Introduction to Computational Nanomechanics

A comprehensive guide on computational nanomechanics discussing basic theoretical concepts and computer modelings in areas such as computational physics, materials, mechanics, and engineering as well as several other interdisciplinary avenues. This book makes the underlying theory accessible to readers without specialized training or extensive background in quantum physics, statistical mechanics, or theoretical chemistry. It combines a careful treatment of theoretical concepts with a detailed tutorial on computer softwares and computing implementations, including multiscale simulation and computational statistical theory. Multidisciplinary perspectives are provided, yielding a deep insight on the applications of computational nanomechanics across diverse engineering fields. The book can serve as a practical guide with step-by-step discussions of coding, example problems, and case studies. This book will be essential reading for students new to the subject, as well as an excellent reference for researchers and developers.

Shaofan Li is a full professor of computational science at the University of California, Berkeley. Doctor Li has been conducting research in atomistic and multiscale simulations since 2000, publishing more than two hundred technical articles in peer-reviewed journals.

Jun Li is a post-doctoral researcher at Wuhan University of Technology. Doctor Li has been conducting research in first-principles modeling and simulations of materials since 2015, publishing over twenty technical papers in peer-reviewed journals.

Introduction to Computational Nanomechanics

Multiscale and Statistical Simulations

SHAOFAN LI
University of California, Berkeley

JUN LI
Wuhan University of Technology

Shaftesbury Road, Cambridge CB2 8EA, United Kingdom

One Liberty Plaza, 20th Floor, New York, NY 10006, USA

477 Williamstown Road, Port Melbourne, VIC 3207, Australia

314–321, 3rd Floor, Plot 3, Splendor Forum, Jasola District Centre, New Delhi – 110025, India

103 Penang Road, #05–06/07, Visioncrest Commercial, Singapore 238467

Cambridge University Press is part of Cambridge University Press & Assessment, a department of the University of Cambridge.

We share the University's mission to contribute to society through the pursuit of education, learning and research at the highest international levels of excellence.

www.cambridge.org
Information on this title: www.cambridge.org/9781107011151

DOI: 10.1017/9780511894770

© Shaofan Li and Jun Li 2022

This publication is in copyright. Subject to statutory exception and to the provisions of relevant collective licensing agreements, no reproduction of any part may take place without the written permission of Cambridge University Press & Assessment.

First published 2022

Printed in the United Kingdom by TJ Books Limited, Padstow, Cornwall

A catalogue record for this publication is available from the British Library

ISBN 978-1-107-01115-1   Hardback

Cambridge University Press & Assessment has no responsibility for the persistence or accuracy of URLs for external or third-party internet websites referred to in this publication and does not guarantee that any content on such websites is, or will remain, accurate or appropriate.

Contents

Preface		*page* ix
Acknowledgments		xi

Part I First-Principles Calculations

1	**A Short Primer on Quantum Mechanics**	3
	1.1 Wave–Particle Duality: Law of Physics	3
	1.2 Schrödinger Equation	8
	1.3 Solution Examples of the Schrödinger Equation	11
	1.4 Interpretations of Quantum Mechanics	28
	1.5 Homework Problems	32
2	**Density Functional Theory**	35
	2.1 Many-body Schrödinger Equation	35
	2.2 Hartree–Fock Approximation	38
	2.3 Hohenberg and Kohn Theorems	39
	2.4 Kohn–Sham Formalism	43
	2.5 Example: DFT Calculation of Silicon Band Structure	46
3	**Quantum Stress**	52
	3.1 Quantum Stress Theory	52
	3.2 Reciprocal-Space Expression for Quantum Stress	55
	3.3 Quantum Stress via DFT	60
	3.4 Quantum Electronic Stress	65
	3.5 Higher-Order Quantum Stress Theory	69
	3.6 Higher-Order Quantum Stress via DFT	70
	3.7 Quantum Couple Stress via DFT	72
4	**Introduction to VASP**	76
	4.1 Files Used by VASP	76
	4.2 Example: Structural Optimization and Self-consistent Charge Density	84
	4.3 Example: DFT Calculation of Si Band Structure	89
	4.4 Example: DFT Modeling of Calcium Silicate Hydrate (C-S-H) Structure	94

Part II Statistical Molecular Dynamics

5 Fundamentals of Statistical Mechanics — 105
 5.1 Lagrangian Mechanics — 105
 5.2 Hamiltonian Mechanics — 106
 5.3 Liouville Theorem — 111
 5.4 Canonical Transformation and Symplectic Condition — 113
 5.5 Laws of Thermodynamics — 115
 5.6 Thermodynamics States — 117
 5.7 Legendre Transformation — 120
 5.8 Statistical Ensembles — 124
 5.9 Homework Problems — 155

6 Fundamentals of Molecular Dynamics — 158
 6.1 How to Derive Molecular Dynamics from Quantum Mechanics — 158
 6.2 Ab-Initio Molecular Dynamics — 162
 6.3 How to Calculate Mechanical Forces in Quantum Mechanical Systems — 163
 6.4 Classical Molecular Dynamics — 166
 6.5 Examples of Atomistic Potentials — 169
 6.6 Periodic Boundary Condition — 180
 6.7 Neighbor Lists — 185
 6.8 Homework Problems — 187

7 Molecular Dynamics Time Integration Techniques — 189
 7.1 Basic Concept of Time Integration — 189
 7.2 Verlet Algorithms — 191
 7.3 Predictor–Corrector Methods — 199
 7.4 Symplectic Algorithm — 201
 7.5 Homework Problems — 203

8 Temperature Control in MD Simulations — 204
 8.1 Velocity Scaling — 204
 8.2 Stochastic Thermostat — 206
 8.3 Nosé–Hoover Thermostat — 208
 8.4 How to Integrate Nosé–Hoover MD? — 216
 8.5 Other Thermostats — 220
 8.6 Homework Problems — 223

9 Andersen–Parrinello–Rahman Molecular Dynamics — 225
 9.1 Andersen's NPH MD — 225
 9.2 Parrinello-Rahman Formulation — 228
 9.3 PR MD for NPH Ensemble — 232
 9.4 Physical Justification of PR MD — 238

	9.5	PR MD for (NσH) or (NτH) Ensemble	242
	9.6	Podio-Guidugli's Interpretation	246
	9.7	Homework Problems	248
10	**Introduction to LAMMPS**		**249**
	10.1	How to Download and Install LAMMPS	249
	10.2	How to Run LAMMPS	251
	10.3	Some Basic LAMMPS Commands	253
	10.4	Case Study (I): Simulation of Three-Dimensional Nano-indentation	282
	10.5	Case Study (II): MD Simulation of Mechanical Properties of Cement	287
	10.6	MD Visualization Software: VMD and OVITO	291
	10.7	Homework Problems	312
11	**Monte Carlo Methods**		**313**
	11.1	Monte Carlo Sampling for Integrations	313
	11.2	Markov Chain Monte Carlo Method	325
	11.3	Hamiltonian (Hybrid) Monte Carlo Method	340
12	**Langevin Equations and Dissipative Particle Dynamics**		**345**
	12.1	Langevin Equation	345
	12.2	LAMMPS Examples for Langevin Dynamics	353
	12.3	Dissipative Particle Dynamics	358
	12.4	Homework Problems	371
13	**Nonequilibrium Molecular Dynamics**		**373**
	13.1	Green–Kubo Relation	373
	13.2	Example: LAMMPS Simulation of Thermal Conductivity	388
	13.3	Fluctuation–Dissipation Theorem (FDT)	392
	13.4	Mori–Zwanzig Formalism	401

Part III Multiscale Modeling and Simulation

14	**Virial Theorem and Virial Stress**		**409**
	14.1	What Is Virial?	411
	14.2	Virial Stress via Tensorial Viral Theorem	413
	14.3	Virial Stress via Liouville Theorem: Irving–Kirkwood Formalism	419
	14.4	Hardy Stress	427
	14.5	Homework Problems	432
15	**Cauchy–Born Rule and Multiscale Methods**		**433**
	15.1	Cauchy–Born Rule	433
	15.2	Higher-Order Cauchy–Born Rule	446
	15.3	Cauchy–Born Rule for Non-Bravais Lattices	450

	15.4 Cauchy–Born Rule for Amorphous Solids	453
	15.5 Homework Problems	458

16 Statistical Theory of Cauchy Continuum — 459
16.1 Quasi-Harmonic Approximation — 459
16.2 Homework Problems — 470

17 Multiscale Method (I): Multiscale Micromorphic Molecular Dynamics — 472
17.1 Multiscale Partition of First-Principles MD Lagrangian — 472
17.2 Multiscale Micromorphic MD — 481
17.3 Numerical Examples — 487
17.4 Multiscale Coupling between MMMD and PD — 495
17.5 Homework Problems — 505

18 Multiscale Methods (II): Multiscale Finite Element Methods — 507
18.1 Multiscale Finite Element Formulation — 507
18.2 MMMD/FEM Coupling Method — 513
18.3 Multiscale Cohesive Interphase Zone Model — 517
18.4 Higher-Order MCZM — 530

Appendix A Crystal Structure — 536
A.1 Lattice and Basis — 536
A.2 Unit Cells — 542
A.3 Miller Indices — 546
A.4 Reciprocal Lattice — 549
A.5 Some Common Crystal Lattice Structures — 551
A.6 Bloch Theorem — 557

Bibliography — 559
Author Index — 565
Subject Index — 567

Preface

This book grew from the lecture notes of the graduate course on Introduction to Computational Nanomechanics at the University of California–Berkeley, which I started teaching in Spring 2011.

Most of the materials are compiled and organized from various textbooks or research papers in quantum mechanics, molecular dynamics, computational statistical mechanics, and the like. However, the contents are edited and streamlined, and the texts are rephrased so that first-year engineering graduate students with different backgrounds but without a formal training in graduate-level physics and chemistry can easily understand them. In fact, the main motivation of the book is to teach computational nanomechanics and computational statistical mechanics to students from various disciplines of engineering, biology, and mathematical sciences, who do not have a formal training in quantum physics and statistical mechanics. Because of continuing developments in nanoscience and nanotechnology and their applications to the broader field of engineering, the needs for in-depth knowledge of nanoscience, especially computational nanoscience, have become more and more urgent and demanding. Usually, conventional wisdom says that acquiring such knowledge would require a career change, which demands so much time and effort that many engineering students or researchers are too intimidated.

There are several books on computational nanomechanics available. To distinguish the present book from the others, I would like to focus more on computational statistical mechanics, multiscale simulation, and its applications to solve engineering problems – a subject that is still in the stage of infancy.

My academic background is not quantum physics or computational chemistry, but applied and computational mechanics, which is a subfield of engineering science or applied physics. I always felt that perhaps this was a disadvantage for me in writing a book on computational nanomechanics; on the other hand, I may be in a unique position to understand how the mind of an engineer works, so that I may have a different perspective to write the book in such a way that most engineers will feel more comfortable to read it. It is only this thought that keeps me thinking that this is a useful endeavor.

One of the main features of the book is that it includes many segments of actual computation scripts and computer codes. It provides step-by-step tutorials to show how to conduct a computer simulation of a first-principle calculation or a molecular dynamics calculation. Most of these scripts are collected from various online resources, as well as private communications and sharings from our research collaborators. Because the computer codes have been migrated from different sources and different versions, we are not able to acknowledge the original sources or developers, and for that we sincerely apologize to the original developers. While we are deeply indebted to these original developers, we are hoping that they also hold same spirits for sharing these information with readers, and especially younger researchers and students. Lastly, in order to help readers have hands-on experience in computer implementation on some of numerical computation examples discussed in this book, computing resource files have been posted on the following website: http://nanomechanics.berkeley.edu/introduction-to-computationalnanomechanics/.

The readers are free to download these resource files.

Shaofan Li

Acknowledgments

My coauthor, Dr. Jun Li, has been working with me on first-principles modeling and simulations of flexoelectricity since 2018. She has provided many inputs and contributed a great deal in the writing of this book. Without her participation and dedication, I would not have been able to finish the book. Finally, I would like to acknowledge and thank the friends, former students, and fellow researchers in computational nanomechanics, who have all helped me gain a better understanding of this topic. In particular, I would like to thank Dr. Shingo Urata, Dr. Lisheng Liu, Dr. Xiaowei Zeng, Dr. Bo Ren, Dr. Houfu Fan, Dr. Tong Qi, Dr. Hiroyuki Miniky, Dr. Dandan Lyu, Dr. Hengameh Shams, Dr. Qingsong Tu, Mr. Qi Zheng, Mr. Caglar Tamur, Dr. Yuxi Xie, Dr. Xin Lai, Dr. Kaiyue Wang, and Dr. Donghoon Kim, among others, who have contributed to the creation and development of this book.

Part I

First-Principles Calculations

1 A Short Primer on Quantum Mechanics

Nanomechanics is part of both quantum physics and molecular physics. As this book is aimed at engineering students and engineers, whom we assume have no formal training in quantum physics, we begin our presentation with a short introduction of quantum mechanics, in order to provide the necessary background for later presentations.

1.1 Wave–Particle Duality: Law of Physics

Light and matter exhibit wave–particle duality, in other words, all matter and light have two manifestations: discreteness as the deterministic being and continuousness in the sense of probabilistic presence. In our current understanding, such wave–particle duality is the law of physics or first principle, because we do not know, at least to date, any other laws of universe that are more fundamental than it.

The relations between wave and particle properties of any object in the universe may be described by the de Broglie relations,

$$E = h\nu, \text{ and } p = \frac{h}{\lambda}, \tag{1.1}$$

where h is the Planck constant, which is a universal constant of nature, and its value is $h = 6.63 \times 10^{-34}$ Js; λ is the matter wavelength; and ν is the matter wave frequency, which is the number of a repeating event, e.g., cycles or temporal wave number per unit time. The unit of frequency is hertz (Hz) (1 Hz means one wave cycle per second). The reciprocal of the frequency is period, which is the time duration of one wave cycle, i.e.,

$$T = \frac{1}{\nu}.$$

At first sight, many of us may experience difficulties understanding such wave–particle proposition because, in our common experience, a finite mass matter is always associated with the discrete particle, whereas the light wave is associated with the continuous electromagnetic field.

However, at the turn of the twentieth century, people had found several counterexamples or evidences that show either (1) light wave behaves like particles, and (2) matter exhibits wave properties.

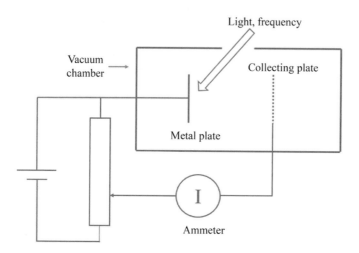

Figure 1.1 Illustration of photoelectric effect experiment

Two famous examples showing that light exhibits particle properties are: (1) photoelectric effect and (2) Compton effect.

1.1.1 Photoelectric Effect

In 1887, Heinrich Hertz found that when ultraviolet (UV) light is shone on a metal plate in a vacuum, and it emits charged particles (see Fig. 1.1), which were later shown to be electrons by J. J. Thomson (1899).

Based on classical electromagnetic theory, electric field $\mathbf{E}$ of light exerts force $F = -e\mathbf{E}$ on electrons. As the intensity of light increases, the input energy to the metal plate increases as well, which may be absorbed by the electrons inside the metal plate, so that the kinetic energy of electrons inside the metal plate increases too. When the kinetic energy of the electrons reach a critical value, they may escape from the metal plate. From this perspective, electrons should be emitted whatever the frequency ν of light is, so long as $\mathbf{E}$ is sufficiently large; and for very low intensity, one may expect a time lag between light exposure and electron emission, because electrons need to absorb enough energy to escape from the metal plate.

The actual experimental observation shows that the maximum kinetic energy of ejected electrons is independent of light intensity, but dependent on the frequency ν of the light. For $\nu < \nu_0$, i.e., for frequencies below a cutoff frequency, no electrons are emitted from the metal plate, and there is no time lag when the light intensity is low. However, the rate of ejection of electrons depends on light intensity.

To interpret the experimental results, Albert Einstein theorized that the energy distribution in light is discrete, or light travels in packets of discrete energy, which are referred to as *quanta*, and they are now called as *photons*,[1]

$$E = h\nu. \tag{1.2}$$

---

[1] Here, we adopt the hypothesis that the group of velocity of light is the velocity of photons.

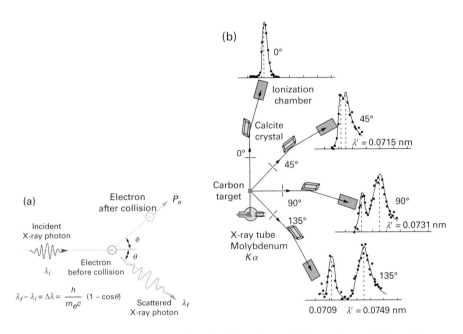

Figure 1.2 Compton scattering: (a) schematic illustration and (b) experimental observation

When an electron absorbs a single photon, it may leave the metal plate. The maximum kinetic energy of an emitted electron can then be expressed as

$$K_{max} = h\nu - \varphi,$$

where φ is the work function, which is the minimum energy needed for an electron to escape from the surface of the metal plate of a given metallic material. It is usually 2~5 eV depending on the type of materials, and it may be written as $\varphi = h\nu_0$, so that we must have $\nu > \nu_0$ for the photoelectric effect to occur. Einstein's theory was later validated by the experiments conducted by Robert Andrews Millikan in 1914.

For his discovery of the law of the photoelectric effect, in 1921 Albert Einstein was awarded the Nobel Prize in Physics.

1.1.2 Compton Scattering

The second example is the so-called Compton scattering or the Compton effect, which is the light scattering due to the inelastic collision of photons and electrons. The experiment is illustrated in Fig. 1.2(a). In the experiment, a high-energy X-ray or gamma ray photon beam hits a target with electrons. In this case, classical theory predicts that when light is scattered on a free electron, the incident electromagnetic (EM) wave will shake the electron transversely, and the oscillating electron then radiates in all directions (except the exact direction of 90°). The classical theory predicts that there may be a change of the wavelength of the colliding photons due to the associated Doppler shift, when the light intensity is large.

However, in the Compton scattering experiment, one can observe the change of the wavelength of the scattering light even when the light intensity is very small, which is called the Compton shift. The shift of the wavelength can be calculated by treating the collision of the photon and electron as the elastic collision of two billiard balls. That is, the photon behaves like a particle and, hence, the photon–electron collision obeys the energy conservation and momentum conservation,

$$h\nu + m_e c^2 = h\nu' + \left(p_e^2 c^2 + m_e^2 c^4\right)^{1/2} \quad \text{and} \quad \mathbf{p}_\nu = \mathbf{p}_{\nu'} + \mathbf{p}_e. \tag{1.3}$$

Note that $(p_e^2 c^2 + m_e^2 c^4)^{1/2} = mc^2$ is Einstein's relativistic energy, which can be derived from Einstein relations,

$$E = mc^2, \; m = \frac{m_e}{\sqrt{1 - v^2/c^2}}, \; \text{and} \; p = mv, \; \rightarrow \; p^2 c^2 = -m_e^2 c^4 + (mc^2)^2$$

and m_e in Eq. (1.3) is the electron's static mass.

From Eq. (1.3), one can find that

$$\lambda' - \lambda = \frac{h}{m_e c}(1 - \cos\theta) \geq 0. \tag{1.4}$$

In Fig. 1.2(b), one finds the shifted wavelength measurement at different angles. Note that for every fixed angle, there is also an unshifted peak, that is due to collision of the X-ray photon and the core of the atom (the nucleus of the atom plus the immobile electrons) because in that case, based on Eq. (1.3), one can find that

$$\lambda' - \lambda = \frac{h}{m_c c}(1 - \cos\theta) \sim 0, \; m_c \gg m_e. \tag{1.5}$$

The Compton effect is a strong evidence that the continuous electromagnetic waves may behave like particles. For the discovery of the Compton effect, Arthur Holly Compton earned the 1927 Nobel Prize in Physics.

On the other hand, discrete matter may also behave like continuous waves. In the following, we consider a well-known double-slit diffraction experiment of matter waves.

1.1.3 Interference of Matter Waves

The double-slit experiment was originally performed by Thomas Young in 1801 in demonstrating the wave nature of light, in which an incoming coherent plane wave is directly hitting a thin plate with two slits, one can observe the wave interference pattern on the screen behind the double-slit plate as shown in Fig. 1.3(b).

On the other hand, if the incoming object is not light, but a beam of particles such as electrons, atoms, or even molecules, what would we expect the measurement result on the back screen to be? A natural expectation on the results of double-slit diffraction of matter waves is depicted in Fig. 1.3(a). However, on the contrary, for matter particle waves, the particle density on the back screen has the same interference pattern as the light wave. Interference pattern produced by a beam of C_{60} molecules is shown in Fig. 1.4, which demonstrates the wave–particle duality of C_{60} molecules. It should be

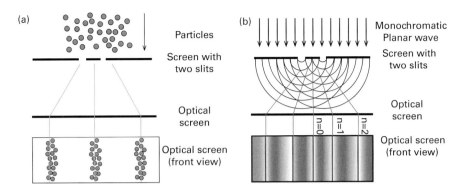

Figure 1.3 Double-slit experiment: (a) expected result for particles and (b) experimental observation

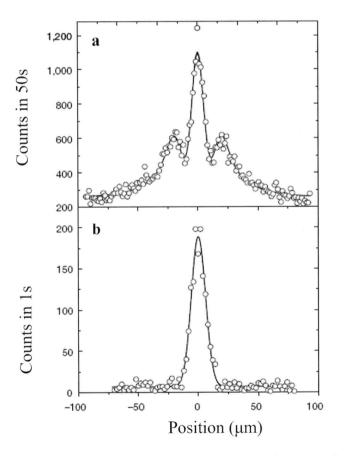

Figure 1.4 Interference pattern produced by C_{60} molecules: (a) experimental recording (open circles) and the fitting curve by using the Kirchhoff diffraction theory (continuous line) – the expected zeroth and first-order maxima can be clearly seen. The details of the theory are discussed in the text; and (b) the molecular beam profile without the grating in the path of the molecules (Arndt et al. (1999))

noted that the position of the matter wave is uncertain, and it is a wave of probability distribution, and it is sometimes called the de Broglie wave. One of the consequences of this probabilistic wave is the uncertainty principle, which is sometimes called the Heisenberg principle. The principle asserts that there is a fundamental limit to the precision with which certain pairs of physical properties of a particle can be simultaneously determined, such as position x and momentum p,

$$\sigma_x \sigma_p \geq \frac{\hbar}{2},$$

where $\hbar = \frac{h}{2\pi} = 1.05457172610^{-34}$ Js is the reduced Planck constant and σ_x, σ_p are standard deviation of position and momentum.

The quantum mechanics uncertainty principle indicates that the more precise the momentum of a particle is determined, the less precise its position can be known, and vice versa. In other words, for a fixed precision of momentum, the precision of the position is bounded below. This is to say that as random variables, position and momentum are intrinsically related, and the product of their variances has a low bound.

To close this section, we note that not only light and matter exhibit wave–particle duality, antimatter also exhibits wave–particle duality.

1.2 Schrödinger Equation

The partial different equation that governs the matter wave motion is called the Schrödinger equation.

1.2.1 A Short Heuristic Derivation

Since this is not a quantum mechanics book but an introduction to nanomechanics to engineers, we derived the Schrödinger equation in a heuristic manner.

Before we get into mathematical derivations, we first make the following assumptions:

1. The total energy E of a particle is

$$E = T + V = \frac{p^2}{2m} + V.$$

This is the energy expression for a classical particle with mass m where the total energy E is the sum of the kinetic energy T, and the potential energy V (which can vary with position, and time). p and m are the momentum and the mass of the particle, respectively.

2. Einstein's light quanta hypothesis (1905) asserts that the energy E of a photon is proportional to the frequency ν (or angular frequency, $\omega = 2\pi\nu$) of the corresponding electromagnetic wave:

$$E = h\nu = \hbar\omega.$$

3. The de Broglie hypothesis (1924) states that any particle can be associated with a wave, and that the momentum p of the particle is related to the wavelength λ (or wave number k) of such a wave by:

$$p = \frac{h}{\lambda} = \hbar k.$$

Expressing p and wavelength k as vectors, we have

$$\mathbf{p} = \hbar \mathbf{k}.$$

4. The three assumptions discussed earlier allow one to derive the governing equation for plane waves only. To extend those assumptions to general situations will require the superposition principle, and thus, one must separately postulate that the Schrödinger equation is linear.

Schrödinger's main idea was to express the phase of the matter wave as a complex phase factor so that the matter wave probability function has the following form:

$$\Psi(\mathbf{r},t) = A \exp i(\mathbf{k} \cdot \mathbf{r} - \omega t), \quad \text{where } \mathbf{r} = x\mathbf{e}_x + y\mathbf{e}_y + z\mathbf{e}_z, \tag{1.6}$$

where $\mathbf{k}$ is the wave number and ω is the angular frequency.

Considering that Eq. (1.6) is the intrinsic form of the wave function, we have

$$\frac{\partial}{\partial t}\Psi = -i\omega\Psi$$

and then

$$E\Psi = h\nu\Psi = \hbar\omega\Psi = i\hbar\frac{\partial}{\partial t}\Psi. \tag{1.7}$$

Similarly, for spatial derivatives, we have

$$\frac{\partial}{\partial x}\Psi = ik_x\Psi, \quad \text{and} \quad \frac{\partial^2}{\partial x^2}\Psi = -k_x^2\Psi.$$

We then have

$$p_x^2\Psi = (\hbar k_x)^2\Psi = -\hbar^2\frac{\partial^2}{\partial x^2}\Psi$$

and hence

$$\mathbf{p}^2\Psi = \left(p_x^2 + p_y^2 + p_z^2\right)\Psi = -\hbar^2\left(\frac{\partial^2}{\partial x^2} + \frac{\partial^2}{\partial y^2} + \frac{\partial^2}{\partial z^2}\right)\Psi = -\hbar^2\nabla^2\Psi.$$

Recalling the Assumption 1 on total energy,

$$E = T + V = \frac{p^2}{2m} + V \Rightarrow E\Psi = (T+V)\Psi = -\frac{\hbar^2}{2m}\nabla^2\Psi + V\Psi \quad (1.8)$$

and combining Eqs. (1.7) and (1.8), we obtain the standard form of time-dependent Schrödinger equation for a single particle as,

$$i\hbar\frac{\partial}{\partial t}\Psi(\mathbf{r},t) = -\frac{\hbar^2}{2m}\nabla^2\Psi(\mathbf{r},t) + V(\mathbf{r},t)\Psi(\mathbf{r},t), \quad (1.9)$$

where m is the mass of the particle, $-\frac{\hbar^2}{2m}\nabla^2$ is said to be the kinetic energy operator, and $V(\mathbf{r},t)$ is the potential energy of the particle at position $\mathbf{r}$ and at time t.

In passing, we note that the Schrödinger equation, i.e., Eq. (1.9), is a second-order, homogeneous, linear partial differential equation.

1.2.2 Wave Function

Further examining the time-dependent Schrödinger equation,

$$i\hbar\frac{\partial\Psi}{\partial t} = -\frac{\hbar^2}{2m}\nabla^2\Psi + V(\mathbf{r},t)\Psi$$

we find that

$$E = T + V = \frac{\mathbf{p}\cdot\mathbf{p}}{2m} + V(\mathbf{r},t) \Rightarrow -\frac{\hbar^2}{2m}\nabla^2 + V(\mathbf{r},t),$$

which may be viewed as a differential operator, and we name the energy differential operator as *Hamiltonian operator* or simply "Hamiltonian,"

$$\hat{H} = \hat{T} + \hat{V} = -\frac{\hbar^2}{2m}\nabla^2 + V(\mathbf{r},t). \quad (1.10)$$

Max Born made a physical interpretation of the wave function $\Psi(\mathbf{r},t)$: The probability of finding the particle in a small volume $\delta\Omega$ at position $\mathbf{r}$ and time t is equal to $|\Psi(\mathbf{r},t)|^2\delta\Omega = \Psi(\mathbf{r},t)\Psi^*(\mathbf{r},t)\delta\Omega$. In other words, $|\Psi(\mathbf{r},t)|^2$ is the probability distribution of finding the particle in the location $\mathbf{r}$ at time t. Since the total probability to find the particle in the space should be one, i.e.,

$$\int_{\mathbb{R}^3} |\Psi(\mathbf{r},t)|^2 d\Omega = 1$$

and a wave function that satisfies this condition is said to be normalized. Suppose that we have a solution of Eq. (1.9), which is not normalized,

$$\int_{\mathbb{R}^3} |\Psi(\mathbf{r},t)|^2 d\Omega = C,$$

we can then normalize it by choosing

$$\Psi(\mathbf{r},t) = \frac{1}{\sqrt{C}}\Psi(\mathbf{r},t).$$

In fact, we can show that the Born interpretation of the wave function

$$P_r = \int_{-\infty}^{\infty} |\Psi(x,t)|^2 dx = const.$$

is correct by proving $\frac{dP_r}{dt} = 0$. We know that

$$i\hbar \frac{\partial \Psi}{\partial t} = -\frac{\hbar^2}{2m} \frac{\partial^2 \Psi}{\partial x^2} + V(x,t)\Psi. \tag{1.11}$$

By taking the complex conjugate of the above equation, we can have the conjugate Schrödinger equation,

$$-i\hbar \frac{\partial \Psi^*}{\partial t} = -\frac{\hbar^2}{2m} \frac{\partial^2 \Psi^*}{\partial x^2} + V(x,t)\Psi^*. \tag{1.12}$$

Then, multiplying Eq. (1.11) with Ψ^* and multiplying Eq. (1.12) with Ψ yield

$$i\hbar \Psi^* \dot{\Psi} = -\frac{\hbar^2}{2m} \Psi^* \frac{\partial^2 \Psi}{\partial x^2} + V(x,t)\Psi^*\Psi, \tag{1.13}$$

$$i\hbar \Psi \dot{\Psi}^* = \frac{\hbar^2}{2m} \Psi \frac{\partial^2 \Psi^*}{\partial x^2} - V(x,t)\Psi\Psi^*. \tag{1.14}$$

By integrating Eqs. (1.13) and (1.14) from $-\infty$ to ∞ and by integration by parts, we obtain

$$i\hbar \frac{\partial}{\partial t} \int_{-\infty}^{\infty} |\Psi|^2 dx = -\frac{\hbar^2}{2m} \Psi^* \frac{\partial \Psi}{\partial x} \Big|_{-\infty}^{\infty} + \frac{\hbar^2}{2m} \Psi \frac{\partial \Psi^*}{\partial x} \Big|_{-\infty}^{\infty}$$
$$+ \frac{\hbar^2}{2m} \int_{-\infty}^{\infty} \left(\Psi^*_{,x} \Psi_{,x} - \Psi_{,x} \Psi^*_{,x} \right) dx = 0,$$

the last equality is derived by considering the fact that wave function should be convergent, which requires $\Psi(x)$ and $\Psi^*(x) \to 0$, as $x \to \infty$. This result leads to Born's statistical interpretation,

$$\int_{-\infty}^{\infty} |\Psi|^2 dx = const.$$

is the probability of finding the particle at time t along the x-axis.

1.3 Solution Examples of the Schrödinger Equation

In this section, we provide a few benchmark solutions, as well as corresponding solution techniques, of the Schrödinger equation.

1.3.1 Time-Independent Schrödinger Equation

There is a large class of problems in which the potential function is independent from time, t. In one-dimensional (1D) cases, we can write such potential function as $V(x,t) = V(x)$, and the corresponding Schrödinger equation can be written as follows:

$$i\hbar \frac{\partial \Psi}{\partial t} = -\frac{\hbar^2}{2m}\frac{\partial^2 \Psi}{\partial x^2}\Psi + V(x)\Psi. \tag{1.15}$$

Using separation of variable and substituting $\Psi(x,t) = \psi(x)T(t)$ into Eq. (1.15), we have

$$i\hbar\psi\frac{dT}{dt} = -\frac{\hbar^2}{2m}T(t)\frac{d^2\psi}{dx^2} + V(x)\psi(x)T(t).$$

Dividing the above equation by $\psi(x)T(t)$, we can obtain

$$i\hbar\frac{1}{T}\frac{dT}{dt} = -\frac{\hbar^2}{2m}\frac{1}{\psi}\frac{d^2\psi}{dx^2} + V(x) = const. = A, \tag{1.16}$$

where A is separation constant.

Equation (1.16) can be separated into two equations, and we can solve the first equation,

$$\frac{dT}{dt} = -i\frac{A}{\hbar}T \quad \rightarrow \quad T(t) = C\exp\left(-i\frac{A}{\hbar}t\right).$$

We know that the matter wave solution should have a factor $\exp(-i\omega t)$ and $\omega = E/\hbar$. Thus, we can now identify that $A = E$.

Subsequently, the two ordinary differential equations of Eq. (1.16) can be written as

$$i\hbar\frac{1}{T}\frac{dT}{dt} = E \quad \text{and} \quad -\frac{\hbar^2}{2m}\frac{1}{\psi}\frac{d^2\psi}{dx^2} + V(x) = E$$

or

$$i\hbar\frac{dT}{dt} = ET \tag{1.17}$$

$$-\frac{\hbar^2}{2m}\frac{d^2\psi}{dx^2} + V(x) = E\psi. \tag{1.18}$$

Equation (1.18) is called the time-independent Schrödinger equation. For time-independent wave function, we always denote $\psi(\mathbf{r})$, which is the stationary part of the total wave function.

In three-dimensional (3D) space, the time-independent Schrödinger equation has the following form

$$-\frac{\hbar^2}{2m}\nabla^2\psi(\mathbf{r}) + V(\mathbf{r})\psi(\mathbf{r}) = E\psi(\mathbf{r}) \quad \text{or} \quad \hat{H}\psi(\mathbf{r}) = E\psi(\mathbf{r}), \tag{1.19}$$

where the energy E is a constant. One may find that Eq. (1.19) is an eigenvalue problem.

The solution of Eq. (1.17) is given as

$$T(t) = a \exp\left(-i\frac{Et}{\hbar}\right), \quad \Rightarrow \quad T(t) = a \exp(-i\omega t), \quad \leftarrow \quad \omega = \frac{E}{\hbar}$$

and the probability density becomes

$$P(x,t) = |\Psi(x,t)|^2 = \psi^*(x)\exp(-i\omega t)\psi(x)\exp(-i\omega t)$$
$$= \psi^*(x)\psi(x) = |\psi(x)|^2 = P(x),$$

which indicates that the probability density only depends on the solution of time-independent Schrödinger equation, and hence the probability distribution is spatial and stationary, if $V(x,t) = V(x)$.

1.3.2 Free Particle Solution

If we let $V(x) = 0$ in Eq. (1.18), this is the case that a particle freely moves in a 1D space. The solution of Eq. (1.18) in this case is:

$$\psi(x) = b \exp\left(\pm i \frac{\sqrt{2mE}}{\hbar} x\right).$$

Consider that, in this case,

$$E = \frac{p^2}{2m} \text{ and } p = \hbar k \quad \Rightarrow \quad \frac{\sqrt{2mE}}{\hbar} = \frac{p}{\hbar} = k, \quad \Rightarrow \quad \psi(x) = b\exp(i \pm kx).$$

Therefore, the total solution of the wave function becomes

$$\Psi(x,t) = (a \cdot b)\exp i(\pm kx - \omega t) = C \exp i(\pm kx - \omega t).$$

In the derivation above, the sign of the time term $(-i\omega t)$ is fixed by the sign adopted in time-dependent Schrödinger equation, while the sign of the position term $\pm ikx$ depends on propagation direction of wave: $+ikx$ term propagates toward $+\infty$, while the term $-ikx$ propagates toward $-\infty$.

In fact, in the above heuristic derivation of the Schrödinger equation we have assumed that the general wave function solution has a plane wave form. We say that the derivation is heuristic, because the derivation does not consider the general cases of potential function $V(\mathbf{r},t) \neq 0$.

1.3.3 Particle in a Finite Potential Well

Assume that a particle is associated with a finite potential well,

$$-\frac{\hbar^2}{2m}\frac{d^2\psi}{dx^2} + V(x)\psi = E\psi, \text{ where } V(x) = \begin{cases} V_0, & \text{if } x < -a \\ 0, & \text{if } |x| < a \\ V_0, & \text{if } x > a \end{cases} \quad (1.20)$$

The eigen solution for the above problem has the form $\exp(kx)$ and

$$\text{Outside the well:} \quad k \sim \pm \frac{\sqrt{2m(V_0 - E)}}{\hbar}, \quad \text{when } V_0 > E,$$

$$\sim \pm i \frac{\sqrt{2m(E - V_0)}}{\hbar} \quad \text{when } V_0 < E,$$

$$\text{Inside the well:} \quad k \sim \pm i \frac{\sqrt{2mE}}{\hbar}.$$

Outside the well, we may discard the eigen solutions

$$\psi \sim \exp\left(\pm i \frac{\sqrt{2m(E - V_0)}}{\hbar}\right)$$

because they are not convergent when $x \to \pm\infty$. However, outside of the well we have convergent solutions,

$$\psi \sim \exp\left(\frac{\sqrt{2m(V_0 - E)}}{\hbar} x\right), \quad \text{when } x < 0,$$

$$\text{and } \exp\left(-\frac{\sqrt{2m(V_0 - E)}}{\hbar} x\right), \quad \text{when } x > 0. \quad (1.21)$$

Now, we even have a solution that corresponds to the case $E < V_0$ outside the well. For classical particles, it is impossible to have a solution when $E < V_0$, which means the classical particles can never escape from the energy well, however, for quantum mechanics, this is a valid possibility. Let $q = \sqrt{2m(V_0 - E)}/\hbar$. Based on the convergence argument, in $x < -a$ we have to discard the solution $\sim \exp(-qx)$ and, for $x > a$, we must discard the solution $\sim \exp(qx)$.

Finally, we have the convergent wave function solution in the whole domain,

$$\psi(x) = \begin{cases} C \exp(qx), & \forall\, x < -a \\ A \cos kx + B \sin kx, & \forall\, -a < x < a \\ D \exp(-qx), & \forall x > a \end{cases} \quad (1.22)$$

where A, B, C, D are coefficient constants determined by boundary conditions. It may be noted again that inside the well $V = 0$ and,

$$\psi(x) = A \cos kx + B \sin kx \quad \text{and} \quad E = \frac{k^2 \hbar^2}{2m}.$$

Using the continuity condition,

$$\psi(-a^-) = \psi(-a^+) \quad \text{and} \quad \frac{d\psi}{dx}(-a^-) = \frac{d\psi}{dx}\psi(-a^+)$$

one can find that the even function solution must satisfy the conditions,

$$C \exp(-qa) = A \cos ka \quad \text{and} \quad \alpha C \exp(-qa) = kA \sin(ka),$$

which lead to

$$q = k\tan(ka). \quad (1.23)$$

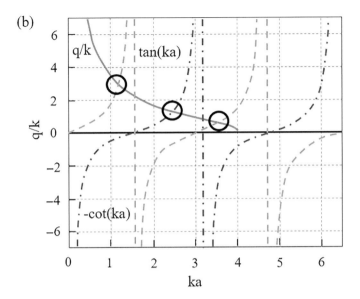

Figure 1.5 Eigenvalue locations for a particle in a finite depth well

While, for the odd function solution

$$\psi(a^-) = \psi(a^+) \text{ and } \frac{d\psi}{dx}(a^-) = \frac{d\psi}{dx}\psi(a^+)$$

it must satisfy the conditions,

$$C\exp(-qa) = B\sin ka \text{ and } -\alpha C\exp(-qa) = kB\cos(ka),$$

which lead to

$$q = -k\cot(ka). \tag{1.24}$$

The graph solution of the two characteristic equations for eigenvalues is shown in Fig. 1.5, in which the intersection points are between $y(ka) = \tan(ka)$ or $-\cot(ka)$ and $y(ka) = q/k = \sqrt{k_0^2/k^2 - 1}$, where $k_0 = \sqrt{2mV_0}/\hbar$. Another approach to find the allowable discrete energy level is utilizing the fact that $q = \sqrt{2m(V_0 - E)}/\hbar$. Let

$$u = qa, \quad v = ka = \frac{\sqrt{2mEa}}{\hbar} \text{ and } u_0 = \frac{\sqrt{2mV_0}a}{\hbar}.$$

Equations (1.23)–(1.24) may be converted to the following energy balance relation:

$$u_0^2 - u^2 = \begin{cases} (v\tan v)^2 & \text{(symmetric case)} \\ (v\cot v)^2 & \text{(antisymmetric case)}. \end{cases}$$

If we choose $X = v$ and $Y = u$, we can find different energy levels for a fixed depth of the well, i.e., V_0, in a plot. In Fig. 1.6, the solid semicircle is the contour for $u_0^2 = 20$, and the dash or dotted curves are functions $u = v\tan v$ or $u = -v\cot v$. In this case,

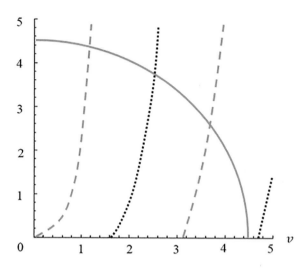

Figure 1.6 Eigenvalues for a particle in a finite depth well

there are exactly three solutions, $v_1 = 1.28, v_2 = 2.54$, and $v_3 = 3.73$, with the corresponding energies

$$E_n = \frac{\hbar^2 v_n^2}{2ma^2}.$$

Consider a special case that $V_0 \to \infty$. The finite depth of potential well becomes a infinite height potential well, and the solution Eq. (1.22) degenerates to

$$\psi(x) = \begin{cases} 0, & \forall\, x < -a \\ A\cos kx + B\sin kx, & \forall\, -a < x < a \,. \\ 0, & \forall x > a \end{cases} \quad (1.25)$$

The wave function solution will then become:

Even solution: $\psi_n(x) = \sqrt{\dfrac{1}{a}} \cos\left(\dfrac{n\pi x}{2a}\right)$, $n = 1, 2, 5, \ldots \infty$ and

Odd solution: $\psi_n(x) = \sqrt{\dfrac{1}{a}} \sin\left(\dfrac{n\pi x}{2a}\right)$, $n = 2, 4, 6, \ldots \infty$,

which satisfy the boundary conditions $\psi(-a) = \psi(a) = 0$, as shown in Fig. 1.7. The energy level for each quantum state n is $E_n = \dfrac{n^2\hbar^2\pi^2}{8ma^2}$. When $n = 1$ $E_1 = \dfrac{\hbar^2\pi^2}{8ma^2}$, it is the minimum quantum energy state in the well, which we call as the *ground state*. In general, the ground state of a quantum mechanical system is its lowest-energy state and the energy of the ground state is known as the zero-point energy of the system. An excited state is any state with energy greater than the ground state. If more than one ground state exist, they are said to be degenerate. Many systems have degenerate ground states. Mathematically speaking, the ground state eigenvalue has multiplicities. According to the third law of thermodynamics, a system at absolute zero temperature only exists in its ground state. At the ground state, the system's

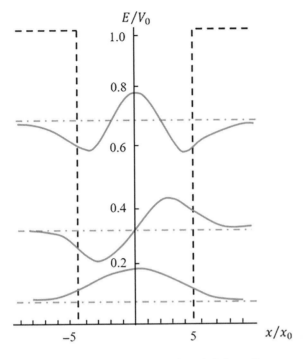

Figure 1.7 Different quantum states in an infinite well

entropy is determined by the degeneracy of the ground state. Many systems, such as a perfect crystal lattice, have a unique ground state and, therefore, have zero entropy at absolute zero temperature.

1.3.4 Harmonic Oscillator

If we choose the potential energy as the quadratic function

$$V(x) = \frac{1}{2}kx^2 = \frac{1}{2}m^2\omega^2x^2, \quad \leftarrow \text{ Recall } \omega^2 = \frac{k}{m},$$

the time-independent Schrödinger equation will take the form

$$-\frac{\hbar^2}{2m}\frac{d^2\psi}{dx^2} + \frac{1}{2}m\omega^2x^2\psi = E\psi. \quad (1.26)$$

Since, in classical mechanics, this is the potential function of the simple harmonic oscillator, we call the solution of Eq. (1.26) the quantum harmonic oscillator.

Choosing $\psi(x) = \exp(-\alpha^2 x^2/2)$ and taking the second-order derivative of the wave function, one may find that

$$\frac{d^2\psi}{dx^2} = -\alpha^2\psi + \alpha^4 x^2\psi,$$

and the Schrödinger equation becomes,

$$-\frac{\hbar^2}{2m}\left(-\alpha^2 + \alpha^4 x^2\right)\psi(x) + \frac{1}{2}m\omega^2 x^2 \psi(x) = E\psi(x).$$

To solve the equation, we obtain

$$\alpha = \sqrt{\frac{m\omega}{\hbar}} \quad \text{and} \quad E_0 = \frac{1}{2}\hbar\omega,$$

which is the lowest achievable energy state or the ground state the particle can occupy.

In general, one can assume the form of solution as $\psi(x) = H(y)\exp(-y^2/2)$, $y = \alpha x$, and Eq. (1.26) becomes,

$$\frac{d^2 H}{dy^2} - 2y\frac{dH}{dy} + (\lambda - 1)H = 0, \quad \text{where } \lambda := \frac{2E}{\hbar\omega}.$$

To obtain a convergent solution for this equation (see, e.g., Pilar 1990), the parameter λ has to satisfy the condition

$$\lambda = 2n + 1, \quad n = 0, 1, 2, \ldots \quad \text{and hence } E_n = \hbar\omega\left(n + \frac{1}{2}\right), \quad n = 0, 1, 2, \ldots$$

and a complete family of solutions may be found as

$$\psi_n(x) = \frac{1}{\sqrt{2^n n!}} \left(\frac{m\omega}{\pi\hbar}\right)^{1/4} H_n\left(\sqrt{\frac{m\omega}{\hbar}}x\right) \exp\left(-\frac{m\omega x}{2\hbar}\right), \quad n = 0, 1, 2, 3, \ldots$$

where the function H_n are the Hermite polynomials

$$H_n(x) = (-1)^n \exp(x^2)\frac{d^n}{dx^n}\left(\exp(-x^2)\right), \quad n = 0, 1, 2, \ldots.$$

For example,

$$H_0(x) = 1, \quad H_1(x) = 2x, \quad H_2(x) = 4x^2 - 2, \quad H_3(x) = 8x^3 - 12x, \ldots.$$

1.3.5 Physical Meaning of the Operators

The operators of the Schrödinger equation have definite physical meanings relating to the physical properties of the particle that is under investigation.

To study the motion of a subatomic particle, the first thing that you may want to do is to measure its position and velocity, which are related with the position operator and momentum operator. The position operator is an algebraic operator that can be expressed as,

$$\hat{x} = x \text{ (in 1D space)} \quad \text{and} \quad \hat{\mathbf{r}} = \mathbf{r}, \text{ (in 3D space)}.$$

The velocity of a particle is related to the linear momentum operator:

$$\hat{p}_x = i\hbar\frac{\partial}{\partial x} \quad \text{and} \quad \hat{\mathbf{p}} = -i\hbar\left[\mathbf{i}\frac{\partial}{\partial x} + \mathbf{j}\frac{\partial}{\partial y} + \mathbf{k}\frac{\partial}{\partial z}\right] = -i\hbar\nabla.$$

Other important operators are:

- Kinetic energy operator

$$T = \frac{\mathbf{p}^2}{2m} \quad \rightarrow \quad \hat{T} = \frac{1}{2m}\left(-i\hbar \frac{\partial^2}{\partial x_i \partial x_i}\right) = -\frac{\hbar^2}{2m}\nabla^2.$$

- Hamiltonian operator

$$E = \frac{\mathbf{p}^2}{2m} + V(\mathbf{r}) \quad \rightarrow \quad \hat{H} = -\frac{\hbar^2}{2m}\nabla^2 + V(\mathbf{r}).$$

- Angular momentum operator

$$\mathbf{L} = \mathbf{r} \times \mathbf{p} \quad \rightarrow \quad \hat{\mathbf{L}} = -i\hbar \mathbf{r} \times \nabla.$$

1.3.6 Dirac's Notation: Bra–Ket Notation

The solution of quantum mechanics belongs to a complex Hilbert space. Paul Dirac developed a set of notations for vector operations in the complex Hilbert space, which are essentially analogs of vector operations in linear vector space.

Dirac denoted that vector basis and its complex conjugate or transpose as ket or bra, i.e.,

$$|e_i> := \mathbf{e}_i \quad \text{and} \quad <e_i| := \mathbf{e}_i^*$$

where the superscript indicates the complex conjugate.

To illustrate the concept here, we slightly alter the notation, so that we may write a vector in a complex finite dimensional vector space by using the bras and kets as

$$\mathbf{A} = A_1 \mathbf{e}_1 + A_2 \mathbf{e}_2 + A_3 \mathbf{e}_3 = A_1|e_1> + A_2|e_2> + A_3|e_3> = \begin{pmatrix} A_1 \\ A_2 \\ A_3 \end{pmatrix}$$

and

$$\mathbf{A}^* = A_1^* \mathbf{e}_1^* + A_2^* \mathbf{e}_2^* + A_3^* \mathbf{e}_3^* = A_1^* <e_1| + A_2^* <e_2| + A_3^* <e_3| = \left(A_1^*, A_2^*, A_3^*\right)$$

and you may see that the bra is basically a complex conjugate of row vector, whereas the ket is a complex column vector. Essentially, the bra may be viewed as the complex conjugate transpose of the ket.

The inner product of finite dimensional vector space may be defined as

$$<e_i|e_j> := \mathbf{e}_i^* \cdot \mathbf{e}_j = \delta_{ij};$$

and the outer product of bra and ket may be viewed as a tensor product, i.e.,

$$|e_i><e_j| := \mathbf{e}_i \otimes \mathbf{e}_j^*.$$

Using engineering notations, we can express the inner product of bra and ket as the scalar dot product of two vectors,

$$<A|B> = (A_1^*, A_2^*, A_3^*) \begin{pmatrix} B_1 \\ B_2 \\ B_3 \end{pmatrix} = A_1^* B_1 + A_2^* B_2 + A_3^* B_3$$

and express the outer product of bra and ket as the following matrix multiplication,

$$|A><B| = \begin{pmatrix} A_1 \\ A_2 \\ A_3 \end{pmatrix} (B_1^* B_2^* B_3^*) = \begin{pmatrix} A_1 B_1^* & A_1 B_2^* & A_1 B_3^* \\ A_2 B_1^* & A_2 B_2^* & A_2 B_3^* \\ A_3 B_1^* & A_3 B_2^* & A_3 B_3^* \end{pmatrix}.$$

Moreover, it may be shown that

$$(|A><B|)|C> = <B|C> |A>.$$

In the complex Hilbert space, we can define the inner product between two function bases as

$$<\psi_i|\psi_j> = \int_{-\infty}^{\infty} \psi_i^*(x)\psi_j(x)dx.$$

As seen previously, the total wave function is the superposition of a complete set of eigenfunctions, which can now be written by using the bra–ket notation,

$$\psi = \sum_{i \in \mathbb{H}} a_i \phi_i \quad \rightarrow \quad |\psi> = \sum_{i \in \mathbb{H}} <\phi_i|\psi> |\phi_i>,$$

where, by definition,

$$<\phi_i|\psi> = \int_{-\infty}^{\infty} \phi_i^* \psi dx = a_i$$

because $\int_{-\infty}^{\infty} \phi_i^*(x)\phi_j(x)dx = \delta_{ij}, \quad \psi(x) = \sum_i a_i \phi_i(x),$

if we view the eigenfunction as the vector(function) basis.

Similarly, one may write,

$$\psi^* = \sum_{i \in \mathbb{H}} a_i^* \phi_i^* \quad \rightarrow \quad |\psi> = \sum_{i \in \mathbb{H}} <\phi_i| <\psi|\phi_i>, \quad \text{where } a_i^* = <\psi|\phi_i>.$$

1.3.7 Measurement and Expectation

In quantum mechanics, a measurable operator, differential or algebraic, is often called the *observable*, by which the state of the physical system can be determined by a sequence of physical measurements. When a measurement of the observable $\hat{Q}$ is made on a normalized wave function ψ, i.e.,

$$\hat{Q}\phi_n = q_n \phi_n, \quad \psi = \sum_n a_n \phi_n(\mathbf{r}),$$

the probability of obtaining the eigenvalue q_n is given by the modulus squared of the overlap integral,

$$P_r(q_n) = |a_n|^2, \quad a_n = \int_{\mathbb{R}^3} \phi_n^*(\mathbf{r})\psi(\mathbf{r})dV.$$

The operator here can be energy, linear momentum, position, angular momentum, and so forth.

Based on the occurrence probability of each eigenvalue, we can define the expectation value of the operator $\hat{Q}$, i.e., in an 1D case,

$$<Q> := \sum_n P_r(q_n)q_n = \sum_n |a_n|^2 q_n, \quad \text{where } a_n = \int_{\mathbb{R}} \phi^*(x)\psi(x)dx.$$

One can readily show that

$$<Q> = \int_{\mathbb{R}} \psi^* \hat{Q} \psi \, dx,$$

this is because

$$<Q> = \int_{\mathbb{R}} \psi^* \hat{Q} \psi \, dx = \int_{real} \left[\sum_i a_i^* \phi_i^*\right] \hat{Q} \left[\sum_j a_j \phi_j\right] dx$$

$$= \int_{real} \left[\sum_i a_i^* \phi_i^*\right]\left[\sum_j a_j q_j \phi_j\right] dx \quad \leftarrow \quad \hat{Q}\phi_j = q_j \phi_j$$

$$= \sum_i \sum_j a_i^* a_j q_j \int_{\mathbb{R}} \phi_i^* \phi_j \, dx$$

$$= \sum_i |a_i|^2 q_i \quad \leftarrow \quad \int_{\mathbb{R}} \phi_i^* \phi_j \, dx = \delta_{ij}.$$

Now, we can write the expectation of a operator $\hat{Q}$ as

$$<Q> = \int_{\mathbb{R}} \psi^* \hat{Q} \psi \, dx = \int_{\mathbb{R}} \left[\sum_i a_i^* \phi_i^*\right] \hat{Q} \left[\sum_j a_j \phi_j\right] dx$$

$$= \sum_i <\phi_i| < \psi|\phi_i> \hat{Q} \sum_j <\phi_j|\psi> |\phi_j> . = <\psi|\hat{Q}|\psi>. \quad (1.27)$$

If the wave function is not normalized, i.e., $\int \psi^* \psi dx \neq 1$, the expectation of the operator $\hat{Q}$ should be written as

$$<\hat{Q}> = \frac{<\psi|\hat{Q}|\psi>}{<\psi|\psi>}.$$

1.3.8 Operator, Commutators, and Uncertainty Principle

All the operators in quantum mechanics have definite physical meanings or properties that may be measurable. The operation order of the two different quantum mechanics operators is not commutable in general, and such incommutable property has profound physical implications To briefly discuss this basic concept of quantum mechanics, we first define the operator commutator. Assume that there are two operators $\hat{A}$ and $\hat{B}$, we define the operator commutator as the difference between the two different orderings,

$$[\hat{A}, \hat{B}] := \hat{A}\hat{B} - \hat{B}\hat{A}.$$

We say that the two operators commute only if the commutator is zero. To better understand this, we now present some examples.

Example 1.1 Let $\hat{A}$ be a position operator and $\hat{B}$ be a momentum operator:

$$\hat{A} = \hat{x} = x \text{ and } \hat{B} = \hat{p}_x = -i\hbar \frac{\partial}{\partial x}.$$

We can find that

$$[\hat{x}, \hat{p}_x]\psi = x\left(-i\hbar \frac{\partial}{\partial x}\right)\psi - \left(-i\hbar \frac{\partial}{\partial x}\right)(x\psi) = -i\hbar x \frac{\partial \psi}{\partial x} + i\hbar x \frac{\partial \psi}{\partial x} + i\hbar \psi = i\hbar \psi, \tag{1.28}$$

which means that the position operator $\hat{x}$ does not commute with the momentum operator $\hat{p}_x$, and based on the calculation,

$$[\hat{x}, \hat{p}_x] = i\hbar. \tag{1.29}$$

However, if we consider $\hat{A} = \hat{x}$ but $\hat{B} = \hat{p}_y$, one may verify that

$$[\hat{x}, \hat{p}_y]\psi = x\left(-i\hbar \frac{\partial}{\partial y}\right)\psi - \left(-i\hbar \frac{\partial}{\partial y}\right)(x\psi) = 0, \quad \rightarrow \quad [\hat{x}, \hat{p}_y] = 0,$$

that is, $\hat{x}$ and $\hat{p}_y$ commute.

Note that any wave function ψ in a complex Hilbert space may be viewed as a vector in a vector space, which may be expressed in terms of eigenfunction expansion. So, a more rigorous way to write Eq. (1.28) is

$$[\hat{x}, \hat{p}_x]|\psi> = i\hbar|\psi>.$$

A profound consequence of $[\hat{x}, \hat{p}_x] = i\hbar$ is the Heisenberg uncertainty principle, which is expressed in the following famous mathematical expression in quantum physics,

$$\sigma_x \sigma_p \geq \frac{\hbar}{2}, \tag{1.30}$$

where

$$\sigma_x := \sqrt{<\hat{x}^2> - <\hat{x}>^2} \text{ and } \sigma_p := \sqrt{<\hat{p}_x^2> - <\hat{p}_x>^2}.$$

We note that, in statistics $\sqrt{<\hat{A}^2> - <\hat{A}>^2}$ is called as the standard deviation of operator $\hat{A}$. If $\hat{A}$ is measurable, σ_A represents the accuracy of the measurement. We note that for an operator both $<\hat{A}^2>$ and $<\hat{A}>^2$ are real numbers, because

$$<\hat{A}> = \frac{<\psi|\hat{A}|\psi>}{<\psi|\psi>} \quad \text{and} \quad <\hat{A}^2> = \frac{<\hat{A}\psi|\hat{A}\psi>}{<\psi|\psi>}$$

if we assume that $\hat{A}$ is self-adjoint and, if the wave function is normalized, we may drop the denominator.

When we say an operator is self-adjoint, what we mean is that

$$<\hat{A}\psi_1|\psi_2> = <\psi_1|\hat{A}\psi_2> \quad \text{or} \quad \int_\mathbb{R} \hat{A}^*\psi_1^*(x)\psi_2(x)dx = \int_\mathbb{R} \psi_1^*(x)\hat{A}\psi_2(x)dx. \quad (1.31)$$

Most operators in quantum mechanics are self-adjoint, for instance, the Hamiltonian operator,

$$\hat{H} = -\frac{\hbar^2}{2m}\nabla^2 + V(\mathbf{r}).$$

In fact, Eq. (1.31) is the definition of symmetric operator. Based on the Hellinger–Toeplitz theorem, a everywhere-defined symmetric operator in the Hilbert space is bounded and self-adjoint, and we call the self-adjoint operators in the Hilbert space as the Hermitian operator. In quantum mechanics, we are mainly dealing with Hermitian operators.

Now, by using Eq. (1.29), we prove Eq. (1.30). We first show that

$$\sigma_A^2 \sigma_B^2 \geq \left(\frac{1}{2}<\{\hat{A},\hat{B}\}> - <\hat{A}><\hat{B}>\right)^2 + \left(\frac{1}{2i}<[\hat{A},\hat{B}]>\right)^2, \quad (1.32)$$

where

$$\{\hat{A},\hat{B}\} := \hat{A}\hat{B} + \hat{B}\hat{A}$$

is called the anticommutator.

Assume that both $\hat{A}$ and $\hat{B}$ are Hermitian. The corresponding standard deviation can be written as

$$\sigma_A^2 = <(\hat{A}-<\hat{A}>)\Psi|(\hat{A}-<\hat{A}>)\Psi> = <f|f>,$$
$$\text{where } |f> = |(\hat{A}-<\hat{A}>)\Psi>$$
$$\sigma_B^2 = <(\hat{B}-<\hat{B}>)\Psi|(\hat{B}-<\hat{B}>)\Psi> = <g|g>,$$
$$\text{where } |g> = |(\hat{B}-<\hat{B}>)\Psi>$$

and, by the complex version of the Cauchy–Schwartz inequality,

$$\sigma_A^2 \sigma_b^2 = <f|f><g|g> \geq <f|g>^2.$$

Note that $<f|f>$ and $<g|g>$ are real numbers, but $<f|g>$ may be a complex number in general, and can be shown that

$$|<f|g>|^2 = \left(\frac{<f|g>+<g|f>}{2}\right)^2 + \left(\frac{<f|g>-<g|f>}{2i}\right)^2 \quad (1.33)$$

and

$$<f|g> = <\Psi|(\hat{A}- <\hat{A}>)(\hat{B}- <\hat{B}>)|\Psi> = <\hat{A}\hat{B}> - <\hat{A}><\hat{B}> \quad (1.34)$$

and

$$<g|f> = <\Psi|(\hat{B}- <\hat{B}>)(\hat{A}- <\hat{A}>)|\Psi> = <\hat{B}\hat{A}> - <\hat{B}><\hat{A}>. \quad (1.35)$$

Substituting Eqs. (1.34) and (1.35) into Eq. (1.33) yields the desired result:

$$\sigma_A^2 \sigma_B^2 \geq <f|g>^2 = \left(\frac{1}{2} <\{\hat{A},\hat{B}\}> - <\hat{A}><\hat{B}>\right)^2 + \left(\frac{1}{2i} <[\hat{A},\hat{B}]>\right)^2.$$

For the case $\hat{A} = \hat{x}$ and $\hat{B} = \hat{p}_x$, we have

$$\sigma_x^2 \sigma_{p_x}^2 \geq \left(\frac{1}{2i} <[\hat{x},\hat{p}_x]>\right)^2 \rightarrow \sigma_x \sigma_{p_x} \geq \left|\frac{1}{2i} <[\hat{x},\hat{p}_x]>\right| = \left|\frac{1}{2i} <i\hbar>\right| = \frac{\hbar}{2}.$$

This is the complete proof of the uncertainty principle.

1.3.9 Hydrogen Atom

Now we consider a 3D example – an electron in hydrogen atom that is floating in free space, and hence, in this case, the potential energy between the electron and nucleus is $V(r) = -Ze^2/(4\pi\epsilon_0 r)$, or in atomic unit $-e^2/r$, where $r = |\mathbf{r}|$. The Schrödinger equation for this problem is

$$-\frac{\hbar^2}{2\mu}\nabla^2 \psi(\mathbf{r}) + V(r)\psi(\mathbf{r}) = E\psi(\mathbf{r}).$$

By symmetry, the problem is being solved in spherical coordinates, and the Schrödinger equation may be cast into the following form:

$$-\frac{\hbar^2}{2\mu}\left\{\frac{1}{r^2}\frac{\partial}{\partial r}\left(r^2\frac{\partial}{\partial r}\right) + \frac{1}{r^2 \sin\theta}\frac{\partial}{\partial\theta}\left(\sin\theta\frac{\partial}{\partial\theta}\right) + \frac{1}{r^2 \sin^2\theta}\frac{\partial^2}{\partial\phi^2}\right\}\psi(\mathbf{r})$$
$$+ (V(r) - E)\psi(\mathbf{r}) = 0, \quad (1.36)$$

where r is the radial distance, θ is the azimuthal angle (longitude), ϕ is the zenith angle (colatitude), and $\mu = m_e m_u/(m_e + m_u)$.

We often write the quantum angular Laplacian operator as

$$\hat{L}^2 = -\hbar^2\left[\frac{1}{\sin\theta}\frac{\partial}{\partial\theta}\left(\sin\theta\frac{\partial}{\partial\theta}\right) + \frac{1}{\sin^2\theta}\frac{\partial^2}{\partial\phi^2}\right], \rightarrow \nabla^2 = \frac{1}{r^2}\frac{\partial}{\partial r}\left(r^2\frac{\partial}{\partial r}\right) - \frac{\hat{L}^2}{\hbar^2 r^2}.$$

The Schrödinger equation can then be written as

$$\hat{H}(\mathbf{r})\psi(\mathbf{r}) = E\psi(\mathbf{r}), \rightarrow -\frac{\hbar^2}{2\mu r^2}\frac{\partial}{\partial r}\left(r^2\frac{\partial\psi(\mathbf{r})}{\partial r}\right) + \frac{\hat{L}^2}{2\mu r^2}\psi(\mathbf{r}) - \frac{Ze^2}{4\pi\epsilon_0 r}\psi(\mathbf{r}) = E\psi(\mathbf{r}).$$

Equation (1.36) can be solved by successive separation of variables. In the first separation of variables, we let $\psi(r,\theta,\phi) = R(r)Y(\theta,\phi)$. After separating r-function part from θ,ϕ-part, we have

$$-\frac{\hbar^2}{2\mu}\frac{1}{R}\frac{\partial}{\partial r}\left(r^2\frac{\partial R}{\partial r}\right) + r^2(V(r) - E)$$

$$= \frac{\hbar^2}{2\mu}\left[\frac{1}{Y\sin\theta}\frac{\partial}{\partial\theta}\left(\sin\theta\frac{\partial Y}{\partial\theta}\right) + \frac{1}{Y\sin^2\theta}\frac{\partial^2 Y}{\partial\phi^2}\right]$$

$$= const. = -\frac{\ell(\ell+1)\hbar^2}{2\mu}. \tag{1.37}$$

The solution of the above equations will correspond to some particular constants (eigenvalues), for reasons which will be discussed later, we choose the constant as $-\ell(\ell+1)\hbar^2/2\mu$, where ℓ is a constant integer. The radial part of Eq. (1.37) is

$$-\frac{\hbar^2}{2\mu r^2}\frac{d}{dr}\left(r^2\frac{dR}{dr}\right) + \frac{\ell(\ell+1)\hbar^2}{2\mu r^2}R - \frac{Ze^2}{4\pi\epsilon_0 r}R = ER \tag{1.38}$$

and the circumference part of Eq. (1.37) is

$$\hat{L}Y(\theta,\phi) = \ell(\ell+1)\hbar^2. \tag{1.39}$$

In the second separation of variables, we can further factor $Y(\theta,\phi) = P(\theta)u(\phi)$. With this form of the solution, we can further separate the second equation of Eq. (1.39) into another two equations,

$$-\frac{1}{u}\frac{d^2u}{d\phi^2} = \frac{\sin\theta}{P}\frac{d}{d\theta}\left(\sin\theta\frac{dP}{d\theta}\right) + \ell(\ell+1)\sin^2\theta = const. = m^2. \tag{1.40}$$

Again, the two equations in Eq. (1.37) equal to a same constant, which for a good reason we may denote it as m^2, and m is a constant.

The separation of variables leads to the following three (eigenvalue) ordinary differential equations,

$$r^2\frac{d^2R}{dr^2} + 2r\frac{dR}{dr} + \frac{2Mr^2}{\hbar^2}(E - V(r)) - \ell(\ell+1) = 0, \tag{1.41}$$

$$\frac{1}{\sin\theta}\frac{d}{d\theta}\left(\sin\theta\frac{dP}{d\theta}\right) + \ell(\ell+1)P - \frac{m^2}{\sin^2\theta}P = 0, \tag{1.42}$$

$$\frac{d^2u}{d\phi^2} = -m^2 u. \tag{1.43}$$

The solution of Eq. (1.41) is

$$R_{n\ell}(r) = \exp(-r/na_0)\left(\frac{r}{na_0}\right)^\ell L_{n-\ell-1}^{2\ell+1}\left(\frac{2r}{na_0}\right),$$

where $L_{n-\ell-1}^{2\ell+1}$ are the associated Laguerre polynomials, and $a_0 = 1$ in atomic units.

The solutions of Eqs. (1.42) and (1.43) are spherical harmonics and exponential functions, which can be expressed as follows:

$$P_{\ell m} = P_\ell^m(\cos\theta) \text{ and } u_m = \exp(im\phi),$$

where $P_{\ell m}(\cos\theta)$ are the associated Legendre polynomials, and the first few associated Legendre polynomials are

$$P_0^0(x) = 1, \ P_1^0(x) = x, \ P_1^1(x) = -(1-x^2)^{1/2},$$
$$P_2^0(x) = \frac{1}{2}(3x^2 - 1), \ P_2^1(x) = -3x(1-x^2)^{1/2}, \ldots$$

and the total stationary wave function solution for a hydrogen atom is

$$\psi_{n\ell m}(\mathbf{r}) = R_{n\ell}(r)Y_{\ell m}(\theta,\phi).$$

1.3.10 Spin

An electron has three basic properties: mass, charge, and spin. To understand what electron spin is, we may imagine that an electron is a charged sphere rotating around the axis of the sphere, which we usually call the Z-axis. According to classical electromagnetics, such a rotating sphere would have an angular momentum S associated with its rotational motion about the Z-axis. Moreover, since the sphere is charged, the rotating charge will give rise to a current loop. According to the classical electromagnetism, such tiny current loop will generate a magnetic dipole, and we denote its magnetic moment as μ_S.

In quantum mechanics, spin angular momentum S can take only certain directions and discrete magnitude. In solving the Schrodinger equation for the hydrogen atom, it is found that the orbital angular momentum is quantized according to the relationship

$$L^2 = \ell(\ell+1)\hbar^2$$

and, hence, the magnitude of the angular momentum in terms of the orbital quantum number is of the form

$$L = \sqrt{\ell(\ell+1)}\hbar$$

and that the z-component of the angular momentum in terms of the magnetic quantum number takes the form:

$$L_z = m_\ell \hbar.$$

The spin angular momentum follows the formula. The magnitude of spin angular momentum S is given by

$$S = \sqrt{s(s+1)}\hbar,$$

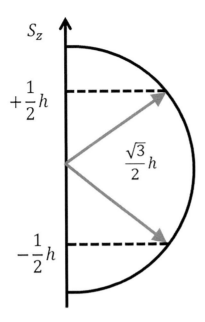

Figure 1.8 $m_s = 1/2$: "spin up" and $m_s = -1/2$: "spin down"

where s is the spin quantum number that equals to $1/2$. The spin direction is specified by the component of S along a Z-axis as shown in Fig. 1.8. The projection of the spin vector to the Z-axis has two possibilities:

$$S_Z = m_s \hbar = \pm \frac{1}{2} \hbar.$$

The associated magnetic moment is

$$\mu_z = \pm \frac{1}{2} g \mu_B,$$

where g is called the gyromagnetic ratio, and the electron spin g-factor has the value $g = 2.00232$. The electron spin can be predicted by the Dirac equation. In 1928, Paul Dirac derived a wave equation for describing the relativistic quantum wave equation of electrons. The Dirac equation is consistent with both the principles of quantum mechanics and the theory of special relativity, and the wave function is in (at least) four-dimensional (4D) space. The additional dimension provides the accommodation for spin and antimatter.

We can then introduce the spin Hermitian operator and its eigenvalues for electron spin as an analog of the orbital angular momentum operator. We denote the spin operator as $\hat{S}$, spin eigenfunction as $\chi_{s,m}$, and two spin quantum numbers as s and m_s, they are

and

$$\hat{\mathbf{S}} = \begin{pmatrix} \hat{S}_x \\ \hat{S}_y \\ \hat{S}_z \end{pmatrix}$$

and

$$\hat{S}_x \chi_{s,m} = m_s \hbar \chi_{s,m} = \pm \frac{1}{2} \hbar \chi_{s,m_s}; \quad \hat{S}^2 \chi_{s,m} = s(s+1)\hbar^2 \chi_{s,m_s}.$$

Note that the spin wave functions χ_{s,m_s} do not depend on the electron spatial coordinates r, θ, ϕ. They represent a purely internal degree of freedom, and we have two choices,

$$\chi_{1/2,1/2} = \begin{pmatrix} 1 \\ 0 \end{pmatrix} \text{ and } \chi_{1/2,-1/2} = \begin{pmatrix} 0 \\ 1 \end{pmatrix}.$$

1.4 Interpretations of Quantum Mechanics

1.4.1 Uncertainty Principle

The uncertainty principle, which is also called Heisenberg's uncertainty principle, was first put forward by Werner Heisenberg see Heisenberg (1985)) in the following expression:

$$\sigma_r \sigma_p \geq \frac{\hbar}{2}.$$

According to Heisenberg's explanation, it indicates that in a quantum system, the more precisely the position of a particle is being measured, the less precisely can one measure its momentum and vice versa. However, from the current understanding, Heisenberg's explanation of the uncertainty principle may be a little bit misleading because it improperly emphasizes the importance of measurement interference and measurement technology. In fact, the uncertainty principle is not about measurement technology, and it is independent from technology; it is a quantum mechanics statement on correlation condition between two observables. The uncertainty principle elucidates the relation of wave–particle duality in clear mathematical terms, it also articulates the necessity of a statistical approach to the study of subatomic particles and, remarkably, it provides an explicit expression on scale transition from determinacy to indeterminacy.

The uncertainty principle provides a lower bound of the position and momentum correlation. However, that is the direct estimate. To make an estimate uncertainty relation, Niels Bohr gave a derivation of the uncertainty relations between position and momentum and between time and energy. Consider

$$E = h\nu \text{ and } p = h/\lambda,$$

which connects the energy E and momentum p from the particle picture with those of frequency ν and wavelength λ from the wave picture. Denoting the spatial and temporal extensions of the wave packet by Δx and Δt, and the extensions in the wave

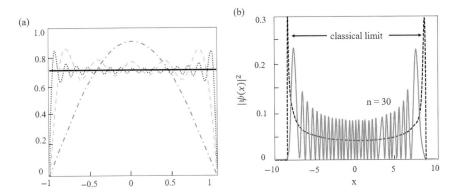

Figure 1.9 Comparison between classical probability density and amplitudes of wave functions: (a) particle in an infinite depth potential well and (b) harmonic oscillator

number and frequency by Δk and $\Delta \nu$, it then follows from Fourier analysis that, in the most favorable case, $\Delta x \Delta k \approx \Delta t \Delta \nu \approx 1$ and one obtains the relations

$$\Delta p = h\Delta k, \quad \Delta E = h\Delta \nu \quad \to \quad \Delta t \Delta E \approx \Delta x \Delta p \approx h.$$

Note that Δx, Δp are not the standard deviations, but unspecified measures of the size of a wave packet.

1.4.2 Correspondence Principle

In quantum physics, the correspondence principle (Bohr) states that the behavior of systems described by the theory of quantum mechanics reproduces classical physics in the limit of large quantum numbers. In other words, it states that, for large orbits and for large energies, quantum calculations must agree with classical calculations.

This can be demonstrated by the previous two examples, i.e., a particle in an infinite well and the quantum harmonic oscillator. From Fig. 1.9, one can find that, as the energy level increases, the profile of amplitudes of the wave function approaches the classical limit. This confirms the correspondence principle. However, for the quantum problems in which particle energy levels does not increase monotonically, how to link the microscale wave function description to the macroscale physical response is often a critical point of quantum mechanics for both theory and applications.

1.4.3 Pauli's Exclusion Principle

For a many-body quantum mechanical system, interchanging two particles occupying two different states should not change the probability density, $|\Psi|^2$, of the system. Consider a two noninteracting identical particle system. The probability density of the two-particle wave function $\Psi(\mathbf{r}_1,\mathbf{r}_2)$ must be identical to that of the the wave function $\Psi(\mathbf{r}_2,\mathbf{r}_1)$, where the particles have been interchanged, i.e.,

$$|\Psi(\mathbf{r}_1,\mathbf{r}_2)| = |\Psi(\mathbf{r}_2,\mathbf{r}_1)|.$$

There are two ways that this can be achieved:

1. Symmetric WF : $\Psi(\mathbf{r}_1,\mathbf{r}_2) = \Psi(\mathbf{r}_2,\mathbf{r}_1)$;
2. Antisymmetric WF : $\Psi(\mathbf{r}_1,\mathbf{r}_2) = -\Psi(\mathbf{r}_2,\mathbf{r}_1)$.

It turns out that particles whose wave functions are symmetric under particle interchange have integral or zero intrinsic spin and are termed "bosons." Particles whose wave functions are antisymmetric under particle interchange have half-integral intrinsic spin and are termed "fermions." Experimentation and quantum theory place electrons in the fermion category. That is the reason why we say that electrons are spin $-1/2$ particles and are described by the antisymmetric wave function:

$$\Psi(\mathbf{r}_1,s_1,\mathbf{r}_2,s_2,\ldots,\mathbf{r}_N,s_N).$$

Consider a two-particle noninteracting fermion system. The "noninteracting" qualifier implies the two-particle wave function can be written as the product of two single-particle wave functions. These can be written as either

$$\Psi_I(\mathbf{r}_1,\mathbf{r}_2) = \psi_a(\mathbf{r}_1)\psi_b(\mathbf{r}_2) \text{ or } \Psi_{II}(\mathbf{r}_2,\mathbf{r}_1) = \psi_a(\mathbf{r}_2)\psi_b(\mathbf{r}_1).$$

where a and b label two different single-particle states. Because we cannot distinguish between the particles, we cannot know which of Ψ_I or Ψ_{II}, describes the system. Consequently, we have to consider the system as being in some linear combination, or superposition of Ψ_I and Ψ_{II}. There are only two correctly normalized combinations possible:

1. Symmetric WF (bosons): $\Psi = \frac{1}{\sqrt{2}}(\Psi_I + \Psi_{II})$;
2. Antisymmetric WF (fermions): $\Psi = \frac{1}{\sqrt{2}}(\Psi_I - \Psi_{II})$.

In the case of fermions, if $a = b$, then $\Psi = 0$, which implies that no two fermions can occupy the same state. By considering the form of wave function for a system of identical particles, we have illustrated Pauli's exclusion principle.

1.4.4 Copenhagen Interpretation

The "Copenhagen Interpretation" of quantum physics refers to the following set of statements for explanation of the meanings of quantum mechanics:

- A system is completely described by a wave function, representing the state of the system, which evolves smoothly in time, except when a measurement is made, at which point it instantaneously collapses to an eigenstate of the observable measured.

- The description of nature is essentially probabilistic, with the probability of a given outcome of a measurement given by the square of the amplitude of the wave function (Born rule, named after Max Born).
- It is not possible to know the value of all the properties of the system at the same time; those properties that are not known exactly must be described by probabilities (Heisenberg's uncertainty principle).
- Matter exhibits a wave–particle duality. An experiment can show the particle-like or wave-like properties of matter. In some experiments, both of these complementary viewpoints must be invoked to explain the results, according to the complementarity principle of Niels Bohr.
- Measuring devices are essentially classical devices and measure only classical properties such as position and momentum.
- The quantum mechanical description of large systems will closely approximate the classical description. This is the correspondence principle of Bohr and Heisenberg.

1.4.5 Schrödinger's Cat and Parallel Universe

The many-worlds interpretation is an interpretation of quantum mechanics that asserts the objective reality of the universal wave function and denies the actuality of wave function collapse. "Many-worlds" implies that all possible alternative histories and futures are real, each representing an actual "world" (or "universe").

Schrödinger's cat is a thought experiment and sometimes it is described as a paradox. It was devised by Erwin Schrödinger in 1935. It illustrates what he saw as the problem of the Copenhagen interpretation of quantum mechanics when applied to everyday objects, resulting in a contradiction with common sense. The scenario presents a cat that might be alive or dead, depending on a previous random event.

One may even set up quite absurd but revealing cases where quantum mechanics, or at least the interpretation of quantum mechanics, conflicts with common sense. For example, we may consider an imaginary experiment where a cat is contained in a closed steel cage, along with a special device, in which there is a Geiger counter with a tiny amount of a radioactive substance. The amount of this radioactive substance is so small that, in the course of one hour, only one of the atoms may decay, but also with the equal probability that no atoms may decay. If the decay happens, a radiation detector will trigger the release of a hammer that will smash a glass container thereby releasing some form of poisonous gas, for instance hydrocyanic acid, killing the cat instantly. Thus, if one has left this entire system to itself for an hour, one may find that the cat may still be alive, if meanwhile no atom has decayed. On the other hand, one may also find the cat is dead, because there is an equal probability that an atom has decayed, and if there is only an atom decay, it would trigger the radioactive device and killed the cat. Therefore, the thought experiment cleverly translates a microscale quantum event described by a wave function into a macroscale event of the life-and-death experience

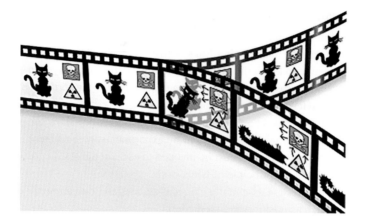

Figure 1.10 Parallel universe interpretation based on Schödinger's cat paradox (Photograph courtesy of Christian Schirm and Wikipedia.org)

of a poor cat. The essence of this thought experiment is that it establishes a direct correlation between a microscale wave function (atom decay or not decay) with a macroscale state of a cat (dead or alive). Since at microscale, there is a superposed state of wave functions, accordingly it will link to a macroscale state of cat, in which both the living and the dead cat are mixed in equal parts. Such a superposed reality at macroscale may be interpreted as the superposition of two events happening in two parallel universes as shown in Fig. 1.10.

1.5 Homework Problems

Problem 1.1 Consider a particle in an one-dimensional (1D) finite depth well, which obeys the Schrödinger equation,

$$\left(-\frac{\hbar^2}{2m}\frac{d^2}{dx^2} + V(x)\right)\psi(x) = E\psi(x),$$

where the potential energy is given as

$$V(x) = \begin{cases} V_0 & x \leq -\frac{L}{2} \\ 0 & |x| < \frac{L}{2} \\ V_0 & x \geq \frac{L}{2} \end{cases}.$$

Find all the eigenvalues (energy levels) and eigenfunctions of the solution. Discuss the solution when $V_0 \to \infty$.

Problem 1.2 Consider the following time-dependent Schrödinger equation,

$$i\hbar\frac{\partial \Psi}{\partial t} = -\left(\frac{\hbar^2}{2m}\right)\nabla^2\Psi + V_{eff}\Psi.$$

Make the following polar decomposition or variable separation,

$$\Psi(\mathbf{x},t) = R(\mathbf{x},t)\exp\left[i\frac{S(\mathbf{x},t)}{\hbar}\right].$$

Show that the time-dependent Schrödinger equation is equivalent to the following quantum hydrodynamic system,

$$\frac{\partial \rho}{\partial t} + \nabla \cdot (\rho \mathbf{v}) = 0 \qquad (1.44)$$

$$\frac{\partial S}{\partial t} + \frac{1}{2m}|\nabla S|^2 + V^{eff} + V^{qu} = 0 \qquad (1.45)$$

where

$$\rho := R^2 \qquad (1.46)$$

$$\mathbf{v} := \nabla\left(\frac{S}{m}\right) \qquad (1.47)$$

$$V_{qu} := -\left(\frac{\hbar^2}{2m}\right)\frac{\nabla^2 R}{R}$$

$$= -\left(\frac{\hbar^2}{2m}\right)\left(\nabla^2 \log R + |\nabla \log R|^2\right) \qquad (1.48)$$

or, in terms of ρ,

$$V^{qu} = \left(-\frac{\hbar^2}{4m\rho}\right)\left[\nabla^2 \rho - \frac{1}{2\rho}(\nabla\rho)\cdot(\nabla\rho)\right]. \qquad (1.49)$$

Equation (1.44) is an analog of conventional mass conservation law of continuum fluid dynamics.

Problem 1.3 Consider a particle in a box (1D), which obeys the following Schrödinger equation,

$$\left(-\frac{\hbar^2}{2m}\frac{d^2}{dx^2} + V\right)\Psi(x,t) = -\hbar\frac{\partial}{\partial t}\Psi(x,t)$$

with the time-independent part,

$$\left(-\frac{\hbar^2}{2m}\frac{d^2}{dx^2} + V\right)\psi(x) = E\psi(x),$$

where

$$V(x) = \begin{cases} \infty, & x \leq -\frac{L}{2} \\ 0, & |x| < \frac{L}{2} \\ \infty, & x \geq \frac{L}{2} \end{cases}.$$

Assume that the initial condition of the wave function is,

$$\Psi(x,0) = A \sin\left(\frac{2\pi x}{L}\right), \quad A \text{ is given.}$$

Find

$$<x>, \quad <p_x>, \quad \text{and} \quad <H>?$$

Problem 1.4 Consider the angular momentum operator, $\hat{\mathbf{L}} = -i\hbar \mathbf{r} \times \nabla$:

$$\hat{L}_x = -i\hbar\left[y\frac{\partial}{\partial z} - z\frac{\partial}{\partial y}\right], \quad \hat{L}_y = -i\hbar\left[z\frac{\partial}{\partial x} - x\frac{\partial}{\partial z}\right], \quad \text{and} \quad \hat{L}_z = -i\hbar\left[x\frac{\partial}{\partial y} - y\frac{\partial}{\partial x}\right].$$

Calculate the operator commutator,

$$[\hat{L}_x, \hat{L}_y] = ?$$

Problem 1.5 Define the standard deviation of two Hermitian operators, $\hat{A}$ and $\hat{B}$, as:

$$\sigma_A = \sqrt{<\hat{A}^2> - <\hat{A}>} \quad \text{and} \quad \sigma_B = \sqrt{<\hat{B}^2> - <\hat{B}>}.$$

Show that

$$\sigma_A^2 = <(\hat{A}- <\hat{A}>)\Psi|(\hat{A}- <\hat{A}>)\Psi> = <f|f>,$$
$$\text{where } |f> := |(\hat{A}- <\hat{A}>)\Psi>$$
$$\sigma_B^2 = <(\hat{B}- <\hat{B}>)\Psi|(\hat{B}- <\hat{B}>)\Psi> = <g|g>,$$
$$\text{where } |g> := |(\hat{B}- <\hat{B}>)\Psi>$$

and

$$<f|g> = <\hat{A}\hat{B}> - <\hat{A}><\hat{B}>$$
$$<g|f> = <\hat{B}\hat{A}> - <\hat{B}><\hat{A}>. \tag{1.50}$$

Let $\hat{A} = \hat{x}$ and $\hat{B} = \hat{p}_x$ and use the Cauchy–Schwartz inequality to show that

$$\sigma_x \sigma_{p_x} \geq \frac{\hbar}{2},$$

considering the fact that, for the complex number $<f|g>$,

$$|<f|g>|^2 = \left(\frac{<f|g> + <g|f>}{2}\right)^2 + \left(\frac{<f|g> - <g|f>}{2i}\right)^2.$$

2 Density Functional Theory

To build the foundation for first-principles computations, in this chapter, we provide a short introduction to the density functional theory (DFT).

2.1 Many-body Schrödinger Equation

The materials that we encounter in science and engineering applications consist of many atoms and electrons. In other words, they are multiatom or multi-electron systems. In principle, to model the material behaviors from first principles, one should consider solving the many-body time-independent Schröinger equation as follows,

$$\hat{H}\Psi(\mathbf{r}_1, \ldots, \mathbf{r}_N; \mathbf{R}_1, \ldots, \mathbf{R}_M) = E\Psi(\mathbf{r}_1, \ldots, \mathbf{r}_N; \mathbf{R}_1, \ldots, \mathbf{R}_M), \tag{2.1}$$

which may be written in details as,

$$\left[\sum_{i=1}^{N} \left(-\frac{\hbar^2}{2m} \nabla_i^2 + \sum_{j>i} \frac{q^2}{|\mathbf{r}_i - \mathbf{r}_j|} + \sum_{A=1}^{M} \frac{Q_A q}{|\mathbf{r}_i - \mathbf{R}_A|} \right) \right.$$
$$\left. + \sum_{A=1}^{M} \left(-\frac{\hbar^2}{2M_A} \nabla_A^2 + \sum_{B>A} \frac{Q_A Q_B}{|\mathbf{R}_A - \mathbf{R}_B|} \right) \right] \Psi(\mathbf{r}_1, \ldots, \mathbf{r}_N; \mathbf{R}_1, \ldots, \mathbf{R}_M)$$
$$= E\Psi(\mathbf{r}_1, \ldots, \mathbf{r}_N; \mathbf{R}_1, \ldots, \mathbf{R}_M), \tag{2.2}$$

where $q = -e$ is the charge of electrons, $Q_A = Z_A e$ is the charge of the nucleus, and M_A is the mass of the nucleus.

In quantum chemistry, the Born–Oppenheimer approximation is adopted under the assumption on the validity of separation of electron variables from nucleus variable in the total wave function of the system. That is the total wave function can be written as the product of the wave function of electron variables and the wave function of nucleus variables. This then leads to a molecular wave function in terms of electron positions $\mathbf{r}_i, i = 1, 2, \ldots, N$ and nucleus positions $\mathbf{R}_A, A = 1, 2, \ldots, M$,

$$\Psi(\mathbf{r}_1, \ldots, \mathbf{r}_N; \mathbf{R}_1, \ldots, \mathbf{R}_M) = \Psi_e(\mathbf{r}_1, \ldots, \mathbf{r}_N; \mathbf{R}_1, \ldots, \mathbf{R}_M)\Psi_u(\mathbf{R}_1, \ldots, \mathbf{R}_M),$$

where $\mathbf{R}_1, \ldots, \mathbf{R}_M$ in Ψ_e are not variables.

Since the electron moves much faster than that of the nucleus, and the electron mass is much smaller than that of the nucleus, so that the two timescales that are associated with electrons and nucleus differ by several orders of magnitude, we can then separate the treatment of the first three terms of the Hamiltonian in Eq. (2.2) with that of the last two terms.

In this chapter we are focusing on the electron and electronic structure problems only, hence we froze the position of the nucleus during a small time interval when we study the electronic structure problems as an approximation due to the big difference of two timescales. Under the Born–Oppenheimer approximation, we then denote the electronic wave function as,

$$\Psi_e(\mathbf{r}_1,\ldots,\mathbf{r}_N;\mathbf{R}_1,\ldots,\mathbf{R}_M) \approx \Psi(\mathbf{r}_1,\ldots,\mathbf{r}_N),$$

so that we only study the following time-independent, nonrelativistic Schrödinger equation,

$$\sum_{i=1}^{N}\left(-\frac{\hbar^2}{2m}\nabla_i^2 + \sum_{j>i}\frac{q^2}{|\mathbf{r}_i-\mathbf{r}_j|} - \sum_{A=1}^{M}\frac{Q_A q}{|\mathbf{r}_i-\mathbf{R}_A|}\right)\Psi(\mathbf{r}_1,\ldots,\mathbf{r}_N) = E\Psi(\mathbf{r}_1,\ldots,\mathbf{r}_N). \quad (2.3)$$

Different from most problems illustrated in Chapter 1, the wave function of the many electron problem contains 3N set of coordinates, and it is a 3N dimensional partial differential equation problem, which is one of the most challenging problems in mathematical physics.

The basic idea of the density functional theory is to reduce the dimension of the problem, so that the problem can be replaced and solved as the approximation of a set of three-dimensional coupled partial differential and integral equation problems. To better understand this process, we shall first understand what is electron density.

2.1.1 Electron Density

Consider an N-electron system. The probabilistic presence of fast-moving electrons may be imagined as a continuous electron cloud as shown in Fig. 2.1, whose probabilistic density to find any electron within a volume is defined as

$$\rho(\mathbf{r}) = N\int\cdots\int |\Psi(\mathbf{r},\mathbf{r}_2,\ldots,\mathbf{r}_N)|^2 d\mathbf{r}_2 d\mathbf{r}_3 \ldots d\mathbf{r}_N, \quad (2.4)$$

and if we integrate it over all the three dimensional space, we should find all N- electrons, i.e.,

$$N = \int_{-\infty}^{\infty} \rho(\mathbf{r})d\mathbf{r} = N[\rho(\mathbf{r})], \quad \text{because of } \sum_i q_i = N, \quad (2.5)$$

here we use the atomic unit in which $e = 1$.

It should be noted that the expression in Eq. (2.5) is actually a functional, i.e., a function of function so to speak, because the integrand, i.e., the electron density ρ,

2 Density Functional Theory

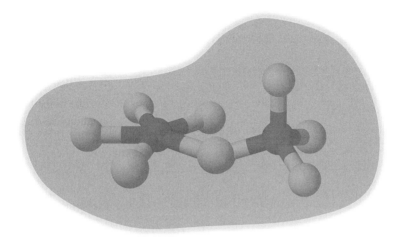

Figure 2.1 Schematic illustration of electron density: $\rho \propto \Psi^*\Psi$

is a function, i.e., $\rho = \rho(\mathbf{r})$. There are many examples of functionals in quantum mechanics as well as in classical mechanics, such as the expectation of the Hamiltonian operator,

$$<H> := \frac{\int \Psi^*(\mathbf{r})\hat{H}\Psi(\mathbf{r})dV}{\int \Psi^*(\mathbf{r})\Psi(\mathbf{r})dV},$$

or the strain energy of an elastic bar,

$$U[\epsilon(x)] = \int_V \frac{E}{2}\epsilon^2(x)A dx,$$

where x is the coordinate the rod and A is section area of the rod.

By definition, a functional is a map from a function space to $\mathbb{R}$, i.e.,

$$N: \rho(\mathbf{r}) \to \mathbb{R},$$

where

$$\rho(\mathbf{r}) \in \mathcal{L} := \{\rho(r) | \rho(r) \in L^1 \text{ and } \rho(r) \geq 0\}.$$

In the following contents, we need to use the calculus of variation to find a particular function that makes a given functional to achieve extreme value, i.e., minimum and maximum.

To get familiarized with calculus of variation, we first recall that in calculus, the first-order derivative of a function $f(x)$, df/dx, measures the first-order change of $y = f(x)$ upon changes of x, i.e., the slope of the function $f(x)$ at x,

$$f(x + dx) = f(x) + \frac{df}{dx}dx + O(dx^2).$$

In calculus of variation, the first-order functional derivative measures, similarly, the first-order change in a functional upon a function variation in its argument:

$$F[f(x) + \delta f(x)] = F[f(x)] + \delta F[f] + O((\delta f)^2),$$

where

$$\delta F[f] := \lim_{\tau \to 0} \frac{F[f + \tau \delta f] - F[f]}{\tau} = \frac{d}{d\tau} F[f + \tau \delta f]\bigg|_{\tau=0} = F'[f]\delta f$$

is called the Gateaux derivative. In many physics literature, people use the notation,

$$F'[f] = \frac{\delta F}{\delta f},$$

where the first functional derivative $\delta F[f]/\delta f(x)$ may be viewed as the first-order coefficient (or "functional slope").

However, there is a subtlety between the notation in physics literature and the notation of functional derivative. To explain it, we consider a functional

$$E[f] = \int_\Omega F(f(x))dx,$$

where $F(f(x))$ is a function of a real-value function $f(x), x \in \Omega$. Since the functional derivative of $E[u]$ is defined as,

$$\delta E = \int_\Omega \frac{\delta E}{\delta f} \delta f dx = \left[\frac{d}{d\tau} \int_\Omega F(f(x) + \tau \delta f)dx\right]_{\tau=0} = \int_\Omega \frac{dF}{df} \delta f dx$$

and thus, by using the definition of functional derivative we may write

$$\frac{\delta E}{\delta f} = F'(f(x)).$$

Furthermore, by using the Gateaux derivative, we can also define the second variation as

$$\delta^2 F = \frac{d^2}{d\tau^2} F[f + \tau \delta f]\bigg|_{\tau=0} = F''[f](\delta f)^2 \tag{2.6}$$

Without going into the details of calculus variation, but as an analog of finding a minimum of a function in calculus, we have the necessary and sufficient condition for a functional $F[f]$ to achieve a minimum, i.e., for a special $f = f^\star(x) \in \mathcal{V}$,

$$\delta F[f^\star(x)] = 0, \quad \text{and} \quad \delta^2 F[f^\star(x)] > 0, \tag{2.7}$$

where $\mathcal{V}$ is a functional space. It may be noted that $\delta^2 F$ is not proportional to F''.

2.2 Hartree–Fock Approximation

To solve many-body electron problems, D. R. Hartree, V.A. Fock, and J. C. Slater adopted the following approaches (see Hartree 1928, 1957; Fock 1930; Slater 1928):

(i) Neglecting the correlation among electrons, i.e., we assume that the electrons are uncorrelated;

(ii) Total wave function can be written as the product of single electron wave functions,

$$\Psi(\mathbf{r}_1, \mathbf{r}_2, \ldots, \mathbf{r}_N) = \frac{1}{\sqrt{N!}} \begin{vmatrix} \psi_1(\mathbf{r}_1) & \psi_1(\mathbf{r}_2) & \cdots & \psi_1(\mathbf{r}_N) \\ \psi_2(\mathbf{r}_1) & \psi_2(\mathbf{r}_2) & \cdots & \psi_2(\mathbf{r}_N) \\ \vdots & \vdots & \ddots & \vdots \\ \psi_N(\mathbf{r}_1) & \psi_N(\mathbf{r}_2) & \cdots & \psi_N(\mathbf{r}_N) \end{vmatrix}$$

Electrons are fermions, and they obey the Pauli exclusion principle and Fermi statistics. Wave function changes sign when the coordinates of two electrons are interchanged, and wave function is zero if two electrons are in the same state. Hence, the total wave function is written to be an antisymmetrized product (Slater determinant). It may be noted that the original Hartree's construction is: $\Psi(\mathbf{r}^N) = \psi_\alpha(\mathbf{r}_1), \psi_\beta(\mathbf{r}_2), \ldots, \psi_\pi(\mathbf{r}_N)$ where $(\alpha, \beta, \ldots, \pi)$ is a combination of states $(1, 2, \ldots N)$, which is called the Hartree product; and $\mathbf{r}^N = \{\mathbf{r}_1, \mathbf{r}_2, \ldots, \mathbf{r}_N\}$.

(iii) The variational principle

$$\delta\left(<\Psi(\mathbf{r}^N)|\hat{H}|\Psi(\mathbf{r}^N)> - \sum_i \epsilon_i <\psi_i(\mathbf{r}_i)|\psi_i(\mathbf{r}_i)> \right) = 0,$$

where

$$\hat{H} = -\sum_i \frac{\hbar^2}{2m}\nabla_i^2 - \sum_i \frac{Ze^2}{4\pi\epsilon_0|\mathbf{r}_i|} + \frac{1}{2}\sum_{i\neq j} \frac{e^2}{4\pi\epsilon_0|\mathbf{r}_i - \mathbf{r}_j|},$$

leads to the Hartree equation

$$\left[-\frac{\hbar^2}{2m}\nabla_i^2 + V_{ext}(\mathbf{r}_i) + V_H(\mathbf{r}_i)\right]\psi_i(\mathbf{r}_i) = \epsilon_i\psi_i(\mathbf{r}_i),$$

where

$$V_{ext} = -\frac{Ze^2}{4\pi\epsilon_0|\mathbf{r}_i|}, \quad \text{and} \quad V_H(\mathbf{r}) = \sum_{j\neq i}\int \frac{|\psi_j(\mathbf{r}_j)|^2}{|\mathbf{r}-\mathbf{r}_j|}dV_j.$$

The second potential is the Hartree potential, which is the average Coulomb potential from other electrons, i.e., $j \neq i$.

(iv) Only consider the ground-state wave function.

2.3 Hohenberg and Kohn Theorems

Before discussing the density functional theory, we need to understand why many-body quantum mechanics problem is so difficult.

This is because that the wave function of N-electron problem

$$\Psi(\mathbf{r}_1(t), \mathbf{r}_2(t), \ldots, \mathbf{r}_N(t)),$$

Figure 2.2 Walter Kohn was an Austrian-born American theoretical physicist and theoretical chemist, and he was a professor at the University of California, Santa Barbara. In 1998, Kohn was awarded the Nobel Prize in Chemistry for his contributions to the understandings of the electronic properties of materials (Photo courtesy of Markus Pössel and Wikipedia.org)

contains 3N degrees of freedom, and hence the computation complexity of wave function renders a 3N-dimensional problem.

Note that here each $\mathbf{r}_i(t)$, $i = 1, 2, \ldots, N$ is not merely or just a particle position, but representing a particle's trajectory or orbital field. One has to remember that this is quantum mechanics problem, we do not know exactly where the particle is (uncertainty principle).

The solution of this problem got a breakthrough in 1960s, starting from the work of Walter Kohn (see Fig. 2.2) and Pierre Hohenberg. In the following, we shall first briefly introduce the Hohenberg–Kohn (HK) theorems.

THEOREM 2.1 (Theorem I) *The external potential is uniquely determined by the electronic charge density, $\rho(\mathbf{r})$, so that the total energy is a unique functional of the density $E[\rho]$.*

Before we present the proof given by Hohenberg and Kohn, we would like to make a few comments.

First, we may define the electron density as

$$\rho(\mathbf{r}) = N \int \prod_{i=2}^{N} d\mathbf{r}_i |\psi(\mathbf{r}_1, \mathbf{r}_2, \ldots, \mathbf{r}_N)|^2.$$

The tricky part is that the order of $\mathbf{r}_1, \mathbf{r}_2, \ldots, \mathbf{r}_N$ can be interchanged arbitrarily. Then we can express E_{ext} as functional of electronic density, i.e., $E_{ext}[\rho]$. Why can we write

$$E_{ext} = <\psi|\hat{V}_{ext}|\psi> = \int V_{ext}(\mathbf{r})\rho(\mathbf{r})dV_r = E_{ext}(\rho) ?$$

This is because

$$E_{ext} = <\psi|\hat{V}_{ext}|\psi> = \sum_i \int <\psi|V_{ext}(\mathbf{r}_i)|\psi> .$$

Assume that all the external forces are due to electrostatic interactions, i.e.,

$$V_{ext}(\mathbf{r}) = -\sum_\alpha \frac{Z_\alpha}{|\mathbf{r} - \mathbf{R}_\alpha|} \quad \text{and} \quad \hat{V}_{ext} = \sum_{i=1}^{N} V_{ext}(\mathbf{r}_i).$$

Hence

$$E_{ext} = \int \sum_i <\psi|V_{ext}(\mathbf{r}_i)|\psi>$$

$$= \int \sum_i \prod_{j=1}^{N} V_{ext}(\mathbf{r}_i)\psi^*(\mathbf{r}_1, \ldots, \mathbf{r}_N)\psi(\mathbf{r}_1, \ldots, \mathbf{r}_N)d\mathbf{r}_1 \cdots d\mathbf{r}_j \cdots d\mathbf{r}_N$$

$$= \int \sum_i \prod_{j=1}^{N} V_{ext}(\mathbf{r}_1)\psi^*(\mathbf{r}_1, \ldots, \mathbf{r}_N)\psi(\mathbf{r}_1, \ldots, \mathbf{r}_N)d\mathbf{r}_1 \cdots d\mathbf{r}_j \cdots d\mathbf{r}_N$$

$$= \int V_{ext}(\mathbf{r}_1)\rho(\mathbf{r}_1)d\mathbf{r}_1$$

in the last line, we used the definition of electron density,

$$\rho(\mathbf{r}) = N \int \prod_{i=2}^{N} d\mathbf{r}_i |\psi(\mathbf{r}_1, \mathbf{r}_2, \ldots, \mathbf{r}_N)|^2.$$

The idea of the proof is to show that the total energy E is unique for fixed ρ, i.e.,

$$E = E[\rho] = T[\rho] + E_{ee}[\rho] + E_{ext}[\rho]$$

So far, we only know that this is the case for $E_{ext}[\rho]$. Below is the proof given by Hohenberg and Kohn (1964).

Proof Suppose that there are two different many-body electron wave functions, Ψ_1 and Ψ_2, that generate the same electronic density, and E_1, E_2 are the two ground-state energies corresponding to two wave functions Ψ_1 and Ψ_2,

$$<\Psi_1|\hat{H}_1|\Psi_1>= E_1, \text{ and } <\Psi_2|\hat{H}_2|\Psi_2>= E_2.$$

Based on the variational principle,

$$<\Psi_2|\hat{H}_1|\Psi_2\gg E_1.$$

However, H_1 and H_2 only differ by the external potentials, i.e., $\hat{H}_1 = \hat{H}_2 + V_2 - V_1$, so that

$$\langle\psi_2|\hat{H}_2|\psi_2\rangle + \langle\psi_2|V_1 - V_2|\psi_2\rangle > E_1 \to E_2 + \langle\psi_2|(V_1(\mathbf{r}) - V_2(\mathbf{r})|\psi_2\rangle > E_1.$$

$$E_2 + \int (V_1(\mathbf{r}) - V_2(\mathbf{r}))\rho(\mathbf{r})dV_r > E_1 \tag{2.8}$$

Similarly

$$E_1 + \int (V_2(\mathbf{r}) - V_1(\mathbf{r}))\rho(\mathbf{r})dV_r > E_2. \tag{2.9}$$

Adding Eqs. (2.8) and (2.9), one has

$$E_1 + E_2 > E_1 + E_2,$$

which is a contradiction, and hence the proof. □

Thus, in principle, the ground-state density determines (to within a constant) the external potential of the Schrödinger equation. Then the external potential and number of electrons determine all the ground-state properties of the system, since the Hamiltonian and ground-state wave function are determined by them.

THEOREM 2.2 (Theorem II) *There is a universal function $E[\rho]$ that can be minimized to obtain the exact ground-state density and energy. The density that minimizes the energy is the ground-state density and the minimum energy is the ground state energy,*

$$\underset{\rho}{\text{Min }} E[\rho] = E_0.$$

The proof of the theorem II is straightforward.

Proof We have already established the fact that $E = E(\rho)$ in Theorem I. For any admissible wave function, ψ, it has the property,

$$<\psi|\hat{H}|\psi>\geq<\psi_0|\hat{H}|\psi_0>, \quad \to \quad E[\rho] \geq E_0.$$

In fact, the variational principle states that the energy computed from a guessed wave function ψ is an upper bound to the true ground-state energy E_0. A full minimization of E with respect to all allowed N-electrons wave functions will give the true ground state,

$$E_0[\Psi_0] = \underset{\psi}{min} E[\psi] = \underset{\psi}{min}\{<\psi|\hat{T} + \hat{V} + \hat{V}_{ee}|\psi>\}$$

and

$$E_0[\rho_0] = \min_{\rho} E[\rho] = \min_{\rho}\{T[\rho] + E_{Ne}[\rho] + E_{ee}[\rho]\}. \qquad \square$$

The first H–K theorem demonstrates that the ground-state properties of a many-electron system are uniquely determined by an electron density that depends on only three spatial coordinates. It establishes the groundwork for reducing the many-body problem of N-electrons with $3N$ spatial coordinates to only three spatial coordinates an electron "continuum," through the use of functionals of the electron density. Theorem I has since been extended to the time-dependent domain to develop time-dependent density functional theory (TDDFT), which can be used to describe excited states. Theorem II defines an energy functional for an electron system, and it proves that the correct ground-state electron density minimizes this energy functional.

On the other hand, the HK theorems do not provide any computational scheme to calculate electron density. It were Kohn and Sham who late developed the so-called Kohn–Sham (K–S) formalism to provide the computational framework to calculate the many-body electron problem.

2.4 Kohn–Sham Formalism

Before proceeding, we first would like to show that we can write the electron density for a set of N noninteracting electron orbitals as follows,

$$\rho(\mathbf{r}) = \sum_{i=1}^{N} |\psi_i(\mathbf{r})|^2. \qquad (2.10)$$

Consider an N-electron system. The probabilistic presence of the fast-moving electrons may be imaged as a continuous cloud, whose probabilistic density to find any electron within a small volume element, $d\mathbf{r}$ ($dv_\mathbf{r}$), is defined as

$$\rho(\mathbf{r}) = N \int \cdots \int |\Psi(\mathbf{r},\mathbf{r}_2,\ldots \mathbf{r}_N)|^2 d\mathbf{r}_2 d\mathbf{r}_3 \cdots d\mathbf{r}_N \qquad (2.11)$$

and if we integrate it over the entire three-dimensional space, we should find all N electrons, i.e.,

$$N = \int_{\mathbf{R}^3} \rho(\mathbf{r}) d\mathbf{r}, \qquad (2.12)$$

which can be seen from Eq. (2.11), if the wave function Ψ is normalized. Physically, in a atom unit $\sum_i q_i = N$. On the other hand, Eq. (2.11) is a very general definition of electron density.

Based on the Hartree–Fock approximation, for a N noninteracting electron system, $\psi_i(\mathbf{r}_i), i = 1, 2, \ldots, N$, we have

$$\Psi(\mathbf{r}_1, \mathbf{r}_2, \ldots, \mathbf{r}_N) = \frac{1}{\sqrt{N!}} \begin{vmatrix} \psi_1(\mathbf{r}_1) & \psi_1(\mathbf{r}_2) & \cdots & \psi_1(\mathbf{r}_N) \\ \psi_2(\mathbf{r}_1) & \psi_2(\mathbf{r}_2) & \cdots & \psi_2(\mathbf{r}_N) \\ \vdots & \vdots & \ddots & \vdots \\ \psi_N(\mathbf{r}_1) & \psi_N(\mathbf{r}_2) & \cdots & \psi_N(\mathbf{r}_N) \end{vmatrix}$$

$$= \frac{1}{\sqrt{N!}} \sum_{\sigma \in S_n} sgn(\sigma) \Pi_{i=1}^N \psi_i(\mathbf{r}_{\sigma_i}) \quad (2.13)$$

where σ is a permutation and S_n is the set of all possible permutations.

If we substitute Eq. (2.13) into the definition of electron density Eq. (2.11), we have

$$\rho = N \int \Psi^*(\mathbf{r}_1, \mathbf{r}_2, \ldots, \mathbf{r}_N) \Psi(\mathbf{r}_1, \mathbf{r}_2, \ldots, \mathbf{r}_N) d\mathbf{r}_2 d\mathbf{r}_3 \cdots d\mathbf{r}_N$$

$$= \int \left(\sum_{\sigma \in S_n} sgn(\alpha) \Pi_{i=1}^N \psi^*(\mathbf{r}_{\alpha i}) \right) \left(\sum_{\sigma \in S_n} sgn(\beta) \Pi_{i=1}^N \psi^*(\mathbf{r}_{\beta i}) \right) d\mathbf{r}_2 d\mathbf{r}_3 \cdots d\mathbf{r}_N.$$

$$(2.14)$$

Because

$$\int \psi_i^* \psi_j d\mathbf{r} = \delta_{ij}$$

the nonzero terms remained in Eq. (2.14) are

$$\int \left(\sum_{\sigma \in S_n} sgn^2(\alpha) \Pi_{i=1}^N \psi_i^*(\mathbf{r}_{\alpha_i}) \psi_i(\mathbf{r}_{\alpha i}) \right) d\mathbf{r}_2 d\mathbf{r}_3 \cdots d\mathbf{r}_N \rightarrow \rho = \sum_{i=1}^N \psi_i^*(\mathbf{r}) \psi_i(\mathbf{r}).$$

$$(2.15)$$

To see this clearly, we consider the example: $N = 2$. Based on the above formula, we have

$$\rho = 2 \int \Psi^*(\mathbf{r}_1, \mathbf{r}_2) \Psi(\mathbf{r}_1, \mathbf{r}_2) d\mathbf{r}_2$$

$$= \int (\psi_1^*(\mathbf{r}_1)\psi_2^*(\mathbf{r}_2) - \psi_2^*(\mathbf{r}_1)\psi_1^*(\mathbf{r}_2))(\psi_1(\mathbf{r}_1)\psi_2(\mathbf{r}_2) - \psi_2(\mathbf{r}_1)\psi_1(\mathbf{r}_2)) d\mathbf{r}_2$$

$$= (\psi_1^*(\mathbf{r}_1)\psi_1(\mathbf{r}_1)) \int \psi_2^*(\mathbf{r}_2)\psi_2(\mathbf{r}_2) d\mathbf{r}_2 + (\psi_2^*(\mathbf{r}_1)\psi_2(\mathbf{r}_1)) \int \psi_1^*(\mathbf{r}_2)\psi_1(\mathbf{r}_2) d\mathbf{r}_2$$

$$= \sum_{i=1}^{2} \psi_i^*(\mathbf{r}_1) \psi_i(\mathbf{r}_1). \quad (2.16)$$

In the above derivation, we used the conditions,

$$\int \psi_1^*(\mathbf{r}_2) \psi_2(\mathbf{r}_2) d\mathbf{r}_2 = 0 \text{ and } \int \psi_1(\mathbf{r}_2) \psi_2^*(\mathbf{r}_2) d\mathbf{r}_2 = 0.$$

For an N noninteracting electron system, the kinetic energy of the system can then be expressed as

$$T_s(\rho) = -\frac{1}{2}\sum_{i=1}^{N} <\psi_i|\nabla^2|\psi_i>, \tag{2.17}$$

and the Hartree energy (the Coulomb interaction between electrons) is

$$E_H(\rho) = \frac{1}{2}\int \frac{\rho(\mathbf{r})\rho(\mathbf{r}')}{|\mathbf{r}-\mathbf{r}'|}d\mathbf{r}d\mathbf{r}'. \tag{2.18}$$

This allows us to write the energy functional of the N noninteracting electron system as

$$E[\rho] = T_s[\rho] + E_H[\rho] + E_{xc}[\rho] + E_{ext}[\rho] + E_{ext}[\rho], \tag{2.19}$$

where $E_{ext}[\rho]$ is the external potential energy, i.e.,

$$E_{ext}[\rho] = \int V_{ext}(\mathbf{r})\rho(\mathbf{r})d\mathbf{r}. \tag{2.20}$$

For example, $E_{ext}[\rho])$ can be induced by a set of fixed nucleuses. In this case,

$$E_{ext}[\rho] = -\sum_A \int \frac{Z_A \rho(\mathbf{r})}{|\mathbf{r}-\mathbf{R}_A|}d\mathbf{r}. \tag{2.21}$$

In Eq. (2.19), we also introduced an exchange-correlation functional,

$$E_{xc}[\rho] = (T[\rho] - T_s[\rho]) + (E_{ee}[\rho] - E_H[\rho]) = E_x[\rho] + E_c[\rho], \tag{2.22}$$

where the exchange-correlation (xc) energy functional $E_{xc}[\rho]$ absorbs all the complicated many-body effects not contained in $T_s[\rho]$, $E_H[\rho]$ and $E_{ext}[\rho]$. $E_{xc}[\rho]$ is a density functional by virtue of the HK theorem for interacting particles. For example, under the local density approximation, it may be explicitly expressed as

$$E_{xc}[\rho] = -\frac{3}{4\pi}(3\pi^2)^{1/3}\int \rho^{4/3}(\mathbf{r})d\mathbf{r}.$$

Moreover, $T_s[\rho]$ is a density functional, by virtue of the HK theorem for noninteracting particles, while $E_H[\rho]$ and $E_{ext}[\rho]$ are explicit density functionals, and thus $E[\rho]$ is a density functional.

We can then set the Kohn–Sham energy density functional as

$$L[\rho] = E[\rho] - \mu\left(\int \rho d\mathbf{r} - N\right) \tag{2.23}$$

where μ is a Lagrangian multiplier.

Applying the stationary condition for the minimum principle of the Kohn–Sham functional, we have

$$\delta\left\{E[\rho] - \mu\left(\int \rho d\mathbf{r} - N\right)\right\}\bigg|_{\rho=\rho_0} = 0. \tag{2.24}$$

In the following, we calculate the variation term by term. Noticing that only the term ψ_i^* is related to the electron density, we have

$$\frac{\delta T_s}{\delta \psi_i^*} = -\frac{\hbar^2}{2m}\nabla^2 \psi_i, \tag{2.25}$$

$$\frac{\delta E_H}{\delta \psi_i^*} = \frac{\delta E_H}{\delta \rho}\frac{\delta \rho}{\delta \psi_i^*} = \int \frac{\rho(\mathbf{r}')}{|\mathbf{r}-\mathbf{r}'|}d\mathbf{r}'\psi_i(\mathbf{r}), \tag{2.26}$$

$$\frac{\delta E_{xc}}{\delta \psi_i^*} = V_{xc}(\mathbf{r})\psi_i(\mathbf{r}), \tag{2.27}$$

$$\frac{\delta E_{ext}}{\delta \psi_i^*} = V_{ext}(\mathbf{r})\psi_i(\mathbf{r}), \quad \leftarrow \quad E_{ext}[\rho] = \int V_{ext}[\mathbf{r}']\rho(\mathbf{r}')d\mathbf{r}' \tag{2.28}$$

and

$$\frac{\delta}{\delta \psi_i^*}\left(\int \rho d\mathbf{r} - N\right) = \mu \psi_i(\mathbf{r}). \tag{2.29}$$

Putting everything together, we obtain the **Kohn–Sham equations**,

$$\left[-\frac{\hbar^2}{2}\nabla^2 + \int \frac{\rho(\mathbf{r}')}{|\mathbf{r}-\mathbf{r}'|}d\mathbf{r}' + V_{xc}(\mathbf{r}) + V_{ext}(\mathbf{r})\right]\psi_i(\mathbf{r}) = \mu\psi_i(\mathbf{r}), \ i=1,2,\ldots,N. \tag{2.30}$$

In the original K–S Lagrangian, the constraint term is defined as

$$\sum_i \epsilon_i <\psi_i^*\psi_i> -\sum_i \epsilon_i.$$

where $\epsilon_i, i = 1, 2, \ldots N$ are the Lagrangian multipliers. Thus the K–S equation have the standard form,

$$\left[-\frac{\hbar^2}{2}\nabla^2 + \int \frac{\rho(\mathbf{r}')}{|\mathbf{r}-\mathbf{r}'|}d\mathbf{r}' + V_{xc}(\mathbf{r}) + V_{ext}(\mathbf{r})\right]\psi_i(\mathbf{r}) = \epsilon_i\psi_i(\mathbf{r}), \ i=1,2,\ldots,N. \tag{2.31}$$

2.5 Example: DFT Calculation of Silicon Band Structure

In solid-state physics, the electronic band structure of a solid describes the range of energies that an electron within the solid may have, which is called allowed bands, and ranges of energy that it may not have (called band gaps or forbidden bands).

In this example, we employ DFT formulation to calculate the electronic band structure for Silicon crystals. Silicon is a typical semiconductor, and the Silicon band structure can be used to explain the electronic material properties of Silicon, such as electrical resistivity and optical absorption, among others, so that it may help us to design or fabricate solid-state devices such as transistors, solar cells, sensors, and so forth.

The DFT formulation is used to solve the following nonlinear eigenvalue equations:

$$\left(-\frac{1}{2}\nabla_i^2 + V_H[\rho] + V_{eN}[\rho] + V_{xc}[\rho]\right)\psi_i = \epsilon_i \psi_i, \quad \rho(\mathbf{r}) = \sum_{i=1}^{N} |\psi_i|^2, \quad (2.32)$$

where the atomic unit is used, i.e., $\hbar = 1, m_e = 1, e = 1$.

For a succinct presentation, we denote

$$V_{eff}[\rho] = V_H[\rho] + V_{ext}[\rho] + V_{xc}[\rho] \rightarrow \left(-\frac{1}{2}\nabla_i^2 + V_{eff}[\rho]\right)\psi_i = \epsilon_i \psi_i. \quad (2.33)$$

The Kohn–Sham equation (2.33) is nonlinear eigenvalue problem, and before you found the wave functions, you do not know the electron density $\rho(\mathbf{r})$. The Kohn–Sham equations must be solved by an iterative solver, which is called the **self-consistent field (SCF) method**, or the **self-consistent solution**. The step to solve this problem is: first guess an electronic density, and then solve a linear eigenvalue problem and obtain ψ_i, subsequently update electron density ρ and hence the Hamiltonian $H[\rho]$,

$$\text{SCF} \quad \psi_i(r) \rightarrow \rho(r) \rightarrow H[\rho] = \left(-\frac{1}{2}\nabla_i^2 + V_{eff}[\rho]\right).$$

When solving the eigenvalue problems, we expand the wave function into a series

$$\psi_i = \sum_{\alpha} a_\alpha \phi_\alpha(\mathbf{r}),$$

and in computations, we only need to determine the coefficient vector $\mathbf{a} = \{a_\alpha\}$. This procedure is outlined in Fig. 2.3.

Since in this example the lattice structure is periodic, the potential energy $V_{eff}[\rho]$ must be a periodic function as well. However, this does not immediately imply that $\psi_i(\mathbf{r})$ is also periodic, or it has the translational symmetry. However, Bloch's theorem (see Appendix A) states that for eigenvalue equation of the form,

$$[\nabla^2 + V(\mathbf{r})]\psi(\mathbf{r}) = E\psi(\mathbf{r}), \quad (2.34)$$

if the potential energy $V(\mathbf{r})$ is periodic, the solution of the equation must have the form,

$$\psi_\mathbf{k}(\mathbf{r}) = \exp(i\mathbf{k} \cdot \mathbf{r})u_\mathbf{k}(\mathbf{r}), \quad (2.35)$$

where $\mathbf{k}$ is the wave number vector and the function $u(\mathbf{r})$ has the same periodicity as the potential function $V(\mathbf{r})$. The proof of the Bloch theorem is given in Appendix A.

In this problem, since the displacement field, $\mathbf{u}(\mathbf{r})$, has the same periodicity of the Silicon lattice, we can expand $\mathbf{u}(\mathbf{r})$ into the Fourier series, i.e.,

$$\mathbf{u}(\mathbf{r}) = c_\mathbf{G} \exp i(\mathbf{G} \cdot \mathbf{r})$$

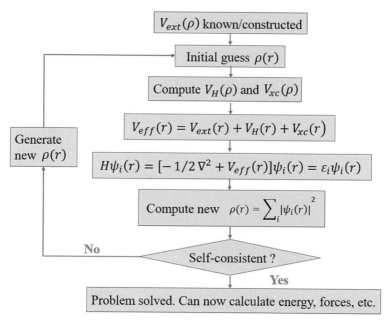

Figure 2.3 Flowchart of self-consistent iterative solution of the Kohn–Sham equation

where **G** is the reciprocal lattice vector, i.e.,

$$\mathbf{G} = g_1 \mathbf{b}_1 + g_2 \mathbf{b}_2 + g_3 \mathbf{b}_3,$$

where g_1, g_2, and g_3 are integers and $\mathbf{b}_i, i = 1, 2, 3$ are the lattice reciprocal bases, i.e.,

$$\mathbf{b}_1 = \frac{2\pi}{\Omega} \mathbf{a}_2 \times \mathbf{a}_3, \quad \mathbf{b}_2 = \frac{2\pi}{\Omega} \mathbf{a}_3 \times \mathbf{a}_1, \quad \text{and} \quad \mathbf{b}_3 = \frac{2\pi}{\Omega} \mathbf{a}_1 \times \mathbf{a}_2,$$

where Ω is the volume of the first Brillouin zone (BZ) that is also called the Wigner–Seitz cell (see Appendix A).

With these preparations, we can express the wave function into the sum of plane wave basis functions, i.e.,

$$\psi_{i,\mathbf{k}}(\mathbf{r}) = \sum_\mathbf{G} c_{i,\mathbf{k}+\mathbf{G}} \exp(i\mathbf{k} \cdot \mathbf{r}) \exp(i\mathbf{G} \cdot \mathbf{r}). \tag{2.36}$$

Therefore, the periodic conditions are following translation symmetry condition

$$\psi_{n,\mathbf{k}+\mathbf{G}}(\mathbf{r}) = \psi_{n,\mathbf{k}}(\mathbf{r}), \quad \epsilon_{n,\mathbf{k}+\mathbf{G}} = \epsilon_{n,\mathbf{k}},$$

Table 2.1 Some common first-principles (including DFT) computation software

ABINIT	Free, open source	www.abinit.org/
Material Studio	Commercial	http://accelrys.com/products/collaborative-science/biovia-materials-studio/
QMCPACK	Free, open source	http://qmcpack.org/
Quantum ESPRESSO	Free, open source	www.quantum-espresso.org/
SIESTA	Free, open source	https://departments.icmab.es/leem/siesta/
VASP	Commercial	www.vasp.at/

where n is an integer. Note that there is a cool way to write down the following expressions,

$$\Psi_{n,\mathbf{k}}(\mathbf{r}) = \exp(i\mathbf{k} \cdot \mathbf{r}) \sum_n C_{\mathbf{G}_n} \exp(i\mathbf{G}_n \cdot \mathbf{r})$$

and $V_{eff}(\mathbf{r}) = \sum_m V_{\mathbf{G}_m} \exp(i\mathbf{G}_m \cdot \mathbf{r})$.

That is,

$$\Psi_{n,\mathbf{k}}(\mathbf{r}) = \sum_n C_{\mathbf{G}_n} |\mathbf{k} + \mathbf{G}_n>, \quad V(\mathbf{r}) = \sum_m V_{\mathbf{G}_m} |\mathbf{G}_m>.$$

Substituting the above expressions into the Schrödinger equation,

$$\left[-\frac{1}{2}\nabla^2 + V_{eff}(\mathbf{r}) \right] \Psi_{n,\mathbf{k}}(\mathbf{r}) = \epsilon_{n,\mathbf{k}} \Psi_{n,\mathbf{k}}(\mathbf{r}).$$

We have the final form of algebraic eigenvalue equation,

$$\frac{\hbar^2}{2m} \sum_n |\mathbf{k} + \mathbf{G}_n|^2 C_{\mathbf{G}_n} e^{i(\mathbf{k}+\mathbf{G}_n)\cdot\mathbf{r}} + \sum_m \sum_n V_{\mathbf{G}_m} C_{\mathbf{G}_n} e^{i(\mathbf{k}+\mathbf{G}_n+\mathbf{G}_m)\cdot\mathbf{r}}$$
$$= \epsilon_{n,\mathbf{k}} \sum_n C_{\mathbf{G}_n} e^{i(\mathbf{k}+\mathbf{G}_n)\cdot\mathbf{r}}. \tag{2.37}$$

In principle, the summation of all the bases of the Fourier series should go over the entire wave number range: $0 \leq |\mathbf{k} + \mathbf{G}| < \infty$. In practical computations, the number of plane waves is determined by the cutoff kinetic energy E_{cut}, i.e.,

$$|\mathbf{k} + \mathbf{G}|^2 < E_{cut},$$

where $|\mathbf{k} + \mathbf{G}|^2$ is proportional to the kinetic energy. It may be noted that the quality of the plane wave solution is critically dependent on this cutoff.

There are several commercial and open source computational chemistry or computational materials software packages available to carry out DFT computations, and we list a few computational chemistry and computational materials software packages in Table 2.1.

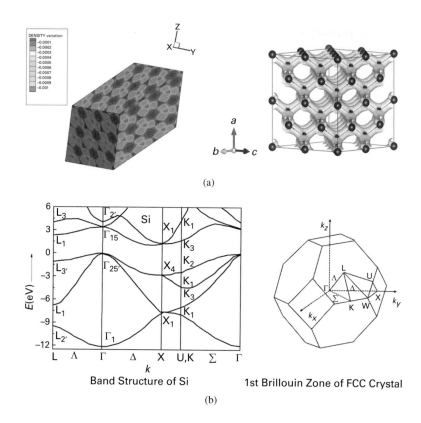

Figure 2.4 (a) Electron density distribution calculated by using ANINIT; (b) Silicon electronic band structure calculated by using Quantum ESPRESSO.

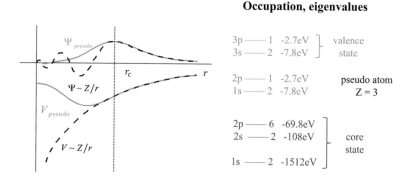

Figure 2.5 The concept of pseudopotential of a pseudo atom with charge number $Z = 3$.

The numerical results of DFT calculations of Silicon electron structure calculations are shown in Fig. 2.4, which were calculated by using both ABINIT and Quantum ESPRESSO software packages, which are open source software.

It may be noted that when using most computational chemistry software packages to do DFT calculations, instead of using the exact potential energy, we often use a so-called **pseudopotential**. This is because when the electron is close to nucleus, due to the strong repulsion, the potential function change dramatically, which causes significant numerical errors resulting slow converge. To remedy this problem, most DFT computation software use the pseudopotential function to replace the real potential function. The pseudopotential replaces the effect of core electrons by a fixed effective potential as shown by the solid lines in Fig. 2.5. Moreover, the pseudopotential removes the core states from the energy spectrum, and only takes into account the valence electron energy state, i.e., the red eigenvalues in Fig. 2.5.

In Chapter 4, as a computation tutorial, we shall demonstrate a step-by-step calculation of electron density in a single crystal Silicon by using another first-principles materials modeling software, The Vienna Ab-Initio Simulation Package, i.e., VASP: www.vasp.at/

3 Quantum Stress

Since the main objective of this book is to study mechanical properties and responses of materials and their microstructure at nanoscale, our focus is on nanoscale mechanical properties of materials. One of such properties is the stress–strain relation. However, stress is a macroscale mechanical property. To study the stress response of materials at nanoscale, we must first define what is stress at nanoscale or at atomistic scale. In fact, there are several physical measures of stress at atomistic scale. In this chapter, we first discuss the material stress as a quantum mechanical quantity.

3.1 Quantum Stress Theory

To begin with, we first consider the following many-body quantum mechanical Hamiltonian form,

$$H = \sum_i \frac{\mathbf{p}_i^2}{2m_i} + V_{int}(\mathbf{r}) + V_{ext}(\mathbf{r}), \tag{3.1}$$

where $\mathbf{p}_i$ is the momentum of particles i, m_i is the mass of electron i, V_{int} and V_{ext} represent the potential energy, which are the function of the positions $\mathbf{r}$. Combined with the Schrödinger equation, $H\Psi = E\Psi$, where Ψ is a wave function, the ground state of the wave function can be obtained by the minimization of the expectation of system's energy, $E = <\Psi|H|\Psi>$. To calculate the quantum stress, an uniform deformation is assumed to be applied on the ground state of the electron system. Based on the first-order Cauchy–Born rule, an deformed particles coordinate $\mathbf{r}$ can be related to the original undeformed coordinate $\mathbf{r}_0$,

$$\mathbf{r} = \mathbf{F} \cdot \mathbf{r}_0 \tag{3.2}$$

where the deformation gradient $\mathbf{F}$ is defined as,

$$\mathbf{F} = \frac{\partial \mathbf{x}}{\partial \mathbf{X}} = \frac{\partial (\mathbf{X} + \mathbf{u})}{\partial \mathbf{X}} = \mathbf{I} + \frac{\partial \mathbf{u}}{\partial \mathbf{X}},$$

where $\mathbf{u}$ is the displacement of particles and $\mathbf{I}$ is the second order unit tensor. Then the wave function $\Psi(\mathbf{r})$ may be written as

$$\Psi_\epsilon(\mathbf{r}) = \frac{1}{(det\mathbf{F})^{1/2}} \Psi(\mathbf{r}_0) = \frac{1}{J^{1/2}} \Psi(\mathbf{F}^{-1}\mathbf{r}), \tag{3.3}$$

where $J = det\mathbf{F}$ and the prefactor $J^{-1/2}$ ensures the normalization of $\Psi_\epsilon(\mathbf{r})$, $\int \Psi_\epsilon \Psi_\epsilon^* d\Omega_\mathbf{r} = 1$. Explicitly, the expectation value of H with respect to $\Psi_\epsilon(\mathbf{r})$ is written as,

$$<\Psi_\epsilon|H|\Psi_\epsilon> = \int \Psi_\epsilon^*(\mathbf{r}) \left[\sum_i \frac{\mathbf{p}_i \otimes \mathbf{p}_i}{2m_i} + V_{int}(\mathbf{r}) + V_{ext}(\mathbf{r}) \right] \Psi_\epsilon(\mathbf{r}) d\Omega_\mathbf{r}. \quad (3.4)$$

Since $\mathbf{r} = \mathbf{F} \cdot \mathbf{r}_0$, $\mathbf{p} \to -i\hbar \nabla$, we then have $\mathbf{p}_i = \mathbf{F}^{-1} \cdot \mathbf{p}_i^0$. By considering the normalization of the wave function, Eq. (3.4) can be rewritten as,

$$<\Psi_\epsilon|H|\Psi_\epsilon> = \int \Psi^* \left[\sum_i \frac{\mathbf{p}_i \otimes \mathbf{p}_i}{2m_i} + V_{int}(\mathbf{r}) + V_{ext}(\mathbf{r}) \right] \Psi d\Omega_{\mathbf{r}_0}$$

$$= \int \Psi^* \left[\sum_i \frac{(\mathbf{F}^{-1}(\mathbf{p}_i^0)) \otimes (\mathbf{F}^{-1}\mathbf{p}_i^0)}{2m_i} + V_{int}(\mathbf{F} \cdot \mathbf{r}_0) + V_{ext}(\mathbf{F} \cdot \mathbf{r}_0) \right] \Psi d\Omega_{\mathbf{r}_0}$$

$$= \int \Psi^* \left[\sum_i \frac{\mathbf{b} : (\mathbf{p}_i^0 \otimes \mathbf{p}_i^0)}{2m_i} + V_{int}(\mathbf{F} \cdot \mathbf{r}_0) + V_{ext}(\mathbf{F} \cdot \mathbf{r}_0) \right] \Psi d\Omega_{\mathbf{r}_0}, \quad (3.5)$$

where $\mathbf{b} = \mathbf{F}^{-T} \cdot \mathbf{F}^{-1}$ is the finger tensor in continuum mechanics.

Considering the quantum virial theorem, we find that the derivative of the expression in Eq. (3.5) with respect to strain $\mathbf{F}$ is zero,

$$\frac{d}{d\mathbf{F}} <\Psi_\epsilon|H|\Psi_\epsilon>$$

$$= \frac{d}{d\mathbf{F}} \int \Psi^* \left[\sum_i \frac{\mathbf{b} : (\mathbf{p}_i^0 \otimes \mathbf{p}_i^0)}{2m_i} + V_{int}(\mathbf{F} \cdot \mathbf{r}_0) + V_{ext}(\mathbf{F} \cdot \mathbf{r}_0) \right] \Psi d\Omega_{\mathbf{r}_0} = 0. \quad (3.6)$$

Noticing that the above integration is with respect to the undeformed lattice configuration. We can then take the derivative of the system's energy with respect to the deformation gradient and find that

$$\frac{d}{d\mathbf{F}} <\Psi_\epsilon|H|\Psi_\epsilon> = \int \Psi^* \frac{d}{d\mathbf{F}} \left[\frac{\sum_i \mathbf{b} : (\mathbf{p}_i^0 \otimes \mathbf{p}_i^0)}{2m_i} + V_{int}(\mathbf{F} \cdot \mathbf{r}_0) + V_{ext}(\mathbf{F} \cdot \mathbf{r}_0) \right] \Psi d\Omega_{\mathbf{r}_0}$$

$$= \sum_i \int \Psi^* \left[-\frac{\mathbf{p}_i \mathbf{F}^{-1} \otimes \mathbf{p}_i}{m_i} + \nabla_i (V_{int} + V_{ext}) \otimes \mathbf{r}_i^0 \right] \Psi d\Omega_{\mathbf{r}_0}$$

$$= \sum_i \left\langle \Psi \left| -\frac{\mathbf{p}_i \mathbf{F}^{-1} \otimes \mathbf{p}_i}{m_i} + \nabla_i (V_{int} + V_{ext}) \otimes \mathbf{r}_i^0 \right| \Psi \right\rangle = 0. \quad (3.7)$$

Using Eq. (3.7), one can derive the first Piola–Kirchhoff quantum stress $\mathbf{P}$ as

$$\mathbf{P} = \sum_i \left\langle \Psi \left| -\frac{\mathbf{p}_i \mathbf{F}^{-1} \otimes \mathbf{p}_i}{m_i} + \nabla_i (V_{int}) \otimes \mathbf{r}_i^0 \right| \Psi \right\rangle. \quad (3.8)$$

One may rewrite Eq. (3.8) as

$$\mathbf{P} = \sum_i \int \Psi^* \left[-\frac{\mathbf{p}_i \otimes \mathbf{p}_i \mathbf{F}^{-T}}{m_i} + \nabla_i(V_{int}) \otimes \mathbf{r}_i \mathbf{F}^{-T} \right] J d\Omega_{\mathbf{r}}$$

$$= \sum_i J \int \Psi^* \left[-\frac{\mathbf{p}_i \otimes \mathbf{p}_i}{m_i} + \nabla_i(V_{int}) \otimes \mathbf{r}_i \right] J d\Omega_{\mathbf{r}} \mathbf{F}^{-T}. \quad (3.9)$$

Considering the relation between the first Piola–Kirchhoff stress and the Cauchy stress

$$\mathbf{P} = J \sigma \mathbf{F}^{-T},$$

we can identify the quantum Cauchy stress tensor from Eq. (3.9) as

$$\sigma = \sum_i \left\langle \Psi \left| -\frac{\mathbf{p}_i \otimes \mathbf{p}_i}{m_i} + \nabla_i(V_{int}) \otimes \mathbf{r}_i \right| \Psi \right\rangle. \quad (3.10)$$

Equations (3.8) and (3.10) are the fundamental results of quantum stress theory.

REMARK 3.1 Nielsen and Martin derived the quantum stress based the assumption of infinitesimal strain deformation (Nielsen and Martin, 1985), and they used the first-order strain transformation on each particle coordinate, i.e., $\mathbf{r} \to (1 + \epsilon)\mathbf{r}$, where ϵ is symmetric strain tensor, to derive the expectation value of H with respect to $\Psi_\epsilon(\mathbf{r})$. That is,

$$< \Psi_\epsilon | H | \Psi_\epsilon > = \int \Psi^* \left[\sum_i \frac{\mathbf{b} : (\mathbf{p}_i^0 \otimes \mathbf{p}_i^0)}{2m_i} + V_{int}(\mathbf{F} \cdot \mathbf{r}) + V_{ext}(\mathbf{F} \cdot \mathbf{r}) \right] \Psi d\Omega_{\mathbf{r}}, \quad (3.11)$$

where $\mathbf{F} = 1 + \epsilon$ and $\mathbf{b} = \mathbf{F}^{-T} \cdot \mathbf{F}^{-1} = 1 - 2\epsilon + \epsilon^2$.

The quantum virial theorem requires that the derivation of Eq. (3.11) with respect to ϵ is zero, that is,

$$\frac{d}{d\epsilon} < \Psi_\epsilon | H | \Psi_\epsilon > = 0 \to \sum_i \left\langle \Psi \left| -\frac{\mathbf{p}_i \otimes \mathbf{p}_i}{m_i} + \nabla_i(V_{int} + V_{ext}) \otimes \mathbf{r}_i \right| \Psi \right\rangle = 0. \quad (3.12)$$

According to Eq. (3.12), Nielsen and Martin also found the quantum Cauchy stress σ_{int} of the many-body system as,

$$\sigma_{int} = \sum_i \left\langle \Psi \left| -\frac{\mathbf{p}_i \otimes \mathbf{p}_i}{m_e} + \nabla_i(V_{int}) \otimes \mathbf{r}_i \right| \Psi \right\rangle, \quad (3.13)$$

which is exactly the same as Eq. (3.10).

Subsequently, we can also define the stress exerted by external environment upon the condensed matter as,

$$\sigma_{ext} = -\sum_i \langle \Psi | \nabla_i V_{ext} \otimes \mathbf{r}_i | \Psi \rangle, \quad (3.14)$$

which leads to the equilibrium condition,

$$\sigma_{int} = \sigma_{ext}. \quad (3.15)$$

For example, if we assume that the internal potential energy has the following form

$$V_{int} = \frac{1}{2}\sum_i \sum_{j \neq i} V(r_{ij}),$$

we can find that

$$\sigma_{int} = -\sum_i \left\{ \left\langle \Psi \left| \frac{\mathbf{p}_i \otimes \mathbf{p}_i}{m_i} \right| \Psi \right\rangle + \sum_{j \neq i} \left\langle \Psi \left| V'(r_{ji}) \frac{\mathbf{r}_{ji} \otimes \mathbf{r}_{ij}}{r_{ij}} \right| \Psi \right\rangle \right\}. \quad (3.16)$$

One may compare this with the classical virial stress derived from molecular physics,

$$\sigma_V = \frac{1}{\Omega}\sum_{k=1}^{N} \left\{ -m_k \left(\frac{\mathbf{p}_k}{m_k} - \mathbf{v} \right) \otimes \left(\frac{\mathbf{p}_k}{m_k} - \mathbf{v} \right) + \frac{1}{2}\sum_{j \neq k} V'(r_{jk}) \frac{\mathbf{r}_{jk} \otimes \mathbf{r}_{jk}}{r_{jk}} \right\}, \quad (3.17)$$

where $\mathbf{v}$ is the average velocity of the atom system. Note that there is a fundamental difference from Eqs. (3.16) and (3.17): The summation in Eq. (3.16) is over electrons whereas the summation in Eq. (3.17) is over the nucleus.

3.2 Reciprocal-Space Expression for Quantum Stress

To evaluate quantum stress, the first approach proposed was using the Fourier transform of the wave functions to find the quantum stress in the reciprocal-space expansions.

In first-principles calculations, it is advantageous to express all quantum mechanical quantities in reciprocal space where a manageable plane wave basis set provides a high computational accuracy. Before we proceed, we first recall the Bloch's theorem.

Instead of being required to consider an infinite number of electrons, it is only necessary to consider the number of electrons within the unit cell (or half of this number if the electrons are spin degenerate). Considering the periodicity of the system, Bloch's theorem states that if the potential $V(\mathbf{r})$ is periodic, the wave function of an electron must have the form,

$$\Psi_{i,\mathbf{k}}(\mathbf{r}) = \exp(i\mathbf{k} \cdot r) u_i(\mathbf{r}), \quad (3.18)$$

where the function $u_i(\mathbf{r})$ has the same periodicity as the potential function $V(\mathbf{r})$, $u_i(\mathbf{r}) = u_i(\mathbf{r} + \mathbf{R})$, i is the band index, and $\mathbf{k}$ is a wave vector confined to the first Brillouin zone. Since $u_i(\mathbf{r})$ is a periodic function, we may expand it in terms of a Fourier series,

$$u_i(\mathbf{r}) = \sum_{\mathbf{G}} c_{i,\mathbf{G}} \exp(i\mathbf{G} \cdot \mathbf{r}), \quad (3.19)$$

where $\mathbf{G}$ is the reciprocal lattice vectors defined through $\mathbf{G} \cdot \mathbf{R} = 2\pi m$, m is an integer, $\mathbf{R}$ is a real space lattice vector, and $c_{i,\mathbf{G}}$ are plane wave expansion coefficients.

The electron wave functions may therefore be written as a linear combination of plane waves,

$$\Psi_{i,\mathbf{k}}(\mathbf{r}) = \exp(i\mathbf{k} \cdot \mathbf{r})u_i(\mathbf{r})$$
$$= \exp(i\mathbf{k} \cdot \mathbf{r}) \sum_{\mathbf{G}} c_{i,\mathbf{G}} \exp(i\mathbf{G} \cdot r)$$
$$= \sum_{\mathbf{G}} c_{i,\mathbf{k}+\mathbf{G}} \exp(i(\mathbf{k}+\mathbf{G}) \cdot \mathbf{r}). \quad (3.20)$$

It is natural to express the wave function as the sum of the Bloch functions,

$$\Psi(\mathbf{r}) = \sum_{i,\mathbf{G},\mathbf{k}} \Psi_{i,\mathbf{k}}(\mathbf{r}) = \sum_{i,\mathbf{G},\mathbf{k}} c_{i,\mathbf{k}+\mathbf{G}} \exp(i(\mathbf{k}+\mathbf{G}) \cdot \mathbf{r}). \quad (3.21)$$

Thus, using the momentum–space representations for the wave function, the charge density, the electronic Coulomb potential, and the exchange-correlation potential, which can be denoted by $\Psi_i(\mathbf{k}+\mathbf{G})$, $\rho(\mathbf{G})$, $V_{Coul}(\mathbf{G})$, and $\mu_{xc}(\mathbf{G})$, the reciprocal space expression for the total energy E_{tot} could be obtained.

(1) The kinetic energy,

$$E_{kin} = \left\langle \Psi_i(\mathbf{k}+\mathbf{G}) \left| -\frac{\hbar^2}{2m_e} \nabla^2 \right| \Psi_i(\mathbf{k}+\mathbf{G}) \right\rangle$$
$$= \sum_{i,\mathbf{k},\mathbf{G}} \frac{\hbar^2}{2m_e}(\mathbf{k}+\mathbf{G})^2 |\Psi_i(\mathbf{k}+\mathbf{G})|^2. \quad (3.22)$$

(2) The Hartree energy
Since the Fourier component of the interelectronic Coulomb potential $V_{Coul}(\mathbf{G})$ can be written as,

$$V_{Coul}(\mathbf{G}) = \frac{8\pi \rho(\mathbf{G})}{G^2}. \quad (3.23)$$

The Coulomb interaction energy, i.e., the Hartree energy, becomes,

$$E_H = \frac{1}{2} \int \int \frac{\rho(\mathbf{r})\rho(\mathbf{r}')}{|\mathbf{r}-\mathbf{r}'|} d^3\mathbf{r} d^3\mathbf{r}' = \frac{1}{2}\Omega_0 \sum_{\mathbf{G}} \frac{1}{2} V_{Coul}(\mathbf{G})\rho(\mathbf{G})$$
$$= \frac{1}{2}4\pi e^2 \sum_{\mathbf{G}} \frac{|\rho(\mathbf{G})|}{G^2}. \quad (3.24)$$

(3) The exchange-correlation energy
We may write the exchange-correlation energy functional as,

$$E_{xc} = \int d\rho(\mathbf{r})\mu_{xc}(\rho) = \sum_{\mathbf{G}} \mu_{xc}(\mathbf{G})\rho(\mathbf{G}). \quad (3.25)$$

(4) The ion–electron energy

$$E_{Ie} = \int V_{Ie}(\mathbf{r})\rho(\mathbf{r}) = \sum_{i,\mu,l} \int \Psi_i^*(\mathbf{r})U_{ps,l}(\mathbf{r}-\mathbf{R}_\mu)\hat{P}_l\Psi_i(\mathbf{r})d^3\mathbf{r}, \quad (3.26)$$

where

$$\Psi_i^*(\mathbf{r}) = \Psi_i^*(\mathbf{r}+\mathbf{R}_\mu) = \Psi_i^*(\mathbf{k}+\mathbf{G})\exp(-i(\mathbf{k}+\mathbf{G})\cdot(\mathbf{r}+\mathbf{R}_\mu)) \quad (3.27)$$

$$\Psi_i(\mathbf{r}) = \Psi_i(\mathbf{r}+\mathbf{R}_\mu) = \Psi_i(\mathbf{k}+\mathbf{G}')\exp(i(\mathbf{k}+\mathbf{G}')\cdot(\mathbf{r}+\mathbf{R}_\mu)). \quad (3.28)$$

Then, it becomes,

$$\begin{aligned} E_{Ie} &= \sum_{i,\mu,l} \int \Psi_i^*(\mathbf{r})U_{ps,l}(\mathbf{r}-\mathbf{R}_\mu)\hat{P}_l\Psi_i(\mathbf{r})d^3\mathbf{r} \\ &= \sum_{i,l,\mathbf{G},\mathbf{G}'} \Psi_i^*(\mathbf{k}+\mathbf{G})\exp(-i(\mathbf{k}+\mathbf{G})\cdot(\mathbf{r}+\mathbf{R}_\mu))U_{ps,l}(\mathbf{r}-\mathbf{R}_\mu) \\ &\quad \times \hat{P}_l\Psi_i(\mathbf{k}+\mathbf{G}')\exp(i(\mathbf{k}+\mathbf{G}')\cdot(\mathbf{r}+\mathbf{R}_\mu)) \\ &= \Omega_0 \sum_{i,l,\mathbf{G},\mathbf{G}'} \Psi_i^*(\mathbf{k}+\mathbf{G})\Psi_i(\mathbf{k}+\mathbf{G}') \sum_\mu \frac{\exp(i(\mathbf{G}-\mathbf{G}')\cdot\mathbf{R}_\mu)}{N} \cdot \\ &\quad \frac{N}{\Omega_0}\int \exp(-i(\mathbf{k}+\mathbf{G})\cdot\mathbf{r})U_{ps,l}(\mathbf{r})\hat{P}_l\exp(i(\mathbf{k}+\mathbf{G}')\cdot\mathbf{r})d^3\mathbf{r} \\ &= \Omega_0 \sum_{i,l,\mathbf{G},\mathbf{G}'} \Psi_i^*(\mathbf{k}+\mathbf{G})\Psi_i(\mathbf{k}+\mathbf{G}')S(\mathbf{G}-\mathbf{G}')U_{ps,l,\mathbf{k}+\mathbf{G},\mathbf{k}+\mathbf{G}'}. \end{aligned} \quad (3.29)$$

The generalized nonlocal form factor is,

$$\begin{aligned} U_{ps,l,\mathbf{k}+\mathbf{G},\mathbf{k}+\mathbf{G}'} &= \frac{N}{\Omega} \int \exp(-i(\mathbf{k}+\mathbf{G})\cdot\mathbf{r})U_{ps,l}(\mathbf{r})\hat{P}_l\exp(i(\mathbf{k}+\mathbf{G}')\cdot\mathbf{r})d^3\mathbf{r} \\ &= \frac{1}{\Omega}(2l+1)4\pi P_l(\cos\gamma)\int U_{ps,l}(r)j_l(|\mathbf{k}+\mathbf{G}|r)j_l(|\mathbf{k}+\mathbf{G}'|r)r^2 dr, \end{aligned} \quad (3.30)$$

where

$$\cos\gamma = \frac{[(\mathbf{k}+\mathbf{G})\cdot(\mathbf{k}+\mathbf{G}')]}{|\mathbf{k}+\mathbf{G}|\cdot|\mathbf{k}+\mathbf{G}'|}.$$

In the local pseudopotential approximation, Eq. (3.29) reduces to the form,

$$\sum_{i,\mu}\int \Psi_i^*(\mathbf{r})U_{ps}(\mathbf{r}-\mathbf{R}_\mu)\Psi_i(\mathbf{r})d^3\mathbf{r} = \Omega_0 \sum_\mathbf{G} S(\mathbf{G})U_{ps}\rho(\mathbf{G}). \quad (3.31)$$

Then the reciprocal expression for the ion–electron (core–valence) interaction becomes,

$$E_{Ie} = \sum_{i,\mu,l} \int \Psi_i^*(\mathbf{r}) U_{ps,l}(\mathbf{r} - \mathbf{R}_\mu) \hat{P}_l \Psi_i(\mathbf{r}) d^3\mathbf{r}$$

$$= \Omega_0 \Big(\sum_G S(\mathbf{G}) U_{ps}(\mathbf{G}) \rho(\mathbf{G})$$

$$+ \sum_{i,l,\mathbf{G},\mathbf{G}'} \Psi_i^*(\mathbf{k}+\mathbf{G}) \Psi_i(\mathbf{k}+\mathbf{G}') S(\mathbf{G}-\mathbf{G}') U_{ps,l,\mathbf{k}+\mathbf{G},\mathbf{k}+\mathbf{G}'} \Big). \quad (3.32)$$

In actual computations, in order to include the contribution from $V_{Coul}(0)$, $U_{ps}(0)$, and $\frac{1}{2}\sum_{I,J,I \neq J} \frac{Z_I Z_J e^2}{|R_I - R_J|}$, we add two additional terms to the total energy,

$$\gamma_{Ewald} + \Big[\sum_\tau \alpha_\tau Z_\tau \Big]. \quad (3.33)$$

Thus, the total energy expression in reciprocal space per unit cell volume Ω_0 becomes,

$$\frac{E_{tot}}{\Omega_0} = \sum_{i,\mathbf{k},\mathbf{G}} \frac{\hbar^2}{2m_e}(\mathbf{k}+\mathbf{G})^2 |\Psi_i(\mathbf{k}+\mathbf{G})|^2 + \frac{1}{2} 4\pi e^2 \sum_G \frac{|\rho(\mathbf{G})|}{G^2} + \sum_G \mu(\mathbf{G})\rho(\mathbf{G})$$

$$+ \sum_G S(\mathbf{G}) U_{ps} \rho(\mathbf{G}) + \sum_{i,l,\mathbf{G},\mathbf{G}'} \Psi_i^*(\mathbf{k}+\mathbf{G}) \Psi_i(\mathbf{k}+\mathbf{G}') S(\mathbf{G}-\mathbf{G}') U_{ps,l,\mathbf{k}+\mathbf{G},\mathbf{k}+\mathbf{G}'}$$

$$+ \Omega_0^{-1} \gamma_{Ewald} + \Big[\sum_\tau \alpha_\tau \Omega_0^{-1} Z_\tau \Big]. \quad (3.34)$$

To calculate the quantum stress, we may apply a constant infinitesimal strain to the unit cell, so that $\mathbf{r} = \mathbf{F} \cdot \mathbf{r}_0$, then the reciprocal wave vector $\mathbf{G}$ and wave number vector $\mathbf{k}$ can be scaled as,

$$\mathbf{G} = \mathbf{F}^{-1} \cdot \mathbf{G}_0 \quad (3.35)$$
$$\mathbf{k} = \mathbf{F}^{-1} \cdot \mathbf{k}_0. \quad (3.36)$$

We can then derive the first Piola–Kirchhoff (PK-I) quantum stress as,

$$\mathbf{P} = \frac{1}{\Omega_0} \frac{\partial E_{tot}}{\partial \mathbf{F}}. \quad (3.37)$$

Considering the relationship between the Cauchy stress σ and PK-I stress,

$$\sigma = J^{-1} \mathbf{P} \mathbf{F}^T,$$

we can obtain the quantum Cauchy stress. Note that $\Omega \rho(\mathbf{G})$, $\Omega |\Psi_i(\mathbf{k}+\mathbf{G})|^2$, and $S_\tau(\mathbf{G})$ are invariant under scaling. Thus, the reciprocal space expression for stress can be expressed as,

$$\sigma_{\alpha\beta} = -\frac{\hbar^2}{m_e} \sum_{\mathbf{k},\mathbf{G},i} |\Psi_i(\mathbf{k}+\mathbf{G})|^2 \cdot (\mathbf{k}+\mathbf{G})_\alpha (\mathbf{k}+\mathbf{G})_\beta$$

$$+ \frac{1}{2} 4\pi e^2 \sum_{\mathbf{G}} \frac{|\rho(\mathbf{G})|^2}{(\mathbf{G})^2} \left(\frac{2\mathbf{G}_\alpha \mathbf{G}_\beta}{\mathbf{G}^2} - \delta_{\alpha\beta} \right)$$

$$+ \delta_{\alpha\beta} \sum_{\mathbf{G}} [\vartheta_{xc}(\rho) - \mu_{xc}(\rho)] \cdot \rho(\mathbf{G})$$

$$- \sum_{\mathbf{G}} S_\tau(\mathbf{G}) \left(\frac{\partial U_{ps}(\mathbf{G})}{\partial \mathbf{G}^2} 2\mathbf{G}_\alpha \mathbf{G}_\beta + U_{ps}(\mathbf{G})\delta_{\alpha\beta} \right) \rho(\mathbf{G})$$

$$+ \sum_{\mathbf{k},\mathbf{G},\mathbf{G}',i,l,\tau} S_\tau(\mathbf{G}-\mathbf{G}') U'_{ps,l,\mathbf{k}+\mathbf{G},\mathbf{k}+\mathbf{G}'} \Psi_i(\mathbf{k}+\mathbf{G}) \Psi_i^*(\mathbf{k}+\mathbf{G}')$$

$$- \delta_{\alpha\beta} \left[\sum_\tau \alpha_\tau \right] \left[\Omega^{-1} \sum_\tau Z_\tau \right] + \Omega^{-1} \gamma'_{Ewald}. \quad (3.38)$$

In the earlier expression, the nonlocal contribution to the quantum stress is,

$$U_{ps,l,\mathbf{k}+\mathbf{G},\mathbf{k}+\mathbf{G}'} = \frac{1}{\Omega}(2l+1) 4\pi P_l(\cos\gamma) F_{\tau l}(\mathbf{k}+\mathbf{G},\mathbf{k}+\mathbf{G}')$$

$$= \frac{1}{\Omega}(2l+1) 4\pi P_l(\cos\gamma) \int U_{ps,l}(\mathbf{r}) j_l(|\mathbf{k}+\mathbf{G}|r) j_l(|\mathbf{k}+\mathbf{G}'|r) r^2 dr, \quad (3.39)$$

where $P_l(\cos\gamma)$ is the Legendre polynomial, γ is the angle between $\mathbf{k}+\mathbf{G}$ and $\mathbf{k}+\mathbf{G}'$, and $j_l(x)$ denotes as spherical Bessel functions. Then we can find that

$$U'_{ps,l,\mathbf{k}+\mathbf{G},\mathbf{k}+\mathbf{G}'} = \frac{4\pi}{\Omega}(2l+1)(-\delta_{\alpha\beta} P_l(\cos\gamma)) F_{\tau l}(\mathbf{k}+\mathbf{G},\mathbf{k}+\mathbf{G}')$$

$$+ P'_{\ell,\alpha\beta} F_{\tau l}(\mathbf{k}+\mathbf{G},\mathbf{k}+\mathbf{G}') + P_l(\cos\gamma) F'_{\tau\ell,\alpha\beta}, \quad (3.40)$$

where $\alpha, \beta = 1, 2, 3$, and

$$P'_{\ell,\alpha\beta} = P_\ell(\cos\gamma) \cdot \cos\gamma \left(\frac{(\mathbf{k}+\mathbf{G})_\alpha (\mathbf{k}+\mathbf{G})_\beta}{(\mathbf{k}+\mathbf{G})^2} + \frac{(\mathbf{k}+\mathbf{G}')_\alpha (\mathbf{k}+\mathbf{G}')_\beta}{(\mathbf{k}+\mathbf{G}')^2} \right)$$

$$- \frac{((\mathbf{k}+\mathbf{G})_\alpha (\mathbf{k}+\mathbf{G}')_\beta + (\mathbf{k}+\mathbf{G}')_\alpha (\mathbf{k}+\mathbf{G})_\beta)}{|\mathbf{k}+\mathbf{G}||\mathbf{k}+\mathbf{G}'|}. \quad (3.41)$$

If $F_{\tau l}(\mathbf{k}+\mathbf{G},\mathbf{k}+\mathbf{G}')$ is calculated from an interpolation table, then we have,

$$F'_{\tau\ell,\alpha\beta} = -(\mathbf{k}+\mathbf{G}) \frac{\partial F_{\tau l}(\mathbf{k}+\mathbf{G},\mathbf{k}+\mathbf{G}')}{\partial (\mathbf{k}+\mathbf{G})} \frac{(\mathbf{k}+\mathbf{G})_\alpha (\mathbf{k}+\mathbf{G})_\beta}{(\mathbf{k}+\mathbf{G})^2}$$

$$- (\mathbf{k}+\mathbf{G}') \frac{\partial F_{\tau l}(\mathbf{k}+\mathbf{G},\mathbf{k}+\mathbf{G}')}{\partial (\mathbf{k}+\mathbf{G}')} \frac{(\mathbf{k}+\mathbf{G}')_\alpha (\mathbf{k}+\mathbf{G}')_\beta}{(\mathbf{k}+\mathbf{G}')^2}. \quad (3.42)$$

For the Madelung energy γ_{Ewald}, which is calculated by the Ewald transformation, its derivative with respect to strain tensor is,

$$\frac{\partial \gamma'_{Ewald}}{\partial \epsilon} = \frac{\pi}{2\Omega\epsilon} \sum_{G\neq 0} \frac{\exp(-G^2/4\epsilon)}{G^2/4\epsilon} \left|\sum_\tau Z_\tau e^{i\mathbf{G}\cdot\mathbf{x}_\tau}\right|^2 \left[\frac{2G_\alpha G_\beta}{G^2}(G^2/4\epsilon + 1) - \delta_{\alpha\beta}\right]$$

$$+ \frac{1}{2}\epsilon^{1/2} \sum_{\tau,\tau',\mathbf{T}} Z_\tau Z'_\tau H'(\epsilon^{1/2}D)\frac{D_\alpha D_\beta}{D^2}\bigg|_{(\mathbf{D}=\mathbf{x}'_\tau - \mathbf{x}_\tau + \mathbf{T} \neq 0)}$$

$$+ \frac{\pi}{2\Omega\epsilon}\left(\sum_\tau Z_\tau\right)^2 \delta_{\alpha\beta}, \tag{3.43}$$

where ϵ is a convergence parameter, which may be chosen for computational performance, T is the lattice translation vectors, and $\mathbf{x}_\tau$ represents the atomic positions in the unit cell. The function $H'(x)$ is,

$$H'(x) = \frac{\partial[\text{erfc}(\mathbf{x})]}{\partial \mathbf{x}} - \mathbf{x}^{-1}\text{erfc}(\mathbf{x}) \tag{3.44}$$

with erfc($\mathbf{x}$) representing the complementary error function.

3.3 Quantum Stress via DFT

The quantum stress formulation derived in the previous section is based on the Fourier transformation. Thus, it is only applicable to systems with linear strain field, and hence the first-order Cauchy–Born rule, because the Fourier transform of a general nonlinear field may become difficult. To obtain the quantum stress field in arbitrary nonlinear strain field, other techniques have to be used.

Within the framework of density functional theory (DFT), the total energy of a many-body system can be written as,

$$E_{tot} = E_{Ion-electron}[\rho] + E_{Ion-Ion}[\rho] + E_{Hartree}[\rho] + F[\rho] \tag{3.45}$$

where $E_{Ion-electron}$ is the ion–electron interaction energy, $E_{Ion-Ion}$ is the direct Coulomb interaction energy of the ions, and $E_{Hartree}$ is the Hartree energy,

$$E_{Ion-electron} = -\frac{1}{2}\int \rho(\mathbf{r})\sum_I V_{Ie}(\mathbf{r} - \mathbf{R}_I)d^3\mathbf{r} \tag{3.46}$$

$$E_{Ion-Ion} = \frac{1}{2}\sum_I \sum_{J,I\neq J} \frac{Z_I Z_J}{|\mathbf{R}_I - \mathbf{R}_J|} \tag{3.47}$$

$$E_{Hartree} = \frac{1}{2}\int\int \frac{\rho(\mathbf{r})\rho(\mathbf{r}')}{|\mathbf{r} - \mathbf{r}'|}d^3\mathbf{r}d^3\mathbf{r}', \tag{3.48}$$

where the ion–electron potential energy V_{Ie} can be defined as $V_{Ie} = \frac{Z_I}{|\mathbf{r}-\mathbf{R}_I|}$, Z_I is the atom's charge number, $\mathbf{r}$ ($\mathbf{r}'$) and $\mathbf{R}_I$ are the electronic and atomic coordinate after deformation, respectively, $F[\rho]$ is a functional that includes the kinetic and exchange-correlation energies of electrons, which can be expressed as,

$$F[\rho] = E_{kin} + E_{xc}(\rho)$$
$$= \sum_i \left\langle \Psi_i \left| \frac{p_i^2}{2m_i} \right| \Psi_i \right\rangle + E_{xc}(\rho). \tag{3.49}$$

The exchange-correlation energy E_{xc} can be estimated by using the local density approximation (LDA) (see Kohn and Sham 1965), which assumes that E_{xc} is a local function of the charge density $\rho(\mathbf{r})$, that is,

$$E_{xc}(\rho) = \int \rho(\mathbf{r}) \vartheta_{xc}(\rho(\mathbf{r})) d^3(\mathbf{r}), \tag{3.50}$$

where $\vartheta_{xc}(\rho(\mathbf{r}))$ represents the exchange-correlation energy density in a homogeneous electron gas state with $\rho(\mathbf{r})$.

To calculate the quantum stress, we first assume that there is an uniform deformation field. Based on the first-order Cauchy–Born rule, the deformed electronic and atomic coordinates $\mathbf{r}$ and $\mathbf{R}_I$ can be scaled from the initial coordinates $\mathbf{r}^0$ and $\mathbf{R}_I^0$, respectively, as shown in Eqs. (3.51) and (3.52),

$$\mathbf{r} = \mathbf{F} \cdot \mathbf{r}^0, \tag{3.51}$$
$$\mathbf{R}_I = \mathbf{F} \cdot \mathbf{R}_I^0. \tag{3.52}$$

Then by using the Cauchy–Born rule, we can find the first Piola–Kirchhoff quantum stress $\mathbf{P}$ based on the total electron density, which can be derived from the following expression,

$$\mathbf{P} = \frac{1}{\Omega_0} \frac{\partial E_{tot}}{\partial \mathbf{F}} = \frac{1}{\Omega_0} \frac{\partial E_{tot}}{\partial \mathbf{r}} \cdot \frac{\partial \mathbf{r}}{\partial \mathbf{F}}, \tag{3.53}$$

where Ω_0 is the volume of original undeformed structure or configuration. Using the transformation between PK-I stress tensor $\mathbf{P}$, Cauchy stress σ, as well as PK-II stress tensor $\mathbf{S}$, the quantum mechanical expressions can be expressed as follows:

(1) Ion–electron quantum stress

$$\mathbf{P}_{Ie} = \frac{1}{\Omega_0} \frac{\partial E_{Ie}}{\partial \mathbf{F}} = \frac{1}{\Omega_0} \frac{\partial}{\partial \mathbf{F}} \left(-\frac{1}{2} \int \rho(\mathbf{r}) \sum_I \frac{Z_I}{|\mathbf{r} - \mathbf{R}_I|} d^3 \mathbf{r} \right)$$
$$= -\frac{1}{2\Omega_0} e^2 \int \rho(\mathbf{r}) \sum_I Z_I \frac{\partial |\mathbf{r} - \mathbf{R}_I|^{-1}}{\partial (\mathbf{r} - \mathbf{R}_I)} \cdot \frac{\partial (\mathbf{r} - \mathbf{R}_I)}{\partial \mathbf{F}} d^3 \mathbf{r}$$
$$= \frac{1}{2\Omega_0} e^2 \int \rho(\mathbf{r}) \sum_I Z_I \frac{(\mathbf{r} - \mathbf{R}_I) \otimes (\mathbf{r}^0 - \mathbf{R}_I^0)}{|\mathbf{r} - \mathbf{R}_I|^3} d^3 \mathbf{r}. \tag{3.54}$$

Subsequently, we may find that

$$\sigma_{Ie} = \frac{1}{2\Omega}e^2 \int \rho(\mathbf{r}) \sum_I Z_I \frac{(\mathbf{r}-\mathbf{R}_I) \otimes (\mathbf{r}-\mathbf{R}_I)}{|\mathbf{r}-\mathbf{R}_I|^3} d^3\mathbf{r}, \qquad (3.55)$$

where $\Omega = J\Omega_0$ is the volume of the deformed lattice, and

$$\mathbf{S}_{Ie} = \frac{1}{2\Omega_0}e^2 \int \rho(\mathbf{r}) \sum_I Z_I \frac{(\mathbf{r}^0-\mathbf{R}_I^0) \otimes (\mathbf{r}^0-\mathbf{R}_I^0)}{|\mathbf{r}-\mathbf{R}_I|^3} d^3\mathbf{r}. \qquad (3.56)$$

(2) Ion–ion quantum stress

$$\mathbf{P}_{II} = \frac{1}{\Omega_0} \frac{\partial E_{II}}{\partial \mathbf{F}} = \frac{1}{\Omega_0} \frac{\partial}{\partial \mathbf{F}} \left(\frac{1}{2} \sum_I \sum_{J, I \neq J} \frac{Z_I Z_J}{|\mathbf{R}_I - \mathbf{R}_J|} \right)$$

$$= \frac{1}{2\Omega_0}e^2 \sum_I \sum_{J, I \neq J} Z_I Z_J \frac{\partial |\mathbf{R}_I - \mathbf{R}_J|^{-1}}{\partial (\mathbf{R}_I - \mathbf{R}_J)} \cdot \frac{\partial (\mathbf{R}_I - \mathbf{R}_J)}{\partial \mathbf{F}}$$

$$= -\frac{1}{2\Omega_0}e^2 \sum_I \sum_{J, I \neq J} Z_I Z_J \frac{(\mathbf{R}_I - \mathbf{R}_J) \otimes (\mathbf{R}_I^0 - \mathbf{R}_J^0)}{|\mathbf{R}_I - \mathbf{R}_J|^3}. \qquad (3.57)$$

Subsequently, we can derive the corresponding Cauchy stress form,

$$\sigma_{II} = -\frac{1}{2\Omega}e^2 \sum_I \sum_{J, I \neq J} Z_I Z_J \frac{(\mathbf{R}_I - \mathbf{R}_J) \otimes (\mathbf{R}_I - \mathbf{R}_J)}{|\mathbf{R}_I - \mathbf{R}_J|^3}, \qquad (3.58)$$

and the corresponding second Piola–Kirchhoff stress (PK-II) form,

$$\mathbf{S}_{II} = -\frac{1}{2\Omega_0}e^2 \sum_I \sum_{J, I \neq J} Z_I Z_J \frac{(\mathbf{R}_I^0 - \mathbf{R}_J^0) \otimes (\mathbf{R}_I^0 - \mathbf{R}_J^0)}{|\mathbf{R}_I - \mathbf{R}_J|^3}. \qquad (3.59)$$

(3) Hartree quantum stress
Based on the definition, the Hartree quantum stress may be derived as,

$$\mathbf{P}_H = \frac{1}{\Omega_0} \frac{\partial E_H}{\partial \mathbf{F}} = \frac{1}{\Omega_0} \frac{\partial}{\partial \mathbf{F}} \left(\frac{1}{2} \int \int \frac{\rho(\mathbf{r})\rho(\mathbf{r}')}{|\mathbf{r}-\mathbf{r}'|} d^3\mathbf{r} d^3\mathbf{r}' \right)$$

$$= \frac{1}{2\Omega_0}e^2 \int \int \rho(\mathbf{r})\rho(\mathbf{r}') \frac{\partial |\mathbf{r}-\mathbf{r}'|^{-1}}{\partial (\mathbf{r}-\mathbf{r}')} \cdot \frac{\partial (\mathbf{r}-\mathbf{r}')}{\partial \mathbf{F}} d^3\mathbf{r} d^3\mathbf{r}'$$

$$= -\frac{1}{2\Omega_0}e^2 \int \int \rho(\mathbf{r})\rho(\mathbf{r}') \frac{(\mathbf{r}-\mathbf{r}') \otimes (\mathbf{r}^0-\mathbf{r}'^0)}{|\mathbf{r}-\mathbf{r}'|^3} d^3\mathbf{r} d^3\mathbf{r}'. \qquad (3.60)$$

Accordingly, we have the corresponding Cauchy–Hartree stress,

$$\sigma_H = -\frac{1}{2\Omega}e^2 \int \int \rho(\mathbf{r})\rho(\mathbf{r}') \frac{(\mathbf{r}-\mathbf{r}') \otimes (\mathbf{r}-\mathbf{r}')}{|\mathbf{r}-\mathbf{r}'|^3} d^3\mathbf{r} d^3\mathbf{r}' \qquad (3.61)$$

and the corresponding PK-II Hartree stress is

$$S_H = -\frac{1}{2\Omega_0} e^2 \int\int \rho(\mathbf{r})\rho(\mathbf{r}') \frac{(\mathbf{r}^0 - \mathbf{r}'^0) \otimes (\mathbf{r}^0 - \mathbf{r}'^0)}{|\mathbf{r} - \mathbf{r}'|^3} d^3\mathbf{r} d^3\mathbf{r}'. \quad (3.62)$$

(4) Kinetic quantum stress

Since $\mathbf{p}_i = \mathbf{F}^{-1}\mathbf{p}_i^0$, it can be expressed as,

$$\mathbf{P}_{kin} = \frac{1}{\Omega_0}\frac{\partial E_{kin}}{\partial \mathbf{F}} = \frac{1}{\Omega_0}\frac{\partial}{\partial \mathbf{F}}\left(\sum_i \left\langle \Psi_i \left| \frac{\mathbf{p}_i^2}{2m_i} \right| \Psi_i \right\rangle\right)$$

$$= \frac{1}{\Omega_0}\frac{\partial\left(\sum_i \left\langle \Psi_i \left| \frac{\mathbf{p}_i^0 \mathbf{F}^{-T}\mathbf{F}^{-1}\mathbf{p}_i^0}{2m_i} \right| \Psi_i \right\rangle\right)}{\partial \mathbf{F}}$$

$$= -\frac{1}{\Omega_0}\sum_i \left\langle \Psi_i \left| \frac{\mathbf{p}_i \mathbf{F}^{-1} \otimes \mathbf{p}_i}{m_i} \right| \Psi_i \right\rangle. \quad (3.63)$$

The corresponding kinetic quantum Cauchy stress is given as,

$$\sigma_{kin} = -\frac{1}{\Omega}\sum_i \left\langle \Psi_i \left| \frac{\mathbf{p}_i \otimes \mathbf{p}_i}{m_i} \right| \Psi_i \right\rangle, \quad (3.64)$$

and the corresponding kinetic quantum PK-II stress is given as

$$S_{kin} = -\frac{1}{\Omega_0}\sum_i \left\langle \Psi_i \left| \frac{\mathbf{p}_i \mathbf{F}^{-1} \otimes \mathbf{p}_i \mathbf{F}^{-1}}{m_i} \right| \Psi_i \right\rangle. \quad (3.65)$$

(5) Exchange-correlation quantum stress

As shown in Eq. (3.50), the exchange-correlation energy density is a local function of $\mathbf{r}$, and thus the exchange-correlation term only gives rise to an isotropic pressure at each particles $\mathbf{r}$. In that case, the pressure per electron is just the volume derivative of $\vartheta_{xc}(\rho(\mathbf{r}))$, i.e.,

$$J\frac{\partial \vartheta_{xc}(\rho(\mathbf{r}))}{\partial J} = J\left(\frac{\partial \vartheta_{xc}}{\partial \mathbf{F}} \cdot \frac{\partial \mathbf{F}}{\partial J}\right)$$

$$= J\left(\frac{\partial\left(\frac{\rho\vartheta_{xc}(\rho)}{\rho}\right)}{\partial \rho} \cdot \frac{\partial \rho}{\partial \mathbf{F}} \cdot \frac{\partial \mathbf{F}}{\partial J}\right). \quad (3.66)$$

Considering the charge equilibrium condition, which can be expressed as,

$$\int_{\Omega_0} \rho_0(\mathbf{r})d\Omega_0 = \int_\Omega \rho(\mathbf{r})d\Omega, \quad (3.67)$$

where the $\rho_0(\mathbf{r})$ is the charge density of structures at ground state. Since $\rho\Omega$ is invariant under scaling and $J = det\mathbf{F} = \frac{\Omega}{\Omega_0}$. Then, Eq. (3.66) can be rewritten as,

$$J\frac{\partial \vartheta_{xc}(\rho(\mathbf{r}))}{\partial J} = J\left(\frac{\partial(\rho\vartheta_{xc}(\rho))}{\partial\rho}\cdot\frac{1}{\rho} - \frac{\vartheta_{xc}(\rho)}{\rho}\right)\cdot\frac{\partial\frac{\rho\Omega}{\Omega}}{\partial F}\cdot\frac{\partial F}{\partial J}$$

$$= J\left(\mu_{xc}(\rho)\cdot\frac{1}{\rho} - \frac{\vartheta_{xc}(\rho)}{\rho}\right)\cdot(\rho\Omega)\frac{1}{\Omega_0}\cdot\frac{\partial J^{-1}}{\partial F}\cdot\frac{\partial F}{\partial J}$$

$$= J\left(\mu_{xc}(\rho)\cdot\frac{1}{\rho} - \frac{\vartheta_{xc}(\rho)}{\rho}\right)\cdot(\rho\Omega)\frac{1}{\Omega_0}\cdot\left(\frac{-1}{J}\right)$$

$$= \vartheta_{xc}(\rho) - \mu_{xc}(\rho), \qquad (3.68)$$

where $\mu_{xc}(\rho) = \frac{\partial(\rho\vartheta_{xc}(\rho))}{\partial\rho}$ is the exchange-correlation energy. Thus, the quantum exchange-correlation Cauchy stress at $\mathbf{r}$ is,

$$\sigma_{xc} = \frac{1}{\Omega}\int I\left[\vartheta_{xc}(\rho(\mathbf{r})) - \mu_{xc}(\rho(\mathbf{r}))\right]\rho(\mathbf{r})\,d\mathbf{r}^3. \qquad (3.69)$$

Consequently, the quantum exchange-correlation PK-I stress will be

$$\mathbf{P}_{xc} = J\sigma_{xc}\mathbf{F}^{-T}, \qquad (3.70)$$

and the quantum exchange-correlation PK-II stress becomes

$$\mathbf{S}_{xc} = J\mathbf{F}^{-1}\sigma_{xc}\mathbf{F}^{-T}. \qquad (3.71)$$

Note that, in this case, the total deformation gradient may not be a spherical tensor, and therefore both the quantum exchange-correlation PK-I stress and PK-II stress may not be spherical.

Recall that $F[\rho] = T_s[\rho] + E_{xc}[\rho]$. In the following, we adopt a LDA form of exchange-correlation function to evaluate

$$\frac{\partial}{\partial \mathbf{F}}F[\rho] = \frac{\partial}{\partial \mathbf{F}}\left(T_s[\rho] + E_{xc}[\rho]\right).$$

By using LDA, we have

$$E_{xc} = -\frac{3}{4\pi}(3\pi^2)^{1/3}\int \rho^{4/3}(\mathbf{r})d^3\mathbf{r}.$$

Then

$$\frac{\partial E_{xc}}{\partial \mathbf{F}} = -\frac{1}{4\pi}(3\pi^2)^{1/3}\int \rho^{1/3}\frac{\partial\rho}{\partial J}\frac{\partial J}{\partial \mathbf{F}}d^3\mathbf{r},$$

where

$$\frac{\partial J}{\partial \mathbf{F}} = J\mathbf{F}^{-T} \quad \text{and}$$

$$\frac{\partial}{\partial J}(\rho J) = \frac{\partial}{\partial J}(\rho_0) = 0 \quad \rightarrow \quad J\frac{\partial\rho}{\partial J} + \rho = 0 \quad \rightarrow \quad \frac{\partial\rho}{\partial J} = -\rho/J.$$

Finally,

$$\frac{\partial E_{xc}}{\partial \mathbf{F}} = \frac{1}{4\pi}(3\pi^2)^{1/3}\int \rho^{4/3}\mathbf{F}^{-T}d^3\mathbf{r}.$$

In general, we may write

$$E_{xc}[\rho] = \int \rho(\mathbf{r})\epsilon_{xc}(\rho(\mathbf{r}))d^3\mathbf{r},$$

and hence

$$\frac{\partial E_{xc}}{\partial \mathbf{F}} = \int J\mathbf{F}^{-T}(\epsilon_{xc}(\rho) - \mu_{xc}(\rho))\rho d^3\mathbf{r},$$

where

$$\mu_{xc}(\rho) = \frac{d(\rho\epsilon_{xc}(\rho))}{d\rho}.$$

3.4 Quantum Electronic Stress

The quantum stress introduced by Nielsen and Martin is due to variation of the quantum potential energy with respect to strain.

In a 2011 paper, Hu et al. considered the problem that when the electron density has a small increment $\Delta\rho$, what would be the system's response in terms of stress (Hu et al. (2012)). In other words, they studied how stress would change when electron density has a sudden change, whereas in the Nielsen–Martin theory, the quantum stress is the effect of potential energy variation due the strain changes.

For instance, if a nanostructure such as a solar cell or a sensor under exposure to photon radiation source or other types of radiation source, the radiation photon, neutron, or other particles knock electrons out from from the nanostructure or nano-device, it would induce mechanical stress inside the nanostructure or device as shown in Fig. 3.1.

To formulate the problem, we first recall that in an N-electron system, the total electron (Kohn–Sham) energy is

$$E[\rho] = T[\rho] + E_H[\rho] + E_{xc}[\rho] + E_{ext}[\rho],$$

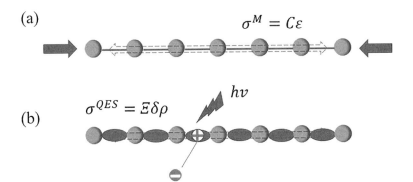

Figure 3.1 Schematic illustration of photon-induced mechanical stress scenarios (a) Quantum stress due to applied load, and (b) Quantum stress due to photon injection.

and the Nielsen–Martin quantum stress is defined as

$$\sigma_{ij}^Q = \frac{de_0}{d\epsilon_{ij}} = \frac{1}{V}\frac{dE}{d\epsilon_{ij}}\bigg|_{\rho^0, \epsilon_{ij}=0},$$

where e_0 is the average energy density defined as $e_0 = E/V$, V is the total volume of the system, and ρ^0 is the ground-state electron density.

For an inhomogeneous system, we may write

$$E = \int e(\rho)d\mathbf{r}, \qquad (3.72)$$

where the electron density is a spatial function, i.e., $\rho(\mathbf{r})$.

Now we want to consider a situation when the electron density has a sudden change or variation from the ground state, i.e.,

$$e[\rho] \approx e_0 + \frac{\partial e}{\partial \rho}\delta\rho = e_0 + \mu\delta\rho,$$

where μ is the electron chemical potential,

$$\mu = \frac{\partial e}{\partial \rho}. \qquad (3.73)$$

Then we have the total energy expressed as

$$E[\rho(\mathbf{r})] = E[\rho^0(\mathbf{r})] + \int_V \left(\frac{\delta e[\rho(\mathbf{r})]}{\delta \rho(\mathbf{r})}\right)\delta\rho(\mathbf{r})d\mathbf{r} = E[n^0(\mathbf{r})] + \int_V \mu\delta\rho(\mathbf{r})d\mathbf{r}.$$

Based on the standard definition, we may evaluate quantum stress in the perturbed electron system, i.e.,

$$\sigma^Q = \frac{1}{V}\int_B \left[\frac{\partial \mu}{\partial \epsilon}\delta\rho + \mu\frac{\partial(\delta\rho)}{\partial \epsilon}\right]d\mathbf{r}\bigg|_{\rho^0, \epsilon=0}, \qquad (3.74)$$

where

$$\mu = \frac{\partial e}{\partial \rho} = \frac{\partial}{\partial \rho}(e_k + e_h + e_{ec} + e_{ext}),$$

and $e = e_k + e_h + e_{ec} + e_{ext}$ is the electron chemical potential. It may be noted that the N-electron system is under the finite perturbation not the infinitesimal perturbation, i.e., $\delta\rho$ is a measurable finite quantity.

We then define the first term inside Eq. (3.74) as the **quantum electron stress** (QES), i.e.,

$$\sigma^{QES} = \frac{1}{V}\int_B \left[\frac{\partial \mu}{\partial \epsilon}\delta\rho\right]d\mathbf{r}\bigg|_{\rho^0, \epsilon=0}, \qquad (3.75)$$

where V is the volume of the unit cell.

In passing, we note that the above quantum electron stress is not the increment of the quantum stress, this is because we assume that the initial stress is zero, i.e.,

$$\frac{\partial^2 e}{\partial \rho \partial \epsilon_{ij}}\bigg|_{\rho^0, \epsilon_{ij}=0} = \frac{\partial \mu}{\partial \epsilon_{ij}}\bigg|_{\rho^0, \epsilon_{ij}=0} = 0.$$

Otherwise, the so-called quantum electron stress is just increment of the quantum stress due to the electron density change. When the strain tensor is spherical, i.e., $\epsilon = \epsilon \mathbf{I}$, we may also define an electron deformation stress as,

$$\Xi = \frac{\partial E}{\partial \epsilon} = \frac{\partial E}{\partial \epsilon} \mathbf{I} = \Xi \mathbf{I}, \tag{3.76}$$

where Ξ is defined as an electron deformation potential,

$$\Xi = \frac{\partial E}{\partial \epsilon} = \frac{1}{V} \int_V \left[\frac{\partial \mu}{\partial \epsilon}\right]\bigg|_{\rho^0, \epsilon=0} dV. \tag{3.77}$$

By doing so, we can then write the quantum electron stress in a simple expression,

$$\sigma^{QES} = \Xi \delta \rho. \tag{3.78}$$

Example 3.2 As an example, we provide the detail computation formulation for Ξ under the LDA.

We first consider

$$\delta E = \delta E_h[\rho] + \delta E_{xc}[\rho] + \delta E_{n-e}[\rho].$$

We then calculate them term-by-term:

$$\delta E_h = \frac{1}{2} \int \frac{d\mathbf{r}}{|\mathbf{r}-\mathbf{r}'|} \delta \rho \int \rho(\mathbf{r}')d\mathbf{r}' + \frac{1}{2} \int \frac{\rho \mathbf{r}}{|\mathbf{r}-\mathbf{r}'|} d\mathbf{r} \int \delta\rho(\mathbf{r}')d\mathbf{r}'$$

$$= \int \left[\int \frac{\rho \mathbf{r}'}{|\mathbf{r}-\mathbf{r}'|} d\mathbf{r}'\right] \delta\rho d\mathbf{r} \tag{3.79}$$

$$\delta E_{n-e} = -\int \sum_I \frac{Z_I}{|\mathbf{r}-\mathbf{R}_I|} \delta\rho d\mathbf{r} \tag{3.80}$$

and

$$\delta E_{xc} = -\frac{1}{\pi} \int (3\pi^2 \rho)^{1/3} \delta\rho d\mathbf{r}. \tag{3.81}$$

Adding them together, we have

$$\mu(\mathbf{r}) = \frac{\delta E}{\delta \rho} = \left(\int \frac{\rho(\mathbf{r}')}{|\mathbf{r}-\mathbf{r}'|} d\mathbf{r}' - \sum_I \frac{Z_I}{|\mathbf{r}-\mathbf{R}_I|} - \frac{(3\pi^2 \rho(\mathbf{r}))^{1/3}}{\pi}\right). \tag{3.82}$$

By chain rule, we can find that

$$\frac{\partial \mu}{\partial \epsilon}\bigg|_{\epsilon=0} = \frac{\partial \mu}{\partial \mathbf{r}} \frac{\partial \mathbf{r}}{\partial \epsilon}\bigg|_{\epsilon=0} + \frac{\partial \mu}{\partial \mathbf{R}_I} \frac{\partial \mathbf{R}_I}{\partial \epsilon}\bigg|_{\epsilon=0}.$$

By using the Cauchy–Born rule,[1] the deformed position vectors and the undeformed position vectors are linked by

$$\mathbf{r} \to \mathbf{Fr} = (1+\epsilon)\mathbf{r}, \quad \mathbf{R} \to \mathbf{FR} = (1+\epsilon)\mathbf{R}, \text{ and } J = det|(\mathbf{I}+\epsilon)|.$$

Thus,

$$\frac{\partial \mathbf{r}}{\partial \epsilon} = \mathbf{I} \otimes \mathbf{r}, \quad \frac{\partial \mathbf{R}_I}{\partial \epsilon} = \mathbf{I} \otimes \mathbf{R}_I, \text{ and } \left.\frac{\partial}{\partial \epsilon} J\right|_{\epsilon=0} = \mathbf{I}.$$

Subsequently, we can find that

$$\left.\frac{\partial \mu}{\partial \epsilon}\right|_{\epsilon=0} = \int \frac{\nabla_{\mathbf{r}'}\rho \otimes \mathbf{r} + \rho \mathbf{I}}{|\mathbf{r}-\mathbf{r}'|} d\mathbf{r}' - \int \frac{(\mathbf{r}-\mathbf{r}') \otimes (\mathbf{r}-\mathbf{r}')}{|\mathbf{r}-\mathbf{r}'|^3} \rho(\mathbf{r}) d\mathbf{r}'$$
$$+ \sum_I Z_I \frac{(\mathbf{r}-\mathbf{R}_I) \otimes (\mathbf{r}-\mathbf{R}_I)}{|\mathbf{r}-\mathbf{r}'|^3} - \frac{(3\pi^2 \rho)^{1/3}}{3\pi\rho} \nabla_{\mathbf{r}} \rho \otimes \mathbf{r}, \quad (3.83)$$

and we can then find Ξ by integration

$$\Xi = \frac{1}{V} \int_V \left.\frac{\partial \mu}{\partial \epsilon}\right|_{\rho_0,\epsilon=0} d\mathbf{r}.$$

It is noted that for semiconductors, there are two types of electron deformation potentials,

$$\Xi^e := \frac{\partial E_{CBM}}{\partial \epsilon} \text{ and } \Xi^h := \frac{\partial E_{VBM}}{\partial \epsilon},$$

where *CBM* refers to *conduction band minimum* and *VBM* stands for *valence band maximum*.

By using DFT formulation, Hu et al. calculated the electron deformation potential values for a group of semiconductor materials, which are listed in Table 3.1 (Hu et al. (2012)).

Table 3.1 Electron deformation potential

Material	Ξ (eV)
Si	$\Xi^e_{Si} = -8.65$
Si	$\Xi^h_{Si} = -9.51$
GaAs	$\Xi^e_{GaAs} = -9.77$
GaAs	$\Xi^h_{GaAs} = -7.33$
$Z_r Q_2$	$\Xi^e_{Z_r Q_2} = -12.36$
$Z_r Q_2$	$\Xi^h_{Z_r Q_2} = -8.87$
Al	$\Xi_{Al} = -10.49$

---

[1] See Chapter 10 for detailed discussions.

3.5 Higher-Order Quantum Stress Theory

To calculate higher-order quantum stress, we first assume that there is an infinitesimal nonuniform deformation field with non-trivial stain gradient, and then by using the second-order Cauchy–Born rule, the deformed particles coordinate **r** can be scaled as,

$$\mathbf{r} \to \mathbf{F} \cdot \mathbf{r} + \frac{1}{2}\mathbf{G} : \mathbf{r} \otimes \mathbf{r} = \mathbf{B} \cdot \mathbf{r}, \qquad (3.84)$$

where $\mathbf{F}$ is the deformation gradient, $\mathbf{B} = \mathbf{F} + \frac{1}{2}\mathbf{G} \cdot \mathbf{r}$, and $\mathbf{G} = \nabla \otimes \mathbf{F}$ is the strain gradient. Then the scaled wave function $\Psi(\mathbf{r})$ becomes

$$\Psi_\mathbf{B}(\mathbf{r}) = \frac{1}{(det\mathbf{B})^{1/2}} \Psi(\mathbf{r}_0) = \frac{1}{L^{1/2}} \Psi(\mathbf{B}^{-1}\mathbf{r}), \qquad (3.85)$$

where $L = det\mathbf{B}$ and the prefactor $L^{-1/2}$ ensures the normalization of $\Psi_\mathbf{B}(\mathbf{r})$, $\int \Psi_\mathbf{B} \Psi_\mathbf{B}^* d\Omega_\mathbf{r} = 1$.

Explicitly, we can write

$$<\Psi_\mathbf{B}|H|\Psi_\mathbf{B}> = \int \Psi_\mathbf{B}^*(\mathbf{r}) \left[\sum_i \frac{\mathbf{p}_i \otimes \mathbf{p}_i}{2m_i} + V_{int}(\mathbf{r}) + V_{ext}(\mathbf{r}) \right] \Psi_\mathbf{B}(\mathbf{r}) d\Omega_\mathbf{r}. \qquad (3.86)$$

Considering normalization of the scaled wave function and using the relations $\mathbf{r} = \mathbf{B} \cdot \mathbf{r}$, $\mathbf{p}_i = \mathbf{B}^{-1} \cdot \mathbf{p}_i$, the expectation value of the Hamiltonian H with respect to $\Psi(\mathbf{r}_0)$ can be found as,

$$<\Psi_\mathbf{B}|H|\Psi_\mathbf{B}> = \int \Psi^* \left[\sum_i \frac{\sum_i \mathbf{p}_i \otimes \mathbf{p}_i}{2m_i} + V_{int}(\mathbf{r}) + V_{ext}(\mathbf{r}) \right] \Psi d\Omega_\mathbf{r}$$

$$= \int \Psi^* \left[\sum_i \frac{(\mathbf{B}^{-1}\mathbf{p}_i \otimes \mathbf{B}^{-1}\mathbf{p}_i)}{2m_i} + V_{int}(\mathbf{B} \cdot \mathbf{r}) + V_{ext}(\mathbf{B} \cdot \mathbf{r}) \right] \Psi d\Omega_\mathbf{r}$$

$$= \int \Psi^* \left[\sum_i \frac{\mathbf{d} : (\mathbf{p}_i \otimes \mathbf{p}_i)}{2m_i} + V_{int}(\mathbf{B} \cdot \mathbf{r}) + V_{ext}(\mathbf{B} \cdot \mathbf{r}) \right] \Psi d\Omega_\mathbf{r} \qquad (3.87)$$

where $\mathbf{d} = \mathbf{B}^{-T} \cdot \mathbf{B}^{-1}$. Assuming that the many-body system is in an equilibrium state, so that the derivative of Eq. (3.87) with respect to higher-order strain gradient $\mathbf{G}$ is zero,

$$\frac{d}{d\mathbf{G}} <\Psi_\mathbf{B}|H|\Psi_\mathbf{B}> = \frac{d}{d\mathbf{G}} \int \Psi^* \left[\sum_i \frac{\mathbf{d} : (\mathbf{p}_i \otimes \mathbf{p}_i)}{2m_i} + V_{int}(\mathbf{B} \cdot \mathbf{r}) + V_{ext}(\mathbf{B} \cdot \mathbf{r}) \right] \Psi d\Omega_\mathbf{r}.$$

$$= 0 \qquad (3.88)$$

Analogous to the generalized Reynolds transport theorem, it may be expressed as,

$$\frac{d}{d\mathbf{G}} <\Psi_\mathbf{B}|H|\Psi_\mathbf{B}> = \int \Psi^* \frac{d}{d\mathbf{G}} \left[\sum_i \frac{\mathbf{d} : (\mathbf{p}_i \otimes \mathbf{p}_i)}{2m_i} + V_{int}(\mathbf{B} \cdot \mathbf{r}) + V_{ext}(\mathbf{B} \cdot \mathbf{r}) \right] \Psi d\Omega_\mathbf{r}$$

$$= \sum_i \left\langle \Psi \left| -\frac{\mathbf{p}_i \otimes \mathbf{p}_i \otimes \mathbf{r}_i}{2m_i} + \frac{1}{2}\nabla_i(V_{int} + V_{ext}) \otimes \mathbf{r}_i \otimes \mathbf{r}_i \right| \Psi \right\rangle$$

$$= 0. \qquad (3.89)$$

Finally, we obtain the following equation,

$$\sum_i \left\langle \Psi \left| -\frac{\mathbf{p}_i \otimes \mathbf{p}_i \otimes \mathbf{r}_i}{2m_i} + \frac{1}{2}\nabla_i(V_{int} + V_{ext}) \otimes \mathbf{r}_i \otimes \mathbf{r}_i \right| \Psi \right\rangle = 0. \qquad (3.90)$$

If one defines the higher-order stress caused by external environment as,

$$\mathbf{Q}_{ext} = \sum_i \left\langle \Psi \left| -\frac{1}{2}\nabla_i V_{ext} \otimes \mathbf{r}_i \otimes \mathbf{r}_i \right| \Psi \right\rangle. \qquad (3.91)$$

Then, based on Eq. (3.91), we obtain the following general expression for the second-order quantum stress:

$$\mathbf{Q}_{int} = \sum_i \left\langle \Psi \left| -\frac{\mathbf{p}_i \otimes \mathbf{p}_i \otimes \mathbf{r}_i}{2m_i} + \frac{1}{2}\nabla_i V_{int} \otimes \mathbf{r}_i \otimes \mathbf{r}_i \right| \Psi \right\rangle. \qquad (3.92)$$

The above expression of the second-order quantum stress is attainable if and only if we have known wave functions of every electrons. In practical calculations of the higher-order stresses for many-body systems, we may only know electron density distribution in ground state. Thus, in the next section, we discuss the higher-order quantum stress in the framework of density functional theory.

3.6 Higher-Order Quantum Stress via DFT

To derive the second-order quantum stress via DFT formulation, we may need to extend the DFT quantum stress formulation to the higher-order stress that is caused by strain gradients, or strain gradient effects. To do so, we first want to use the second-order Cauchy–Born rule as discussed in the previous section in conjunction with DFT formulation.

First, we assume that there is an infinitesimal nonuniform strain with a nontrivial strain gradient is applied in an electron unit cell, so that we can have the following quadratic mapping in the space,

$$\mathbf{r} \to \mathbf{F} \cdot \mathbf{r} + \frac{1}{2}\mathbf{G} : (\mathbf{r} \otimes \mathbf{r}) \text{ and} \qquad (3.93)$$

$$\mathbf{R}_I \to \mathbf{F} \cdot \mathbf{R}_I + \frac{1}{2}\mathbf{G} : (\mathbf{R}_I \otimes \mathbf{R}_I). \qquad (3.94)$$

Taking into account the total quantum energy and second-order coordinates transformation, the second-order quantum stress can be defined as

$$\mathbf{Q} = \frac{1}{\Omega}\frac{\partial E_{tot}}{\partial \mathbf{G}} = \frac{1}{\Omega}\frac{\partial E_{tot}}{\partial \mathbf{r}} \cdot \frac{\partial \mathbf{r}}{\partial \mathbf{G}}. \qquad (3.95)$$

For clarity, we provide the contribution of each part to higher-order quantum stress one by one as follow:

(1) Kinetic higher-order quantum stress

Since $\mathbf{p}_i \to \mathbf{B}^{-1} \cdot \mathbf{p}_i$, so the kinetic higher-order quantum stress $\mathbf{Q}_{kin}$ can be derived as,

$$\begin{aligned}
\mathbf{Q}_{kin} &= \frac{1}{\Omega} \frac{\partial E_{kin}}{\partial \mathbf{G}} \\
&= \frac{1}{\Omega} \frac{\partial \sum_i \left\langle \Psi_i \left| \frac{\mathbf{p}_i^2}{2m_e} \right| \Psi_i \right\rangle}{\partial \mathbf{G}} \\
&= -\frac{1}{2m_e \Omega} \sum_i \left\langle \Psi_i \left| \mathbf{p}_i \otimes \mathbf{p}_i \otimes \mathbf{r}_i \right| \Psi_i \right\rangle.
\end{aligned} \qquad (3.96)$$

(2) Ion–electron higher-order quantum stress

$$\begin{aligned}
\mathbf{Q}_{Ie} &= \frac{1}{\Omega} \frac{\partial E_{Ie}}{\partial \mathbf{G}} = \frac{1}{\Omega} \frac{\partial}{\partial \mathbf{G}} \left(-\frac{1}{2} \int \rho(\mathbf{r}) \sum_I V_{Ie}(\mathbf{r} - \mathbf{R}_I) d^3 \mathbf{r} \right) \\
&= -\frac{1}{2\Omega} e^2 \int \rho(\mathbf{r}) \sum_I \frac{\partial V_{Ie}(\mathbf{r} - \mathbf{R}_I)}{\partial (\mathbf{r} - \mathbf{R}_I)} \cdot \frac{\partial (\mathbf{r} - \mathbf{R}_I)}{\partial \mathbf{G}} d^3 \mathbf{r} \\
&= \frac{1}{2\Omega} e^2 \int \rho(\mathbf{r}) \sum_I Z_I \frac{(\mathbf{r} - \mathbf{R}_I)}{|\mathbf{r} - \mathbf{R}_I|^3} \cdot \frac{\partial (\mathbf{r} - \mathbf{R}_I)}{\partial \mathbf{G}} d^3 \mathbf{r} \\
&= \frac{e^2}{4\Omega} \sum_I \int \rho(\mathbf{r}) Z_I \frac{(\mathbf{r} - \mathbf{R}_I) \otimes (\mathbf{r} \otimes \mathbf{r} - \mathbf{R}_I \otimes \mathbf{R}_I)}{|\mathbf{r} - \mathbf{R}_I|^3} d^3 \mathbf{r}.
\end{aligned} \qquad (3.97)$$

(3) Ion–ion higher-order quantum stress

$$\begin{aligned}
\mathbf{Q}_{II} &= \frac{1}{\Omega} \frac{\partial E_{II}}{\partial \mathbf{G}} = \frac{1}{\Omega} \frac{\partial}{\partial \mathbf{G}} \left(\frac{1}{2} \sum_{I,J,I \neq J} \frac{Z_I Z_J}{|\mathbf{R}_I - \mathbf{R}_J|} \right) \\
&= \frac{1}{2\Omega} e^2 \sum_{I,J,I \neq J} Z_I Z_J \frac{\partial |\mathbf{R}_I - \mathbf{R}_j|^{-1}}{\partial (\mathbf{R}_I - \mathbf{R}_J)} \cdot \frac{\partial (\mathbf{R}_I - \mathbf{R}_J)}{\partial \mathbf{G}} \\
&= -\frac{1}{2\Omega} e^2 \sum_{I,J,I \neq J} Z_I Z_J \frac{(\mathbf{R}_I - \mathbf{R}_J)}{|\mathbf{R}_I - \mathbf{R}_J|^3} \cdot \frac{\partial (\mathbf{R}_I - \mathbf{R}_J)}{\partial \mathbf{G}} \\
&= -\frac{e^2}{4\Omega} \sum_{I,J,I \neq J} Z_I Z_J \frac{(\mathbf{R}_I - \mathbf{R}_J) \otimes (\mathbf{R}_I \otimes \mathbf{R}_I - \mathbf{R}_J \otimes \mathbf{R}_J)}{|\mathbf{R}_I - \mathbf{R}_J|^3}.
\end{aligned} \qquad (3.98)$$

(4) Hartree higher-order quantum stress

$$\begin{aligned}\mathbf{Q}_H &= \frac{1}{\Omega}\frac{\partial E_H}{\partial \mathbf{G}} = \frac{1}{\Omega}\frac{\partial}{\partial \mathbf{G}}\left(\frac{1}{2}\int\int\frac{\rho(\mathbf{r})\rho(\mathbf{r}')}{|\mathbf{r}-\mathbf{r}'|}d^3\mathbf{r}d^3\mathbf{r}'\right)\\ &= \frac{e^2}{2\Omega}\int\int\rho(\mathbf{r})\rho(\mathbf{r}')\frac{\partial|\mathbf{r}-\mathbf{r}'|^{-1}}{\partial(\mathbf{r}-\mathbf{r}')}\cdot\frac{\partial(\mathbf{r}-\mathbf{r}')}{\partial\mathbf{G}}\\ &= -\frac{e^2}{2\Omega}\int\int\rho(\mathbf{r})\rho(\mathbf{r}')\frac{(\mathbf{r}-\mathbf{r}')}{|\mathbf{r}-\mathbf{r}'|^3}\cdot\frac{\partial(\mathbf{r}-\mathbf{r}')}{\partial\mathbf{G}}\\ &= -\frac{e^2}{4\Omega}\int\int\rho(\mathbf{r})\rho(\mathbf{r}')\frac{(\mathbf{r}-\mathbf{r}')\otimes(\mathbf{r}\otimes\mathbf{r}-\mathbf{r}'\otimes\mathbf{r}')}{|\mathbf{r}-\mathbf{r}'|^3}d^3\mathbf{r}d^3\mathbf{r}'. \end{aligned} \quad (3.99)$$

(5) Exchange-correlation higher-order quantum stress

$$\mathbf{Q}_{xc} = 0. \quad (3.100)$$

This is true, if LDA is used. This is because that under LDA the exchange-correlation energy density will be a local function of $\mathbf{r}$, and it only contributes to an isotropic pressure.

3.7 Quantum Couple Stress via DFT

The higher-order stress is usually referred to the third-order force-couple intensity tensors, or even higher-order stress tensors.

In engineering practice, there is a special higher-order stress that is particularly useful in material property analysis, which is called the couple stress. The couple stress is a distributed force couple in the continuum, which is present in many condensed matters such as dielectric and magnetic materials as well as defect-rich media, such as crystals with large amount of dislocations.

Just as the higher-order stress is the material responses of strain gradients, the couple stress is the material response to curl of lattice rotations, which correspond to local deformation curvatures or material rotations. However, unlike the third-order or even higher-order stress tensors, the couple stress is a second-order stress tensor, whose self-equilibrium diagram in a unit cell is shown in Fig. 3.2.

This is because that the curl of material rotation is a second-order tensor, which is often called the lattice curvature, and the couple stress is a pair of the conjugate variables in mechanical work.

To derive the quantum couple stress by using DFT, we first apply an infinitesimal nonuniform strain with a strain gradient to the many-body unit cell. Thus, based on the couple stress theory and the higher-order Cauchy–Born rule, the electronic and atomic coordinates are scaled by considering the second-order transformation with respect to deformation gradient $\mathbf{F} = \mathbf{I} + \boldsymbol{\epsilon}$ and strain gradient $\mathbf{G} = \nabla \times \mathbf{F}$,

$$\mathbf{r} = \mathbf{F}\cdot\mathbf{r}^0 + \frac{1}{2}\mathbf{G}:(\mathbf{r}^0\otimes\mathbf{r}^0), \text{ and} \quad (3.101)$$

$$\mathbf{R}_I = \mathbf{F}\cdot\mathbf{R}_I^0 + \frac{1}{2}\mathbf{G}:(\mathbf{R}_I^0\otimes\mathbf{R}_I^0), \quad (3.102)$$

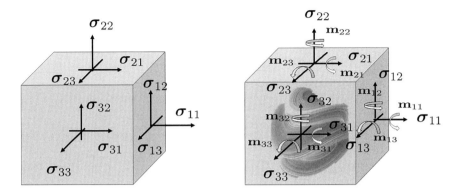

Figure 3.2 Illustration and comparison of self-equilibrium states of stress and couple stress in an infinitesimal element

where $\mathbf{r}$ and $\mathbf{R}_I$ are the electronic and atomic coordinates of deformed material configuration, respectively, and $\mathbf{r}^0$ and $\mathbf{R}_I^0$ are the electronic and atomic coordinates of initial configuration, respectively. Note that here ϵ is the infinitesimal strain, and we neglect the effect of small rotation. According to the couple stress theory, e.g., Fleck and Hutchinson (1997), the strain gradient tensor $\mathbf{G}$ can be decomposed into a symmetric part $\mathbf{G}^S$ representing stretch gradients, and an antisymmetric part $\mathbf{G}^A$, which is a measure of lattice curvature.

A similar decomposition is applied on the higher-order quantum stress $\mathbf{Q}$, which leads to a symmetric tensor $\mathbf{Q}^S$, and an antisymmetric tensor $\mathbf{Q}^A$.

Since $\mathbf{Q}^S$ is orthogonal to $\mathbf{G}^A$ and $\mathbf{Q}^A$ is orthogonal to $\mathbf{G}^S$, the density of the internal virtual work may be expressed as,

$$\delta w = \sigma : \delta \epsilon + \mathbf{Q}^S : \delta \mathbf{G}^S + \mathbf{Q}^A : \delta \mathbf{G}^A. \tag{3.103}$$

In this study, we mainly focus on the antisymmetric work term $\mathbf{Q}^A : \delta \mathbf{G}^A$, which represents the work done by the couple stress $\mathbf{m}$ acting through the curvature increment $\delta \chi$, i.e.,

$$\mathbf{Q}^A : \delta \mathbf{G}^A = \mathbf{m} : \delta \chi. \tag{3.104}$$

The curvature tensor is defined as the antisymmetric part of the strain gradient tensor, $\chi_{ij} = \frac{1}{2} e_{iqr} G^A_{jqr}$, where e_{iqr} is the permutation symbol. Then the antisymmetric part of strain gradient can be expressed as $G^A_{ijk} = \frac{2}{3}(e_{ikp}\chi_{pj} + e_{jkp}\chi_{pi})$. We may write the second-order coordinate transformations in indicial notation,

$$r_i = F_{ij} r_j^0 + \frac{1}{3}(e_{ikp}\chi_{pj} + e_{jkp}\chi_{pi}) r_k^0 r_j^0 \quad \text{and} \tag{3.105}$$

$$R_{Ii} = F_{ij} R_{Ij}^0 + \frac{1}{3}(e_{ikp}\chi_{pj} + e_{jkp}\chi_{pi}) R_{Ik}^0 R_{Ij}^0. \tag{3.106}$$

By virtue of Eqs. (3.103) and (3.104), the quantum couple stress can be derived as follows:

$$\mathbf{m} = \frac{1}{\Omega_0} \frac{\partial E_{tot}}{\partial \chi}, \tag{3.107}$$

where Ω_0 is the volume of initial unit cell. In the following, we present the detailed expressions for each part of the total quantum couple stress **m** in terms of DFT formalism:

(1) Kinetic quantum couple stress

Since $p_i = \left[F_{ij} + \frac{1}{3}\left(e_{ikp}\chi_{pj} + e_{jkp}\chi_{pi}\right)r_k\right]^{-1} p_j$, so that the kinetic quantum couple stress $\mathbf{m}_{kin}$ can be derived as

$$\mathbf{m}_{kin} = \frac{1}{\Omega_0}\frac{\partial E_{kin}}{\partial \chi} = \frac{1}{\Omega_0}\frac{\partial}{\partial \chi}\sum_i \left\langle \Psi_i \left| \frac{\mathbf{p}_i^2}{2m_e} \right| \Psi_i \right\rangle$$

$$= -\frac{1}{3m_e\Omega_0}\sum_i \left\langle \Psi_i \left| (\mathbf{p}_i \times \mathbf{r}_i^0) \otimes \mathbf{p}_i^0 + (\mathbf{p}_i^0 \times \mathbf{r}_i^0) \otimes \mathbf{p}_i \right| \Psi_i \right\rangle. \quad (3.108)$$

(2) Ion–electron quantum couple stress

$$\mathbf{m}_{Ie} = \frac{1}{\Omega_0}\frac{\partial E_{Ie}}{\partial \chi} = \frac{1}{\Omega_0}\frac{\partial}{\partial \chi}\left(-\frac{1}{2}\int \rho(\mathbf{r})\sum_I V_{Ie}(\mathbf{r} - \mathbf{R}_I)d^3\mathbf{r}\right)$$

$$= -\frac{e^2}{2\Omega_0}\int \rho(\mathbf{r})\sum_I \frac{\partial V_{Ie}(\mathbf{r} - \mathbf{R}_I)}{\partial(\mathbf{r} - \mathbf{R}_I)} \cdot \frac{\partial(\mathbf{r} - \mathbf{R}_I)}{\partial \chi} d^3\mathbf{r}$$

$$= \frac{e^2}{2\Omega_0}\int \rho(\mathbf{r})\sum_I Z_I \frac{(\mathbf{r} - \mathbf{R}_I)}{|\mathbf{r} - \mathbf{R}_I|^3} \cdot \frac{\partial(\mathbf{r} - \mathbf{R}_I)}{\partial \chi} d^3\mathbf{r}$$

$$= \frac{e^2}{6\Omega_0}\sum_I \int \rho(\mathbf{r})Z_I \frac{(\mathbf{r} - \mathbf{R}_I) \times (\mathbf{r}^0 \otimes \mathbf{r}^0 - \mathbf{R}_I^0 \otimes \mathbf{R}_I^0)}{|\mathbf{r} - \mathbf{R}_I|^3} d^3\mathbf{r}. \quad (3.109)$$

(3) Ion–ion quantum couple stress

$$\mathbf{m}_{II} = \frac{e^2}{\Omega_0}\frac{\partial E_{II}}{\partial \chi} = \frac{e^2}{\Omega_0}\frac{\partial}{\partial \chi}\left(\frac{1}{2}\sum_{I,J,I \neq J}\frac{Z_I Z_J}{|\mathbf{R}_I - \mathbf{R}_J|}\right)$$

$$= \frac{e^2}{2\Omega_0}\sum_{I,J,I \neq J} Z_I Z_J \frac{\partial |\mathbf{R}_I - \mathbf{R}_J|^{-1}}{\partial(\mathbf{R}_I - \mathbf{R}_J)} \cdot \frac{\partial(\mathbf{R}_I - \mathbf{R}_J)}{\partial \chi}$$

$$= -\frac{e^2}{2\Omega_0}\sum_{I,J,I \neq J} Z_I Z_J \frac{(\mathbf{R}_I - \mathbf{R}_J)}{|\mathbf{R}_I - \mathbf{R}_J|^3} \cdot \frac{\partial(\mathbf{R}_I - \mathbf{R}_J)}{\partial \chi}$$

$$= -\frac{e^2}{6\Omega_0}\sum_{I,J,I \neq J} Z_I Z_J \frac{(\mathbf{R}_I - \mathbf{R}_J) \times (\mathbf{R}_I^0 \otimes \mathbf{R}_I^0 - \mathbf{R}_J^0 \otimes \mathbf{R}_J^0)}{|\mathbf{R}_I - \mathbf{R}_J|^3}. \quad (3.110)$$

(4) Hartree quantum couple stress

$$\begin{aligned}
\mathbf{m}_H &= \frac{1}{\Omega_0}\frac{\partial E_H}{\partial \chi} = \frac{e^2}{\Omega_0}\frac{\partial}{\partial \chi}\left(\frac{e^2}{2}\int\int\frac{\rho(\mathbf{r})\rho(\mathbf{r}')}{|\mathbf{r}-\mathbf{r}'|}d^3\mathbf{r}d^3\mathbf{r}'\right)\\
&= \frac{e^2}{2\Omega_0}\int\int \rho(\mathbf{r})\rho(\mathbf{r}')\frac{\partial|\mathbf{r}-\mathbf{r}'|^{-1}}{\partial(\mathbf{r}-\mathbf{r}')}\cdot\frac{\partial(\mathbf{r}-\mathbf{r}')}{\partial\chi}\\
&= -\frac{e^2}{2\Omega_0}\int\int \rho(\mathbf{r})\rho(\mathbf{r}')\frac{(\mathbf{r}-\mathbf{r}')}{|\mathbf{r}-\mathbf{r}'|^3}\cdot\frac{\partial(\mathbf{r}-\mathbf{r}')}{\partial\chi}\\
&= -\frac{e^2}{6\Omega_0}\int\int \rho(\mathbf{r})\rho(\mathbf{r}')\frac{(\mathbf{r}-\mathbf{r}')\times(\mathbf{r}^0\otimes\mathbf{r}^0-\mathbf{r}'^0\otimes\mathbf{r}'^0)}{|\mathbf{r}-\mathbf{r}'|^3}d^3\mathbf{r}d^3\mathbf{r}'. \quad (3.111)
\end{aligned}$$

(5) Exchange-correlation quantum couple stress,

$$\mathbf{m}_{xc} = 0. \quad (3.112)$$

This is because the exchange-correlation energy density under the LDA is a local function of $\mathbf{r}$, and it only gives rise to an isotropic pressure without higher-order couple stress effect.

4 Introduction to VASP

In this chapter, we introduce a popular first-principles-based computation software package, Vienna Ab initio Simulation Package (VASP), and provide a few examples to help readers understand DFT calculations.

VASP is one of the most widely used commercial software programs for performing ab initio quantum mechanical molecular dynamics (MD) simulations by using pseudopotentials or the projector-augmented wave method and a plane wave basis set. VASP adopts the finite-temperature local-density approximation with the free energy as a variational quantity, and it can evaluate an exact and instantaneous electronic ground state at each MD time step. VASP can also calculate forces and a full quantum stress tensor, and these quantities are used when relaxing atoms into their instantaneous ground state. In VASP, the interaction between ions and electrons is described by ultrasoft Vanderbilt pseudopotentials (US-PPs) or by the projector-augmented wave (PAW) method, so that the number of plane waves per atom for transition metals and first-row elements can be reduced greatly.

4.1 Files Used by VASP

Various input and output files are used in VASP, as shown in Table 4.1. For example, INCAR, POSCAR, KPOINTS, and POTCAR are necessary input files required for all calculations in VASP, and CHGCAR/CHG and CONCAR are important output files of VASP. We introduce and explain in detail the main input and output files subsequently.

4.1.1 INCAR File

INCAR is a main input file of VASP, which determines what things to do and how to do them. This file contains a large number of parameters, and the main parameters are described here.

1. **SYSTEM**

    SYSTEM defines the computational cell, which helps users identify what they want to do with this specific input file.

Table 4.1 Examples of input and output files in VASP

File name		File name	
INCAR	in	CONTCAR	out
POSCAR	in	CHG	out
KPOINTS	in	EIGENVAL	out
POTCAR	in	DOSCAR	out
STOPCAR	in	PROCAR	out
EXHCAR	in	OSZICAR	out
CHGCAR	in/out	PCDAT	out
WAVECAR	in/out	XDATCAR	out
TMPCAR	in/out	LOCPOT	out
stout	out	ELFCAR	out
IBZKPT	out	PROOUT	out

2. ENCUT

ENCUT is the cutoff energy for the plane wave basis set in eV. Its default value is the largest ENMAX in the POTCAR file. But it is strongly recommended to specify the cutoff energy manually in the INCAR file, which can be determined by convergence tests. For consistency reasons, it is better to keep the ENCUT value constant throughout the entire calculations.

3. PREC

PREC is an important parameter for controlling computational accuracy. It determines the default value for ENCUT, FFT mesh, and ROPT, if no value is given for them in the INCAR file.

FFT mesh is determined by flags NGX (NGY, NGZ) and NGXF (NGYF, NGZF). NGX (NGY, NGZ) controls the number of grid points in the FFT mesh along the direction of three lattice vectors. NGXF (NGYF, NGZF) controls the number of grid points for a second, finer FFT mesh. ROPT controls the number of grid points within the integration sphere around each ion, if real space projectors are used.

The influences of PREC parameters on the default value for ENCUT, FFT mesh, and ROPT are shown in Table 4.2, where $\frac{\hbar^2}{2m_e}|G_{cut}|^2 = $ ENCUT, and $\frac{\hbar^2}{2m_e}|G_{aug}|^2 = $ ENAUG. ENMAX/ENMIN corresponds to the maximum ENMAX/ENMIN found in POTCAR; ENAUG is the kinetic energy cutoff for the augmentation charges, and its default value is the largest value of ENMAX on the POTCAR file.

In general, it is recommended to use

```
PREC = Normal
```

for calculation in VASP.4.5 and higher (default in VASP.5.X).

Table 4.2 The influence of PREC parameters on the default for ENCUT, FFT mesh, and ROPT

PREC	ENCUT	NGX	NGXF	ROPT
Normal	ENMAX	$\frac{3}{2}G_{cut}$	2NGX	−5E-4
Single	ENMAX	$\frac{3}{2}G_{cut}$	NGX	−5E-4
Accurate	ENMAX	$2G_{cut}$	2NGX	−2.5E-4
Low	ENMIN	$\frac{3}{2}G_{cut}$	$3G_{aug}$	−1E-2
Medium	ENMAX	$\frac{3}{2}G_{cut}$	$4G_{aug}$	−2E-3
High	ENMAX ×1.3	$2G_{cut}$	$\frac{16}{3}G_{aug}$	−4E-4

4. ISTART and ICHARG

ISTART determines whether to read the file WAVECAR or not. If WAVECAR exists, the default value of ISTART is 1, which means orbitals could be read from WAVECAR. Else, the default value of ISTART is 0, where the orbitals can be initialized according to the flag INIWAV.

ICHARG determines how to construct the "initial" charge density. If ISTART = 0, the default value of ICHARG is 2, which means the initial charge density is from overlapping atoms. Otherwise, the default value of ICHARG is 0, which means that the charge density is calculated from the initial orbital. There exist other meanings for ISTART and ICHARG. Readers may consult the VASP manual for more details.

5. EDIFF and EDIFFG

EDIFF defines the global break condition for the electronic SC loop. If the changes in the total energy and the band structure energy between two electronic steps are both smaller than EDIFF, the relaxation process will be stopped. The default value of EDIFF is 10^{-4}, and there is no need to use a much smaller number in general. If EDIFF = 0, NELM (the maximum number of electronic SC steps) will always be performed.

EDIFFG defines the break condition for the ionic relaxation loop. The ionic relaxation will be stopped if the total energy change between two steps is smaller than EDIFFG. Its default value is EDIFF ×10. If EDIFFG is negative, it means the relaxation will be stopped if all forces are smaller than the absolute value of EDIFFG. If EDIFFG = 0, the ionic relaxation is stopped after NSW (the maximum number of ionic steps).

6. IBRION

IBRION determines how the ions are updated and moved. The meaning of IBRION is shown in Table 4.3. For relaxation problems, it is recommended to use the conjugate gradient algorithm, that is,

```
IBRION = 2
```

Table 4.3 The meanings of IBRION index

IBRION	Algorithm for ions relaxation
−1	No update
0	Standard ab initio molecular dynamics
1	Quasi-Newton (variable metric) algorithm
2	Conjugate-gradient algorithm
3	Damped molecular dynamics
5,6	Finite differences
7,8	Density functional perturbation theory
44	Improved dimer method

Table 4.4 The meaning of ISIF

ISIF	Calculate force	Calculate stress tensor	Relax ions	Change cell shape	Change cell volume
0	yes	no	yes	no	no
1	yes	trace only	yes	no	no
2	yes	yes	yes	no	no
3	yes	yes	yes	yes	yes
4	yes	yes	yes	yes	no
5	yes	yes	no	yes	no
6	yes	yes	no	yes	yes
7	yes	yes	no	no	yes

7. **ISIF**

ISIF is a very useful parameter for determining which degrees of freedom (ions, cell volume, cell shape) are allowed to change. The meaning of ISIF is shown in Table 4.4, where "trace only" means that only the sum of the diagonal components or total pressure is given.

8. **ISMEAR** and **SIGMA**

ISMEAR determines how the partial occupancies $f_{n\mathbf{k}}$ are set for each orbital. For the finite temperature, **SIGMA** determines the width of the smearing in eV. The meaning of ISMEAR is displayed in Table 4.5. There is also a guideline for choosing the smearing method:

(i) For semiconductors or insulators, ISMEAR = −5 is recommended to use. If the cell is too large, ISMEAR = 0 with a small SIGMA = 0.05 is used. In general, one should avoid using ISMEAR > 0 for semiconductors and insulators.
(ii) For metals, ISMEAR = 1 or ISMEAR = 2 with an appropriate SIGMA value is used. Generally, the value of SIGMA is chosen as 0.2.
(iii) For the calculation of density of states (DOS) and high-accuracy total energy calculations, it is recommended to use ISMEAR = −5.

Table 4.5 The meaning of ISMEAR

ISMEAR	Smearing method
N	Method of Methfessel–Paxton order N
0	Gaussian smearing
−1	Fermi smearing
−2	Partial occupancies are read in from WAVECAR (or INCAR), and kept fixed throughout the run
−3	Perform a loop over smearing parameters supplied in the INCAR file
−4	Tetrahedron method without Blöchl corrections
−5	Tetrahedron method with Blöchl corrections

There are some other parameters used in INCAR files. Readers can find the details in the VASP manual. In the following, we present an example of an INCAR file:

```
SYSTEM = Si-diamond
ISTART = 0
ICHGCAR = 2
ENCUT = 245
ISMEAR = -5
PREC = Accurate
```

4.1.2 POSCAR File

A POSCAR file contains the lattice constants, atom numbers, and ionic positions, and optionally also starting velocities and predictor–corrector coordinates for a MD run. Its usual format is:

```
Si-diamond
5.43
    0.00    0.50    0.50
    0.50    0.00    0.50
    0.50    0.50    0.00
    2
Direct
    0.00    0.00    0.00
    0.25    0.25    0.25
```

The first line is a comment line, which can be used to describe the calculation system. The second line provides a universal scaling factor, which is used to scale all lattice vectors and all atomic coordinates. It may also be regarded as the lattice constant sometimes. In the example discussed earlier, "5.43" is the lattice constant of Si with cubic diamond symmetry. The next three lines represent the first, the second, and the third lattice vectors that define the unit cell of the system. The sixth line defines the number of atoms per atomic species. The seventh line specifies whether the atomic positions are given in cartesian coordinates or in direct (fractional) coordinates. The next lines give the three coordinates for each atom. If direct coordinates are adopted, the atomic positions are given by,

$$\mathbf{R} = x_1\mathbf{a}_1 + x_2\mathbf{a}_2 + x_3\mathbf{a}_3 \tag{4.1}$$

where $\mathbf{a}_i$, $i = 1, 2, 3$ are the three basis vectors, and x_i, $i = 1, 2, 3$ are the values of atomic coordinates. If the cartesian coordinates are used, atomic positions will be scaled by the universal scaling factor supplied in the second line.

4.1.3 KPOINTS File

KPOINTS file determines the k-points setting, including k-point coordinates and weights or the mesh size for creating the k-point grid. There are several methods for generating k-points mesh.

1. The coordinates and weights of k-points can be listed explicitly. The first line is a comment line. The second line provides the number of k-points. The third line specifies whether the atomic positions are given in cartesian or in reciprocal coordinates. The next lines give the three coordinates for each k-point and its weight.

```
Entering all k-points explicitly
4
Cartesian
0.0    0.0    0.0    1
0.0    0.0    0.5    1
0.0    0.5    0.5    2
0.5    0.5    0.5    4
```

2. "Line mode" for generating "strings" of k-points connecting specific points of the Brillouin zone. As shown in the following example, VASP will generate 10 k-points with the weight of 1 per line in cartesian coordinates.

```
K-points along high symmetry lines
40   ! 40 intersections
Line-mode
Cartesian
0.0    0.0    0.0    1 ! gamma
0.0    0.0    1.0    1 ! X

0.0    0.0    1.0    1 ! X
0.5    0.0    1.0    1 ! W

0.5    0.0    1.0    1 ! W
0.0    0.0    1.0    1 ! gamma
```

3. The k-points mesh could be generated automatically. It only requires the input of subdivisions of the Brillouin zone in each direction and the origin ("shift") of the k-mesh. The first line is a comment line. In the second line, the number 0 activates the automatic generation scheme. The third line provides the way to generate k-points mesh. In the example, k-points are generated automatically using the Monkhorst–Pack's technique. The number of subdivisions along each reciprocal lattice vector are shown in the fourth line. The fifth line is optional, which provides an additional shift of the k-mesh.

```
Automatic generation
0
Monkhorst-pack
9   9   9
0.0    0.0    0.0
```

4. For hexagonal lattices, it is strongly recommended to use only Gamma-centered grids.

```
Automatic generation
0
Gamma
4   4   4
0.0    0.0    0.0
```

4.1.4 POTCAR File

POTCAR files contain pseudopotential information for each atomic type or species used in the calculation, as well as the information about the atoms. Although the standard pseudopotentials supplied with VASP are among the best pseudopotentials presently available, it is strongly recommended to use better electronic structure methods, such as the PAW method. In addition, if there are more than one atomic species, one should concatenate the POTCAR files of all atom species. There is a one-to-one correspondence between the atomic species in POSCAR file and in POTCAR file.

4.1.5 OUTCAR File

OUTCAR file contains most calculation results and more detailed information for every electronic and ionic step. One can find some useful information from OUTCAR file, such as,

1. Volume of cell:

```
energy-cutoff   :   600.00
volume of cell  :   157.91
```

2. Basis vectors:

```
   direct lattice vectors                    reciprocal lattice vectors
     5.405044237  0.000000000  0.000000000    0.185012362  0.000000000  0.000000000
     0.000000000  5.405044237  0.000000000    0.000000000  0.185012362  0.000000000
     0.000000000  0.000000000  5.405044237    0.000000000  0.000000000  0.185012362

   length of vectors
     5.405044237  5.405044237  5.405044237    0.185012362  0.185012362  0.185012362
```

3. Energy:

```
   free  energy   TOTEN  =    -47.57238622 eV
   energy  without entropy=   -47.57238622   energy(sigma->$0) =   -47.57238622
```

If ISMEAR = −5, "free energy TOTEN" is equal to "energy without entropy," otherwise they are different.

4.1.6 CHGCAR and CHG Files

CHGCAR and CHG files contain the lattice vectors, atomic coordinates, and the total charge density multiplied by the volume $\rho(r) \times V_{cell}$ on the fine FFT grid (NG(X,Y,Z)F). CHGCAR also contains PAW one-center occupancies, and it can be used to restart VASP from existing charge density. The data arrangements of CHGCAR and CHG files are similar, except the PAW one-center occupancies. The charge density in CHGCAR and CHG files is written by using the following commands in Fortran, where the x index is the fastest index and the z index the slowest index.

```
WRITE(IU,FORM) (((C(NX,NY,NZ),NX=1,NGXF),NY=1,NGYF),NZ=1,NGZF)
```

The following is an example of CHGCAR file. The first fifteen lines are similar to that of POSCAR file, which provides lattice vectors and atomic coordinates. The three integers observed in the seventeen lines are NGXF, NGYF, and NGZF, respectively, which are the number of grid points in the finer FFT mesh along the directions of three lattice vectors. The charge density is stored as a 3D (NGXF, NGYF, NGZF) matrix. It should be noted that only self-consistent charge density could be used for accurate DOS and band structure calculations.

```
Si-diamond
1.00000000000000
    5.405044    0.000000    0.000000
    0.000000    5.405044    0.000000
    0.000000    0.000000    5.405044
   8
Direct
   0.000000  0.000000  0.000000
   0.000000  0.500000  0.500000
   0.500000  0.500000  0.000000
   0.500000  0.000000  0.500000
   0.750000  0.250000  0.750000
   0.250000  0.250000  0.250000
   0.250000  0.750000  0.750000
   0.750000  0.750000  0.250000

 80 80 80
 -.23057754672E+01 -.77932705631E+00 0.36228445125E+01 0.10394537811E+02 0.18775581373E+02
 0.27869725282E+02 0.36785096251E+02 0.44763551966E+02 0.51264147413E+02 0.55982774117E+02
 . . . . . . . . . . . .
```

4.1.7 CONTCAR File

CONTCAR file is written after the end of each ionic step. Its format is almost the same as that of POSCAR file. For relaxation calculation, CONTCAR file contains lattice vectors and atomic coordinates of the last ionic step of relaxation. If the relaxation has converged, the information in CONTCAR is optimized into lattice constants and atomic positions. If the relaxation has not converged yet, CONTCAR can be copied as POSCAR to continue the job. For static calculations, CONTCAR file is identical to POSCAR file.

4.2 Example: Structural Optimization and Self-consistent Charge Density

In this section, we present an example of first-principles calculations for $SrTiO_3$ crystal structure, as implemented in VASP. The task of the example includes finding the optimized $SrTiO_3$ crystal structure, and its selfconsistent charge density.

4.2.1 Structural Optimization

The relaxation calculation is first employed to obtain optimized $SrTiO_3$ structure. The entered command is,

```
mpirun -np 16 vasp
```

The input files and results are shown as follows.

1. POSCAR File

It stores the initial structural information about $SrTiO_3$, including lattice constants and ionic positions, which can be obtained from experimental data.

```
SrTiO3-cubic
1.00
    3.9049     0.0000     0.0000
    0.0000     3.9049     0.0000
    0.0000     0.0000     3.9049
     Sr    Ti    O
     1     1     3
Direct
    0.00   0.00   0.00
    0.50   0.50   0.50
    0.50   0.50   0.00
    0.00   0.50   0.50
    0.50   0.00   0.50
```

2. POTCAR File

The POTCAR file of $SrTiO_3$ crystal structure can be obtained directly from VASP package. Here, we use recommended GW PAW potentials for Sr, Ti, and O atoms, that is, Sr_sv_GW, Ti_sv_GW, and O_GW potentials. Since there are three atomic species in $SrTiO_3$, the POTCAR file should include the three atomic PAW potentials. The atomic species in POTCAR file and in POSCAR file must correspond with each other, which means that the order of the atomic potentials in POTCAR file should be Sr, Ti, and O.

3. INCAR File

Before the relaxation calculation, it is strongly recommended to specify the plane wave cutoff energy manually in the INCAR file. Thus, the value of cutoff energy shall be first determined by convergence tests. In the convergence tests, the experimental structural information of $SrTiO_3$ is used in POSCAR file, and the k-points mesh remains unchanged. The bash script loop.sh runs INCAR at several different cutoff energies (400–1000 eV), and it saves the ground-state energy versus cutoff energy into the file ENCUT.

```
\#! /bin/bash
BIN=/path/to/your/vasp/executable
rm WAVECAR
for i in 400 450 500 550 600 650 700 750 800 850 900 950 10000
do
cat $>$ INCAR $<<$ !
SYSTEM = SrTiO3-cubic
ENCUT = \$i
ISTART = 0
ICHARG = 2
ISMEAR = -5
PERC = Accurate
!
echo "ENCUT= \$i eV" ; mpirun -np 2 \$BIN
E= 'grep ''entropy'' OUTCAR $\mid$ tail -1 $\mid$ awk'
echo \$i \$E  $>>$ ENCUT
done
```

Figure 4.1 illustrates the variation of ground-state energy of $SrTiO_3$ as a function of plane wave cutoff energy. It indicates that the total energy has converged when the plane wave cutoff energy reaches 850 eV.

In the following, we show the INCAR file for the relaxation calculation of $SrTiO_3$, in which conjugate–gradient algorithm is adopted for ionic relaxation. In addition, ISIF = 3 means internal parameters, volume of the cell, and cell shape can be changed simultaneously.

```
SYSTEM = SrTiO3-cubic
ENCUT = 850
ISTART = 0
ICHGCAR = 2
ISIF = 3
IBRION = 2
EDIFF = 1.0E-6
EDIFFG = -0.001
ISMEAR = -5
NSW = 100
PREC = Accurate
```

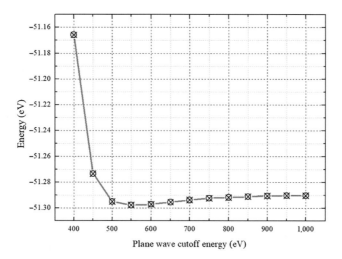

Figure 4.1 Convergence tests with respect to plane wave cutoff energy for $SrTiO_3$

4. KPOINTS File

Similarly, some convergency tests shall also be performed on the $SrTiO_3$ structure to determine the number of k-points required. Generally, in the convergence tests, INCAR file is set as static calculation, that is,

```
SYSTEM = SrTiO3-cubic
ENCUT = 850
ISTART = 0
ICHGCAR = 2
ISMEAR = -5
PREC = Accurate
```

The bash script loop.sh runs KPOINTS at several different number of k-points. The variation of ground-state energy of $SrTiO_3$ as a function of the number of k-points is shown in Fig. 4.2. The results show that the total energy has converged when the $(9 \times 9 \times 9)$ k-points mesh is adopted.

```
\#! /bin/bash
BIN=/path/to/your/vasp/executable
rm WAVECAR
for i in  3 4 5 6 7 8 9 10 11 12 13
do
cat $>$ KPOINTS $<<$ !
Automatic generation
0
Monkhorst-pack
\$i   \$i  \$i
0.0   0.0   0.0
!
echo "k-mesh= \$i x \$i x \$i" ; mpirun -np 2 \$BIN
```

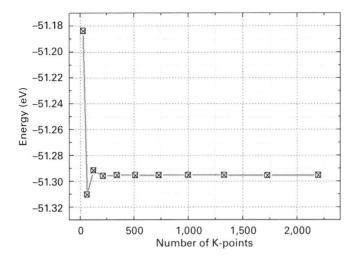

Figure 4.2 Convergence tests with respect to the number of k-points for $SrTiO_3$

```
E= 'grep ''entropy'' OUTCAR $\mid$ tail -1 $\mid$ awk'
KP= 'grep ''irreducible'' OUTCAR $\mid$ tail -1 $\mid$ awk'
echo \$i \$KP \$E   $>>$ Kpoints
done
```

In the following, we show the KPOINTS file for $SrTiO_3$.

```
Automatic generation
0
Monkhorst-pack
9    9    9
0.0    0.0    0.0
```

5. Optimized Structure

After the relaxation, the optimized $SrTiO_3$ structure can be obtained, as displayed in Fig. 4.3. The optimized lattice constants and atomic positions of $SrTiO_3$ are listed in CONTCAR output file.

```
SrTiO3-cubic
1.000000000000
   3.8557968342891606    0.0000000000000000    0.0000000000000000
   0.0000000000000000    3.8557968342891606    0.0000000000000000
   0.0000000000000000    0.0000000000000000    3.8557968342891606
   Sr   Ti   O
   1    1    3
Direct
   0.0000000000000000    0.0000000000000000    0.0000000000000000
   0.5000000000000000    0.5000000000000000    0.5000000000000000
   0.5000000000000000    0.5000000000000000    0.0000000000000000
   0.0000000000000000    0.5000000000000000    0.5000000000000000
   0.5000000000000000    0.0000000000000000    0.5000000000000000
```

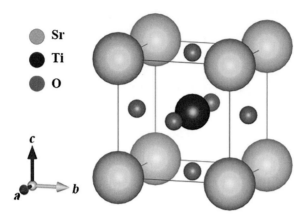

Figure 4.3 The optimized crystal structure of $SrTiO_3$

4.2.2 Self-consistent Calculation

Subsequently, based on the optimized $SrTiO_3$ structure, the self-consistent charge density of $SrTiO_3$ can be obtained by a static calculation. The execution command is,

```
mpirun -np 16 vasp
```

In the static calculation, POTCAR and KPOINTS files remain unchanged, and the obtained CONTCAR file after relaxation calculation is saved, and renamed as POSCAR file. The INCAR file is shown as follows.

```
SYSTEM = SrTiO3-cubic
ENCUT = 850
ISTART = 0
ICHGCAR = 2
ISMEAR = -5
PREC = Accurate
```

After the static calculation, the self-consistent charge density of $SrTiO_3$ is listed in the CHGCAR output file.

```
SrTiO3-cubic
1.00000000000000
    3.855797      0.000000      0.000000
    0.000000      3.855797      0.000000
    0.000000      0.000000      3.855797
     Sr    Ti    O
      1     1    3
Direct
  0.000000  0.000000  0.000000
  0.500000  0.500000  0.500000
  0.500000  0.500000  0.000000
  0.000000  0.500000  0.500000
  0.500000  0.000000  0.500000

  56    56    56
  0.15719145600E+03  0.15928624202E+03  0.16476888629E+03  0.17150243021E+03  0.17670683782E+03
  0.17780108673E+03  0.17310158669E+03  0.16218700942E+03  0.14588362542E+03  0.12593859930E+03
  0.10451281379E+03  0.83651482745E+02  0.64882465405E+02  0.49038884853E+02  0.36312685822E+02
  0.54396148020E+01  0.42542926437E+01  0.34676832913E+01  0.29535130106E+01  0.26230141462E+01
  0.24151212334E+01  0.22897898043E+01  0.22228397388E+01  0.22017847171E+01  0.22228397388E+01
  ............
```

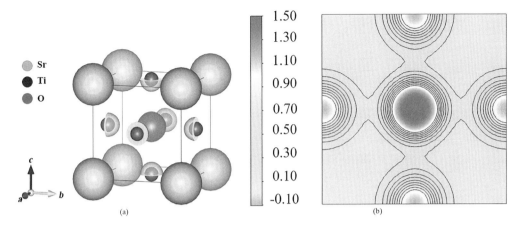

Figure 4.4 Electron charge density distribution of $SrTiO_3$: (a) 3D model and (b) [110] plane

The charge density of $SrTiO_3$ can also be drawn via VESTA software, as illustrated in Fig. 4.4.

4.3 Example: DFT Calculation of Si Band Structure

The band structure describes the quantum mechanical behavior of electrons in solids, which can be used to explain the electronic material properties of crystal structures. In the following example, we present a detailed step-by-step DFT calculation of the band structure of Si by using VASP.

4.3.1 Self-consistent Calculation

Before calculating the band structure of Si, the ground-state charge density of Si should be first obtained by the self-consistent calculation. As shown in Section 4.2, the ground-state charge density of Si is calculated based on its optimized structure by using a static calculation. The input files POSCAR, POTCAR, INCAR, and KPOINTS for relax and static calculations are shown as follows.

1. POSCAR File
   This file contains the lattice geometry and the ionic positions of Si with diamond structure.

```
Si
5.430
    0.50    0.50    0.00
    0.00    0.50    0.50
    0.50    0.00    0.50
    Si
    2
```

```
Cartesian
  0.00  0.00  0.00
  0.25  0.25  0.25
```

Note that the above POSCAR file is only for the relax calculation. In the static calculation, the POSCAR file is the saved copy of the CONCAR file, which is the output file after the relax calculation.

2. POTCAR File

The recommended pseudopotential Si_GW is used for the calculation, which can be obtained from VASP package directly.

3. INCAR File

First, to ensure calculation accuracy and efficiency, the required cutoff energy is determined by convergence tests. The INCAR files for the relax and static calculations are shown as follows.

```
! relax calculation
SYSTEM = Si
ENCUT = 800
ISTART = 0
ICHARG = 2
ISMEAR = -5
IBRION = 2
ISIF = 3
EDIFF = 1E-6
EDIFFG = -0.001
NSW = 200
PERC = Accurate

! self-consistent calculation
SYSTEM = Si
ENCUT =800
ISTART = 0
ICHARG = 2
ISMEAR = 0
SIGMA = 0.1
PERC = Accurate
```

4. KPOINTS File

According to the convergence tests, the required k-points mesh is displayed as follows.

```
Auto generation
0
Monkhorst-Pack
4 4 4
0.0 0.0 0.0
```

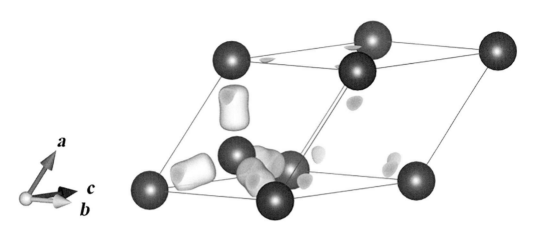

Figure 4.5 The charge density distribution in a silicon unit cell

The ground-state charge density of Si is listed in the CHGCAR file after self-consistent calculation.

```
Si
5.43000000000000
   0.497497      0.497497      0.000000
   0.000000      0.497497      0.497497
   0.497497      0.000000      0.497497
   Sr
   2
Direct
  0.000000  0.000000  0.000000
  0.250000  0.250000  0.250000

56    56    56
 -.57924515671E+00 -.19018275071E+00  0.93485251921E+00  0.26744834492E+01  0.48437827150E+01
  0.72204606841E+01  0.95783481822E+01  0.11720694367E+02  0.13503956727E+02  0.14845826244E+02
  0.15718142084E+02  0.16131842046E+02  0.16123018609E+02  0.15745618836E+02  0.15070200819E+02
  0.14183665984E+02  0.13184690900E+02  0.12172060367E+02  0.11210963426E+02  0.10331485945E+02
  0.95493916565E+01  0.88689310548E+01  0.82875950596E+01  0.78009431667E+01  0.74055197214E+01
. . . . . . . . . . . .
```

The charge density of Si can also be plotted by VESTA software, as shown in Fig. 4.5.

4.3.2 Non-self-consistent Calculation

In the non-self-consistent calculation, the POSCAR and POTCAR files are similar to that used in the self-consistent calculation. The INCAR and KPOINTS files are shown as follows.

1. INCAR File

The non-self-consistent calculation needs a converged charge density as input. Thus, you may need to use the CHGCAR file of the earlier self-consistent calculation.

```
! non-self-consistent calculation
SYSTEM = Si
ENCUT =800
```

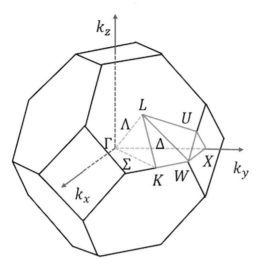

Figure 4.6 The first Brillouin zone of Si crystal structure

```
ISTART = 1
ICHARG = 11
ISMEAR = 0
SIGMA  = 0.1
LORBIT = 11
PERC   = Accurate
```

2. KPOINTS File

To do band structure calculation, one should select high symmetry points and link them along edges of irreducible Brillouin zone, which is the first Brillouin zone reduced by all of the symmetries in the points group of the lattice. The first Brillouin zone of Si crystal structure is shown in Fig. 4.6. The KPOINTS file in the self-consistent calculation is shown as follows:

```
k-points for bandstructure
  10
line
reciprocal
    0.50000   0.50000   0.50000    1
    0.00000   0.00000   0.00000    1

    0.00000   0.00000   0.00000    1
    0.00000   0.50000   0.50000    1

    0.00000   0.50000   0.50000    1
    0.25000   0.62500   0.62500    1

    0.37500   0.7500    0.37500    1
    0.00000   0.00000   0.00000    1
```

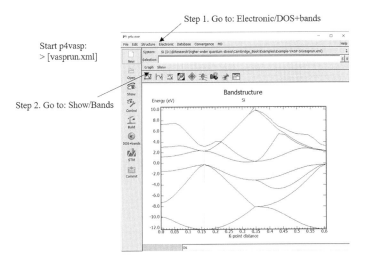

Figure 4.7 The band structure of Si via p4VASP

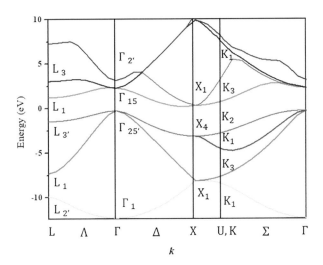

Figure 4.8 The band structure of Si via Origin

In non-self-consistent calculations, the execution command is:

```
mpirun -np 16 vasp
```

The band structure describes the variation of the energy E with the wave vector **k**. In the simulation, the band structure of Si is plotted by using p4VASP (*www.p4vasp.at*), which is a visualization tool for the VASP, and the visual result is shown in Fig. 4.7.

If you want to draw the band structure of Si by using other data post-processing software, such as Origin, the data of band structure of Si can be exported from p4VASP (Graph > Export). Figure 4.8 shows the band structure of Si plotted by Origin.

4.4 Example: DFT Modeling of Calcium Silicate Hydrate (C-S-H) Structure

Concrete is one of the most important engineering materials in the field of civil and construction engineering. Calcium silicate hydrate (C-S-H) is the main product of the hydration of portland cement, which serves as the binder to hold the different constituents of the concrete material together. Thus, it is primarily responsible for the strength in cement-based materials. C-S-H is a nano-sized material with some degree of crystallinity as observed by using X-ray diffraction imaging. The underlying atomic structure of C-S-H is similar to the microstructure of the naturally occurring mineral tobermorite.

In the following example, we present a detailed step-by-step DFT modeling and simulation of C-S-H structure by using VASP. In this example, we conduct a uniaxial compression test for a DFT C-S-H model to find its stress–strain relation at atomistic scale.

4.4.1 Structural Optimization

Before applying compressive strains, structural optimization of C-S-H should be employed first. The input files POSCAR, POTCAR, INCAR, and KPOINTS are displayed as follows.

1. POSCAR File
   This file contains the lattice geometry and the ionic positions of C-S-H ($[Ca_6Si_6O_{18}]$ $2H_2O$).

```
C-S-H model
1.00
    6.6900000572        0.0000000000        0.0000000000
   -4.0777384113        6.1631280593        0.0000000000
    0.0000000000        0.0000000000       22.7700004578
     Ca   Si    O    H
     12   12   40    8
Direct
    0.750000000         0.750000000         0.000000000
    0.250000000         0.250000000         0.500000000
    0.750000000         0.250000000         0.000000000
    0.250000000         0.750000000         0.500000000
    0.750000000         0.750000000         0.412999988
    0.250000000         0.250000000         0.912999988
    0.750000000         0.250000000         0.412999988
    0.250000000         0.750000000         0.912999988
    0.505999982         0.379999995         0.197999999
    0.494000018         0.620000005         0.698000014
```

0.505999982	0.879999995	0.197999999
0.494000018	0.120000005	0.698000014
0.250000000	0.287000000	0.056000002
0.750000000	0.713000000	0.555999994
0.250000000	0.707000017	0.056000002
0.750000000	0.292999983	0.555999994
0.068000004	0.908999979	0.141000003
0.931999981	0.091000021	0.641000032
0.250000000	0.207000002	0.372999996
0.750000000	0.792999983	0.873000026
0.083999999	0.416999996	0.282000005
0.916000009	0.583000004	0.782000005
0.250000000	0.787000000	0.372999996
0.750000000	0.213000000	0.873000026
0.250000000	0.170000002	0.119999997
0.750000000	0.829999983	0.620000005
0.015000000	0.136999995	0.018900000
0.985000014	0.863000035	0.518899977
0.483999997	0.372000009	0.018900000
0.516000032	0.628000021	0.518899977
0.250000000	0.500000000	0.077000000
0.750000000	0.500000000	0.577000022
0.015000000	0.621999979	0.018900000
0.985000014	0.378000021	0.518899977
0.483999997	0.856000006	0.018900000
0.516000032	0.143999994	0.518899977
0.250000000	0.829999983	0.119999997
0.750000000	0.170000017	0.620000005
0.068000004	0.908999979	0.210999995
0.931999981	0.091000021	0.710999966
0.819999993	0.785000026	0.112999998
0.180000007	0.214999974	0.612999976
0.250000000	0.000000000	0.347999990
0.750000000	0.000000000	0.847999990
0.015000000	0.122000001	0.410800010
0.985000014	0.878000021	0.910799980
0.483999997	0.356999993	0.410800010
0.516000032	0.643000007	0.910799980
0.250000000	0.335000008	0.310000002
0.750000000	0.664999962	0.810000002
0.100000001	0.425000012	0.213000000
0.899999976	0.574999988	0.713000000
0.250000000	0.644999981	0.310000002
0.750000000	0.355000019	0.810000002

0.015000000	0.638000011	0.410800010
0.985000014	0.361999989	0.910799980
0.483999997	0.871999979	0.410800010
0.516000032	0.128000021	0.910799980
0.824999988	0.287999988	0.305999994
0.174999997	0.712000012	0.805999994
0.750000000	0.750000000	0.303000003
0.250000000	0.250000000	0.802999973
0.750000000	0.250000000	0.109999999
0.250000000	0.750000000	0.610000014
0.767880023	0.870069981	0.335500002
0.232119977	0.129930019	0.835500002
0.915920019	0.750000000	0.303000003
0.084079981	0.250000000	0.802999973
0.767880023	0.370070010	0.142499998
0.232119977	0.629930019	0.642499983
0.915920019	0.250000000	0.109999999
0.084079981	0.750000000	0.610000014

2. POTCAR File

The POTCAR file contains the pseudopotential information for each atomic species used in the calculation. Here, the frozen-core all-electron projector augmented wave (PAW) method is used to model the ion–electron interactions. The generalized gradient approximation (GGA) with the Perdew–Burke–Ernzerhof (PBE) functional is employed as the exchange correlation potential. That is, the recommended Ca_sv_GW, Si_GW, O_GW, and H_GW are used in the C-S-H simulations.

3. INCAR File

To ensure accuracy and efficiency, convergence tests are made before calculations to determine the cutoff energy required. After convergence tests, the INCAR file is listed as follows:

```
! relax calculation
SYSTEM = C-S-H
ENCUT = 900
ISTART = 0
ICHARG = 2
ISMEAR = -5
ISIF = 3
IBRION = 2
EDIFF = 1E-5
EDIFFG = -0.02
PERC = Accurate
```

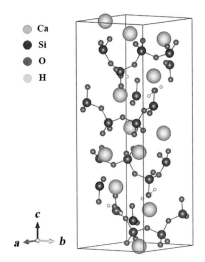

Figure 4.9 The optimized crystal structure of C-S-H

```
! static calculation
SYSTEM = C-S-H
ENCUT = 900
ISTART = 0
ICHARG = 2
ISMEAR = -5
PERC = Accurate
```

4. KPOINTS File

Similarly, some convergence tests shall also be first performed on the C-S-H to determine the number of k-points required. The determined KPOINTS file is displayed as follows:

```
Auto generation
0
Monkhorst-Pack
3 3 1
0.0 0.0 0.0
```

After structural optimization, the optimized C-S-H model could be obtained, as shown in Fig. 4.9. Readers could find the optimized lattice constants and atomic positions of C-S-H in CONCAR file.

4.4.2 Uniaxial Compression Test

In this section, we discuss the actual DFT uniaxial compression test. Uniaxial compressive strains are applied on the optimized C-S-H structure along the c-axis, while allowing full structure relaxation of the other five strain components and requiring the residual stresses after relaxation are less than 0.1 GPa. At each compression deformation step, a small increment (a 0.02 compressive strain) is applied sequentially to the relaxed structure in the previous step. The input files are listed as follows.

1. POSCAR File

The following calculations are based on the optimized C-S-H structure, whose lattice constants and atomic positions are included in the CONCAR file after relaxation calculation. Then, the c-axis compressive strain is applied by changing the lattice constant c. In the following, we list the POSCAR file for C-S-H structure under 0.02 c-axis compressive strain.

```
C-S-H 0.02 c-axis compressive strain
1.00
    6.6463890076        0.0000000000        0.0000000000
   -4.0943219015        6.1640420625        0.0000000000
    0.0000000000        0.0000000000       23.2739644623
    Ca   Si   O    H
    12   12   40   8
Direct
   0.772674263          0.761398137         0.014303752
   0.227325723          0.238601804         0.514303744
   0.745510399          0.249186292         0.997438729
   0.254489601          0.750813723         0.497438788
   0.749451578          0.749009669         0.428005725
   0.250548452          0.250990331         0.928005695
   0.776404202          0.261113077         0.411106706
   0.223595783          0.738886893         0.911106706
   0.478128821          0.195940837         0.258199543
   0.521871150          0.804059148         0.758199513
   0.477994382          0.698649347         0.166793540
   0.522005558          0.301350653         0.666793525
   0.263675749          0.295957953         0.054283667
   0.736324251          0.704042017         0.554283619
   0.266196311          0.720813930         0.050405238
   0.733803749          0.279186100         0.550405204
   0.113610581          0.928846121         0.140249670
   0.886389375          0.071153902         0.640249670
   0.270038337          0.220592052         0.374924749
   0.729961693          0.779407918         0.874924779
   0.116665363          0.428362817         0.285086602
   0.883334696          0.571637154         0.785086632
   0.267363310          0.795674026         0.371142894
```

0.732636750	0.204325974	0.871142983
0.255116999	0.176053688	0.112609021
0.744883001	0.823946297	0.612609029
0.025547411	0.151777983	0.015333158
0.974452615	0.848222017	0.515333116
0.498658299	0.372684360	0.015161173
0.501341701	0.627315700	0.515161157
0.288516551	0.521750569	0.077071838
0.711483419	0.478249401	0.577071846
0.028506635	0.630651712	0.012059865
0.971493363	0.369348317	0.512059867
0.508870721	0.882758081	0.015363868
0.491129279	0.117241919	0.515363872
0.265944749	0.826039195	0.112786613
0.734055281	0.173960805	0.612786651
0.163807169	0.952928245	0.207111254
0.836192787	0.047071781	0.707111239
0.831930280	0.776437104	0.120097004
0.168069690	0.223562896	0.620096982
0.292137891	0.021352604	0.348246843
0.707862079	0.978647411	0.848246813
0.032637008	0.130548164	0.413374364
0.967363000	0.869451880	0.913374305
0.513096273	0.382785559	0.409797132
0.486903816	0.617214441	0.909797072
0.268937021	0.325395912	0.312540531
0.731062949	0.674604058	0.812540591
0.167283699	0.452534646	0.218246952
0.832716286	0.547465324	0.718246996
0.257667810	0.675277829	0.312808186
0.742332160	0.324722141	0.812808216
0.029497735	0.651721060	0.410194188
0.970502257	0.348278970	0.910194159
0.502660930	0.872465551	0.410149336
0.497339100	0.127534464	0.910149336
0.834629655	0.275650829	0.305067986
0.165370345	0.724349141	0.805067956
0.748203635	0.888204157	0.241049081
0.251796365	0.111795887	0.741049051
0.749418676	0.378469169	0.183353052
0.250581324	0.621530890	0.683353066
0.772543728	0.998871267	0.268295288

```
0.227456257          0.001128757          0.768295288
0.914945304          0.920876324          0.231682450
0.085054681          0.079123653          0.731682420
0.772087157          0.489127398          0.156458035
0.227912858          0.510872662          0.656458020
0.917105198          0.413531363          0.193001926
0.082894832          0.586468637          0.693001926
```

2. POTCAR File

This file is the same as the POTCAR file in the above structural optimization calculations.

3. INCAR File

In order to apply the compressive strain in c-axis on C-S-H structure, the c-axis is fixed during relaxation calculations. Thus, we modify the code $constr\_cell\_relax.F$ of VASP and recompile the VASP as VASP-z. Here are the INCAR files.

```
! relax calculation
SYSTEM = C-S-H
ENCUT = 900
ISTART = 0
ICHARG = 2
ISMEAR = -5
IBRION = 2
ISIF = 3
EDIFF = 1E-5
EDIFFG = -0.02
NSW = 200
PERC = Accurate
```

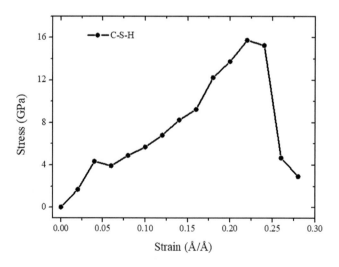

Figure 4.10 The stress–strain curve of C-S-H under the c-axis uniaxial compression

```
! static calculation
SYSTEM = C-S-H
ENCUT = 900
ISTART = 0
ICHARG = 2
ISMEAR = -5
PERC = Accurate
```

Note that in relax calculations you should enter the command:

`mpirun -np 16 vasp-z`

And in static calculations, the execution command is:

`mpirun -np 16 vasp`

4. KPOINTS Files

This file is the same as the KPOINTS file used in the earlier structural optimization calculations.

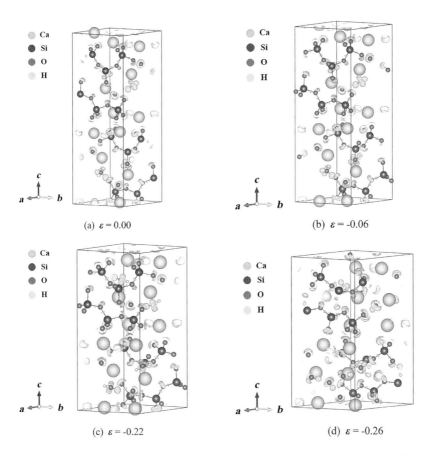

Figure 4.11 The structural changes at various critical strains for C-S-H structure under the c-axis uniaxial compression: (a) Compression strain $\epsilon = 0$; (b) Compression strain $\epsilon = 0.06$; (c) Compression strain $\epsilon = 0.22$, and (d) Compression strain $\epsilon = 0.26$.

```
Auto generation
0
Monkhorst-Pack
3 3 1
0.0 0.0 0.0
```

The stress–strain curve of C-S-H under the c-axis uniaxial compression is shown in Fig. 4.10. It should be noted that the compression is denoted as positive in DFT calculations, which is opposite to that of classical mechanics notation.

In order to further understand the deformation mechanism of C-S-H under the c-axis uniaxial compression, structural changes and isosurface of the electron localization function (ELF) in C-S-H are also examined, as displayed in Fig. 4.11.

Part II

Statistical Molecular Dynamics

5 Fundamentals of Statistical Mechanics

In this chapter, we review and outline the basic theories of statistical mechanics, including Lagrangian and Hamiltonian mechanics of particle systems.

5.1 Lagrangian Mechanics

Consider an N-particle system. The motion of each particle is described by its position coordinates and velocity, $\mathbf{q}_i$ and $\dot{\mathbf{q}}_i, i = 1, 2 \ldots N$. We denote the Lagrange function, which is often called the Lagrangian, as

$$L(\mathbf{q}_1, \mathbf{q}_2, \ldots, \mathbf{q}_N; \dot{\mathbf{q}}_1, \dot{\mathbf{q}}_2, \ldots \dot{\mathbf{q}}_N; t) = T - V, \tag{5.1}$$

where T denotes the kinetic energy, and V denotes the potential energy of the system.

5.1.1 Least Action (Hamilton's) Principle

For a fixed time interval $[t_1, t_2]$, we may consider the following action integral:

$$S = \int_{t_1}^{t_2} L(\{\mathbf{q}_i, \dot{\mathbf{q}}_i, t\}) dt, \tag{5.2}$$

in which $\mathbf{q}_i, i = 1, 2, \ldots N$, and they have fixed boundary conditions at the two ends of the time interval as shown in Fig. 5.1. They can be written as,

$$\mathbf{q}_i(t_1) = const. \text{ and } \mathbf{q}_i(t_2) = const., i = 1, 2, \ldots N,$$

which means that the displacements of a particle at the beginning and end of its trajectory are fixed in space, or $\delta \mathbf{q}_i(t_1) = \delta \mathbf{q}_i(t_2) = 0$.

The least action, or the Hamilton, principle states that the actual trajectories of the particles will make the action integral-stationary, i.e.,

$$\delta S = \delta \int_{t_1}^{t_2} L(\{\mathbf{q}_i, \dot{\mathbf{q}}_i, t\}) dt = 0. \tag{5.3}$$

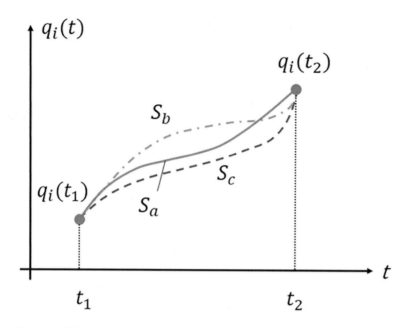

Figure 5.1 Illustration of the Hamilton principle and its boundary conditions.

Using the general form of the Lagrange function, one derives the following expressions based on the least action principle:

$$\delta S = \delta \int_{t_1}^{t_2} L(\mathbf{q}_i, \dot{\mathbf{q}}_i, t) dt = \int_{t_1}^{t_2} \delta L(\mathbf{q}_i, \dot{\mathbf{q}}_i, t) dt$$

$$= \int_{t_1}^{t_2} \left(\frac{\partial L}{\partial \mathbf{q}_i} \delta \mathbf{q}_i + \frac{\partial L}{\partial \dot{\mathbf{q}}_i} \delta \dot{\mathbf{q}}_i \right) dt = \frac{\partial L}{\partial \dot{\mathbf{q}}_i} \delta \mathbf{q}_i \Big|_{t_1}^{t_2} + \int_{t_1}^{t_2} \left(\frac{\partial L}{\partial \mathbf{q}_i} - \frac{d}{dt} \frac{\partial L}{\partial \dot{\mathbf{q}}_i} \right) \delta \mathbf{q}_i dt. \tag{5.4}$$

Considering the boundary conditions, we have

$$\delta \mathbf{q}_i(t_1) = \delta \mathbf{q}_i(t_2) = 0 \rightarrow \frac{\partial L}{\partial \dot{\mathbf{q}}_i} \delta \mathbf{q}_i \Big|_{t_1}^{t_2} = 0,$$

and the second term in Eq. (5.4) yields the Euler–Lagrange equations

$$\frac{d}{dt} \left(\frac{\partial L}{\partial \dot{\mathbf{q}}_i} \right) - \frac{\partial L}{\partial \mathbf{q}_i} = 0, \quad i = 1, 2, \ldots N. \tag{5.5}$$

The Lagrangian equations are based on the least action principle only and are valid for all coordinate systems.

5.2 Hamiltonian Mechanics

The description of a mechanical system in terms of generalized coordinates and velocities is not unique. Alternative formulations in terms of different variables can be used for convenience. In statistical mechanics, it is often convenient to use the generalized coordinates and momenta to describe particle trajectories in the phase space rather than using the coordinates and velocities in the physical space.

One can use the Legendre transformation, which is the passage from one set of independent variables to another, to transform the variable from $\{\mathbf{q}_i, \dot{\mathbf{q}}_i\}$ to $\{\mathbf{q}_i, \mathbf{p}_i\}, i = 1, 2, \ldots N$.

To illustrate this process, we first consider a general form of a Lagrangian

$$L = T - V = \sum_\alpha \frac{m_\alpha}{2} \dot{q}_\alpha \dot{q}_\alpha - V(q_\alpha) \tag{5.6}$$

and its related Lagrangian equation

$$\frac{d}{dt}\frac{\partial L}{\partial \dot{q}_\alpha} = \frac{\partial L}{\partial q_\alpha} \quad \rightarrow \quad \frac{d}{dt}(m_\alpha \dot{q}_\alpha) = \frac{d}{dt}(p_\alpha) = \dot{p}_\alpha = \frac{\partial L}{\partial q_\alpha}. \tag{5.7}$$

Inspired by Eq. (5.7), it is natural for us to define the generalized momentum

$$\mathbf{p}_i = \frac{\partial L}{\partial \dot{\mathbf{q}}_i}, \tag{5.8}$$

which leads to another form of the Lagrangian equation,

$$\dot{\mathbf{p}}_i = \frac{\partial L}{\partial \mathbf{q}_i}. \tag{5.9}$$

Example 5.1 Consider

$$L = T - V = \sum_i \frac{m_i}{2} \dot{\mathbf{r}}_i \cdot \dot{\mathbf{r}}_i - V(\{\mathbf{r}_i\}).$$

For the material points in Cartesian coordinates, we can define the momenta as

$$\mathbf{p}_i = \frac{\partial L}{\partial \dot{\mathbf{r}}_i} = m_i \dot{\mathbf{r}}_i.$$

In component form, we have

$$p_{ix} = m_i \dot{x}_i, \quad p_{iy} = m_i \dot{y}_i, \quad p_{iz} = m_i \dot{z}_i.$$

Then the alterative Lagrangian equations are

$$\dot{\mathbf{p}}_i = \frac{\partial L}{\partial \mathbf{r}_i} \quad \rightarrow \quad m_i \ddot{\mathbf{r}}_i = -\frac{\partial V}{\partial \mathbf{r}_i},$$

which recovers the standard form of the Lagrangian equations.

To illustrate the Legendre transformation, we first differentiate a general form of the Lagrange function and obtain

$$L = L(\mathbf{q}_i, \dot{\mathbf{q}}_i) \Rightarrow dL = \sum_i \frac{\partial L}{\partial \mathbf{q}_i} d\mathbf{q}_i + \sum_i \frac{\partial L}{\partial \dot{\mathbf{q}}_i} d\dot{\mathbf{q}}_i, \quad i = 1, 2, \ldots N. \tag{5.10}$$

We then substitute the definition of momenta (5.8) and the alternative form of the Lagrangian equation (5.9) into (5.10). Equation (5.10) becomes

$$dL = \sum_i \dot{\mathbf{p}}_i d\mathbf{q}_i + \sum_i \mathbf{p}_i d\dot{\mathbf{q}}_i. \tag{5.11}$$

Using the product rule, we may rewrite Eq. (5.11) as

$$dL = \sum_i \dot{\mathbf{p}}_i d\mathbf{q}_i + d\left(\sum_i (\mathbf{p}_i \cdot \dot{\mathbf{q}}_i)\right) - \sum_i \dot{\mathbf{q}}_i d\mathbf{p}_i, \tag{5.12}$$

which leads to the following differential form of the Legendre transformation:

$$d\left(\sum_i \mathbf{p}_i \cdot \dot{\mathbf{q}}_i - L\right) = \sum_i \dot{\mathbf{q}}_i d\mathbf{p}_i - \sum_i \dot{\mathbf{p}}_i d\mathbf{q}_i. \tag{5.13}$$

We can then pair $(\mathbf{q}_i, \mathbf{p}_i)$ as conjugate variables.
The conjugate variable transforms

$$(\mathbf{q}_i, \dot{\mathbf{q}}_i) \rightarrow (\mathbf{q}_i, \mathbf{p}_i)$$

with respect to the Lagrangian in the Legendre transformation above, which leads to a new state function (or variable) that we call the *Hamiltonian*, i.e.,

$$H(\{\mathbf{q}_i\}, \{\mathbf{p}_i\}) = \sum_i \mathbf{p}_i \cdot \dot{\mathbf{q}}_i - L(\{\mathbf{q}_i\}, \{\dot{\mathbf{q}}_i\}) = \sum_i \frac{\partial L}{\partial \dot{\mathbf{q}}_i} \cdot \dot{\mathbf{q}}_i - L(\{\mathbf{q}_i\}, \{\dot{\mathbf{q}}_i\}). \tag{5.14}$$

Based on Eq. (5.12), the differential of the Hamiltonian is

$$dH = \sum_i \dot{\mathbf{q}}_i d\mathbf{p}_i - \sum_i \dot{\mathbf{p}}_i d\mathbf{q}_i. \tag{5.15}$$

For a conservative system, the differential of the Hamiltonian is a total differential, which implies that it is a state variable. Thus, we also should have

$$dH = \sum_i \left(\frac{\partial H}{\partial \mathbf{p}_i} d\mathbf{p}_i + \frac{\partial H}{\partial \mathbf{q}_i} d\mathbf{q}_i\right). \tag{5.16}$$

Comparing Eqs. (5.15) and (5.16), one can identify the *Hamiltonian equation of motion* or Hamilton's equation as

$$\dot{\mathbf{q}}_i = \frac{\partial H}{\partial \mathbf{p}_i}, \tag{5.17}$$

$$\dot{\mathbf{p}}_i = -\frac{\partial H}{\partial \mathbf{q}_i}. \tag{5.18}$$

The above equations are the governing equations of a Hamiltonian system. Moreover, for a time-dependent Lagrangian, we have

$$dL = \sum_i \dot{\mathbf{p}}_i d\mathbf{q}_i + d\left(\sum_i (\mathbf{p}_i \cdot \dot{\mathbf{q}}_i)\right) - \sum_i \dot{\mathbf{q}}_i d\mathbf{p}_i + \frac{\partial L}{\partial t} dt, \tag{5.19}$$

which may be rewritten as

$$d\left(\sum_i \mathbf{p}_i \cdot \dot{\mathbf{q}}_i - L\right) = \sum_i \dot{\mathbf{q}}_i d\mathbf{p}_i - \sum_i \dot{\mathbf{p}}_i d\mathbf{q}_i - \frac{\partial L}{\partial t} dt \qquad (5.20)$$

or

$$dH = \sum_i \dot{\mathbf{q}}_i d\mathbf{p}_i - \sum_i \dot{\mathbf{p}}_i d\mathbf{q}_i - \frac{\partial L}{\partial t} dt. \qquad (5.21)$$

On comparison with the corresponding time-dependent Hamiltonian

$$dH = \left(\sum_i \frac{\partial H}{\partial \mathbf{p}_i} d\mathbf{p}_i + \frac{\partial H}{\partial \mathbf{q}_i} d\mathbf{q}_i\right) + \frac{\partial H}{\partial t} dt, \qquad (5.22)$$

we should have

$$\boxed{\frac{\partial H}{\partial t} = -\frac{\partial L}{\partial t}. \qquad (5.23)}$$

Example 5.2 Find the Hamiltonian and Hamiltonian equations for the following Lagrangian:

$$L = \frac{C}{2}\left(\dot{r}^2 + r^2\dot{\theta}^2\right) - \frac{k}{2}r^2.$$

First, we can derive the two Lagrangian equations associated with variable pairs $(\theta, \dot{\theta})$ and $(r, \dot{r})$:

$$\frac{d}{dt}\left(\frac{\partial L}{\partial \dot{\theta}}\right) - \frac{\partial L}{\partial \theta} = 0, \qquad (5.24)$$

$$\frac{d}{dt}\left(\frac{\partial L}{\partial \dot{r}}\right) - \frac{\partial L}{\partial r} = 0. \qquad (5.25)$$

Since in Eq. (5.24)

$$\frac{\partial L_{rel}}{\partial \theta} = 0,$$

it leads to the following Lagrangian equation of motion

$$\frac{d}{dt}\left(\frac{\partial L}{\partial \dot{\theta}}\right) = 0 \quad \rightarrow \quad \frac{d}{dt}\left(Cr^2\dot{\theta}\right) = 0. \qquad (5.26)$$

On the other hand, Eq. (5.25) gives the equation of motion,

$$C\ddot{r} - Cr\dot{\theta}^2 + \frac{\partial V}{\partial r} = 0. \qquad (5.27)$$

However, based on the definition of conjugate momentum variable, one can find that

$$p_\theta = \frac{\partial L}{\partial \dot{\theta}} = Cr^2\dot{\theta}, \text{ and } p_r = \frac{\partial L}{\partial \dot{r}} = C\dot{r}$$

and hence the Hamiltonian is

$$H = (p_r \dot{r} + p_\theta \dot{\theta}) - L = \frac{C}{2}(\dot{r}^2 + r^2 \dot{\theta}^2) + V.$$

Then, after a simple calculation, we may find that

$$\frac{\partial H}{\partial \theta} = 0 \rightarrow \frac{d}{dt} p_\theta = \frac{d}{dt}(Cr^2 \dot{\theta}) = 0. \tag{5.28}$$

Considering the fact that

$$\frac{\partial H}{\partial p_\theta} = \frac{1}{Cr^2} p_\theta = \dot{\theta} \rightarrow \frac{1}{2} p_\theta \dot{\theta} = \frac{p_\theta^2}{2Cr^2},$$

one may find that

$$\frac{\partial H}{\partial r} = -\frac{1}{Cr^3} p_\theta^2 + \frac{\partial V}{\partial r},$$

which then leads to

$$\frac{d}{dt} p_r = C\ddot{r} = -\left(-\frac{p_\theta^2}{Cr^3} + \frac{\partial V}{\partial r}\right). \tag{5.29}$$

One can find that Eqs. (5.28) and (5.29) are exactly the same as Eqs. (5.26) and (5.27).

REMARK 5.3 A natural question is why we are interested in the Hamiltonian description of an atomistic or molecular system. There are several good reasons and advantages why we would like to use the Hamiltonian description to study the motions of a molecular system.

(1) In statistical physics, the probability density of a given thermodynamics system is proportional to its energy states according to

$$Prob(state) \propto \exp\left(\frac{-H(state)}{k_B T}\right).$$

(2) Geometrical elegance: Hamilton equations basically imply that the flow of a Hamiltonian system in space-time is equivalent to its flow along a vector field in the phase space as shown in Fig. 5.2.

(3) Hamiltonian equations are symmetric with respect to both its primary variables $\eta = (\mathbf{q}, \mathbf{p})$. Thus, the computational algorithm of a Hamiltonian system should also possess corresponding symmetry, which leads to easy, fast, and accurate computations.

If a particle system is a Hamiltonian system, by using the Hamiltonian equation, we can express the rate of time change of the probability distribution of the system as,

$$\frac{d\mathbf{f}}{dt} = \sum_i \left(\frac{\partial \mathbf{f}}{\partial \mathbf{q}_i} \dot{\mathbf{q}}_i + \frac{\partial \mathbf{f}}{\partial \mathbf{p}_i} \dot{\mathbf{p}}_i\right) = \sum_i \left(\frac{\partial \mathbf{f}}{\partial \mathbf{q}_i} \cdot \frac{\partial H}{\partial \mathbf{p}_i} - \frac{\partial \mathbf{f}}{\partial \mathbf{p}_i} \cdot \frac{\partial H}{\partial \mathbf{q}_i}\right). \tag{5.30}$$

Introduce a differential operator

$$[\mathbf{f}, H] := \left(\frac{\partial \mathbf{f}}{\partial \mathbf{q}} \cdot \frac{\partial H}{\partial \mathbf{p}} - \frac{\partial \mathbf{f}}{\partial \mathbf{p}} \cdot \frac{\partial H}{\partial \mathbf{q}}\right), \tag{5.31}$$

which is called the Poisson's bracket in physics literature.

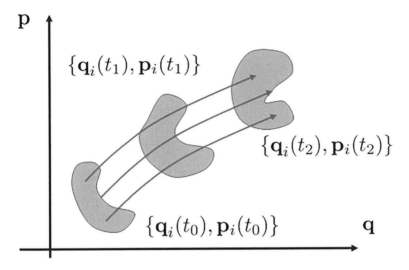

Figure 5.2 Motion flow of a Hamiltonian system in the phase space

With the notation of Poisson's bracket, we can rewrite Eq. (5.30) succinctly as

$$\frac{d\mathbf{f}}{dt} = [\mathbf{f}, H]. \tag{5.32}$$

5.3 Liouville Theorem

As shown in Eqs. (5.30) and (5.32), the motion of a Hamiltonian system may be described by its evolution of volume flow in the phase space. The Liouville theorem is a theorem about the phase space volume change of Hamiltonian systems.

For Hamiltonian systems, the control volume is time invariant, i.e.,

$$\frac{\partial}{\partial t} dV_\eta = 0.$$

Denote the control volume in the phase space as ω. The balance of the probability flow $\mathbf{f}$ for a controlled volume is

$$\frac{\partial}{\partial t} \int_\omega f \, dV_\eta = -\oint_{\partial \omega} f \dot{\boldsymbol{\eta}} \cdot \mathbf{n} \, dS_\eta, \tag{5.33}$$

where $\eta = (\mathbf{q}, \mathbf{p})$. By the divergence theorem,

$$\int_\omega \frac{\partial f}{\partial t} dV_\eta = -\int_\omega \frac{\partial}{\partial \eta} \cdot (f \dot{\boldsymbol{\eta}}) \, dV_\eta. \tag{5.34}$$

Let $\eta = (\mathbf{q}, \mathbf{p})$. Since ω is arbitrary, we can derive the local form of balance (conservation) of probability density in the phase space (the Liouville theorem):

$$\frac{\partial f}{\partial t}\bigg|_\eta = \mathcal{L}f = -\left(f \frac{\partial}{\partial \eta} \cdot \dot{\boldsymbol{\eta}} + \dot{\boldsymbol{\eta}} \cdot \frac{\partial f}{\partial \eta}\right). \tag{5.35}$$

By chain rule, the total derivative of the probability density can be written as,

$$\frac{df}{dt} = \frac{\partial f}{\partial t} + \dot{\boldsymbol{\eta}} \cdot \frac{\partial f}{\partial \boldsymbol{\eta}} = -f \frac{\partial}{\partial \boldsymbol{\eta}} \dot{\boldsymbol{\eta}} =: -f \Lambda(\boldsymbol{\eta}), \qquad (5.36)$$

where $\Lambda(\boldsymbol{\eta})$ is called the *phase-space compression factor* and $\Lambda(\boldsymbol{\eta}) = 0$ is called the *adiabatic incompressibility condition* (AIη).

For a Hamiltonian system,

$$\frac{df}{dt} = 0 \;\rightarrow\; \Lambda(\boldsymbol{\eta}) = 0,$$

we have

$$\Lambda(\boldsymbol{\eta}) = \sum_{i=1}^{N} \left(\frac{\partial}{\partial \mathbf{q}_i} \cdot \dot{\mathbf{q}}_i + \frac{\partial}{\partial \mathbf{p}_i} \cdot \dot{\mathbf{p}}_i \right) = \sum_{i=1}^{N} \left(\frac{\partial}{\partial \mathbf{q}_i} \cdot \frac{\partial H}{\partial \mathbf{p}_i} - \frac{\partial}{\partial \mathbf{p}_i} \cdot \frac{\partial H}{\partial \mathbf{q}_i} \right) \equiv 0. \qquad (5.37)$$

The last equation holds because

$$\frac{\partial H}{\partial \mathbf{p}_i} = \dot{\mathbf{q}}_i, \quad \frac{\partial H}{\partial \mathbf{q}_i} = -\dot{\mathbf{p}}_i.$$

Conversely, one may say that the condition for existence of a Hamiltonian is that the phase space is incompressible, i.e.,

$$\frac{df}{dt} = 0.$$

Now we formally state the Liouville theorem.

THEOREM 5.4 (Liouville theorem) *For a Hamiltonian system, the probability distribution function f is constant along any trajectory in the phase space, which is expressed in the following Liouville equation,*

$$\frac{df}{dt} = 0 \;\rightarrow\; \frac{df}{dt} = \frac{\partial f}{\partial t} + \sum_i \left(\frac{\partial f}{\partial \mathbf{q}_i} \dot{\mathbf{q}}_i + \frac{\partial f}{\partial \mathbf{p}_i} \dot{\mathbf{p}}_i \right)$$

$$= \frac{\partial f}{\partial t} + \sum_i \left(\frac{\partial f}{\partial \mathbf{q}_i} \frac{\partial H}{\partial \mathbf{p}_i} - \frac{\partial f}{\partial \mathbf{p}_i} \frac{\partial H}{\partial \mathbf{q}_i} \right) = 0, \qquad (5.38)$$

which may be written in a more compact form,

$$\frac{\partial f}{\partial t} + [f, H] = 0, \qquad (5.39)$$

where $[,]$ *is the Poisson bracket.*

The Liouville equation may also be expressed in terms of the Liouville operator or Liouvillian,

$$i\mathbf{L} = \sum_i \left[\frac{\partial H}{\partial \mathbf{p}_i} \frac{\partial}{\partial \mathbf{q}_i}, -\frac{\partial H}{\partial \mathbf{q}_i} \frac{\partial}{\partial \mathbf{p}_i} \right] = \{\cdot, H\} \qquad (5.40)$$

or

$$\frac{\partial f}{\partial t} + i\mathbf{L}f = 0. \qquad (5.41)$$

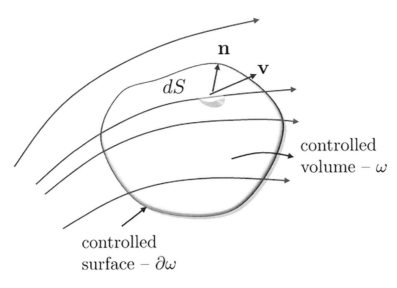

Figure 5.3 Controlled volume of a Hamiltonian system in the phase space

REMARK 5.5 Similar to the conservation of mass in continuum mechanics,

$$\frac{\partial \rho}{\partial t} + \frac{\partial \rho v_j}{\partial x_j} = 0 \quad \rightarrow \quad \frac{d\rho}{dt} + \rho \nabla \cdot \mathbf{v} = 0.$$

where ρ is mass density and $\mathbf{v}$ is the velocity field as shown in Fig. 5.3, the *conservation of probability* of a mechanical system in the phase space reads as

$$\frac{\partial f}{\partial t} + \frac{\partial (f\dot{\eta})}{\partial \eta} = \frac{\partial f}{\partial t} + \left(f \frac{\partial}{\partial \eta} \cdot \dot{\eta} + \dot{\eta} \cdot \frac{\partial f}{\partial \eta} \right) = \frac{df}{dt} + f \frac{\partial \dot{\eta}}{\partial \eta} = 0.$$

For the Hamiltonian systems,

$$\frac{\partial \dot{\eta}}{\partial \eta} = 0 \quad \rightarrow \quad \frac{df}{dt} = 0.$$

This is because for a Hamiltonian system we can show that

$$\Lambda(\eta) = \frac{\partial \dot{\eta}}{\partial \eta} = \sum_{i=1}^{N} \left(\frac{\partial}{\partial \mathbf{q}_i} \cdot \dot{\mathbf{q}}_i + \frac{\partial}{\partial \mathbf{p}_i} \cdot \dot{\mathbf{p}}_i \right) = \sum_{i=1}^{N} \left(\frac{\partial}{\partial \mathbf{q}_i} \cdot \frac{\partial H}{\partial \mathbf{p}_i} - \frac{\partial}{\partial \mathbf{p}_i} \cdot \frac{\partial H}{\partial \mathbf{q}_i} \right) \equiv 0.$$

(5.42)

5.4 Canonical Transformation and Symplectic Condition

In this section, we would like to explore the mathematical structure of the Hamiltonian equations,

$$\dot{\mathbf{q}} = \frac{\partial H(\mathbf{q}, \mathbf{p})}{\partial \mathbf{p}}, \quad \dot{\mathbf{p}} = -\frac{\partial H(\mathbf{q}, \mathbf{p})}{\partial \mathbf{q}}.$$

Let $\eta = (\mathbf{q}, \mathbf{p})$. The Hamiltonian equations can be written in the following form:

$$\dot{\eta} = [\eta, H] \quad \rightarrow \quad \dot{\eta} = \Omega \frac{\partial H(\eta)}{\partial \eta}, \quad \text{where} \quad \Omega = \begin{bmatrix} 0 & \mathbf{I} \\ -\mathbf{I} & 0 \end{bmatrix}, \quad (5.43)$$

where $\mathbf{I}$ is the 3×3 identity matrix.

For Hamiltonian systems,

$$\dot{\eta} = \Omega \frac{\partial H(\eta)}{\partial \eta},$$

we consider a general variable transform,

$$\xi = \xi(\eta), \quad \text{and} \quad \frac{\partial \xi}{\partial \eta} = \mathbf{M}, \quad d\xi = \mathbf{M} d\eta \quad \text{and} \quad \dot{\xi} = \mathbf{M}\dot{\eta}. \quad (5.44)$$

A transformation is called canonical if after the transform we still have

$$\dot{\xi} = \Omega \frac{\partial H(\xi)}{\partial \xi}, \quad \text{where} \quad \Omega = \begin{bmatrix} 0 & \mathbf{I} \\ -\mathbf{I} & 0 \end{bmatrix}. \quad (5.45)$$

Because of $\dot{\xi} = \mathbf{M}\dot{\eta}$,

$$\frac{d\eta}{dt} = \mathbf{M}^{-1} \frac{d\xi}{dt}, \quad \text{and} \quad \frac{\partial H}{\partial \eta} = \frac{\partial H}{\partial \xi} \frac{\partial \xi}{\partial \eta} = \frac{\partial H}{\partial \xi} \mathbf{M} = \mathbf{M}^T \frac{\partial H}{\partial \xi}.$$

Hence

$$\dot{\eta} = \Omega \frac{\partial H(\eta)}{\partial \eta}, \quad \rightarrow \quad \mathbf{M}^{-1} \dot{\xi} = \Omega \mathbf{M}^T \frac{\partial H}{\partial \xi}, \quad \rightarrow \quad \dot{\xi} = \mathbf{M} \Omega \mathbf{M}^T \frac{\partial H}{\partial \xi}.$$

Comparing the definition of canonical transformation

$$\dot{\xi} = \Omega \frac{\partial H(\xi)}{\partial \xi}, \quad \text{where} \quad \Omega = \begin{bmatrix} 0 & \mathbf{I} \\ -\mathbf{I} & 0 \end{bmatrix}$$

with the actual transformation equations,

$$\dot{\eta} = \Omega \frac{\partial H(\eta)}{\partial \eta}, \quad \rightarrow \quad \mathbf{M}^{-1} \dot{\xi} = \Omega \mathbf{M}^T \frac{\partial H}{\partial \xi}, \quad \rightarrow \quad \dot{\xi} = \mathbf{M} \Omega \mathbf{M}^T \frac{\partial H}{\partial \xi}.$$

We conclude the condition for canonical transformation is,

$$\mathbf{M} \Omega \mathbf{M}^T = \Omega. \quad (5.46)$$

which is called the symplectic condition. This means that under this condition, the system remains a *Hamiltonian system* after the transformation.

The physical implication of the symplectic condition

$$det\{\mathbf{M} \Omega \mathbf{M}^T\} = det\{\mathbf{M}\}^2 det\{\Omega\} = 1$$

recalling

$$\Omega = \begin{bmatrix} 0 & \mathbf{I} \\ -\mathbf{I} & 0 \end{bmatrix} \quad (5.47)$$

which leads to

$$(det\mathbf{M})^2 = 1, \quad \rightarrow \quad \det \mathbf{M} = 1. \tag{5.48}$$

Consider

$$d\xi = \mathbf{M}d\eta$$

and the symplectic condition $\det \mathbf{M} = 1$ means that the phase space is invariant, i.e.,

$$|d\xi| = |d\eta|. \tag{5.49}$$

Recall that $\Lambda(\eta)$ is called the *phase-space compression factor*, and we call $\Lambda(\eta) = 0$ the *adiabatic incompressibility condition* (AIη).

5.5 Laws of Thermodynamics

Before we start, we would like to ask a fundamental question: *What is objective molecular dynamics?*

The answer to this question is: To find the statistical average of macroscale or mesoscale quantities at given thermodynamics conditions, such as stresses, strains, temperature, and the like. In fact, one may say that molecular dynamics is an abbreviation of statistical molecular dynamics and statistical molecular mechanics.

In statistical physical, an intensive variable is a physical property of a system that does not depend on the system size or the amount of material in the system: it is scale invariant, e.g., density, temperature, stress, concentration, and so on.

By contrast, an extensive parameter of a system is directly proportional to the system size or the amount of material in the system, e.g., volume, mass, energy, and so on.

A state variable is one of the set of variables that are used to describe the physical "state" of a dynamical system. Intuitively, the state of a system represents and describes the existence condition of the system in the absence of any external forces affecting the system.

A state function has the property that depends solely on the state of the system. It does not depend on how the system was brought to that state. When a system is brought from an initial to a final state, the change in a state function is independent of the path followed.

5.5.1 The First Law of Thermodynamics

The first law of thermodynamics, about the conservation of energy, states that, the change in the internal energy of a closed thermodynamic system is equal to the sum of the amount of heat energy supplied to or removed from the system and the work done on or by the system. It may be expressed as follows:

$$dE = đQ + đW \tag{5.50}$$

where dE is an infinitesimal change of internal energy of a thermodynamics system that is a state variable. Moreover, dE is an exact differential, which means that its reverse operation, i.e. the following integral

$$\int_i^f dE = E_f - E_i = \Delta E \text{ is path-independent.}$$

where $đQ$ is increment of heat, or heat (flux) input, and $đw$ is infinitesimal work done by the external agent. The differential operator $đ$ denotes the inexact differential or imperfect differential that is a specific type of differential used in thermodynamics to express the path dependence of a particular differential. It is contrasted with the concept of the exact differential in calculus, which can be expressed as the gradient of another function and is therefore path independent.

5.5.2 Entropy

Before we discuss the concept of entropy, we first define the isolated system. An isolated system satisfies the condition $đQ = 0$, which is also called the *adiabatic condition*.

Entropy is a measure of disorder of a system, and we denote it by the letter S. At macroscale for an isolated system, entropy, S, is monotonously increasing in time, and it is convex and position. Thus many call it *"the arrow of time,"* because it breaks the time symmetry or T-symmetry. However, at microscale, it may be possible for a local entropy density increment become negative.

Generally speaking in thermodynamics, we have two processes:

1. Reversible process

$$dS = \frac{đQ}{T},$$

where T is the system's temperature. In fact, the quantity, $đQ$, becomes an exact differential in a reversible process.

2. Spontaneous (irreversible) process

$$dS > \frac{đQ}{T}.$$

Since in an isolated system, $đQ = 0$, we say that the entropy in an isolated system is a non-negative function, i.e.,

$$dS \geq 0. \tag{5.51}$$

In particular, a reversible adiabatic process is an isentropic process, in which $dS = 0$.

In summary, in thermodynamics, we postulate that there exists a state function S called the entropy. It is such that, for a reversible process,

$$dS = \frac{đQ}{T} = \frac{dQ}{T} \quad \rightarrow \quad \oint dS = 0. \tag{5.52}$$

Here the fact $1/T$ is the integration factor of $đQ$. The unit of S is k_B.

For a reversible process, we can rewrite the first law of thermodynamics as

$$dE = dE = đQ + đW \rightarrow dE = dQ + dW = TdS - PdV, \tag{5.53}$$

if the external mechanical work is caused by pressure $\sigma = -P$ and strain $\epsilon = dV/V$. Since internal energy increment is an extensive variable, the volumetric change $dV = \epsilon V$ may be viewed as an extensive volumetric strain.

5.5.3 The Second Law of Thermodynamics

The second law of thermodynamics states that the entropy of an isolated system never decreases, because isolated systems always evolve toward thermodynamic equilibrium, a state with maximum entropy. That is in an isolated system,

$$\frac{dS}{dt} \geq 0.$$

The equality holds when the system is in reversible process or equilibrium state.

In nature, physical processes evolve toward minimum energy and maximum disorder states. The second law is about the change of entropy and its role in determining whether a process will proceed spontaneously.

5.6 Thermodynamics States

There are two types of thermodynamics: *equilibrium thermodynamics* and *nonequilibrium thermodynamics*. Today, most molecular dynamics simulations are conducted under equilibrium conditions. On the other hand, nonequilibrium thermodynamics and nonequilibrium molecular dynamics (NEMD) have been active research fields in computational chemistry and computational materials science.

In this section, we briefly review some basic concepts and methods related to thermodynamic equilibrium state.

In all systems there is a tendency to evolve toward states whose properties are determined by intrinsic factors and not by applied external influences. Such simple states are, by definition, time-independent. They are called equilibrium states. Based on this argument, we postulate that there exist particular states (called equilibrium states) of simple systems that, macroscopically, are characterized completely by state variables such as internal energy E, volume V, and mole numbers $N_1, \ldots, N_r$ of the chemical components.

There are many types of thermodynamics equilibrium. For example, a system is in **thermal equilibrium**, if the temperature is the same throughout the entire system. On the other hand, a system is in *mechanical equilibrium* state if the pressure of the system is the same at any spatial point of the system with time. We may say that a system is in a *chemical equilibrium*, if its chemical composition does not change with time, or when no chemical reaction occurs.

5.6.1 Example: Thermal Equilibrium

Consider an isolated system chamber with rigid wall all around that is also subjected to the adiabatic condition. Inside the chamber, there is diathermal wall that is impermeable to matter, and it divides the system into two parts: part A and part B, as shown in Fig. 5.4. The system is subjected to the following constraints:

$$V^A = const., \quad V^B = const., \quad E = E^A + E^B = const. \tag{5.54}$$

Then, the increment of internal energy E^A is related to the increment of internal energy E^B by,

$$dE^B = -dE^A.$$

For an isolated reversible system, we have

$$E^A = E^A(S^A, V^A), \text{ and } S = S^A(E^A, V^A) + S^B(E^B, V^B).$$

Since the volumes of all chambers cannot change, we have

$$dS = \frac{\partial S^A}{\partial E^A}\bigg|_{V_A} dE^A + \frac{\partial S^B}{\partial E^B}\bigg|_{V_B} dE^B = \left(\frac{\partial S^A}{\partial E^A}\bigg|_{V_A} - \frac{\partial S^B}{\partial E^B}\bigg|_{V_B}\right) dE^A.$$

From $dE = TdS - PdV$, we find

$$\frac{\partial S}{\partial E}\bigg|_V = \frac{1}{T} \rightarrow dS = \left(\frac{1}{T^A} - \frac{1}{T^B}\right) dE^A.$$

At equilibrium $dS = 0$ for any dE^A. It follows that $T^A = T^B$.

In molecular dynamics modeling, the earlier discussed system is an example of so-called NVT system, i.e., the number of atoms is fixed, the volume of the system is fixed, and the temperature of the system is fixed.

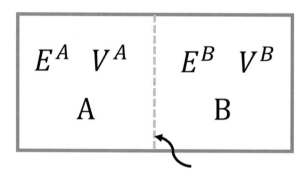

Figure 5.4 A thermal equilibrium system with a rigid diathermal wall in the middle of an isolated system

5.6.2 Thermal and Mechanical Equilibrium

Figure 5.5 shows another isolated system with rigid wall all around. Different from the previous system, the rigid diathermal wall in the previous system is replaced with a movable diathermal wall.

The independent internal variables in this case are: E^A, V^A and E^B, V^B. However, the constraint conditions change to

$$V = V^A + V^B = const., \quad E = E^A + E^B = const.,$$

which imply

$$dV_B = -dV_A \text{ and } dE_B = -dE_A.$$

Now we have

$$dS = \frac{\partial S}{\partial E^A}\bigg|_{V_A} dE^A + \frac{\partial S}{\partial V^A}\bigg|_{E^A} dV^A + \frac{\partial S}{\partial E^B}\bigg|_{V^B} dE^B + \frac{\partial S}{\partial V^B}\bigg|_{E^B} dV^B$$

$$= \left(\frac{\partial S}{\partial E^A}\bigg|_{V_A} - \frac{\partial S}{\partial E^B}\bigg|_{E^B}\right) dE^A + \left(\frac{\partial S}{\partial V^A}\bigg|_{E^A} - \frac{\partial S}{\partial V^B}\bigg|_{E^B}\right) dV^A.$$

From $dE = TdS - PdV$, we find that

$$dS = \frac{1}{T} dE + \frac{P}{T} dV$$

and hence

$$\frac{\partial S}{\partial E}\bigg|_V = \frac{1}{T} \text{ and } \frac{\partial S}{\partial V}\bigg|_E = \frac{P}{T},$$

and therefore at equilibrium $dS = 0$ for any dE^A and dV^A. It then follows that $T^A = T^B$ and $P^A = P^B$, which are the thermal–mechanical equilibrium conditions.

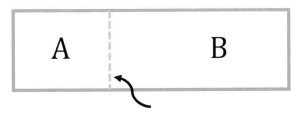

Movable diathermal wall impermeable to matter

Figure 5.5 A thermal–mechanical equilibrium system with a movable diathermal wall in the middle of an isolated system

5.7 Legendre Transformation

Given a function $f(x)$, we seek a transformed but equivalent function, $\psi(\chi)$, that contains the same information as $f(x)$. The variable of the transformed function $\psi(\chi)$ is determined by,

$$\chi = \frac{df}{dx}. \tag{5.55}$$

Equation (5.55) is the first part of the relation of the conjugate variable pair (x, χ). The second part of the relation of the conjugate variable pair (x, χ) is

$$x = -\frac{d\psi}{d\chi}, \tag{5.56}$$

and we shall make this clear next.

To construct the complementary function in the Legendre transformation, $\psi(\chi)$, we let

$$\psi(\chi) = f(x) - x\chi. \tag{5.57}$$

Based on the complementary function $\psi(\chi)$, we can reconstruct $f(x)$ geometrically as the envelope of a family of tangent lines of $f(x)$, i.e.,

$$\psi(\chi) = f(x) - x\chi \quad \rightarrow \quad f(x) = \chi x + \psi,$$

where ψ is the intercept and χ is the conjugate variable,

$$\chi = f' = \frac{f - \psi}{x}$$

that represent the tangent or slope at x. This is illustrated in Fig. 5.6.

We now define the Legendre transformation of $f(x)$ as $\psi(\chi)$,

$$\psi(\chi) = f(x) - \chi x. \tag{5.58}$$

The meaning of the function $\psi(\chi)$ is: for fixed χ and for $x \in I$ where I is the interval or the range of x, $\psi(\chi)$ is the minimum or infimum of $f(x) - \chi x$, i.e.,

$$\psi(\chi) = \inf_{x \in I} (f(x) - x\chi).$$

Interested readers may consult specialized articles such as Nielsen (2010).

Note that $\psi = \psi(\chi)$. We may solve $f' = \chi(x)$ for x and substitute it into the above relation to obtain $\psi = \psi(x)$. This is only possible if f' is a single-value function, i.e., $f'' \neq 0$, $\forall x$.

Example 5.6 In Lagrangian mechanics, we denote the Lagrangian of a single particle system as

$$L(\mathbf{r}, \dot{\mathbf{r}}) = \frac{1}{2} m \dot{\mathbf{r}}^2 - V(\mathbf{r}),$$

where $\mathbf{r}(t)$ is the position of the particle, $\dot{\mathbf{r}}(t)$ is the velocity of the particle, and $V(\mathbf{r})$ is the potential energy.

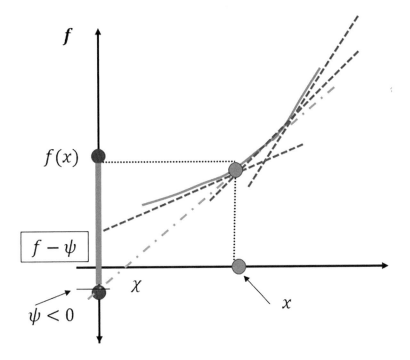

Figure 5.6 Geometrical interpretation of the Legendre transformation

Consider the dual variable transformation

$$\mathbf{p} = \frac{\partial L(\mathbf{r}, \dot{\mathbf{r}})}{\partial \dot{\mathbf{r}}} = m\dot{\mathbf{r}}.$$

By Legendre transformation, we construct the Hamiltonian of the system

$$\psi = f - xf'(x) \quad \rightarrow \quad H = \dot{\mathbf{r}} \cdot \mathbf{p} - L(\mathbf{r}, \dot{\mathbf{r}}) = \frac{\mathbf{p}^2}{m} - \left(\frac{1}{2}m\dot{\mathbf{r}}^2 - V(\mathbf{r})\right) = \frac{1}{2m}\mathbf{p}^2 + V(\mathbf{r}).$$

Note that the Hamiltonian is equivalent to $H \sim -\psi$. The reason why we make such choice is because the dual variable transformation will be symmetric, i.e.,

$$\dot{\mathbf{r}} = \frac{\partial H}{\partial \mathbf{p}}.$$

Example 5.7 In solid mechanics, we often define the strain energy as

$$U(\epsilon) = \int_0^\epsilon \tilde{\sigma}(\tilde{\epsilon}) d\tilde{\epsilon} \sim f(\epsilon),$$

as shown in Fig. 5.7. Thus, we may define the conjugate variable of strain as

$$\sigma = \frac{dU}{d\epsilon}$$

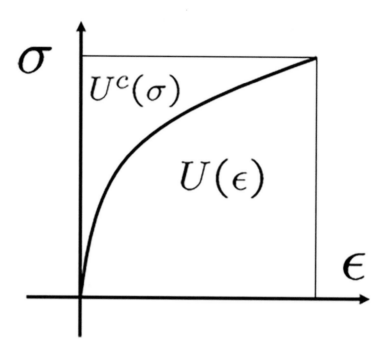

Figure 5.7 Geometrical interpretation of strain energy and complementary strain energy

and hence construct complementary strain energy through a Legendre transformation, i.e.,

$$-\psi(x) = xf' - f(x) \quad \rightarrow \quad U^c(\sigma) = \epsilon\sigma - U(\epsilon),$$

where the complementary energy $U^c(\sigma) \sim -\psi$. This transform is apparent because

$$U(\epsilon) + U^c(\sigma) = \epsilon\sigma$$

as the area of the rectangular box in Fig. 5.7.

Again by choosing such arrangement, we have a symmetric conjugate variable transformation

$$\epsilon = \frac{dU^c}{d\sigma}.$$

In thermodynamics, we often carry out multivariable Legendre transformation.

Let $f = f(x_1, x_2, x_3, \ldots, x_n)$, where $x_1, x_2, \ldots, x_n$ are state variables and $f = f(x_i)$ is a state variable function. Now we would like to transform the variable set $(x_1, \ldots, x_r, x_{r+1}, \ldots, x_n)$ to its conjugate set $(x_1, \ldots, x_r, u_{r+1}, \ldots, u_n)$ where $r = 0, 1, \ldots n-1$, where

$$u_i = \frac{\partial f}{\partial x_i}, \quad i = r+1, \ldots, n,$$

5 Fundamentals of Statistical Mechanics

and hence the Legendre transformation is

$$\psi((x_1, \ldots, x_r, u_{r+1}, \ldots, u_n) = f(x_1, x_2, x_3, \ldots, x_n) - \sum_{i=r+1}^{n} u_i x_i.$$

In thermodynamics literature, we often write $g = \psi$,

$$g = f - \sum_{i=r+1}^{n} u_i x_i, \quad g = g(x_1, \ldots, x_r, u_{r+1}, \ldots u_n),$$

where

$$u_{r+1} = \frac{\partial f}{\partial x_{r+1}}, \ldots, u_n = \frac{\partial f}{\partial x_n};$$

while

$$x_{r+1} = -\frac{\partial g}{\partial u_{r+1}}, \ldots, x_n = -\frac{\partial g}{\partial u_n}.$$

Example 5.8 We start with the internal energy function $E = E(S, V)$ and

$$dE = TdS - PdV, \text{ with } T = \left(\frac{\partial E}{\partial S}\right)_V, P = -\left(\frac{\partial E}{\partial V}\right)_S,$$

which provides the conjugate variable relations, because E is a state variable and

$$dE = \frac{\partial E}{\partial S} dS + \frac{\partial E}{\partial V} dV.$$

1. The Helmholtz free energy: Switch $S \to T$.
By the Legendre transformation of the conjugate variable (S, T):

$$F(T, V) = E - \left(\frac{\partial E}{\partial S}\right) S \Rightarrow F(T, V) = E - TS;$$

or

$$dF = -SdT - PdV, \text{ with } S = -\frac{\partial F}{\partial T}, P = -\frac{\partial F}{\partial V}.$$

Sometimes in literature, we denote the Helmholtz free energy $F(T, V)$ as $A(T, V)$.

2. Enthalpy: Switch from $V \to P$ or $(-P)$.
Using the conjugate variable $-P = \frac{\partial E}{\partial V}$, Legendre transformation gives

$$H(S, P) = E - \left(\frac{\partial E}{\partial V}\right) V, \Rightarrow H = E + PV.$$

One may verify that the enthalpy function has the relation:
$dH = dE + PdV + VdP = dQ + VdP$. In molecular dynamics, we often use enthalpy to calculate the heat capacity at constant pressure, i.e.,

$$c_p = \left(\frac{\partial H}{\partial T}\right)_T.$$

3. Gibbs free energy: Switch from $S \to T$ and $V \to -P$. By using the conjugate variable pairs,

$$S = \frac{\partial E}{\partial T} \quad \text{and} \quad -P = \frac{\partial E}{\partial V},$$

we can construct the Gibbs free energy via the following Legendre transformation,

$$G(T, P) = E - \left(\frac{\partial E}{\partial S}\right) S - \left(\frac{\partial E}{\partial V}\right) V \quad \rightarrow \quad G(T, P) = E - (TS - PV).$$

One can find that Legendre transformation provides a convenient tool to construct state variable functions under different sets of independent state variables.

5.8 Statistical Ensembles

5.8.1 How to Calculate Entropy

The definition of entropy by Boltzmann for a given macrostate is linked to the total number of microstates or quantum states in that macrostate of the molecular system.

To quantify entropy, according to Boltzmann, we need to calculate how many ways to arrange A number particles into Ω number of microstates or quantum states, such that a_1 number of particles are in the first quantum state, a_2 number of particles are in the second quantum state, and so on, and a_Ω number of particles are in the Ωth quantum state, where $a_1, a_2, \ldots, a_\Omega$ are called occupation numbers, i.e., how many particles in each microstate. For a fixed set of occupation numbers $\{a_1, a_2, \ldots, a_\Omega\}$, the total energy or the macroscale energy is fixed, while there are many different ways to put particles into different quantum states, and each of these arrangement represents a microstate. To find out how many possible microstates existing in the system, we need to calculate how many different ways to arrange particles into these quantum states, which is expressed as

$$W(a_1, a_2, \ldots, a_\Omega) = \frac{A!}{a_1! a_2! \ldots a_\Omega!} = \frac{A!}{\prod_j a_j!}, \tag{5.59}$$

where A is the total number of particles and a_j is the *occupation number* in the jth energy level, or quantum state. Note that the energy level, or quantum state, is not microstate, and a microstate is a particular arrangement to put the A particles into different energy levels.

We assume that there are Ω quantum states. For a fixed partition, or a fixed set of occupation numbers $\{a_1, a_2, \ldots, a_\Omega\}$, it will be subjected to a constraint

$$\sum_{j=1}^{\Omega} a_j = A. \tag{5.60}$$

Example 5.9 We label this example as Case I, in which we have a Boltzmann box that has 36 cells, and each of them represents an energy level as shown in in Fig. 5.8. If we have 10 distinguishable particles, for the macroscale represented by the Box 1, we shall have

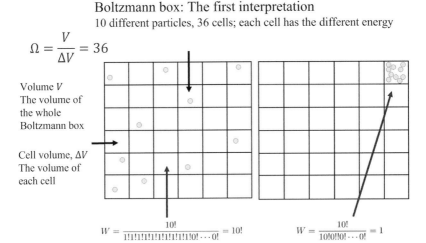

This is a macrostate with W many microstates

Figure 5.8 Illustration example I of microstates and macrostate

$$W = \frac{10!}{1!\,1!\,1!\,1!\,1!\,1!\,1!\,1!\,1!\,1!\,0!\cdots 0!} = 10!,$$

many ways to arrange the particles into these energy partition. Whereas in Box 2 of Fig. 5.8, all 10 particles are within the same cell. Hence the energy partition is:

$$a_1 = 0,\ a_2 = 0,\ a_3 = 0,\ a_4 = 0,\ a_5 = 10,\ a_6 = 0, \ldots, a_{36} = 0.$$

Therefore

$$W = \frac{10!}{0!\,0!\,0!\,0!\,10!\,0!\,0!\cdots 0!} = 1.$$

Ludwig Boltzmann proposed to use the total number of different ways to put all particles into a given energy partition as the base to calculate the disorder of the system, and hence the measure of the entropy of the system. Boltzmann's ensemble entropy formula is given as,

$$S_B = k_B \ln W = k_B \left(\ln A! - \sum_{j=1}^{\Omega} a_j! \right), \tag{5.61}$$

in which the constant, k_B, is named as the Boltzmann constant by Max Planck, and it equals to 1.38055×10^{-23} J/K; A is the total number of atoms or base objects in the system; Ω is number of quantum energy state or energy level, while $a_j, j = 1, \ldots, \Omega$ are the occupation number in each quantum state or energy level.

Example 5.10 We label this example as Case II, and we calculate the entropy for the Boltzmann box shown in Fig. 5.9. Based on Boltzmann's formula,

Case II: We have 10 particles and 36 cells with different energy levels, and each energy level with an occupation number a_j

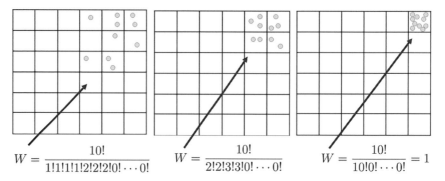

$$W = \frac{10!}{1!1!1!1!2!2!2!0!\cdots 0!} \qquad W = \frac{10!}{2!2!3!3!0!\cdots 0!} \qquad W = \frac{10!}{10!0!\cdots 0!} = 1$$

Figure 5.9 Illustration example II of microstates and macrostate

$$S_B = k_B \ln W = k_B \left(\ln A! - \sum_{i=1}^{\Omega} \ln a_i! \right).$$

For case 1

$$S_B = k_B(\ln(10!) - (2\ln 2)) = 13.718 k_B.$$

For case 2

$$S_B = k_B(\ln(10!) - (2\ln 2 + 2\ln 3)) = 11.520 k_B,$$

and for case 3

$$S_B = k_B(\ln 1) = 0.$$

Note that each box in Fig. 5.9 may be only a part of a macrostate, so that the entropy of the macrostate is actually far bigger than the number that we just calculated. Note that Boltzmann's entropy is an extensive state variable, meaning that it depends on the volume of the system.

For instance, depending on particular energy combinations, three different boxes may have the same energy level, and hence belong to a same macrostate as shown in Fig. 5.10. Thus, the total number of microstates in such a macrostate is the summation of the microstates in these three Boltzmann boxes,

$$W = W_1 + W_2 + W_3.$$

To understand the case of a fixed number of particle configuration that may have different microstates because of different types of particles, we illustrate the concept in Fig. 5.11.

One may ask a question that for a given set of particles A and a given energy partition Ω, which specific occupation number set $\{a_1, a_2, \ldots a_\Omega\}$ maximizes the multinominal coefficient

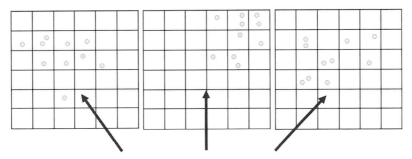

Depending on the energy level, it could be true that all these three box configurations have the same total energy, and they are under the same (one) macroscale state. Hence

$$W = W_1 + W_2 + W_3$$

Figure 5.10 Illustration example III of microstates and macrostate

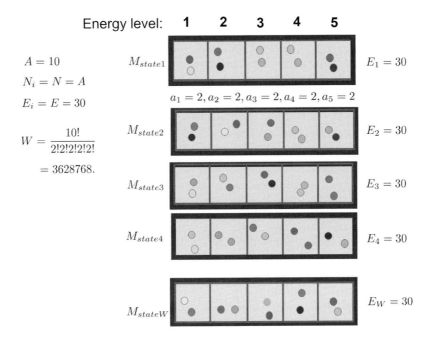

$A = 10$
$N_i = N = A$
$E_i = E = 30$

$$W = \frac{10!}{2!2!2!2!2!}$$
$$= 3628768.$$

Figure 5.11 Illustration of multiplicity of the microstate in a same energy configuration for a macrostate

$$W(a_1, a_2, \ldots, a_\Omega) = \frac{A!}{a_1! a_2! \ldots a_\Omega!} = \frac{A!}{\Pi_j a_j!}?$$

The answer is that the following set of the occupation number

$$a_1 = a_2 = \ldots a_\Omega = a = \frac{A}{\Omega} \rightarrow W(a, a, \ldots, a) = W_{max}.$$

Obviously, this particular set of the occupation number maximizes the Boltzmann entropy as well, which implies that it is in equilibrium state. Therefore, the maximum of the Boltzmann entropy associates with the maximum number of microstates, which is the maximum disorder state of the system, and hence the definition of the Boltzmann entropy is consistent with the second law of thermodynamics.

On the other hand, from Boltzmann's entropy definition (Eq. (5.61)), one can see that for a completely ordered ensemble, i.e.,

$$a_1 = A, a_2 = a_3 = \cdots = 0 \quad \rightarrow \quad S_{ensemble} = 0,$$

we have

$$W = 1 \quad \rightarrow \quad S_B = 0.$$

When every particles stays in a same energy level, i.e., all atoms stay at the ground state, the thermodynamic system has zero temperature, and Boltzmann's entropy definition marks a zero entropy at such energy partition. Thus, by Boltzmann's entropy definition we get the third law of thermodynamics by free!

Ludwig Boltzmann formulated his entropy formula (Eq. (5.61)) in 1872, when he was studying the kinetic theory of gases. In Boltzmann's entropy definition, he assumes that every microstate in the system has the same probability to occur in the statistical ensemble, which implies that the thermodynamic system in an equilibrium state, or near an equilibrium state, because the condition that every microstates of the macrosystem has the same likelihood to maximize the multinomial coefficient W.

Example 5.11 We consider the Boltzmann box illustrated in Fig. 5.12, in which we assume that each box is a microstate. Under the equilibrium condition, we assume that the probability that a given particle may get into a specific cell is $1/36$. If we neglect the interactions among particles, i.e., we assume that particle motions are independent (Boltzmann studies the motion of ideal gas at the time). Then the joined probability of the 10 particles may get into a given energy partition will be $(1/36)^{10}$.

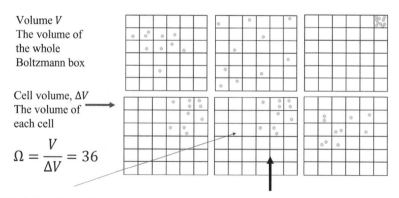

Figure 5.12 The second interpretation of the Boltzmann box

Assume that under the equilibrium condition, each microstate has the same probability to occur in the statistical ensemble. We then have

$$p_1 = p_2 = \cdots = p_W = \frac{1}{W} = \left(\frac{1}{36}\right)^{10} \rightarrow W = (36)^{10}.$$

That is: each particle has 36 ways to be arranged into the Boltzmann box, and then the ways that all 10 particles can be put into the various cells are $W_{max}(30)^{10}$, or there are total $W = (36)^{10}$ microstates.

Therefore, the ensemble entropy will be

$$S_B = k_B \ln((36)^{10}) = 10 k_B \ln 36 = 35.845 k_B.$$

This particular statistical thermodynamics system is connected to the statistical thermodynamics system that is related the so-called microcanonical ensemble, which will be discussed soon after.

In the previous several examples, we illustrated the differences between a microstate and a macrostate. In general, a macrostate can correspond to many microstates, and Boltzmann's definition of entropy is basically using the number of microstates corresponding to a macrostate to calculate the entropy of a macrostate.

Boltzmann's definition of entropy was further generalized by an American chemist Josiah Willard Gibbs, who called the set of all microstates corresponding to a given macrostate as the *microstate ensemble* of the given macrostate, and the probability of the occurrence or emergence of a microstate in a given ensemble may be different. In other words, for a thermodynamics system, microstates of the system may not have equal probabilities, even in the equilibrium state. Suppose that we have total N possible microstates, Gibbs proposed to use the following expression to calculate entropy:

$$S_G = -k_B \sum_{i=1}^{N} p_i \ln p_i, \quad \text{and} \quad \sum_{i=1}^{N} p_i = 1, \tag{5.62}$$

where p_i is the probability of the emergence frequency of the ith microstate in the given macrostate, while the term $-\ln p_i$ may be interpreted as the *information* gained upon the measurement of a microstate in the energy level ε_i or in the cell i. The Gibbs entropy is not a simple extension from the Boltzmann entropy, but makes a significant generalization of the Boltzmann original concept of the entropy.

However, in a special case, we can relate the Boltzmann entropy with the Gibbs entropy. To elucidate such connection, we first consider the following example.

Example 5.12 In fact, Boltzmann's original definition of entropy is slightly more complicated than the previously discussed examples. In Boltzmann's original idea each particle in Fig. 5.13 represents a microstate without assigning energy as what was called a *single particle microstate*. When a particle moves into a specific cell in the phase space, it acquires a specific energy level becoming a specific microstate. In each cell, the energy level is fixed, but it may allow multiple microstate particles to move in, depending on its given probabilities in different microstate ensembles. To

Boltzmann box: The third interpretation
36 microstate configurations, 36 cells with specific energy level

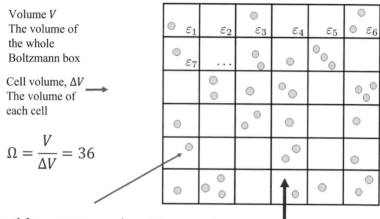

Volume V
The volume of the whole Boltzmann box

Cell volume, ΔV
The volume of each cell

$$\Omega = \frac{V}{\Delta V} = 36$$

Particle represents a microstates

When a particle moves to the $i\,th$ cell, the microstate has energy ε_i

Figure 5.13 The third interpretation of the Boltzmann box

illustrate this idea, in this example, the Boltzmann box has $N = 36$ particles and 36 cells with specific energies: $\varepsilon_1, \varepsilon_2, \ldots, \varepsilon_N$, where $N = 36$ as shown in Fig. 5.13. Thus Boltzmann's entropy is given as

$$S_B = k_B \ln \frac{N!}{a_1! a_2! \ldots a_N!} = k_B \ln \frac{36!}{1!\,0!\,0!\,0!\,2! \cdots 1!\,1!\,2!}.$$

To relate the Gibbs entropy with the Boltzmann entropy, one may express the probability of the occurrence of a microstate in terms of occupation number,

$$p_i = \frac{a_i}{N}, \quad i = 1, 2, \ldots, N.$$

By using Stirling's approximation $\ln(x!) = x \ln x - x + O(\ln x)$, we can write the Boltzmann entropy in the previous special case as

$$S_B = k_B \ln \frac{N!}{\prod_{i=1}^{N}(a_i!)} \approx N \ln N - \sum_{i=1}^{N}(a_i \ln a_i a_i)$$

$$= k_B \left(N \ln N - N \sum_{i=1}^{N} \frac{a_i}{N} \ln \left(N \frac{a_i}{N} \right) \right)$$

$$= k_B \left(N \ln N - N \sum_{i=1}^{N} p_i \ln N - N \sum_{i=1}^{N} p_i \ln p_i \right)$$

$$= -k_B N \sum_{i=1}^{N} p_i \ln p_i = N S_G. \tag{5.63}$$

One can see that in this special case

$$S_B = \frac{1}{N} S_G,$$

which indicates that the Gibbs entropy is an intensive state variable, whereas the Boltzmann entropy is an extensive state variable.

The above probabilistic formula was given by Gibbs in 1878, which was only a few years after Boltzmann's initial proposal. Boltzmann himself recognized the Gibbs entropy as a more general definition of entropy. That is, if S_G is valid, S_B is also valid, but not vice versa.

5.8.2 Microcanonical Ensemble

To motivate the discussion on the microcanonical ensemble, we first recall the example shown in Fig. 5.12. In the Boltzmann box shown in Fig. 5.12, all microstates have the same probability of occurrence, i.e., $p_i = (1/36)^{10}$. That is, (1) all 10 particles have the same chance or probability getting into any of the cells in a box, and (2) all 10 particles are independent or weakly interactive.

In a microcanonical ensemble, all the microstates have the same number of particles (N), the same volume (V), and the same total energy (E). If N is the number of accessible microstates including degeneracies, the probability of any microstate occupied at random from the ensemble is simply $1/N$. For some historical reasons, in the literature, many authors use the letter Ω to denote the total number of microstates mixing with the total number of quantum states. In this book, we still follow this confusing notation $N \to \Omega$ but warn the readers at the outset.

In a microcanonical ensemble of a N microstate ensemble, when the macrostate is in equilibrium, which means the maximum disorder, occupation number in each microstate will be the same, i.e.,

$$a_1 = a_2 = \ldots a_i = \ldots a_N = \frac{N}{N} = \frac{\Omega}{\Omega} = 1 \to p_i = \frac{a_i}{\Omega} = \frac{1}{\Omega}, \ i = 1, \ldots, \Omega. \tag{5.64}$$

This is to say that given an isolated system in equilibrium, it is found with equal probability or "opportunity" in each of its accessible quantum state.

Since the probability of each microstate occurring is

$$p_i = \frac{1}{\Omega},$$

the entropy of the system is

$$S = -k_B \sum_{i=1}^{\Omega} \frac{1}{\Omega} \ln \frac{1}{\Omega} = k_B \ln \Omega,$$

where Ω is the total number of microstate states.

The earlier result can also be rigorously derived by maximizing the system's entropy. For an (N, V, E) system, by definition, we can assume that

$$<E> = \sum_j p_j E_j, \quad <N> = \sum_j p_j N_j, \text{ and } \sum_j p_j = 1.$$

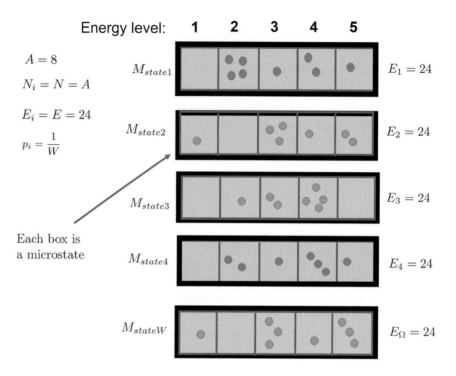

Figure 5.14 Illustration of the concept of microcanonical ensemble

Note that in a microcanonical ensemble (N, V, E), all the accessible microstates have the same energy value $E_j = E = const.$, and the same number of particles, $N_j = N = const.$, with the same volume, but they are different microstates as shown in Fig. 5.14.

When an (N, V, E) ensemble reaches an equilibrium state, its entropy becomes maximum, which requires that

$$\delta S_{(N,V,E)} = 0.$$

Using a Lagrange multiplier γ, we consider the following stationary condition:

$$\delta \left(S + \gamma \left(\sum_j p_j - 1 \right) \right) = 0.$$

Then we have

$$\delta \left(-k_B \sum_j p_j \ln p_j + \gamma \left(\sum_j p_j - 1 \right) \right)$$

$$= \sum_j \delta p_j \left(-k_B \ln p_j - k_B + \gamma \right) + \delta \gamma \left(\sum_j p_j - 1 \right) = 0.$$

Hence,
$$\ln p_j = \frac{\gamma - k_B}{k_B} = const. \rightarrow p_j = \exp(const.) = const.$$

Since
$$\sum_j p_j = 1 \rightarrow \sum_{j=1}^{\Omega} \frac{1}{\Omega} = 1 \rightarrow p_j = \frac{1}{\Omega},$$

where Ω is the number of states with the energy E.

Finally, we have
$$S = -k_B \sum_j p_j \ln p_j = -k_B \sum_{j=1}^{\Omega} \frac{1}{\Omega} \ln \frac{1}{\Omega} = k_B \ln \Omega,$$

where Ω is the total number of energy states.

Since the microcanonical ensemble is an isolated system as shown in Fig. 5.15, there is no external interaction with its environment, and if the initial conditions are fixed, the total energy of the system
$$\mathcal{H}(\mathbf{q}^N, \mathbf{p}^N) = \sum_{i=1}^{N} \left(\frac{\mathbf{p}_i \cdot \mathbf{p}_i}{2m_i} + U(q_i) \right) = E = const.$$

is a constant.

For an (N, V, E) system,
$$p_j = \begin{cases} \frac{1}{\Omega}, & for\ E_j = E \\ 0, & for\ E_j \neq E \end{cases}.$$

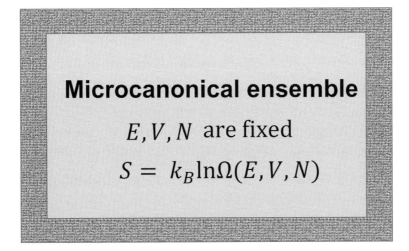

Figure 5.15 Schematic illustration of the adiabatic bridge of microcanonical ensemble

A microcanonical ensemble molecular dynamics provides a natural setting to consider the ergodic hypothesis, i.e., the long time average coincides with the ensemble average

$$\bar{f} = \lim_{T \to \infty} \frac{1}{T} \int_0^T f(x(t))dt = <f> = \int_\Gamma f(x)d\mu(x), \quad (5.65)$$

where $f(x)$ is an arbitrary field of a random variable.

Since the distribution function in the phase space for a microcanonical ensemble (N, V, E) system is,

$$f_{NVE}(\mathbf{q}^N, \mathbf{p}^N) = \begin{cases} 1/\Omega, & \forall \mathcal{H} \in [E, E+\delta E] \\ 0, & \text{elsewhere} \end{cases} \quad (5.66)$$

If $\delta E \to 0$, the above expression can be written as

$$f_{NVE}(\mathbf{q}^N, \mathbf{p}^N) = \delta(\mathcal{H}(\mathbf{q}^N, \mathbf{p}^N) - E)/\Omega.$$

If the system is *ergodic*, the time average will equal to ensemble average, i.e., for a given field $A(\mathbf{q}^N, \mathbf{p}^N)$,

$$\bar{A} = <A>,$$

where

$$\bar{A} := \lim_{T \to \infty} \frac{1}{T} \int_0^T A(\mathbf{q}^N, \mathbf{p}^N) dt$$

$$<A>_{NVE} := \frac{\int_V d\mathbf{q}^N \int_{V_p} d\mathbf{p}^N A(\mathbf{q}^N, \mathbf{p}^N) \delta(\mathcal{H} - E)}{\int_V d\mathbf{q}^N \int_{V_p} d\mathbf{p}^N \delta(\mathcal{H} - E)}. \quad (5.67)$$

5.8.3 How to Find Ω ?

In microcanonical ensemble, the total energy of the system

$$\mathcal{H}(\mathbf{q}^N, \mathbf{p}^N) = \sum_{i=1}^N \left(\frac{\mathbf{p}_i \cdot \mathbf{p}_i}{2m_i} + U(q_i) \right) = E = const.$$

is a constant.

Considering the probability distribution function of a microcanonical ensemble (N, V, E) system in Eq. (5.66), we can find the total volume of the ensemble as

When $\delta E \to 0$, $\quad V := \int_{E \leq \mathcal{H} \leq E+\delta E} d\mathbf{q}^N d\mathbf{p}^N = \int \delta(\mathcal{H} - E) d\mathbf{q}^N d\mathbf{p}^N.$

On the other hand, when $\delta E \neq 0$ but $\delta E \ll 1$ and $N \gg 1$, the total phase-space volume V when $\mathcal{H} = E$ be approximated as,

$$V := \int_{E \leq \mathcal{H} \leq E+\delta E} d\mathbf{q}^N d\mathbf{p}^N = \int d\mathbf{q}^N d\mathbf{p}^N H[E - \mathcal{H}].$$

Note that V is insensitive to δE. This is because in mathematics for a N-dimensional space ($N \gg 1$), the volume of the N-dimensional sphere $\mathcal{H} \leq E$ is very close to the volume of its outer layer $E \leq \mathcal{H} \leq E + \delta E$ when $N \gg 1$.

Since V is the total volume of the ensemble in the phase space, we can find the total number of microstates by dividing it with the volume of a microstate,

$$\Omega := \frac{V}{\Delta V} = \frac{1}{\Delta V} \int d\mathbf{q}^N d\mathbf{p}^N H[E - \mathcal{H}], \tag{5.68}$$

where H is the heaviside function. Based on Heisenberg's uncertainty principle, the volume of any atomic or subatomic particle size is $|\Delta \mathbf{r} \Delta \mathbf{p}| \sim h^3$. Thus for a N-particle microstate, its volume would correspond to the volume of the quantum state h^{3N}. In the above integral, every set of distinct coordinates occurs $N!$. If we assume that the particles are all of the same type as indistinguishable particles, we need to take away a factor $N!$. Thus, we can calculate the total number of the microstates in a microcanonical ensemble as

$$\Omega = \frac{1}{h^{3N} N!} \int d\mathbf{q}^N d\mathbf{p}^N H[E - \mathcal{H}]. \tag{5.69}$$

where $\mathcal{H}$ is the Hamiltonian of the system. Note that the factor $N! h^{3N}$ converting the phase-space volume to the number of microstates that was established through the Bohr's correspondence principle.

Accordingly, the entropy of the microcanonical ensemble is

$$S(N, V, E) = k_B \ln \Omega = k_B \ln \left\{ \frac{1}{h^{3N} N!} \int d\mathbf{q}^N d\mathbf{p}^N H(E - \mathcal{H}) \right\}. \tag{5.70}$$

One of the advantages to have the above definition of Ω is that we can find the local density of microstates by using the following elegant formula:

$$\omega = \left(\frac{\partial \Omega}{\partial E} \right)_{N,V} = \frac{1}{h^{3N} N!} \int d\mathbf{q}^N d\mathbf{p}^N \delta[E - \mathcal{H}]. \tag{5.71}$$

Then according to Eq. (5.67), we have

$$<A>_{NVE} := \frac{\int_V d\mathbf{q}^N \int_{V_p} d\mathbf{p}^N A(\mathbf{q}^N, \mathbf{p}^N) \delta(\mathcal{H} - E)}{\int_V d\mathbf{q}^N \int_{V_p} d\mathbf{p}^N \delta(\mathcal{H} - E)}$$

$$= \frac{1}{h^{3N} N! \omega} \int d\mathbf{q}^N d\mathbf{p}^N A(\mathbf{q}^N, \mathbf{p}^N) \delta[E - \mathcal{H}]. \tag{5.72}$$

Applying the following formula of Laplace transformation of the Heaviside function,

$$L[H(t - a)] = \int_0^\infty H(t - a) e^{-st} dt = \frac{1}{s} e^{-sa}$$

to Ω and changing of variables ($t \to E$, $a \to \mathcal{H}$), we have

$$L[\Omega] = \frac{1}{h^{3N} N!} \int d\mathbf{q}^N d\mathbf{p}^N \frac{1}{s} \exp(-s\mathcal{H})$$

$$= \frac{1}{h^{3N} N!} \int d\mathbf{q}^N d\mathbf{p}^N \frac{1}{s} \exp(-s E_k(\mathbf{p}^N)) \exp(-s U(\mathbf{q}^N))$$

in which

$$\int d\mathbf{p}^N \exp\left(-\frac{s}{2m}\sum_i \mathbf{p}_i^2\right) = \left(\frac{2\pi m}{s}\right)^{3N/2}$$

by using the identity

$$\int_{-\infty}^{\infty} \exp(-sx^2)dx = \frac{\sqrt{\pi}}{\sqrt{s}}.$$

Hence, we find that

$$L[\Omega] = \left(\frac{2\pi m}{sh^2}\right)^{3N/2} \frac{1}{N!} \int d\mathbf{q}^N \frac{1}{s} \exp\left(-sU(\mathbf{q}^N)\right). \quad (5.73)$$

Recall the inverse Laplace transform,

$$L^{-1}\left(\frac{\exp[-sk]}{s^n}\right) = H[t-k]\frac{(t-k)^{n-1}}{\Gamma[n]}.$$

where $\Gamma[n+1] = n\Gamma[n] = n!$.
Let $t \to E$, $k \to U$, and $n \to \frac{3N}{2} + 1$. We can then find the total number of microstates as

$$\Omega = L^{-1}(L[\Omega]) = \left(\frac{2\pi m}{h^2}\right)^{3N/2} \frac{1}{N!\Gamma[3N/2+1]} \int d\mathbf{q}^N (E-U)^{3N/2} H[E-\mathcal{H}]. \quad (5.74)$$

Considering the definition $\omega = \frac{\partial \Omega}{\partial E}$ and $\Gamma(n) = (n-1)!$, we have

$$\omega = \left(\frac{2\pi m}{h^2}\right)^{3N/2} \frac{1}{N!\Gamma[3N/2]} \int d\mathbf{q}^N (E-U)^{3N/2-1} H[E-U]$$

Considering for any field function in the phase space that only depends on position $\mathbf{q}^N$, i.e., $A(\mathbf{q}^N)$, we can simplify the ensemble average in the microcanonical ensemble

$$<A> = \frac{1}{h^{3N}N!\omega} \int d\mathbf{q}^N d\mathbf{p}^N A(\mathbf{q}^N)\delta[E-\mathcal{H}]$$

by substituting into the above expression

$$\omega = \frac{1}{h^{3N}N!} \int d\mathbf{q}^N d\mathbf{p}^N \delta[E-\mathcal{H}],$$

and finally we obtain

$$<A> = \frac{\int d\mathbf{q}^N (E-U)^{3N/2-1} H(E-U) A(\mathbf{q}^N)}{\int d\mathbf{q}^N (E-U)^{3N/2-1} H(E-U)}. \quad (5.75)$$

We can use Eq. (5.75) to calculate temperature. Based on the definition,

$$\frac{1}{T} = \left(\frac{\partial S}{\partial E}\right)_{N,V},$$

we have

$$\frac{1}{T} = k_B \left(\frac{\partial \ln \Omega}{\partial E}\right)_{N,V} = \frac{k_B}{\Omega}\left(\frac{\partial \Omega}{\partial E}\right)_{N,V} = \frac{k_B}{\Omega}\omega.$$

Hence,

$$k_B T = \frac{\Omega}{\omega}. \tag{5.76}$$

Substituting

$$\Omega = \left(\frac{2\pi m}{h^2}\right)^{3N/2} \frac{1}{N!\Gamma[3N/2+1]} \int d\mathbf{q}^N (E-U)^{3N/2} H[E-\mathcal{H}]$$

and

$$\omega = \left(\frac{2\pi m}{h^2}\right)^{3N/2} \frac{1}{N!\Gamma[3N/2]} \int d\mathbf{q}^N (E-U)^{3N/2-1} H[E-U]$$

into Eq. (5.76) and noting that $\Gamma[\frac{3}{2}N+1] = \left(\frac{3}{2}N\right)\Gamma[\frac{3}{2}N]$, we have

$$k_B T = \frac{2}{3N}\frac{\int d\mathbf{q}^N (E-U)^{3N/2-1} H(E-U)(E-U)}{\int d\mathbf{q}^N (E-U)^{3N/2-1} H(E-U)}. \tag{5.77}$$

Since $\mathcal{H} = E \to E - U = E_k$, we reach a familiar result

$$<E_k>_{V_c} = \frac{3N}{2}k_B T, \tag{5.78}$$

where V_c denotes the configurational space.

This indicates that one can calculate the temperature by averaging kinetic energy in configuration space rather than the whole phase space. However, one may note that $E_k(\mathbf{q}^N) \neq E_k(\mathbf{p}^N)$!

In the earlier derivation, we implicitly assume that

$$E_k = E - U(\mathbf{q}^N) = E_k(\mathbf{q}^N).$$

This is only true for the microcanonical ensembles. To make a distinction between two definitions of temperature, i.e.,

$$k_B T = \frac{2}{3N}<E_k(\mathbf{q}^N)> \quad \text{and} \quad k_B T = \frac{2}{3N}<E_k(\mathbf{p}^N)>.$$

We call

$$T = \frac{2}{3Nk_B}<E_k(\mathbf{q}^N)>_{v_c}$$

as the *configurational temperature*.

5.8.4 Canonical Ensemble

In statistical mechanics, a canonical ensemble is the statistical equilibrium state ensemble that has thermal equilibrium, i.e., it can go through all possible microstates that have the same temperature. For a finite system to achieve thermal equilibrium, an ideal model is to put the finite system within a heat bath, or thermal reservoir, that has a fixed temperature, so that any thermal fluctuation in the finite system will be calmed down through the heat exchange with the infinite large thermal reservoir, or molecule collisions with the heat bath at microscale, to maintain the thermal equilibrium, as shown in Fig. 5.16.

In molecular dynamics modelings and simulations, the canonical ensemble is also referred to as an NVT ensemble: that is the system has the fixed number of particles (N), fixed volume (V), and fixed temperature (T), which is given by the temperature of the heat bath with which it would be in equilibrium. In a canonical ensemble or an NVT ensemble, each microstate has the same number of particles, i.e., $N_i = N$, $i = 1, 2, \ldots, \Omega$, but the energy of each microstate is not the same, i.e., $E_i \neq E_j$, if $i \neq j$. Nevertheless, the ensemble energy $<E> = \sum_{i=1}^{\Omega} p_i E_i$ is fixed, i.e., $<E> = \bar{E}$. What this means is that the energy of system will fluctuate around $\bar{E}$.

For canonical ensembles, the thermal equilibrium condition implies that the entropy reaches maximum under the constraints $<E> = \sum_j p_j E_j = \bar{E}$ and $\sum_j p_j = 1$. We consider equilibrium condition (maximize entropy),

$$\delta\left(S + \alpha(<E> - \bar{E}) + \gamma\left(\sum_j p_j - 1\right)\right) = 0, \qquad (5.79)$$

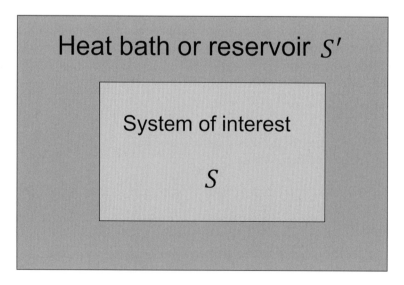

Figure 5.16 Schematic of the concept of canonical ensemble

where α and γ are Lagrange multipliers.

Using the Gibbs entropy formula, we have

$$\sum_j \left[-k_B \ln p_j - k_B + \alpha E_j + \gamma\right] \delta p_j + \delta\gamma\left(\sum_j p_j - 1\right) + \delta\alpha(<E> - \bar{E}) = 0, \quad (5.80)$$

where p_j is no longer a constant.

Solving p_j from Eq. (5.80), we have

$$\ln p_j = \frac{\alpha E_j - k_B + \gamma}{k_B} \rightarrow p_j = \exp\left(\frac{\alpha E_j - k_B + \gamma}{k_B}\right). \quad (5.81)$$

To determine Lagrangian multipliers, α and γ, we use identity,

$$T = \frac{\partial <E>}{\partial S}\bigg|_{V,N}.$$

Consider

$$(\delta <E>)_{V,N} = \sum_j E_j \delta p_j \quad (5.82)$$

and

$$(\delta S)_{V,N} = -k_B \sum_j \delta p_j \ln p_j = -k_B \sum_j \delta p_j \left[\frac{\alpha E_j - k_B + \gamma}{k_B}\right]$$

$$= -k_B \sum_j \delta p_j E_j \alpha / k_B,$$

in which we use the identity $\sum_j p_j = 1 \rightarrow \sum_j \delta p_j = 0$.

Hence,

$$T = \left[\frac{\delta <E>}{\delta S}\right]_{V,N} = -\frac{1}{\alpha} \rightarrow \alpha = -\frac{1}{T}.$$

Substituting $\alpha = 1/T$ into Gibbs' entropy formula, we have

$$S = -k_B \sum_j p_j \ln p_j = -k_B \sum_j p_j \left[\frac{\alpha E_j - k_B + \gamma}{k_B}\right]$$

$$= \sum_j p_j \frac{E_j + k_B T - \gamma T}{T} = \frac{1}{T}(<E> + k_B T - \gamma T).$$

Because of $F = <E> - ST$, we find that

$$\gamma T = F + k_B T, \text{ and } p_j = \exp\left(\frac{k_B T - E_j - \gamma T}{k_B T}\right). \quad (5.83)$$

Finally, we obtain the probability for every microstates with different energies

$$p_j = \exp\left((-\beta(E_j + F))\right), i = 1, 2, \ldots, \Omega, \text{ where } \beta = \frac{1}{k_B T}. \quad (5.84)$$

Since p_j is normalized,

$$\sum_j p_j = 1 = \frac{1}{\exp(-\beta F)} \sum_j \exp(-\beta E_j).$$

We call the normalization factor, $Z = \exp(-\beta F) = \sum_j e^{-\beta E_j}$, as the **partition function**. Therefore, the partition function of the canonical ensemble is related to the Helmholtz free energy:

$$Z = \exp(-\beta F) \rightarrow -kT \ln Z = F.$$

Equation (5.84) tells us that for the canonical ensemble the probability distribution of microstates obeys the following Boltzmann distribution,

$$p_i = \frac{1}{Z} \exp(-E_i/(k_B T)), \quad (5.85)$$

where E_i is the energy level of the energy state i and Z is the partition function,

$$Z = \sum_i \exp(-E_i/(k_B T)) = \exp(-\beta F). \quad (5.86)$$

Figure 5.17 provides an illustration example of canonical ensemble. One may find that in the canonical ensemble the energy of the system fluctuates, but the temperature of the system is fixed, and the ensemble energy is fixed, i.e., $<E> = \bar{E}$.

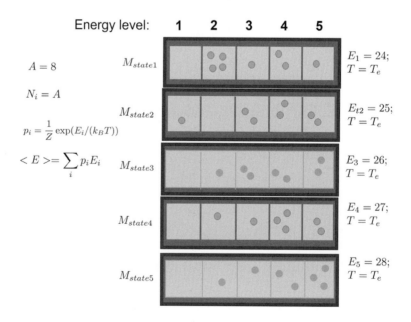

Figure 5.17 Illustrative example of canonical ensemble

For classical systems, the partition function of canonical ensembles of N-particles in the phase space may be written as,

$$Z = \int_\Gamma \exp\left(-\frac{H(p,q)}{k_B T}\right) dp^N dq^N. \tag{5.87}$$

Hence, the ensemble average for any function of the state variables, $A(\mathbf{p,q})$, in the canonical ensemble can be defined as

$$<A(\mathbf{p,q})> = \frac{\int_\Gamma A(\mathbf{p,q}) \exp\left(-\frac{H(\mathbf{p,q})}{k_B T}\right) d\mathbf{p}^N d\mathbf{q}^N}{\int_\Gamma \exp\left(-\frac{H(\mathbf{p,q})}{k_B T}\right) d\mathbf{p}^N d\mathbf{q}^N}, \tag{5.88}$$

where $\Gamma = V_q \cup V_p$ is the volume of the phase space.

For example, we can use the above formula to calculate the average kinetic energy of a canonical ensemble,

$$<E_{kin,j}> = \left\langle \frac{|\mathbf{p}_j|^2}{2m} \right\rangle = \frac{1}{Z} \int d\mathbf{q}^N d\mathbf{p}^N \frac{|\mathbf{p}_j|^2}{2m} \exp\left[-\beta H(\{\mathbf{r}_j, \mathbf{p}_j\})\right]$$

$$= \frac{\int d\mathbf{p}^N \frac{|\mathbf{p}_j|^2}{2m} \exp\left[\frac{-|\mathbf{p}_j|^2}{2mk_B T}\right]}{\int d\mathbf{p}^N \exp\left[\frac{-|\mathbf{p}_j|^2}{2mk_B T}\right]} = \frac{3}{2} k_B T, \tag{5.89}$$

in which we used the identities:

$$\int_0^\infty x^2 \exp(-\beta x^2) dx = \frac{1}{4\beta}\sqrt{\frac{\pi}{\beta}} \quad \text{and} \quad \int_0^\infty \exp(-\beta x^2) dx = \frac{1}{2}\sqrt{\frac{\pi}{\beta}}.$$

Equation (5.89) yields a familiar result that the kinetic energy for each atom is $3/2k_B T$ as stated by the equipartition theorem.

Then we can calculate the temperature of a canonical ensemble by summing its total (ensemble) kinetic energy,

$$E_{kin} = E_{kin}(\mathbf{p}^N) = \sum_{j=1}^N <E_{kin,j}> = \frac{3}{2} N k_B T. \tag{5.90}$$

For the canonical ensemble, the ensemble average is the observable internal energy,

$$<E> = \sum_i p_i E_i = \frac{1}{Z} \sum_i \exp(-\beta E_i) E_i = U, \tag{5.91}$$

in which

$$Z = \sum_i \exp(-\beta E_i), \quad \beta = \frac{1}{k_B T}.$$

Thus, we have

$$-\ln Z = \sum_i \beta E_i \rightarrow \frac{\partial}{\partial \beta}(-\ln Z) = \frac{1}{Z} \sum_i E_i \exp(-\beta E_i) = U.$$

Changing of variable, we can further derive that

$$U = -\left(\frac{\partial \ln Z}{\partial \beta}\right)_{N,V} = -\left(\frac{\partial \ln Z}{\partial T}\frac{\partial T}{\partial \beta}\right)_{N,V} = k_B T^2 \left(\frac{\partial \ln Z}{\partial T}\right)_{N,V}, \quad (5.92)$$

in which we use the identities

$$\beta = \frac{1}{k_B T}, \quad \frac{\partial \beta}{\partial T} = -\frac{1}{k_B T^2}, \quad \text{and} \quad \frac{\partial T}{\partial \beta} = -k_B T^2.$$

Moreover, we can also express the specific heat as

$$C_V = \left(\frac{\partial U}{\partial T}\right)_{N,V} = -\frac{1}{k_B T^2}\left(\frac{\partial U}{\partial \beta}\right)_{N,V} = k_B \beta^2 \left(\frac{\partial^2 \ln Z}{\partial \beta^2}\right)_{N,V}. \quad (5.93)$$

in which we use the chain rule

$$\frac{\partial}{\partial T} = \frac{\partial}{\partial \beta}\frac{\partial \beta}{\partial T} = -\frac{1}{k_B T^2}\frac{\partial}{\partial \beta}.$$

From Eq. (5.92), we have

$$<E> = U = -\frac{\partial \ln Z}{\partial \beta}. \quad (5.94)$$

On the other hand,

$$<E^2> = \frac{1}{Z}\sum_i E_i^2 \exp(-\beta E_i) = \frac{1}{Z}\frac{\partial}{\partial \beta^2}\sum_i \exp(-\beta E_i) = \frac{1}{Z}\frac{\partial^2 Z}{\partial \beta^2}. \quad (5.95)$$

The energy fluctuation can be calculated as

$$<(E-<E>)^2> = <E^2 - 2E<E> + <E>^2> = <E^2> - <E>^2$$
$$= \frac{1}{Z}\frac{\partial^2 Z}{\partial \beta^2} - \left(-\frac{1}{Z}\frac{\partial Z}{\partial \beta}\right)^2 = \frac{\partial^2 \ln Z}{\partial \beta^2} = -\frac{\partial <E>}{\partial \beta}. \quad (5.96)$$

Thus, we find that the heat capacity or the specific heat under the constant volume can be measured using energy fluctuation,

$$C_V = \frac{\partial <E>}{\partial T} = -\frac{1}{k_B T^2}\frac{\partial <E>}{\partial \beta} = \frac{1}{k_B T^2}<(E-<E>)^2>. \quad (5.97)$$

The objective of molecular dynamics is to calculate macroscale properties of a thermodynamics ensemble, and the specific heat C_V is a macroscale property, or a property of a canonical ensemble. This is why we often use Eq. (5.97) to calculate the specific heat by using molecular dynamics.

REMARK 5.13 Since

$$C_V = \frac{\partial <E>}{\partial T} = -\frac{1}{k_B T^2}\frac{\partial <E>}{\partial \beta} \rightarrow C_V = \frac{1}{k_B T^2}(\Delta E)^2,$$

we have

$$(\Delta E)^2 = k_B T^2 C_V = N k_B T^2 c_V \quad \leftarrow \text{ specific heat per atom.}$$

Therefore,
$$\Delta E = \sqrt{Nk_B T^2 c_V} \sim \sqrt{N} \text{ as } N \to \infty$$

and then for ideal gas
$$E \sim \frac{3}{2} N k_B T \sim N \text{ as } N \to \infty.$$

Thus,
$$\frac{\Delta E}{<E>} \sim \frac{1}{\sqrt{N}} \to 0, \text{ as } N \to \infty.$$

Therefore, for large (NVT) systems, the relative energy fluctuation diminishes at the thermodynamics limit, which means that in the thermodynamic limit (NVT) ensemble behaves like a (NVE) ensemble. This is the reason why we also say at the thermodynamic limit (NVE) ensemble behaves like a (NVT) ensemble, which is the consequence of the "central limit theorem" in statistics theory.

5.8.5 Probability Distributions in the Canonical Ensemble

The canonical ensemble is the most typical case of a thermodynamics equilibrium ensemble. This is because when the number of particles of a molecular system increase to infinite, many other equilibrium ensembles will approach to or behave the same as the canonical ensemble. This is guaranteed by the central limited theorem of statistics. Therefore, it is meaningful for us to examine the various probability distributions in the canonical ensemble.

1. Probability Distribution of Velocity

In the canonical ensemble, the occupation number is,
$$p_j(E_j) = \exp(-\beta E_j)/Z, \quad \beta = \frac{1}{k_B T}.$$

Consider a molecular system of ideal gas, in which $U_{potential} = 0$ and $E_j = \frac{1}{2} m_j \mathbf{v}_j^2$. The probability that a particle in the microstate E_j in the phase space is
$$\frac{1}{Z} \exp\left(-\frac{1}{2}\beta m_j \mathbf{v}_j^2\right) d\mathbf{p}_j \sim C \exp\left(-\frac{1}{2}\beta m_j \mathbf{v}_j^2\right) d\mathbf{v}_j.$$

Normalizing it
$$\int C \exp\left(-\frac{1}{2}\beta m \mathbf{v}^2\right) d\mathbf{v} = 1 \to C = \left(\frac{\beta m}{2\pi}\right)^{3/2}.$$

We then find that the atom velocity distribution in a canonical ensemble is a Gaussian distribution:

$$g(v) = \left(\frac{\beta m}{2\pi}\right)^{3/2} \exp\left(-\frac{1}{2}\beta m \mathbf{v}^2\right). \tag{5.98}$$

For a one-dimensional (1D) component, we have

$$g(v_x) = \left(\frac{\beta m}{2\pi}\right)^{1/2} \exp\left(-\frac{1}{2}\beta m v_x^2\right).$$

Note that Eq. (5.98) is the velocity distribution, but not the velocity square distribution. In molecular dynamic simulations, the velocity of atom is a random variable under a given temperature for the ensemble. If in an NVT MD simulation, every atoms are the same type, and this simulation may be viewed as an ensemble MD simulation. By finding the velocity distribution of all atoms, we can find the probability distribution of atoms. On the other hand, the velocity square is a random variable function, i.e., $f(v) = v^2$, the probability distributions of a random variable and its functions are different in general, and we shall discuss the probability distribution of the velocity square next.

Before we discuss the velocity square distribution, we first calculate the average velocity, which can be found as

$$\bar{v} = \int_{-\infty}^{\infty} v g(v) dv = \left(\frac{\beta m}{2\pi}\right)^{1/2} \int_{-\infty}^{\infty} v \exp\left(-\frac{1}{2}\beta m v^2\right) dv = 0, \tag{5.99}$$

because the Gaussian distribution is symmetric.

However, the velocity probability distribution is not a standard Gaussian distribution, because the variance of the velocity distribution is not 1. In fact,

$$<v_x^2> = \left(\frac{\beta m}{2\pi}\right)^{1/2} \int_{-\infty}^{\infty} v_x^2 \exp\left(-\frac{1}{2}\beta v_x^2\right) dv_x$$

$$= \left(\frac{\beta m}{2\pi}\right)^{1/2} \frac{\sqrt{\pi}}{2} \left(\frac{2}{\beta m}\right)^{3/2} = \frac{1}{\beta m} = \frac{k_B T}{m}. \tag{5.100}$$

In 1D case, the standard deviation of the velocity distribution is then given as,

$$\sqrt{v_x^2} = \sqrt{\frac{1}{\beta m}} = \sqrt{\frac{k_B T}{m}}.$$

This is shown in Fig. 5.18. In fact, Eq. (5.100) has a profound consequence, i.e.,

$$<v^2> = <v_x^2> + <v_y^2> + <v_z^2> = \frac{3k_B T}{m}. \tag{5.101}$$

The physical meaning of this expression is that the root mean square speed

$$v_{rms} = \left(<v^2>\right)^{1/2} = \sqrt{\frac{3k_B T}{m}}.$$

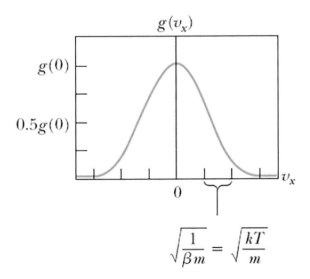

Figure 5.18 Velocity probability distribution and its standard deviation

In consequence, we have the relationship between the kinetic energy and temperature,

$$<K> = \frac{m}{2}(<v_x>^2 + <v_y>^2 + <v_z>^2) = \frac{1}{2}m\left(\frac{3k_BT}{m}\right) = \frac{3}{2}k_BT. \quad (5.102)$$

Equation (5.102) is related to the so-called **equal-partition theorem**, which states that each quadratic term in the kinetic energy of a molecular system contributes $1/2k_BT$ to the internal energy U and $1/2k_B$ to the heat capacity C_V of monoatomic system at high temperature. However, things become different for diatomic molecular system, in which

- Translation in 1D: $\hat{H} = \frac{1}{2}mv_i^2$ (one quadratic term) $\Rightarrow \frac{1}{2}k_BT$;
- Rotation about an axis: $\hat{H} = \frac{1}{2}J_i\omega_i^2$ (one quadratic term) $\Rightarrow \frac{1}{2}k_BT$
  (linear molecules: 2 such terms, nonlinear molecules: ?3 such terms)
- Vibration: $\hat{H} = \frac{1}{2}\frac{\hat{p}^2}{m} + \frac{1}{2}m\omega^2\hat{x}^2$ (two quadratic terms) $\Rightarrow k_BT$

For the diatomic gas molecule, it has three translational degrees of freedom or three translation quadratic terms in kinetic energy, two rotational quadratic terms in kinetic energy, and two vibrational quadratic terms in kinetic energy. Thus, its average thermal energy per gas molecule is

$$\bar{\epsilon} = 7 \times \left(\frac{1}{2}k_BT\right) = \frac{7}{2}k_BT,$$

while it contributes

$$U = \frac{7}{2}k_BT$$

to the internal energy.

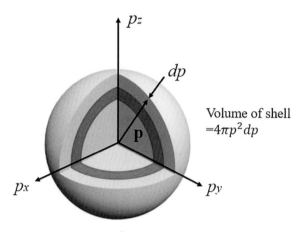

Figure 5.19 Number of energy states is a thin shell

1. Probability Distribution of Energy: Maxwell–Boltzmann Distribution

For a canonical ensemble system, the probability distribution density function in the phase is

$$p = \frac{1}{Z}\exp(-\beta H),$$

which is a relative probability that an energy state is occupied at a given temperature.

To find the energy probability distribution in a classical system, we first define the following quantities:

- $n(E)dE$ = the number of particles with energies between $[E, E+dE]$;
- We define the density of states, $\omega(E)$, as the number of states available per unit energy range

$$n(E) = \omega(E)p(E).$$

For ideal gas particles, we have

$$E = K = \frac{1}{2}m\mathbf{p}^2$$

$$|\mathbf{p}| = \sqrt{2mE} = \sqrt{p_x^2 + p_y^2 + p_z^2}$$

$$dp = \sqrt{\frac{2m}{E}}dE.$$

The number of momentum states in a spherical shell from p to $p+dp$ is proportional to $4\pi p^2 dp$ (the volume of the shell). Thus, we can write the number of states having momentum between p and $p+dp$ as

$$\omega_p(\mathbf{p})d\mathbf{p} = B\mathbf{p}^2 d\mathbf{p},$$

where B is a proportionality constant as shown in Fig. 5.19.

Since the energy in ideal gas is totally contributed from kinetic energy, each $\mathbf{p}$ can be directly linked to energy E, i.e.,

$$\omega(E)dE = \omega_p(\mathbf{p})d\mathbf{p} = B\mathbf{p}^2 d\mathbf{p}.$$

Considering

$$\mathbf{p}^2 = 2mE \quad \rightarrow \quad d|\mathbf{p}| = \sqrt{\frac{m}{2}} E^{-1/2} dE,$$

we have

$$\mathbf{p}^2 d\mathbf{p} = (2mE)\left(\sqrt{m/2} E^{-1/2} dE\right) \sim E E^{-1/2} dE \quad \rightarrow \quad \omega(E)dE \sim E^{1/2} dE.$$

Thus,

$$n(E)dE = \omega(E) \frac{\exp(-\beta E)}{Z} dE \sim C\sqrt{E} \exp\left(-\frac{E}{k_B T}\right) dE$$

$$\rightarrow \quad n(E) \sim C\sqrt{E} \exp\left(-\frac{E}{k_B T}\right).$$

We find the density state in terms of energy as

$$\omega(E) = C\sqrt{E},$$

where C is a constant.

To identify the constant C, we can evaluate the following integration,

$$N = \int_0^\infty n(E)dE = C \int_0^\infty \sqrt{E} \exp(-E/k_B T)\, dE,$$

where N is the total number of particles in the system. Using the integration identity,

$$\int_0^\infty x^{1/2} \exp(-\beta x) dx = \frac{1}{2\beta}\sqrt{\frac{\pi}{\beta}}, \quad \beta = \frac{1}{k_B T},$$

we find that

$$N = \frac{C}{2}\sqrt{\pi}\,(k_B T)^{3/2} \quad \rightarrow \quad C = \frac{2N}{\sqrt{\pi}(k_B T)^{3/2}}$$

and

$$\omega(E) = \frac{2N}{\sqrt{\pi}(k_B T)^{3/2}} \sqrt{E}.$$

Finally, we obtain the density function,

$$n(E) = \frac{2N}{\sqrt{\pi}(k_B T)^{3/2}} \sqrt{E} \exp(-E/k_B T). \tag{5.103}$$

Equation (5.103) is called the Maxwell–Boltzmann distribution for kinetic energy (see Fig. 5.20). For a single atom, the energy distribution may be written in the following standard form,

$$n(\epsilon_k) = \frac{2}{\sqrt{\pi}(k_BT)^{3/2}} \sqrt{\epsilon_k} \exp(-\epsilon_k/(k_BT)), \tag{5.104}$$

which is shown in Fig. 5.20. We can then calculate the expectation of internal energy of the molecular system, and by definition it is given as,

$$U = <E> = \int_0^\infty En(E)dE = \frac{2N}{\sqrt{\pi}(k_BT)^{3/2}} \int_0^\infty E^{3/2} \exp(-E/(k_BT))dE. \tag{5.105}$$

Evaluation or integration of Eq. (5.105) yields,

$$U = \frac{3Nk_BT}{2}.$$

This is the total energy for N molecules, so that the average (kinetic) energy per molecule is

$$<\epsilon> = \frac{3}{2}k_BT,$$

which recovers the result from the equipartition theorem of ideal gas.

Based on the earlier expressions, the speed of a molecule can be solved by using the average energy, i.e.,

$$<\epsilon> = \frac{m<v^2>}{2} = \frac{3}{2}k_BT \rightarrow v_{rms} = \sqrt{<v^2>} = \sqrt{\frac{3k_BT}{m}}, \tag{5.106}$$

where

$$<v^2> := <v_x^2> + <v_y^2> + <v_z^2> = \frac{3k_BT}{m} \text{ and}$$

v_{rms} is the speed of a molecule having the average energy $\bar{E}$. The v_{rms} speed is the square root of mean square velocity.

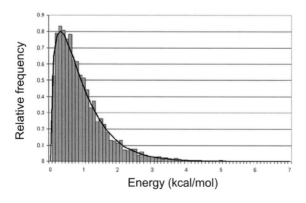

Figure 5.20 Maxwell–Boltzmann distribution for kinetic energy

Furthermore, by substituting: $\epsilon = \dfrac{mv^2}{2}$ and $d\epsilon = mvdv$ into the density function expression $n(\epsilon)d\epsilon$, i.e.,

$$n(\epsilon)d\epsilon = \dfrac{2}{\sqrt{\pi}(k_B T)^{3/2}}\sqrt{\epsilon}\exp(-\epsilon/k_B T)(mvdv).$$

We can identify the speed probability distribution as

$$p(v) = 4\pi \left(\dfrac{m}{2\pi k_B T}\right)^{3/2} v^2 \exp\left(-\dfrac{mv^2}{2k_B T}\right), \qquad (5.107)$$

which is often called the Maxwell speed distribution. Figure 5.21 is a graphic illustration of the Maxwell speed distribution.

In fact, the Maxwell speed distribution can be directly derived from the Maxwell–Boltzmann distribution of kinetic energy, i.e., Eq. (5.104). Consider that the random variable speed v is a function of the random variable energy ϵ_k, i.e.,

$$\epsilon_k = \dfrac{mv^2}{2} \quad \to \quad v = g(\epsilon_k) = \sqrt{\dfrac{2\epsilon_k}{m}}.$$

Using the probability distribution function formula (see Roussas 2003), we have the following expression,

$$f_V(v) = f_E(g^{-1}(v))\left|\dfrac{dg^{-1}(v)}{dv}\right| \quad \to \quad p(v) = n(\epsilon_k)\left|\dfrac{d\epsilon_k}{dv}\right|,$$

where $f_E(\epsilon_k) = n(\epsilon_k)$ is the probability distribution of energy ϵ_k and $f_V(v)$ is the probability distribution of speed v.

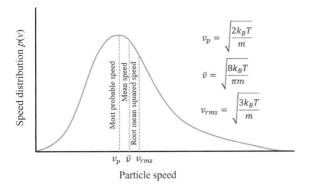

Figure 5.21 Maxwell speed distribution for particles in a canonical ensemble

Then we can readily find that

$$p(v) = \frac{2}{\sqrt{\pi}(k_BT)^{3/2}} \sqrt{\epsilon_k} \exp(-\epsilon_k/(k_BT)) \left| mv \right|$$

$$= 4\pi \left(\frac{m}{2\pi k_B T}\right)^{3/2} v^2 \exp\left(-\frac{mv^2}{2k_BT}\right), \tag{5.108}$$

which recovers Eq. (5.107).

The difference between the velocity distribution (Eq. (5.98)) and the speed distribution (Eq. (5.107)) is that the former is the probability distribution for physical velocity as a vector that has direction, or plus/minus signs, and the latter is the probability distribution for the velocity value (without sign).

Based on this interpretation, we can calculate the average speed value $\bar{v}$ by carrying the following integration,

$$\bar{v} = \frac{\int_0^\infty v p(v) dv}{\int_0^\infty p(v) dv} \quad \rightarrow \quad \bar{v} = \sqrt{\frac{8k_BT}{\pi m}} \tag{5.109}$$

Comparing with v_{rms}, one may find that $v_{rms} = 1.09\bar{v}$.

Moreover, we can also find the so-called most probable speed v_p by solving the following stationary point condition,

$$\frac{dp(v)}{dv} = 0 \quad \rightarrow \quad v_p = \sqrt{\frac{2k_BT}{m}}, \tag{5.110}$$

where the subscript p means the most probable. From Fig. 5.21, one can compare v_p, $\bar{v}$, and v_{rms}.

2. Temperature Fluctuation Distribution

In the canonical ensemble, the temperature of the system is fixed, while the energy of the system fluctuates. However, since the temperature is a macroscale quantity, it will also fluctuate around the expected value.

There are different ways to derive the temperature fluctuation formula. In the following, we derive the temperature fluctuation formula based on the central limit theorem in statistics.

In statistics, the **central limit theorem** states that for large sample size ($N \to \infty$), no matter what form of the actual probability distribution function is for a random variable x, the distribution of the sample mean

$$\bar{x} = \frac{1}{N} \sum_i x_i$$

will approach a normal distribution with the sample mean as $\mu(\bar{x}) = \mu(x)$ and the sample variance as $\sigma^2(\bar{x}) = \sigma^2(x)/N$.

We can rephrase this statement as: no matter what form the distribution $p(x)$ is, the distribution of the sample mean will approach a normal (Gaussian) distribution with certain parameters,

$$p(x) \to P(\bar{x}) = \frac{1}{\sqrt{2\pi}} \exp\left(-\left(\frac{\sum_j x_j - N <x>}{\sqrt{N}\sigma(x)}\right)^2\right).$$

Let $x = \epsilon_k$. We can apply the central limit theorem to the Maxwell–Boltzmann energy distribution

$$n(\epsilon_k) = \frac{2}{\sqrt{\pi}(k_B T)^{3/2}} \sqrt{\epsilon_k} \exp(-\epsilon_k/k_B T) \to \frac{1}{\sqrt{2\pi}} \exp\left(-\left(\frac{E_k - <E>}{\sqrt{N}\sigma(\epsilon)}\right)^2\right), \quad (5.111)$$

where $\sigma(\epsilon_k) := <\epsilon_k^2> - <\epsilon_k>^2$, $E_k = N\epsilon_k$, and $<E> = N<\epsilon_k>$. Considering the energy distribution of the ideal gas

$$n(\epsilon_k) = \frac{2}{\sqrt{\pi}(k_B T)^{3/2}} \sqrt{\epsilon_k} \exp(-\epsilon_k/k_B T)$$

and assuming that the NVT system with target temperature T_0 and using the following identities,

$$\int_0^\infty x^{1/2} \exp(-\beta x) dx = \frac{1}{2\beta}\sqrt{\frac{\pi}{\beta}}$$

$$\int_0^\infty x^{3/2} \exp(-\beta x) dx = \frac{3}{4\beta^2}\sqrt{\frac{\pi}{\beta}}$$

$$\int_0^\infty x^{5/2} \exp(-\beta x) dx = \frac{15}{8\beta^3}\sqrt{\frac{\pi}{\beta}}$$

we can find that

$$<\epsilon_k> = \frac{3}{2}k_B T_0, \quad <\epsilon_k^2> = \frac{15}{4}(k_B T_0)^2$$

and

$$\sigma^2 = <\epsilon_k^2> - <\epsilon_k>^2 = \frac{15}{4}(k_B T_0)^2 - \frac{9}{4}(k_B T_0)^2 = \frac{3}{2}(k_B T_0)^2. \quad (5.112)$$

On the other hand, we have

$$E_k = \frac{1}{2}\sum_{j=1}^N m_j v_j^2 = \frac{3N}{2}k_B T \quad (5.113)$$

and

$$<E> = N<\epsilon_k> = \frac{3N}{2}k_B T_0. \quad (5.114)$$

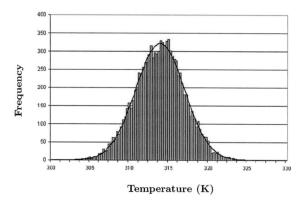

Figure 5.22 Probability distribution for temperature fluctuation

Substituting Eqs. (5.112), (5.113), and (5.114) into (5.111) yields,

$$p(\bar{\epsilon}_k) = \frac{1}{\sqrt{2\pi}} \exp\left(-\left(\frac{E_k - <E>}{\sqrt{N}\sigma}\right)^2\right) = \frac{1}{\sqrt{2\pi}} \exp\left(-\left(\frac{3N\Delta T^2}{2T_0^2}\right)\right),$$

which leads to the distribution for the temperature fluctuation,

$$p(\Delta T) = \frac{1}{\sqrt{2\pi}} \exp\left(-\left(\frac{\Delta T^2}{2\sigma_T^2}\right)\right), \quad (5.115)$$

where $\Delta T = T - T_0$ and $\sigma_T^2 = T_0^2/3N$.

Note that Eq. (5.115) is the temperature fluctuation distribution not the temperature distribution. In the canonical ensemble, the target temperature is fixed as T_0. Figure 5.22 displays such probability distribution for temperature fluctuation.

5.8.6 Grand Canonical Ensemble (μVT)

A grand canonical ensemble may be thought as an open system inside a canonical ensemble. Thus its temperature will be a constant or fixed. Since the system is open, the number of particles inside the grand canonical ensemble varies as shown in Fig. 5.23.

Even though both energy and the number of particles inside the grand canonical ensemble change from time to time, both the ensemble energy as well as the ensemble particle number are fixed. Thus, for a grand canonical ensemble, we have

$$<E> = \sum_i p_i E_i = \bar{U} = const., \text{ as well as } <N> = \sum_i p_i N_i = \bar{N} = const.$$

The previous thermal reservoir for the canonical ensemble will also act as a particle reservoir to feed or absorb the particles coming in and out of the grand canonical

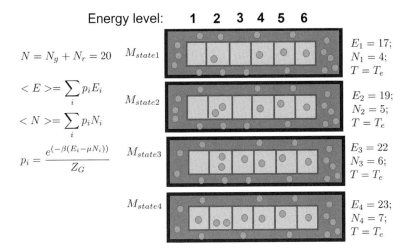

Figure 5.23 Illustrative example of the grand canonical ensemble

ensemble. Since the particle type in the reservoir and in the grand canonical ensemble should be the same, its chemical potential should be fixed as well, i.e., $\mu = const$. This is the reason why we also call the grand canonical ensemble as the μVT ensemble.

Considering the equilibrium condition, we set the condition for maximum entropy, i.e.,

$$\delta\left(S - \gamma\left(\sum_i p_i E_i - \bar{U}\right) + \lambda\left(\sum_i p_i - 1\right) - \eta\left(\sum_i p_i N_i - \bar{N}\right)\right) = 0. \tag{5.116}$$

Solving Eq. (5.116), we obtain,

$$p_j = \exp(-(1 + \lambda/k_B))\exp(-\gamma E_j/k_B - \eta N_j/k_B) = \frac{1}{Z_G}\exp(-\beta(E_j - \mu N_j)), \tag{5.117}$$

where $\beta = (k_B T)^{-1}$ and

$$Z_G = \exp((1 + \lambda/k_B)) = \sum_j \exp(-\gamma E_j/k_B - \eta N_j/k_B)$$

is the partition function of the grand canonical ensemble.

Substituting

$$p_i = \frac{1}{Z_G}\exp(-\gamma E_i/k_B - \eta N_i)$$

into Gibbs' entropy formula, we have

$$S = -k_B \sum_i p_i \ln p_i \rightarrow S = -k_B \sum_i p_i\left(-\gamma E_i - \eta N_i\right) - \ln Z_G)$$
$$= \gamma <E> + \eta <N> + k_B \ln Z_G = \gamma U + \eta N + k_B \ln Z_G. \tag{5.118}$$

Taking derivatives of Eq. (5.118) yields

$$\left(\frac{\partial S}{\partial U}\right)_{V,N} = \gamma, \text{ and } \left(\frac{\partial S}{\partial N}\right)_{V,U} = \eta,$$

and comparing them with

$$\left(\frac{\partial S}{\partial U}\right)_{V,N} = \frac{1}{T} \text{ and } \left(\frac{\partial S}{\partial N}\right)_{V,U} = -\frac{\mu}{T},$$

it leads to

$$\gamma = \frac{1}{T} \text{ and } \eta = -\frac{\mu}{T}.$$

Considering

$$p_j = \frac{1}{Z_G} \exp(-\beta(E_j - N_j\mu)) \rightarrow \ln p_j = -\beta E_j + \beta N_j \mu - \ln Z_G$$

and recalling $F = U - ST = <E> - ST$ and in the canonical ensemble

$$Z = \exp(-\beta F) \rightarrow -kT \ln Z = F,$$

we have

$$\ln Z_G = \ln Z + \beta \mu <N>. \tag{5.119}$$

One may rewrite the partition function of the grand canonical ensemble as

$$Z_G(z, V, T) = \sum_i \exp(-\beta(E_i + \mu N_i)) = \sum_i z^{N_i} \exp(-\beta E_i),$$

where $z = \exp(\beta\mu)$ is called the *fugacity*.
Then Eq. (5.119) becomes

$$\ln Z_G = \ln Z + \ln z <N>, \tag{5.120}$$

which leads to the expression of the average number of particles in the grand canonical ensemble,

$$<N> = z\frac{\partial}{\partial z} \ln Z_G(z, V, T).$$

Recalling in the canonical ensemble

$$<E> = U = -\frac{\partial}{\partial \beta} \ln Z,$$

we have

$$<E> - \mu N = -\frac{\partial}{\partial \beta} \ln Z_G.$$

In canonical ensemble, we have

$$F(T, V, N) = -k_B T \ln Z \text{ or } Z = \exp(-\beta F).$$

Table 5.1 Summary of three thermodynamics equilibrium ensembles

Ensemble	State variable	Probability density	Potential/partition function
Microcanonical	E, V, N	$p_i = 1/\Omega$	$S(E, V, N) = k_B \ln \Omega$
Canonical	T, V, N	$p_i = \exp(-\beta E_i)/Z$	$F(T, V, N) = -k_B T \ln Z$
Grand canonical	T, V, μ	$p_i = \exp(-\beta(E_i - \mu N_i))/Z_G$	$\Phi(T, V, \mu) = -k_B T \ln Z_G$

We can define the grand canonical potential as

$$\Phi(T, V, \mu) = -k_B T \ln Z_G \quad \text{or} \quad Z_G = \exp(-\beta \Phi),$$

thus,

$$\Phi(T, V, \mu) = -k_B T \ln Z_G = -k_B T (\ln Z + \beta \mu <N>)$$
$$= -k_B T \ln Z - \mu N) = F - \mu <N>.$$

We summarize the three thermodynamics equilibrium ensembles in Table 5.1.

5.9 Homework Problems

Problem 5.1 Consider the following total Lagrangian for a single particle,

$$L(x, y, \dot{x}, \dot{y}) = \frac{m}{2}(\dot{x}^2 + \dot{y}^2) - \alpha \exp(-\beta(R - r)),$$

where $r = \sqrt{x^2 + y^2}$, and α, β are constants. Derive the Euler–Lagrangian equations for the particle.

Problem 5.2 Consider a heteronuclear diatomic molecule (in two dimension) in the gas phase whose two atoms have masses m_1 and m_2. The Lagrangian for the molecule is given by

$$L(\mathbf{r}_1, \mathbf{r}_2, \dot{\mathbf{r}}_1, \dot{\mathbf{r}}_2) = \frac{1}{2} m_1 \dot{\mathbf{r}}_1^2 + \frac{1}{2} m_2 \dot{\mathbf{r}}_2^2 - U(|\mathbf{r}_1 - \mathbf{r}_2|),$$

where

$$U(r) = \frac{kr^2}{2}, \quad r = |\mathbf{r}_1 - \mathbf{r}_2|.$$

Define the center of mass and relative coordinate as

$$\mathbf{R} = \frac{m_1 \mathbf{r}_1 + m_2 \mathbf{r}_2}{m_1 + m_2}, \quad \mathbf{r} = \mathbf{r}_1 - \mathbf{r}_2.$$

1. Rewrite the Lagrangian in terms of $\mathbf{R}$ and $\mathbf{r}$;
2. Just consider the relative part of Lagrangian, L_{rel}, and assume that the two atoms only move in a plane, so that we can transform the Cartesian coordinate into polar coordinates,

$$x_1 = r \sin \theta$$
$$x_2 = r \cos \theta$$

Derive equations of motion in terms of r and θ;

3. Solve for r by assuming that $r - r_0 \ll 1$, where r_0 is the equilibrium position of the total potential V (which is not just U), i.e.,

$$F = -\frac{dV}{dr}\bigg|_{r=r_0} = 0,$$

where

$$V(r) = \frac{k}{2}r^2 + \frac{B}{2r^2} \approx \frac{V''(r_0)}{2}(r - r_0)^2.$$

Note that you have to identify constant B.

Reference (see): https://en.wikipedia.org/wiki/Lagrangian_mechanics

Problem 5.3 Let f be the probability density distribution function in the phase space $\eta = (\mathbf{q}^N, \mathbf{p}^N)$, where $\mathbf{q}^N = [\mathbf{q}_1, \ldots \mathbf{q}_i, \ldots \mathbf{q}_N]$ and $\mathbf{p}^N = [\mathbf{p}_1, \ldots \mathbf{p}_i, \ldots \mathbf{p}_N]$. Consider the balance of probability density function f in an arbitrary volume of the phase space, ω, i.e.,

$$\frac{\partial}{\partial t} \int_\omega f d\eta = -\oint_{\partial \omega} f \dot{\eta} \cdot \mathbf{n} dS.$$

Show

(1)

$$f \sum_i \frac{\partial}{\partial \eta} \dot{\eta} = f \left(\frac{\partial}{\partial \mathbf{q}_i} \dot{\mathbf{q}}_i + \frac{\partial}{\partial \mathbf{p}_i} \dot{\mathbf{p}}_i \right) = 0,$$

where $\eta = (\mathbf{q}^N, \mathbf{p}^N)$ satisfy the Hamiltonian equation,

$$\frac{\partial H}{\partial \mathbf{p}_i} = \dot{\mathbf{q}}_i, \text{ and } \frac{\partial H}{\partial \mathbf{q}_i} = -\dot{\mathbf{p}}_i.$$

(2) The Liouville equation

$$\frac{df}{dt} = \frac{\partial f}{\partial t} + \dot{\eta} \cdot \frac{\partial f}{\partial \eta} = 0.$$

Problem 5.4 Based on definition,

$$c_v = \left(\frac{\partial E}{\partial T} \right)_{N,V},$$

show that

$$<E> = -\frac{\partial \ln Z}{\partial \beta} \text{ and } <E^2> = \frac{1}{Z} \frac{\partial^2 Z}{\partial \beta^2}, \quad \beta = \frac{1}{k_B T},$$

and

$$c_v = \frac{<E^2> - <E>^2}{k_B T^2}.$$

Problem 5.5 Show that in the microcanonical ensemble,

$$k_B T = \frac{2}{3N} \frac{\int d\mathbf{q}^N (E-U)^{3N/2-1} H(E-U)(E-U)}{\int d\mathbf{q}^N (E-U)^{3N/2-1} H(E-U)},$$

where U is the potential energy, and $H[\cdot]$ is the Heaviside function. N is the number of particles.

Problem 5.6 Prove that for a given field distribution $A(\mathbf{q},\mathbf{p}))$ the following equality holds,

$$<A(\mathbf{q},\mathbf{p}')>_{NVE,Nosé} = \frac{\int A(\mathbf{q},\mathbf{p}') \exp\left[-\frac{H(\mathbf{q},\mathbf{p}')}{k_B T}\right] d\mathbf{q}^N d\mathbf{p}'^N}{\int \exp\left[-\frac{H(\mathbf{q},\mathbf{p}')}{k_B T}\right] d\mathbf{q}^N d\mathbf{p}'^N} =: <A(\mathbf{q},\mathbf{p}')>_{NVT},$$

where $A(\mathbf{q},\mathbf{p})$ is an arbitrary field in the phase space.

Problem 5.7 Assume that an ensemble system contains nine particles ($A = 9$), and having three quantum energy levels with the occupation number (a_1, a_2, a_3). Calculate both the maximum and the minimum numbers of microstates under this condition,

$$W = \frac{A!}{a_1! a_2! a_3!}.$$

Problem 5.8 Consider a canonical ensemble system with N identical atoms. Show that the kinetic energy

$$E_{kin} = \frac{1}{2} \sum_{i=1}^{N} m \dot{r}_i^2.$$

Show that

$$<E_{kin}>_{NVT} = \frac{3}{2} N k_B T.$$

6 Fundamentals of Molecular Dynamics

In this chapter, we discuss how to compute motions of atoms and molecules by modeling them as the rigid particle motion in classical Newtonian mechanics. Before we proceed to discuss the classical molecular dynamics (MD), we first examine the relationship between the MD and quantum mechanics, and then the briefly introduce ab-initio MD.

6.1 How to Derive Molecular Dynamics from Quantum Mechanics

Our starting point is under the framework of the non-relativistic quantum mechanics, which is formulated via the following time-dependent Schrödinger equation for an N nuclear ($I = 1, 2, \ldots, N$) and an n electron ($i = 1, 2, \ldots, n$) system (see Fig. 6.1),

$$i\hbar \frac{\partial}{\partial t} \Phi(\{\mathbf{r}_i\}, \{\mathbf{R}_I\}; t) = \mathcal{H} \Phi(\{\mathbf{r}_i\}, \{\mathbf{R}_I\}; t), \qquad (6.1)$$

where $\mathbf{r}_i, i = 1, 2, \ldots n$ are electron position coordinates, and we call $\{\mathbf{r}_i\}$ the electron degrees of freedom; $\mathbf{R}_I, I = 1, 2, \ldots N$ are nuclear coordinates, and we call $\{\mathbf{R}_I\}$ the nuclear degrees of freedom; $\Phi(\{\mathbf{r}_i\}, \{\mathbf{R}_I\}; t)$ is the total wave function of the system; and $\mathcal{H}$ is the Hamiltonian of the system, which may be expressed as

$$\mathcal{H} = -\sum_I \frac{\hbar^2}{2M_I} \nabla_I^2 - \sum_i \frac{\hbar^2}{2m_e} \nabla_i^2 + \sum_{i<j} \frac{e^2}{|\mathbf{r}_i - \mathbf{r}_j|} - \sum_{I,i} \frac{e^2 Z_I}{|\mathbf{R}_I - \mathbf{r}_j|} + \sum_{I<J} \frac{e^2 Z_I Z_J}{|\mathbf{R}_I - \mathbf{R}_J|}$$

$$= -\sum_I \frac{\hbar^2}{2M_I} \nabla_I^2 - \sum_i \frac{\hbar^2}{2m_e} \nabla_i^2 + V_{n-e}((\{\mathbf{r}_i\}, \{\mathbf{R}_I\})$$

$$= -\sum_I \frac{\hbar^2}{2M_I} \nabla_I^2 + \mathcal{H}_e(\{\mathbf{r}_i\}, \{\mathbf{R}_I\} \qquad (6.2)$$

in terms of electronic coordinate $\{\mathbf{r}_i\}$ and nucleus position $\{\mathbf{R}_I\}$ degrees of freedom. In Eq. (6.2), the electronic Hamiltonian is defined as

$$\mathcal{H}_e := -\sum_i \frac{\hbar^2}{2m_e} \nabla_i^2 + V_{n-e}((\{\mathbf{r}_i\}, \{\mathbf{R}_I\}$$

$$= -\sum_i \frac{\hbar^2}{2m_e} \nabla_i^2 + \sum_{i<j} \frac{e^2}{|\mathbf{r}_i - \mathbf{r}_j|} - \sum_{I,i} \frac{e^2 Z_I}{|\mathbf{R}_I - \mathbf{r}_j|} + \sum_{I<J} \frac{e^2 Z_I Z_J}{|\mathbf{R}_I - \mathbf{R}_J|}. \qquad (6.3)$$

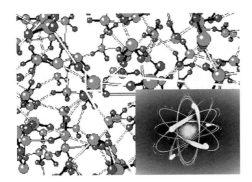

Figure 6.1 Atoms and their electronic structures

To derive the governing equation of the MD from quantum mechanics, the trick is "separation of variable"! We separate the electronic coordinates with the nuclear coordinates by writing the total wave function into the product of the two wave functions with an additional phase angle,

$$\Phi(\{\mathbf{r}_i\}, \{\mathbf{R}_I\}; t) \approx \Psi(\{\mathbf{r}_i\}; t) \chi(\{\mathbf{R}_I\}; t) \exp\left[\frac{i}{\hbar} \int_{t_0}^{t} d\tau \bar{E}_e(\tau)\right], \quad (6.4)$$

where Ψ is the electron wave function and χ is nucleus wave function. In Eq. (6.4), we require the following normalization conditions being satisfied

$$<\Psi|\Psi> = 1 \text{ and } <\chi|\chi> = 1,$$

so that we can define

$$\bar{E}_e := <\Psi\chi|\mathcal{H}_e|\Psi\chi>, \text{ and } S(t) := \int_{t_0}^{t} d\tau \bar{E}_e(\tau). \quad (6.5)$$

Thus,

$$i\hbar \frac{\partial \Phi}{\partial t} = i\hbar \chi(\{\mathbf{R}_I\}) \frac{\partial \Psi(\{\mathbf{r}_i\})}{\partial t} \exp \frac{i}{\hbar} \left[\int d\tau E_e\right]$$
$$+ i\hbar \frac{\partial \chi(\{\mathbf{R}_I\})}{\partial t} \Psi(\{\mathbf{r}_i\}) \exp \frac{i}{\hbar} \left[\int d\tau E_e\right]$$
$$- E_e(t) \Psi(\{\mathbf{r}_i\}) \chi(\{\mathbf{R}_I\}) \exp \frac{i}{\hbar} \left[\int d\tau E_e\right].$$

Canceling the phase factor $\exp \frac{i}{\hbar} \left[\int d\tau E_e\right]$, we can rewrite Eq. (6.1) as (separation of variable),

$$i\hbar \chi(\{\mathbf{R}\}) \frac{\partial \Psi}{\partial t} + i\hbar \frac{\partial \chi}{\partial t} \Psi(\{\mathbf{r}_i\}) - \Psi(\{\mathbf{r}_i\}) \chi(\{\mathbf{R}\}) E_e(t) = \left(-\sum_I \frac{\hbar^2}{2M_I} \nabla_I^2 \chi\right) \Psi(\{\mathbf{r}_i\})$$
$$+ \left(-\sum_i \frac{\hbar^2}{2m_e} \nabla_i^2 \Psi\right) \chi(\{\mathbf{R}_I\}) + V_{n-e}(\{\mathbf{r}_i\}; \{\mathbf{R}\}) \Psi(\{\mathbf{r}_i\}) \chi(\{\mathbf{R}_I\}). \quad (6.6)$$

Multiplying $<\chi|$ and $<\Psi|$ successively on Eq. (6.6) and considering the normalization conditions: $<\chi|\chi>=1$ and $<\Psi|\Psi>=1$, we have,

$$<\Psi|i\hbar\frac{\partial}{\partial t}|\Psi> + <\chi|i\hbar\frac{\partial}{\partial t}|\chi> = \left\langle \chi \left| <\Psi|\hat{H}_e|\Psi> \right| \chi \right\rangle$$

$$= \left\langle \chi \left| \sum_I -\left(\frac{\hbar^2}{2M_I}\nabla_I^2\right) \right| \chi \right\rangle + \left\langle \Psi \left| \sum_i -\left(\frac{\hbar^2}{2M_e}\nabla_i^2\right) \right| \Psi \right\rangle$$

$$+ \left\langle \Psi \left| <\chi|V|\chi> \right| \Psi \right\rangle, \quad (6.7)$$

which can be rearranged as

$$\mathcal{H}_\Psi := <\Psi|i\hbar\frac{\partial}{\partial t}|\Psi> - \left\langle \Psi \left| \sum_i -\left(\frac{\hbar^2}{2M_e}\nabla_i^2\right) \right| \Psi \right\rangle - \left\langle \Psi \left| <\chi|V_{n-e}|\chi> \right| \Psi \right\rangle$$

$$\equiv \mathcal{H}_\chi := -<\chi|i\hbar\frac{\partial}{\partial t}|\chi> + \left\langle \chi \left| <\Psi|\hat{H}_e|\Psi> \right| \chi \right\rangle + \left\langle \chi \left| \sum_I -\left(\frac{\hbar^2}{2M_I}\nabla_I^2\right) \right| \chi \right\rangle$$

$$= const. \quad (6.8)$$

because $\{\mathbf{R}_I\}$ and $\{\mathbf{r}_i\}$ are different variables, and in fact, they are in different scales too! It may be noted that the inner bracket has been integrated out as constants. Thus the only way that left-hand side and right-hand side of Eq. (6.8) can only be equal to a same constant, based on the standard argument in separation of variables.

Since both sides of Eq. (6.8) equal to a same constant, it is straightforward to derive the stationary conditions: $\delta\mathcal{H}_\Psi = 0$ and $\delta\mathcal{H}_\chi = 0$, which yields the following two sets of equations:

$$i\hbar\frac{\partial}{\partial t}\Psi = -\sum_i \frac{\hbar^2}{2m_i}\nabla_i^2\Psi + <\chi|V_{n-e}|\chi>\Psi \quad (6.9)$$

$$i\hbar\frac{\partial}{\partial t}\chi = -\sum_I \frac{\hbar^2}{2M_I}\nabla_I^2\chi + <\Psi|H_e|\Psi>\chi, \quad (6.10)$$

where

$$V_{n-e} = +\sum_{i<j}\frac{e^2}{|\mathbf{r}_i - \mathbf{r}_j|} - \sum_{I,i}\frac{e^2 Z_I}{|\mathbf{R}_I - \mathbf{r}_j|} + \sum_{I<J}\frac{e^2 Z_I Z_J}{|\mathbf{R}_I - \mathbf{R}_J|} \quad (6.11)$$

$$\mathcal{H}_e = -\sum_I \frac{\hbar^2}{2M_I}\nabla_I^2 + V_{n-e}(\{\mathbf{r}_i\},\{\mathbf{R}_I\}). \quad (6.12)$$

Equations (6.9) and (6.10) are a set of coupled time-dependent self-consistent (TDSCF) Schrödinger equations, which were introduced as early as 1930s by Paul Dirac.

One can see that Eq. (6.9) is essentially the Schrödinger equation for electrons, and Eq. (6.10) is the Schrödinger equation for nucleus. Taking the approach in quantum hydrodynamics, we can rewrite the nuclear wave function as

$$\chi(\{\mathbf{R}_I\};t) = A(\{\mathbf{R}_I\},t)\exp\left(\frac{i}{\hbar}S(\{\mathbf{R}_I\},t)\right), \quad (6.13)$$

where the function A is the amplitude factor of the wave function and the function S is the phase factor of the wave function. By doing so, the complex wave function χ is now expressed by two real functions, in which A is a position function, i.e., $A > 0$.

Based on the quantum fluid dynamics representation, we can calculate the following quantities,

$$i\hbar\frac{\partial \chi}{\partial t} = i\hbar\frac{\partial A}{\partial t}\exp\left(\frac{i}{\hbar}S\right) - A\exp\left(\frac{i}{\hbar}S\right)\frac{\partial S}{\partial t} \quad (6.14)$$

$$-\frac{\hbar^2}{2M_I}\nabla^2\chi = -\frac{\hbar^2}{2M_I}\left(\nabla^2 A + \frac{i}{\hbar}\nabla A\cdot\nabla S - \frac{1}{\hbar^2}A(\nabla S)^2 + \frac{i}{\hbar}A\nabla^2 S\right)\exp\left(\frac{i}{\hbar}S\right) \quad (6.15)$$

$$<\Psi|H_e|\Psi> \chi = <\Psi|H_e|\Psi> A\exp\left(\frac{i}{\hbar}S\right). \quad (6.16)$$

Substituting Eqs. (6.14)–(6.16) into (6.13) and separating its real and imaginary parts, one may obtain

$$\frac{\partial S}{\partial t} + \sum_I \frac{1}{2M_I}(\nabla_I S)^2 + <\Psi|H_e|\Psi> = \hbar^2\sum_I \frac{1}{2M_I}\frac{\nabla_I^2 A}{A} \quad (6.17)$$

$$\frac{\partial A}{\partial t} + \sum_I \frac{1}{M_I}(\nabla_I A)\cdot(\nabla_I S) + \sum_I \frac{1}{2M_I}A(\nabla_I^2 S) = 0. \quad (6.18)$$

Considering the classical limit $\hbar \to 0$, Eq. (6.17) becomes

$$\frac{\partial S}{\partial t} + \sum_I \frac{1}{2M_I}(\nabla_I S)^2 + <\Psi|H_e|\Psi> = 0. \quad (6.19)$$

Equation (6.19) is an analog to the Hamilton–Jacobi equation in classical Newtonian mechanics, i.e.,

$$\frac{\partial S}{\partial t} + \mathcal{H} = 0,$$

here the analog Hamiltonian is

$$\mathcal{H}(\{\mathbf{R}_I\},\{\mathbf{P}_I\}) \to \sum_I \frac{1}{2M_I}(\nabla_I S)^2 + <\Psi|H_e|\Psi>. \quad (6.20)$$

To conform with the description of classical Newtonian mechanics, we define

$$\mathbf{P}_I = \nabla_I S, \text{ and } V_e^{Eff} := <\Psi|H_e|\Psi>, \quad (6.21)$$

where $\mathbf{P}_I$ is the momentum of the Ith nucleus and V_e^{Eff} is the atomistic potential or potential energy.

Equation (6.19) can be written as

$$\frac{\partial S}{\partial t} + \sum_I \frac{1}{2M_I}(\mathbf{P}_I)^2 + V_e^{Eff}(\{\mathbf{R}_I\}) = 0. \quad (6.22)$$

Taking a gradient operation over Eq. (6.22), it reads as

$$\nabla_I \left(\frac{\partial S}{\partial t} + \sum_J \frac{1}{2M_J}(\mathbf{P}_J)^2 + V_e^{Eff}(\{\mathbf{R}_I\}) \right) = 0$$

$$\rightarrow \frac{\partial \mathbf{P}_I}{\partial t} + \mathbf{V}_I \cdot \nabla_I \otimes \mathbf{P}_I = -\nabla_I V_e(\{\mathbf{R}_I\}), \qquad (6.23)$$

where we used the identity $\frac{\mathbf{P}_I}{M_I} = \mathbf{V}_I$.

To this end, one may realize that Eq. (6.23) is in fact the Newton's second law, i.e., the equation of motion of classical particles,

$$M_I \frac{d^2 \mathbf{R}_I}{dt^2} = -\nabla_I V_e^{Eff}(\{\mathbf{R}_I\}). \qquad (6.24)$$

Now, we have shown that under the classical limit, the wave equation of the nucleus is reduced to the governing equation of the classical particle under the Newton's second law. Equations (6.17)–(6.18) become a set of coupled equations,

$$i\hbar \frac{\partial \Psi}{\partial t} = \left(-\sum_i \frac{\hbar^2}{2m_e} \nabla_i^2 + V_{n-e}(\{\mathbf{r}_i\},\{\mathbf{R}_I(t)\}) \right) \Psi(\{\mathbf{r}_i\},\{\mathbf{R}_I(t)\})$$

$$= \hat{H}_e(\{\mathbf{r}_i\},\{\mathbf{R}_I(t)\}) \Psi(\{\mathbf{r}_i\},\{\mathbf{R}_I(t)\}) \qquad (6.25)$$

and

$$M_I \ddot{\mathbf{R}}_I = -\nabla_I V_e^{Eff}(\{\mathbf{R}_I(t)\}) \qquad (6.26)$$

Equations (6.25)–(6.26) are the governing equations of the Ehrenfest molecular dynamics.

6.2 Ab-Initio Molecular Dynamics

6.2.1 Born–Oppenheimer Molecular Dynamics

In the Ehrenfest MD, the electronic wave function is solved by using a time-dependent Schrödinger equation. If we only consider the ground state of the electronic energy, which is called the Born–Oppenheimer approximation, we can replace the time-dependent Schrödinger equation with the time-independent Schrödinger equation,

$$\left(-\sum_i \frac{\hbar^2}{2m_e} \nabla_i^2 + V_{n-e}(\{\mathbf{r}_i\},\{\mathbf{R}_I\}) \right) \Psi_0(\{\mathbf{r}_i\},\{\mathbf{R}_I\}) = E(\{\mathbf{R}_I\}) \Psi_0(\{\mathbf{r}_i\},\{\mathbf{R}_I\}) \quad (6.27)$$

$$M_I \ddot{\mathbf{R}}_I = -\nabla_I V_e^{Eff}(\{\mathbf{R}_I\}), \qquad (6.28)$$

where $V_e^E := <\Psi_0|H_e|\Psi_0>$. Equations (6.27)–(6.28) are called the Born–Oppenheimer (OP) MD (OPMD). An immediate benefit of OPMD is that one can use DFT to calculate V_e^E.

Figure 6.2 Paul Ehrenfest (18 January 18, 1880–September 25, 1933) was an Austrian and Dutch theoretical physicist, who made major contributions to the field of statistical mechanics and its relations with quantum mechanics, including the theory of phase transition, the Ehrenfest theorem, and Ehrenfest molecular dynamics. The portrait is from the Dibner Library of the History of Science and Technology (reworked) (Photo courtesy of Wikipedia.org)

Recall that in Eq. (6.21) we define

$$V_e^{Eff} := <\Psi|H_e|\Psi>.$$

The next question is how to calculate $\mathbf{F}_I = -\nabla_I V_e^{Eff}$?

6.3 How to Calculate Mechanical Forces in Quantum Mechanical Systems

6.3.1 Hellmann–Feynman Theorem

In quantum mechanics, the Hellmann–Feynman theorem relates the derivative of the total energy with respect to a parameter, to the expectation value of the derivative of the Hamiltonian with respect to that same parameter. According to the theorem, once the spatial distribution of the electrons has been determined by solving the Schrödinger equation, all the forces in the system can be calculated using classical electrostatics.

The theorem has been proven independently by many authors, including Paul Güttinger (1932), Wolfgang Pauli (1933), Hans Hellmann (1937), and Richard Feynman (1939).

The theorem states that

$$\frac{dE}{d\lambda} = <\psi(\lambda)|\hat{H}_\lambda|\psi(\lambda)> \quad \rightarrow \quad \frac{dE}{d\lambda} = \int \psi^*(\lambda) \frac{d\hat{H}_\lambda}{d\lambda} \psi(\lambda)\, dr,$$

where $\hat{H}_\lambda$ is a Hamiltonian operator depending upon a continuous parameter λ, $\psi(\lambda)$ is a wavefunction of the Hamiltonian, depending implicitly upon λ, E is the energy (eigenvalue) of the wavefunction, and dr implies an integration over the domain of the wave function.

Proof Using Dirac's bra–ket notation we have the following two conditions,

$$\hat{H}_\lambda|\psi(\lambda)\rangle = E_\lambda|\psi(\lambda)\rangle \text{ and } \langle\psi(\lambda)|\psi(\lambda)\rangle = 1 \Rightarrow \frac{\partial}{\partial\lambda}\langle\psi(\lambda)|\psi(\lambda)\rangle = 0.$$

The proof then follows through an application of the derivative product rule to the expectation value of the Hamiltonian viewed as a function of λ:

$$\begin{aligned}
\frac{dE_\lambda}{d\lambda} &= \frac{d}{d\lambda}\langle\psi(\lambda)|\hat{H}_\lambda|\psi(\lambda)\rangle \\
&= \left\langle \frac{d\psi(\lambda)}{d\lambda}\middle|\hat{H}_\lambda\middle|\psi(\lambda)\right\rangle + \left\langle\psi(\lambda)\middle|\hat{H}_\lambda\middle|\frac{d\psi(\lambda)}{d\lambda}\right\rangle + \left\langle\psi(\lambda)\middle|\frac{d\hat{H}_\lambda}{d\lambda}\middle|\psi(\lambda)\right\rangle \\
&= E_\lambda\left\langle\frac{d\psi(\lambda)}{d\lambda}\middle|\psi(\lambda)\right\rangle + E_\lambda\left\langle\psi(\lambda)\middle|\frac{d\psi(\lambda)}{d\lambda}\right\rangle + \left\langle\psi(\lambda)\middle|\frac{d\hat{H}_\lambda}{d\lambda}\middle|\psi(\lambda)\right\rangle \\
&= E_\lambda \frac{d}{d\lambda}\left\langle\psi(\lambda)\middle|\psi(\lambda)\right\rangle + \left\langle\psi(\lambda)\middle|\frac{d\hat{H}_\lambda}{d\lambda}\middle|\psi(\lambda)\right\rangle \\
&= \left\langle\psi(\lambda)\middle|\frac{d\hat{H}_\lambda}{d\lambda}\middle|\psi(\lambda)\right\rangle.
\end{aligned}$$

□

6.3.2 Force on Nucleus

The most common application of the Hellmann–Feynman theorem is the calculation of intramolecular forces in molecules. This allows for the calculation of equilibrium geometries of the nuclear coordinates where the forces acting upon the nuclei, due to the electrons and other nuclei, vanish. The parameter λ corresponds to the coordinates of the nuclei. For a molecule with $1 \leq i \leq N$ electrons with coordinates $\{\mathbf{r}_i\}$, and $1 \leq \alpha \leq M$ nuclei, each located at a specified point $\mathbf{R}_\alpha = \{X_{\alpha 1}, X_{\alpha 2}, X_{\alpha 3}\}$ and with nuclear charge Z_α, the clamped nucleus Hamiltonian is

$$\hat{H} = \hat{T} + \hat{V} - \sum_{i=1}^{N}\sum_{\alpha=1}^{M} \frac{Z_\alpha}{|\mathbf{r}_i - \mathbf{R}_\alpha|} + \sum_{\alpha}^{M}\sum_{\beta>\alpha}^{M} \frac{Z_\alpha Z_\beta}{|\mathbf{R}_\alpha - \mathbf{R}_\beta|}.$$

The force acting on the x-component of a given nucleus is equal to the negative of the derivative of the total energy with respect to that coordinate,

$$F = -\frac{\partial E}{\partial X}.$$

The nucleus Hamiltonian is given as

$$\hat{H} = \hat{T} + \hat{V} - \sum_{i=1}^{N}\sum_{\alpha=1}^{M} \frac{Z_\alpha}{|\mathbf{r}_i - \mathbf{R}_\alpha|} + \sum_{\alpha}^{M}\sum_{\beta>\alpha}^{M} \frac{Z_\alpha Z_\beta}{|\mathbf{R}_\alpha - \mathbf{R}_\beta|}.$$

The force acting on the x-component of a given nucleus is equal to the negative of the derivative of the total energy with respect to that coordinate. Employing the Hellmann–Feynman theorem this is equal to

$$F_{X_\gamma} = -\frac{\partial E}{\partial X_\gamma} = -\left\langle \psi \left| \frac{\partial \hat{H}}{\partial X_\gamma} \right| \psi \right\rangle.$$

Only two components of the Hamiltonian contribute to the required derivative the electron–nucleus and nucleus–nucleus terms. Differentiating the Hamiltonian yields

$$\frac{\partial <\hat{H}>}{\partial X_\gamma} = \frac{\partial}{\partial X_\gamma}\left(-\sum_{i=1}^{N}\sum_{\alpha=1}^{M} \frac{Z_\alpha}{|\mathbf{r}_i - \mathbf{R}_\alpha|} + \sum_{\alpha}^{M}\sum_{\beta>\alpha}^{M} \frac{Z_\alpha Z_\beta}{|\mathbf{R}_\alpha - \mathbf{R}_\beta|}\right),$$

$$= Z_\gamma \sum_{i=1}^{N} \frac{x_i - X_\gamma}{|\mathbf{r}_i - \mathbf{R}_\gamma|^3} - Z_\gamma \sum_{\alpha \neq \gamma}^{M} Z_\alpha \frac{X_\alpha - X_\gamma}{|\mathbf{R}_\alpha - \mathbf{R}_\gamma|^3}.$$

Only two components of the Hamiltonian contribute to the required derivative – the electron–nucleus and nucleus–nucleus terms. Differentiating the Hamiltonian yields

$$\frac{\partial <\hat{H}>}{\partial X_\gamma} = \frac{\partial}{\partial X_\gamma}\left(-\sum_{i=1}^{N}\sum_{\alpha=1}^{M} \frac{Z_\alpha}{|\mathbf{r}_i - \mathbf{R}_\alpha|} + \sum_{\alpha}^{M}\sum_{\beta>\alpha}^{M} \frac{Z_\alpha Z_\beta}{|\mathbf{R}_\alpha - \mathbf{R}_\beta|}\right),$$

$$= Z_\gamma \sum_{i=1}^{N} \frac{x_i - X_\gamma}{|\mathbf{r}_i - \mathbf{R}_\gamma|^3} - Z_\gamma \sum_{\alpha \neq \gamma}^{M} Z_\alpha \frac{X_\alpha - X_\gamma}{|\mathbf{R}_\alpha - \mathbf{R}_\gamma|^3}.$$

Insertion of this in to the Hellmann–Feynman theorem returns the force on the x-component of the given nucleus in terms of the electronic density $\rho(\mathbf{r})$ and the atomic coordinates and nuclear charges:

$$F_{X_\gamma} = -Z_\gamma \left(\int d\mathbf{r}\, \rho(\mathbf{r}) \frac{x - X_\gamma}{|\mathbf{r} - \mathbf{R}_\gamma|^3} - \sum_{\alpha \neq \gamma}^{M} Z_\alpha \frac{X_\alpha - X_\gamma}{|\mathbf{R}_\alpha - \mathbf{R}_\gamma|^3}\right). \quad (6.29)$$

6.3.3 Car–Parrinello Molecular Dynamics

In 1985, Car and Parrinello proposed the following first-principles Lagrangian

$$\mathcal{L}_{cp} = \frac{1}{2}\left(\sum_{I}^{\text{nuclei}} M_I \dot{\mathbf{R}}_I^2 + \mu \sum_{i}^{\text{orbitals}} \int d\mathbf{r}\, |\dot{\psi}_i(\mathbf{r},t)|^2 \right) - E_{KH}\left[\{\psi_i\},\{\mathbf{R}_I\}\right], \quad (6.30)$$

where $E[\{\psi_i\},\{\mathbf{R}_I\}] = E[\rho]$ is the Kohn–Sham energy density functional, which outputs energy values when Kohn–Sham orbitals and nuclear positions are given.

It has additional orthogonality constraints,

$$\int d\mathbf{r}\, \psi_i^*(\mathbf{r},t)\psi_j(\mathbf{r},t) = \delta_{ij} \quad \rightarrow \quad \sum_{i,j} \Lambda_{ij}\left(<\psi_i|\psi_j> -\delta_{ij}\right),$$

where δ_{ij} is the Kronecker delta.

The equations of motion are obtained by finding the stationary point of the Lagrangian functional under variations of ψ_i and $\mathbf{R}_I$, i.e., the use of Hamiltonian principle. Precisely speaking, by using the stationary variation condition,

$$\delta \int_0^t \left[\mathcal{L}_{cp}(\{\mathbf{R}_I\}, \{\dot{\mathbf{R}}_I\}, \{\psi_i\}, \{\dot{\psi}_i\}) + \sum_{i,j} \Lambda_{ij}\left(<\psi_i|\psi_j> -\delta_{ij}\right)\right] dt = 0$$

one can derive that

$$M_I \ddot{\mathbf{R}}_I = -\nabla_I E\left[\{\psi_i\}, \{\mathbf{R}_J\}\right] \quad (6.31)$$

$$\mu \ddot{\psi}_i(\mathbf{r},t) = -\frac{\delta E}{\delta \psi_i^*(\mathbf{r},t)} + \sum_j \Lambda_{ij} \psi_j(\mathbf{r},t), \quad (6.32)$$

where Λ_{ij} is a Lagrangian multiplier matrix to comply with the orthonormal constraint.

One of the striking features of the Car–Parrinello molecular dynamics (CPMD) is that both its equations of motions for electrons and atoms (nucleus) have accelerations, i.e., the second-order time derivative terms. This is in contrast with the Schrödinger equation. When $\mu \to 0$, CPMD degenerates to the Born–Oppenheimer MD.

6.4 Classical Molecular Dynamics

Discussed in the previous sections, as a good approximation, we can model the atom motion by focusing on the motions of their nucleuses, and the interaction among electrons and nucleuses can be described by the *internal* force field. Because the nucleuses of the atoms may be approximated as the coarse-grained rigid particles, the motion or the dynamics of an atomistic system in many-body quantum mechanics may be simplified as the many-body dynamics in classical Newtonian mechanics.

To describe their motions, we may first set a Cartesian coordinate, so that we can make the position of each atom at any time instance, say t. Assume that in an atomistic system that has N atoms. We denote the position of the representative atom i as $\mathbf{r}_i, i = 1, 2, \ldots N$ as shown in Figure 6.3. The positions of all other adjacent atoms in the system are denoted as $\mathbf{r}_j, j \neq i$. Thus we may define the relative position vectors as $\mathbf{r}_{ij} = \mathbf{r}_j - \mathbf{r}_i$.

The basic governing equation of classical MD can now be written as

$$m_i \frac{\partial^2 \mathbf{r}_i}{\partial t^2} = \mathbf{F}_i, \ i = 1, 2, \ldots, N, \quad (6.33)$$

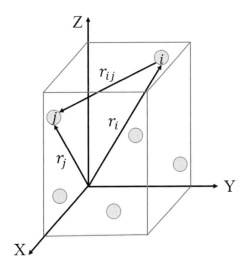

Figure 6.3 Atoms as classical particles and their relative positions

where $\mathbf{F}_i$ is the force acting on the atom i by the rest of atoms $j = 1, 2, \ldots \neq i, N$, which are the source of the internal force, even though there may be as possible external forces as well. However, at the atomistic scale, we mainly focus on atomistic interactions, i.e., the internal force. Therefore, in the rest of this book, we treat $\mathbf{F}_i$ as the internal force unless it is otherwise noted.

Since $\mathbf{F}_i$ is the internal force of the system, and it should be a conservative force. Thus, in physical principle, we can write

$$\mathbf{F}_i = -\frac{\partial V}{\partial \mathbf{r}_i}, \tag{6.34}$$

where V is a potential function that depends on the interatomic position, i.e., $r_{ij} = |\mathbf{r}_{ij}|$ and their orientations. The parameters that are used in the atomistic potential are often characterized by quantum or first-principles calculations, and they determine the *force field*.

In general, we may symbolically express the many-body atomistic potential depending on how many numbers of atoms interact with each other, i.e.,

$$V(\mathbf{r}^N) = \sum_i V_1(\mathbf{r}_i) + \sum_i \sum_{j>i} V_2(\mathbf{r}_i, \mathbf{r}_j) + \sum_{i, j>i, k>j} V_3(\mathbf{r}_i, \mathbf{r}_j, \mathbf{r}_k) + \cdots, \tag{6.35}$$

where in the notation $\mathbf{r}^N = (\mathbf{r}_1, \mathbf{r}_2, \ldots, \mathbf{r}_N)$.

In literature, $V_2(\mathbf{r}_i, \mathbf{r}_j) = V(r_{ij})$ is often called as pair potential, and $V(\mathbf{r}_i, \mathbf{r}_j, \mathbf{r}_k)$ is the three-body potential, and so forth. Figure 6.4 illustrates the interaction patterns in both pair potentials (a) and three-body potentials (b). One may find that in three-body potential, not only the relative positions, i.e., r_{ij}, r_{ik}, and r_{jk} matter, but also depend on the relative orientations or angles as well (see Figure 6.4(b)). For example, we have a special form of the three-body potential that may be expressed as

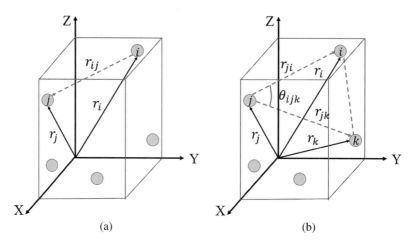

Figure 6.4 Interaction among different atoms: (a) pair interaction and (b) three atom interaction

$$V_3(\mathbf{r}_i, \mathbf{r}_j, \mathbf{r}_k) = V(\cos\theta_{ijk}), \quad \cos\theta_{ijk} = \frac{\mathbf{r}_{ji} \cdot \mathbf{r}_{jk}}{|\mathbf{r}_{ji}||\mathbf{r}_{jk}|}. \quad (6.36)$$

In actual atomistic modeling and simulations, we may have many-body potentials that involve number of atoms more than three or even a dozen, which will be discussed in later chapters.

For pedagogical and learning purpose, we first illustrate how to calculate interatomic forces from pair potentials. If we only consider the pair potential, the total potential in an N-atom system may be written as,

$$V(\mathbf{r}^N) = \sum_i \sum_{j>i} V(r_{ij}) = \frac{1}{2} \cdot \sum_i \sum_j V(r_{ij}). \quad (6.37)$$

The factor $1/2$ is because of the double counting of the part bonds. Thus,

$$\mathbf{F}_i = \sum_{j \neq i} \mathbf{F}_{ji} = -\frac{\partial V}{\partial \mathbf{r}_i} = \sum_{j \neq i} V'(r_{ij}) \frac{\mathbf{r}_{ij}}{r_{ij}} = -\sum_{j \neq i} V'(r_{ij}) \frac{\mathbf{r}_{ji}}{r_{ji}}, \quad (6.38)$$

where $\mathbf{F}_{ji}$ is the force from the jth atom to the ith atom, and

$$\mathbf{F}_{ji} = -V'(r_{ij}) \frac{\mathbf{r}_{ji}}{r_{ji}}, \quad (6.39)$$

where $\mathbf{r}_{ij} = \mathbf{r}_j - \mathbf{r}_i$.

In the above derivation, we use the chain rule, i.e.,

$$\mathbf{F}_{ji} = -\frac{\partial}{\partial \mathbf{r}_i} V(r_{ij}) = -\frac{\partial V}{\partial r_{ij}} \frac{\partial r_{ij}}{\partial \mathbf{r}_i} = -V'(r_{ij}) \frac{\mathbf{r}_{ji}}{r_{ij}} = V'(r_{ij}) \frac{\mathbf{r}_{ij}}{r_{ij}}. \quad (6.40)$$

Considering the fact that

$$r_{ij}^2 = (\mathbf{r}_j - \mathbf{r}_i) \cdot (\mathbf{r}_j - \mathbf{r}_i) \quad \rightarrow \quad \frac{\partial r_{ij}}{\partial \mathbf{r}_i} = \frac{\mathbf{r}_i - \mathbf{r}_j}{r_{ij}} = \frac{\mathbf{r}_{ji}}{r_{ij}},$$

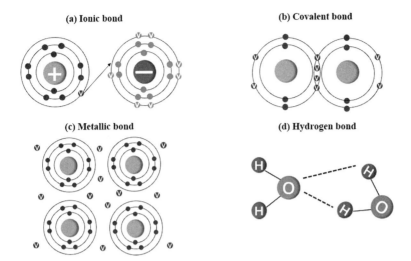

Figure 6.5 Different atomistic bonding types: (a) ionic bonds, (b) covalent bond, (c) metallic bond, and (d) hydrogen bond

and summing the force contribution from all the adjacent atoms, we have the desired results (6.39) (see: Fig. 6.4(a)).

In some literatures, you may see the expression like

$$\mathbf{F}_i = -\frac{\partial V}{\partial \mathbf{r}_i} = -\sum_{j \neq i} V'(r_{ij}) \frac{\mathbf{r}_{ij}}{r_{ij}} = \sum_{j \neq i} V'(r_{ij}) \frac{\mathbf{r}_{ji}}{r_{ij}},$$

where a different definition is used $\mathbf{r}_{ij} = \mathbf{r}_i - \mathbf{r}_j$. The advantage of such expression is that all the forces actually point to atom i. Whereas in this book, we follow the standard notation.

6.5 Examples of Atomistic Potentials

The next question is: What is the atomistic potential or how do we construct the atomistic potential?

In the current knowledge of physics, the universe has four forces: gravitational force, electromagnetic force, weak nuclear force, and strong nuclear force. The interatomic force among atoms are *electromagnetic force*, which depends on the atomistic or chemical bonding types among atoms and long-range intermolecular force and electrostatic force. In chemistry and physics, different atoms and molecules have different types of bonding structures, such as ionic bond, covalent bond, metallic bond, hydrogen bond, the van der Waals force interaction, and so forth (see: Fig. 6.5 and Fig. 6.6), which result in different types of atomistic potentials. Many atomistic potentials used in actual computational chemistry or computational materials modeling are fitted with the data obtained from first-principles or ab-initio MD calculations

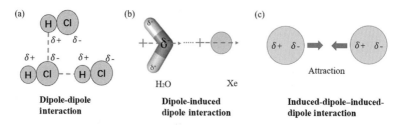

Figure 6.6 The van der Waals forces: (a) the Keesom force, (b) the Debye force, and (c) the London (dispersion) force

or electronic structure calculations by using DFT, Car–Parrinello MD, and other quantum mechanics calculations. However, there are several generic types of atomistic potentials that are suitable for a group-specific bonds. For instance, the embedded atom method (EAM) is specifically designed for metallic bonds, and the many-body Stillinger–Weber potential is for the materials with covalent bonds, and so forth. The construction of most many-body potentials are based on geometric configurations of an atom cluster. For example, the three-body atomistic potential has the generic form,

$$V_3(r_{ij}, r_{ik}, r_{jk}) = h(r_{ij}, r_{ik}, \theta_{jik}) + h(r_{ji}, r_{jk}, \theta_{ijk}) + h(r_{ki}, r_{kj}, \theta_{ikj}), \qquad (6.41)$$

where θ_{ijk} is the angle defined in Eq. (6.36). A simple example of h function may be given as follows:

$$h(r_{ij}, r_{ik}, \theta_{jik}) = \lambda \exp\left[\frac{\gamma}{r_{ij} - r_c} + \frac{\gamma}{r_{ik} - r_c}\right] (\cos \theta_{jik} - \beta)^2, \qquad (6.42)$$

where r_c is a cutoff distance.

More flexibly, the so-called bond-order potentials have been developed for group IV elements (S_i, G_e, C) based on the concept of bond strength dependence on local environment, such as the Tersoff potential for S_i and G_e, the Brenner potential for C, and so forth. We shall discuss them separately in later chapters, because of the specific applications.

6.5.1 Lennard-Jones Potential

The Lennard-Jones potential is designed for intermolecular interactions, mainly the van der Waals interaction, which is a long-range weak interaction among different molecules and atoms. There are three types of van der Waals interactions: (1) dipole–dipole interaction, (2) dipole–induced dipole interaction, and (3) induced dipole–induced dipole interaction. In 1930s, Fritz London conducted a quantum mechanics calculation of the dispersion force due to the induced dipole–induced dipole interaction, and he found that the so-called London force (a type of van der Waals forces) is proportional to r^{-6}, where r is the distance between two atoms or molecules. In Table 6.1, we list all three van der Waals potentials, where ε_o is the permittivity of free space, ε_r is the dielectric constant of surrounding material, T is the temperature, k_b is the Boltzmann constant, and r is distance between two molecules. In Table 6.1, $m_i, i = 1, 2$ are dipole moment for corresponding atoms,

Table 6.1 The van der Waals forces

Interaction	Origin of interactions	Equation
Keesom	Dipole–dipole interaction	$V(r) = \dfrac{-m_1^2 m_2^2}{24\pi^2 \epsilon_0^2 \epsilon_r^2 k_b T r^6}$
Debye	Dipole–induced dipole interaction	$V(r) = \dfrac{-m_1^2 \alpha_2}{16\pi^2 \epsilon_0^2 \epsilon_r^2 r^6}$
London (Dispersion)	Induced dipole–induced dipole interaction	$V(r) = -\dfrac{3}{2}\dfrac{\alpha_1 \alpha_2 I_1 I_2}{((4\pi\epsilon_0)^2 (I_1 + I_2) r^6}$

$\alpha_i, i = 1, 2$ are dipole polarizability for corresponding atoms, and $I_i, i = 1, 2$ are the first ionization potential for corresponding atoms. A representative atomistic potential of this type is the Lennard-Jones (LJ) potential, which was proposed by Sir John Edward Lennard-Jones (see Fig. 6.7), and was named after him. It is often expressed in the following form:

$$V(r) = 4\epsilon \left[\left(\frac{\sigma}{r}\right)^{12} - \left(\frac{\sigma}{r}\right)^6 \right] = \epsilon \left[\left(\frac{r_0}{r}\right)^{12} - 2\left(\frac{r_0}{r}\right)^6 \right]. \quad (6.43)$$

where ϵ is the depth of the potential well, σ is the (finite) distance at which the inter-particle potential is zero, and r is the distance between the particles. These parameters can be fitted to reproduce experimental data or accurate quantum chemistry calculations. The r^{-12} term describes the Pauli repulsion at short ranges due to overlapping electron orbitals and the r^{-6} term describes attraction at long ranges due to van der Waals force (see Table 6.1), or dispersion force as shown in Fig. 6.8.

As one can see from Eq. (6.43), when $r = \sigma$, $V(\sigma) = 0$. So $r = \sigma$ is the position with separate positive potential range with the negative potential range. To find the equilibrium distance, we set

$$F(r_0) = -\frac{dV}{dr}\bigg|_{r=r_0} = 0 \quad r_0 = 2^{1/6}\sigma \rightarrow V(r_0) = V_{min} = -\epsilon,$$

that is why ϵ is the depth of potential well and r_0 also separates the repulsion area with the attraction area. To find the maximum interaction force, we use the following expressions:

$$\frac{d^2 V}{dr^2}\bigg|_{r=r_1} = 0 \rightarrow r_1 = (26/7)^{1/6}\sigma \approx 2r_0 \rightarrow F(r_1) = F_{max} = \frac{2436\epsilon}{169\sigma}.$$

As $r \rightarrow 0$, $V(r) \sim r^{-12} \rightarrow +\infty$. In fact, the potential approaches to infinity so fast, seldom an atom can penetrate to the region $r < 0.9\sigma$. Therefore, the LJ potential can be viewed having a "hard core." It may be noted that although the LJ potential originated from the modeling of the van der Waals interactions, e.g., the induced dipole–induced dipole interaction with a characteristic $r^{-1/6}$ term, it is also used in calculations of other types of atomistic interactions, such in metallic bonds as well as covalent bonds. In those cases, the LJ potential parameters becomes simply a "best

Sir John Edward Lennard-Jones (1894–1954)

Figure 6.7 John Edward Lennard-Jones was a British mathematician and was a professor of theoretical physics at University of Bristol, and then of theoretical science at the University of Cambridge. Lennard-Jones studied math at Manchester, then joined the Royal Flying Corp. He flew during World War-I. Following the war, he completed doctorate in Cambridge, and also he gave us the most recognized atomistic potential expression for molecular forces. It was later called the Lennard-Jones potential. That was after he had married Kathleen Lennard and added her name to his own. (The copyright of the portrait is from Computer Laboratory, University of Cambridge. Reproduced by permission) (Photo courtesy of Wikipedia.org)

fit" to the actual potential energy curve. The depth of the potential is then related to the dissociation energy of the bond. It is fair to say that in computational chemistry and computational materials science, the LJ potential is a primary atomistic potential used in modeling and simulations.

6.5.2 Cutoff Distance and Cutoff Method

Ideally, every atom should interact with every other atom in the system. However, this creates a force calculation algorithm of quadratic order of N, i.e., $O(N^2)$. We may be able to ignore atoms at large distances from each other without suffering too much loss of accuracy but with only computation effort of order $O(N)$. To do so, we ignore the atoms with the distance larger than a *cutoff* distance from the atom where the force is being calculated or exerted.

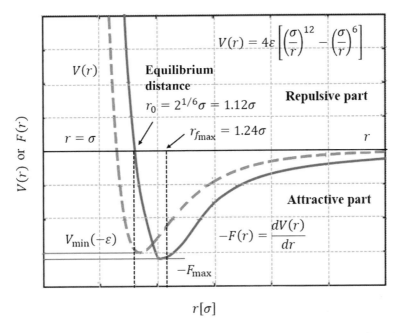

Figure 6.8 The Lennard-Jones potential (dotted line) and the corresponding force between two particles (the solid line)

We demonstrate the cutoff method through the LJ potential. Considering the generic form of the LJ potential,

$$V_{LJ}(r) = 4\epsilon \left[\left(\frac{\sigma}{r}\right)^{12} - \left(\frac{\sigma}{r}\right)^{6} \right],$$

we may adopt the following truncated LJ potential,

$$V_t(r) = \begin{cases} V^{LJ}(r), & r \leq r_c \\ 0, & r > r_c \end{cases}, \quad (6.44)$$

where r_c is the cutoff distance. The problem with the truncated LJ potential (6.44) is that the potential is discontinuous at $r = r_c$, which has the issue of evaluating force at $r = r_c$. In computation practice, a so-called truncated and shifted LJ potential is used, which is defined as

$$V_{ts}(r) = \begin{cases} V^{LJ}(r) - V_{LJ}(r_c), & r \leq r_c \\ 0, & r > r_c \end{cases}. \quad (6.45)$$

Obviously, Eq. (6.45) is continuous at $r = r_c$. To make the force continuous, one can adopt the following shifted potential,

$$V_{shift}(r) = \begin{cases} V_{LJ}(r) - V_{LJ}(r_c) - (r - r_c)V'_{LJ}(r_c), & r \leq r_c \\ 0 & r > r_c \end{cases}. \quad (6.46)$$

The truncated and shifted LJ potentials are displayed in Fig. 6.9.

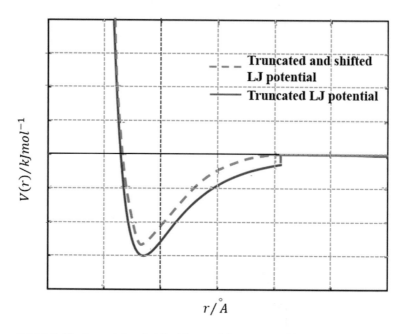

Figure 6.9 The truncated and shifted Lennard-Jones potentials

6.5.3 Some Commons on Atomistic Potentials

There are several other well-known atomistic potentials that are either important during the development of MD or still extensively used in MD.

1. Hard Sphere Model

The hard sphere model was the first atomistic potential used in MD modeling and simulation. In this model, the particle is not a zero-volume particle, but a finite volume rigid sphere. Assume that the radius of the hard sphere is σ, the hard sphere model adopts the following pairwise interaction potential,

$$V(\mathbf{r}_1, \mathbf{r}_2) = \begin{cases} 0 & \text{if } |\mathbf{r}_1 - \mathbf{r}_2| \geq \sigma \\ \infty & \text{if } |\mathbf{r}_1 - \mathbf{r}_2| < \sigma \end{cases}. \quad (6.47)$$

This potential is still used in some scientific and engineering modeling and simulation today. In fact, the hard sphere model is the primary prototype of discrete element method (DEM), which is extensively used in granular material modeling and simulation.

2. Soft Sphere Model

The soft sphere potential may be defined as

$$V_{soft}(r) = \begin{cases} \epsilon \left(\dfrac{\sigma}{r}\right)^n & ; \quad r \leq \sigma \\ 0 & ; \quad r > \sigma \end{cases},$$

where $V_{soft}(r)$ is the intermolecular pair potential between two soft spheres separated by a distance $r := |\mathbf{r}_1 - \mathbf{r}_2|$. For the soft sphere model, $r := |\mathbf{r}_1 - \mathbf{r}_2| < \sigma_1 + \sigma_2$,

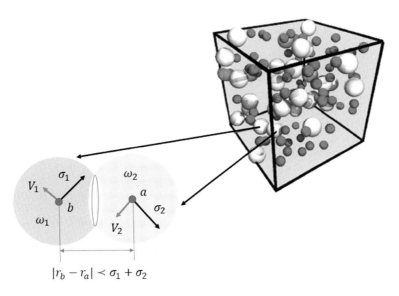

Figure 6.10 The soft sphere model

where $\sigma_i, i = 1, 2$ are the radius of the two spheres, ϵ is the interaction strength, and σ is the radius of the sphere. Frequently the value of n is taken to be 12, thus the model effectively becomes the high temperature limit of the LJ model. If $n \to \infty$ one recovers the hard sphere model. For $n \leq 3$, the material model is not thermodynamically stable. An illustration of the soft sphere MD model is shown in Fig. 6.10.

3. Coulomb Potential

The Coulomb potential is mainly used to model electrostatic interactions. For point charge interaction, it is expressed as

$$V_C(r) = \frac{Q_1 Q_2}{4\pi\epsilon_0 r}. \tag{6.48}$$

The Coulomb potential is often used when atoms are inside the liquids, or when electrostatic charge are present.

4. Morse Potential

The simplest bonded potential is the harmonic potential that is a quadratic well, which is named after the physicist Philip M. Morse; it can capture effects of bond breaking during molecular vibrations.[1] Thus, the Morse potential is suitable for modeling vibrations of diatomic molecules, because it yields a harmonic oscillator solution with anharmonic corrections, i.e., the Morse potential accounts for the effects of bond dissociation, making it more useful than the Harmonic potential. The typical form of the Morse potential may be written as

---

[1] By the way, the LJ potential is a nonbonded potential.

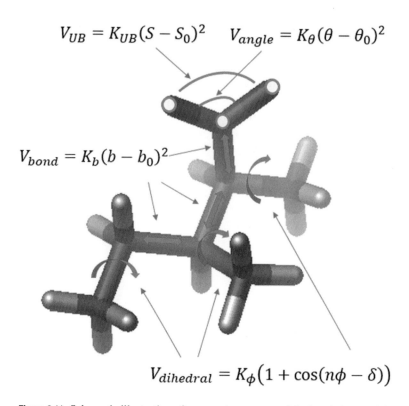

Figure 6.11 Schematic illustration of geometric structure of the bonded potentials

$$V_M(r) = \epsilon \left[1 - \exp(-\beta(r - r_e))\right]^2, \quad (6.49)$$

where $\epsilon = V(r \to \infty)$ is the well depth, r_0 is the equilibrium distance is the equilibrium distance, because both $V(r_e) = 0, F(r_e) = -V'(r_e) = 0$, and the parameter β controls the width of the potential well.

Since the zero location of the potential is arbitrary, the potential energy of the Morse potential can be rewritten by adding or subtracting a constant value. Commonly, the energy zero may be redefined so that the Morse potential may be rewritten as

$$V(r) = \epsilon \left((1 - \exp^{-\beta(r-r_e)})^2 - 1\right) = \epsilon \left(\exp(-2\beta(r - r_e)) - \exp(-\beta(r - r_e))\right).$$

This form of the Morse potential approaches zero when $r \to \infty$, and equals $-\epsilon$ at its minimum, when $r = r_e$. It clearly shows that the Morse potential is the combination of a short-range repulsion term (the first term) and a long-range attractive term (the second term), making it analogous to the LJ potential.

6.5.4 Semiempirical Force Fields of Molecular Dynamics

Most of the potentials used in large-scale simulations of organic and bio-organic systems (proteins, polymers, etc.) are based on ideas similar to what discussed for

Stillinger–Weber potential, i.e., the potential is defined through geometrical parameter fittings, such as bond lengths and bond angles. The potentials for organic systems are typically much more complex as compared to Stillinger–Weber potential, and they have many more parameters that are fitted based on experimental data or first-principles calculations. The approach is to set up a few intrinsic bond types based on geometry of the bond structures, and then use either experimental data or first-principles calculations to fit or calibrate the parameters, and the interatomic force generated by such approach is generally referred to as the *force field*. Typically, the total potential energy consists of the following parts:

$$V = V_{bonds} + V_{angle} + V_{dihedral} + V_{nonbonded},$$

in which the part of the nonbonded potential has two parts:

$$V_{nonbonded} = V_{electrostatic} + V_{van\ der\ Waals}.$$

For example, the generic form of CHARMM (Chemistry at Harvard Macromolecular Mechanics) potential energy equation has the following form:

$$V(\mathbf{r}) = \sum_{bonds} \frac{K_d}{2}(b-b_0)^2 + \sum_{UB}(S-S_0)^2 + \sum_{angle} K_\theta(\theta-\theta_0)^2$$

$$+ \sum_{dihedrals} K_\chi(1+\cos(n\chi - \delta)) + \sum_{impropers} K_{impropers}(\phi - \phi_0)^2$$

$$+ \sum_{\substack{nonbond \\ van\ der\ Waals}} 4\epsilon_{ij}\left[\left(\frac{\sigma_{ij}}{r_{ij}}\right)^{12} - \left(\frac{\sigma_{ij}}{r_{ij}}\right)^{6} + \frac{q_i q_j}{\epsilon_1 r_{ij}}\right], \quad (6.50)$$

where $K_b, K_{UB}, K_\theta, K_\chi, K_{imp}$ are force field constants, b is the bond length, b_0 is the equilibrium bond length, UB refers to the stretch part of the Urey–Bradley bond, S is the stretch of 1,3 atom pair, and S_0 is the equilibrium distance of UB 1–3 pair bond. The bending part of the Urey–Bradley bond is labeled as *angle*, and θ is the angle value. θ_0 is the equilibrium angle value, χ is the dihedral angle value, n is the periodicity, ϕ is the improper angle value, and ϕ_0 is the equilibrium or ideal improper angle value. Since the bonded potential is based on the change of geometric position or kinematics of molecular structures, we illustrate the geometric structures of these bonds in Fig. 6.11. Among these bonds, the bond structure that is most difficult to understand is the so-called dihedral bond, which is essentially a torsional bond. However, it has two variances: dihedral bonds and improper dihedral bonds. In both cases, the torsional angle is the angle between two atomic planes. Nevertheless, in former case, the two planes share two atoms; and in latter case, the two planes share only one atom, as shown in Fig. 6.12.

For nonbond potential energies, ϵ is the LJ well depth, σ_{ij} is the zero potential position, r_{ij} is the distance between atom i and atom j, q_i and q_j are the atoms' charges, and ϵ_1 is the effective dielectric constant.

There are a dozen open source and commercial MD software packages available, which equip various useful atomistic potentials. In Table 6.2, we list several most

Table 6.2 Some common molecular dynamics softwares

Molecular dynamics software		
GROMACS	Free, open source	www.gromacs.org/
LAMMPS	Free, open source	http://lammps.sandia.gov/
NAMP+VMD	Free, open source	www.ks.uiuc.edu/Research/vmd/
MDynaMix	Free, open source	www.fos.su.se/~sasha/mdynamix/
CHARMM	Proprietary, commercial	www.charmm.org/

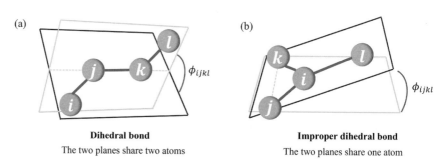

Figure 6.12 Schematic illustration of dihedral bonds: (a) dihedral bonds and (b) improper dihedral bonds

popular MD software, readers can easily download them to your own computer and follow the instruction to install and use them to solve various scientific and engineering problems.

6.5.5 Reduced Unit

In MD simulations, it is often convenient to use a so-called reduced unit. The reduced unit is an extension of the reduced properties in thermodynamics to numerical calculations. In thermodynamics, the reduced properties of a material are a set of state variables normalized by the material's state properties at its critical point. These dimensionless thermodynamic properties provide the basis for the simplest form of the theorem of corresponding states.

While in MD, if we choose the characteristic length scale, timescale, and energy scale to normalize all other dependent and independent variables, we can express both potential energy and equation of motion in the simplest and dimensionless form, which can reduce the round-off numerical error and enhance computational stability. This is because by doing so one synchronizes or scales the numerical computation to the correct temporal and spatial regions of the actual physical process and avoids the numerical pollution-induced artificial units. For example, if a physical event lasts only for a few nanoseconds, while the time unit used in the computation to simulate the event is light year. you would not expect a convenient computation, if though in principle there is nothing wrong with it.

Table 6.3 A list of reduced units

Variable	Reduced units	Relation to SI units
Length	r^*	r/σ
Volume	V^*	V/σ^3
Density	ρ^*	$N/V^* = N/(V/\sigma^3) = \rho\sigma^3$
Energy	E^*	E/ϵ
Temperature	T^*	$T/(\epsilon/k_B)$
Time	t^*	$t/(\sigma\sqrt{M/\epsilon})$
Pressure	P^*	$P/(\epsilon/\sigma^3) = P\sigma^3/\epsilon$
Force	F^*	$F/(\epsilon/\sigma) = F\sigma/\epsilon$
Surface tension	γ^*	$\gamma/(\epsilon/\sigma^2) = \gamma\sigma^2/\epsilon$

In the literature, people often refer the reduced unit as the application of the principle of corresponding state. This is because that one set of the computation or simulation results can actually correspond to many different simulations of different materials. However, one simulation is equivalent to one state defined by a set of fixed characteristic state variable values or scales. There is no direct way to extrapolate the results of one simulation to other states. Moreover, the principle of corresponding states is only an approximation and works best for gases composed of spherical particles.

In the following, we illustrate how to use the reduced unit to convert all computation variables into dimensionless forms. For example, for a fixed LJ potential of certain van der Waals interaction,

$$V(r) = 4\epsilon\left[\left(\frac{\sigma}{r}\right)^{12} - \left(\frac{\sigma}{r}\right)^6\right]. \tag{6.51}$$

In this case, we can select the following five characteristic scales:

length scale: σ, energy scale: ϵ, temperature scale: ϵ/k_B, and timescale: $\sigma\sqrt{M/\epsilon}$.

and convert all other dependent and independent variables into dimensionless forms, as shown in Table 6.3.

Once we convert all the variables into the reduced unit, Eq. (6.51) becomes the following form:

$$V^*(r^*) = 4\left[\left(\frac{1}{r^*}\right)^{12} - \left(\frac{1}{r^*}\right)^6\right], \tag{6.52}$$

and this is the form that we shall use in the MD simulations.

Example 6.1 How Much Is the Real Temperature If the Reduced Unit Temperature is 2? Assume that the MD simulation is conducted for Argon. The reduced unit can be converted into SI unit based on the following formula.

$$T^* = k_B T/\epsilon \;\;\Rightarrow\;\; T = T^*\epsilon/k_B.$$

For Argon,
$$\epsilon/k_B = 120 \Rightarrow T = 2 \times 120 = 240K,$$
here we used
$$k_B = 1.3806488 \times 10^{-23} J K^{-1}, \text{ or } 8.6173324 \times 10^{-5} eV K^{-1}.$$

Example 6.2 How Much Is the Real-Time Step If the Reduced Unit Time Step is: 0.0032? The reduced unit can be converted into SI unit based on the following formula:
$$t^* = t\sigma^{-1}\sqrt{\epsilon/M} \rightarrow t = t^*\sigma\sqrt{M/\epsilon} = 0.0032 \times 0.34 \times 10^{-9}.$$
For Argon,
$$\epsilon = 120 k_B = 165.72 \times 10^{-23} J K^{-1}; M = 0.03994 \, kg/mol, \, 1mol = 6.02214 \times 10^{23}.$$
Thus,
$$\Delta t = 0.69 fs.$$

6.6 Periodic Boundary Condition

There is a major difference between computational materials sciences and computational nanomechanics. One of the objectives of the nanomechanics is to study the mechanics of a particular material microstructure at nanoscale, whereas the main objective of computational materials science is to model and simulate the material properties at atomistic scale. In order to exclude the size effect, we always assume that the material under simulation occupies the entire material space, and on the other hand, however, we are not able to simulate the entire material space because of the lacking of computational power.

To resolve the issue, in computational materials science as well as computational chemistry, the so-called *periodic boundary condition* is extensively and almost ubiquitously used. By doing so, even if we only simulate the material in a small unit cell, at certain level we may believe that the simulation results may reflect the bulk material properties. To impose the periodic boundary condition, we assume that the entire simulation space consists many periodic cells that are identical with the original unit cell. In Fig. 6.13, we show an example of two-dimensional (2D) periodically distributed unit cells. If the interaction of cutoff r_{cut} between two atoms is much smaller than the cell size L, we only need to consider the eight adjacent cells to the original unit cell as shown in Fig. 6.13. In the three-dimensional cases, to consider the periodic boundary condition, we need to consider the adjacent 26 cells in which all the atom configurations are identical to the original unit cell.

There are several features of the periodic boundary: (1) As shown in Fig. 6.13, if an atom at the upper-right corner of the original unit cell (shaped square) is about to leave the original unit cell, then another atom in the adjacent cell number (VII) is about to

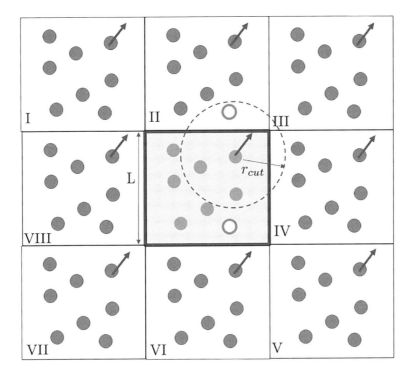

Figure 6.13 Schematic illustration of two-dimensional periodic unit cell distribution and the periodic boundary condition: (1) the number of atoms in the original simulation cell is conserved and (2) in calculating interactions within the cutoff range, the information of an image of a neighboring atom can be found in a real atom that is its identical twin (the light-colored atoms)

enter the original unit cell, so that we can always maintain a fixed number of atoms in the original unit cell. This is important, because the number of atoms in some of the most common equilibrium ensemble MD simulations, e.g., NVE, NVT, NPT, and the like, are always fixed. (2) If we want to calculate the interaction of atoms with the atom in the upper right corner of the unit cell that is about to leave, we have to consider all the atoms inside its cutoff region. By examining Fig. 6.13, one may find that there is an atom (light color) in the adjacent cell number (II), which is outside of the original unit cell. However, because of the periodic boundary condition, there is an atom inside the original unit cell that is exactly the same as the light-colored atom in the cell (II). Therefore, we can calculate the interaction between the light-colored atom and the atom that is about to leave by using its copy in the original unit cell, although we may have to "shift" it to the adjacent cell number (II) in order to keep the distance of the two atom the same. By doing so, at any instance of the time, we only need to calculate trajectories of all atoms inside the original unit cell, no more and no less. This is the so-called periodic boundary condition. Because of this, we usually call the atoms in the original unit cell as the *real atoms*, whereas the atoms in the adjacent cells as the *image atoms*, which can be mapped from the original unit cell.

6.6.1 Minimum Image Criterion

In the above example, the light-colored image atom exists in all adjacent unit cells. In computations, how to identify the corresponding light-colored real atom and put them into calculation needs additional cares.

A principle to do so is called the *minimum-image criterion*. In the following, we illustrate how this works for the pair potential case in the 2D unit cell setting.

In a periodic system, interatomic forces can be exerted by image atoms, as well as by real atoms, so that when we calculate interaction force for an atom

$$\mathbf{F}_{ji} = -\frac{\partial V(r_{ij})}{\partial \mathbf{r}_i},$$

where the atom j may be the image atoms located in the adjacent cells. To distinguish the real atom j, we denote the relative position vector $\mathbf{r}_{ij}$ as $\mathbf{r}_{ij}^*$ when atom j is an image atom.

To conduct the computation, as mentioned earlier, we first need to shift a real atom in the original unit cell into the image atom position in the adjacent cell. To do so, we need to have a systematic notation and approach. Consider a 2D original simulation cell with eight adjacent cells shown in Fig. 6.14.

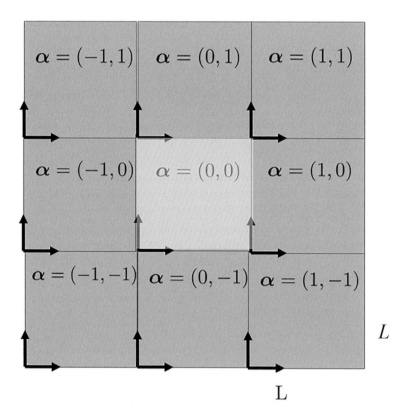

Figure 6.14 Schematic illustration of the minimum-image criterion in periodic boundary condition calculations

Denote $\alpha = (i, j), i, j = -1, 0, 1$ as the origins of the unit cell α. As shown in Fig. 6.14, the origin of the original unit cell is $\alpha = (0,0)$, and all the other adjacent cells have their own coordinate systems of different origins as indicated in Fig. 6.14. In fact, we may use the vector α to denote a specific cell.

By doing so, we can find the coordinates of an arbitrary cell α as

$$\mathbf{R}_{(\alpha)} = L\alpha,$$

where L is the length of the squared unit cell.

Thus, if we have an atom i with the position vector $\mathbf{r}_i^{(0,0)}$ in the original unit cell, its image in the surrounding adjacent cells are

$$\mathbf{r}_i^\alpha = \mathbf{r}_i^{(0,0)} + L\alpha, \tag{6.53}$$

where the position vector for the image atom i in the local frame α is denoted as $\mathbf{r}_i^\alpha$ and the atom i measured in the original frame $\alpha = (0,0)$ is denoted as $\mathbf{r}_i^{(0,0)}$ or simply $\mathbf{r}_i$.

Thus, if we want to calculate the force from an image atom i acting on an insider atom j in the force field calculation, we have

$$\mathbf{F}_{ji} = -\sum_\alpha \frac{\partial V(r_{ij}^\star)}{\partial \mathbf{r}_i},$$

where $\mathbf{r}_{ij}^\star = \mathbf{r}_j^{(\alpha)} - \mathbf{r}_i^{(0,0)} = \mathbf{r}_{ij} + L\alpha$.

In actual computations, we often use cubic simulation cells and calculate force interaction between the real atom and the image atom based on one-dimensional (1D) shifting rule of the Cartesian product. Therefore, if we understand how to shift the image atom interaction with the real atom interaction in 1D, we can find the interaction force in any dimension.

For simplicity of the discussion and without loss generality, we can always assume that $r_{cut} < L/2$. Thus the force interaction between an image atom and a real atom can happen if and only if $x_{ij} > L/2$, where L is the dimension of the 1D unit cell as shown in Fig. 6.15.

As shown in Fig. 6.15, if we assume $|x_{ij}| > L/2$, there are only two cases, i.e.,

(1) $x_{ij}^\star = x_{ij} + L,\ x_{ij} < 0$, and (6.54)
(2) $x_{ij}^\star = x_{ij} - L,\ x_{ij} > 0$. (6.55)

The two cases can be combined into one unified expression, i.e.,

If $|x_{ij}| > \frac{L}{2}$, replace $x_{ij} \rightarrow x_{ij}^\star = x_{ij} - sign(x_{ij})L$. (6.56)

In fact, one may also show that

If $|x_{ij}| < \frac{L}{2}$, replace $x_{ij} \rightarrow x_{ij}^\star = x_{ij} - sign(x_{ij})L$, (6.57)

even though it is irrelevant.

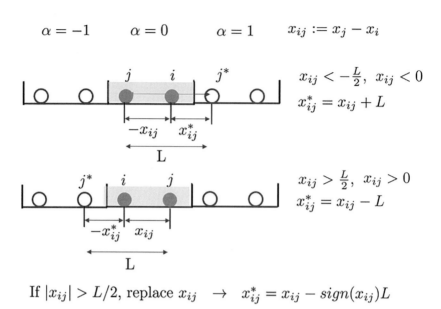

Figure 6.15 Schematic illustration of the minimum-image criterion in the one-dimensional case

The following is an example that is exerted from an actual Fortran 90 MD code, in which Eqs. (6.54) and (6.55) are used to implement the periodic boundary in the real computations. Note that the box dimension is normalized into 1.

```
subroutine Refold_Positions
!
! Particles that left the box are refolded back into the box by
! periodic boundary conditions
!
use Particles
implicit none
where (pos >  0.5d0) pos = pos - 1.0d0
where (pos < -0.5d0) pos = pos + 1.0d0
end subroutine Refold-Position
```

The following is another computation example that is exerted from an actual MD computer program (Fortran 90), in which Eq. (6.56) is used to implement the periodic boundary for a LJ potential–based force field. The part of the code for the periodic boundary is given as follows:

```
subroutine Compute_Forces
!
! Compute forces on atoms from the positions, using the Lennard-Jones
! potential.
!
use Particles
use Potential
implicit none
```

```fortran
      double precision, dimension(DIM) :: Sij,Rij
      double precision :: Rsqij,phi,dphi
      double precision :: rm2,rm6,rm12
      integer :: i,j
!
!  Reset to zero potential energies, forces, virial term
!
      ene_pot = 0.d0
      acc = 0.d0
      virial = 0.d0
!
!  Loop over all pairs of particles
!
      do i = 1,N-1                           ! looping an all pairs
         do j = i+1,N
            Sij = pos(:,j) - pos(:,i)        ! distance vector between i,j
            where ( abs(Sij) > 0.5d0 )       ! (in box scaled units)
               Sij = Sij - sign(1.d0,Sij)    ! periodic boundary condition
            end where                        ! applied where needed.
            Rij = BoxSize*Sij                ! go to real space units
            Rsqij = dot_product(Rij,Rij)     ! compute square distance
            if ( Rsqij < Rcutoff**2 ) then   ! particles are interacting
                                             ! compute Lennard-Jones pot.
               rm2 = 1.d0/Rsqij              !   1/r^2
               rm6 = rm2**3                  !   1/r^6
               rm12= rm6**2                  !   1/r^12
               phi = 4.d0*(rm12 - rm6) - phicutoff   ! 4[1/r^12-1/r^6]-phi(Rc)
                                             ! The following is dphi
                                             ! = -(1/r)(dV/dr)
               dphi = 24.d0*rm2*( 2.d0*rm12 - rm6 )  ! =24[2/r^14 - 1/r^8]
               ene_pot(i) = ene_pot(i) + 0.5d0*phi  ! sum: energy
               ene_pot(j) = ene_pot(j) + 0.5d0*phi  ! (i and j share it)
               virial = virial - dphi*Rsqij         ! sum: virial=sum r(dV/dr)
               acc(:,i) = acc(:,i) + dphi*Sij       ! sum: forces
               acc(:,j) = acc(:,j) - dphi*Sij       !    (Fji = -Fij)
            endif
         enddo
      enddo
      virial = - virial/DIM                  ! definition of virial term
      end subroutine Compute_Forces
```

6.7 Neighbor Lists

As mentioned in the previous sections, in MD simulations the interaction between two atoms has a cutoff, and by using the cutoff we do not need to calculate interaction of two atoms that are far away from each other. Thus, for the convenience of the computation, we always create a neighbor list for each atom so that if we want to calculate the force acting on an atom we only need to calculate the interaction of the atom with the other atoms in its neighbor list.

However, the challenge is that as the simulation proceeds the neighboring atoms change their positions: some atoms may leave the cutoff region and some other atoms may enter the cutoff region. Thus, we need to update the neighbor list from time to time. In principle, when each time one updates the neighbor list one has to conduct

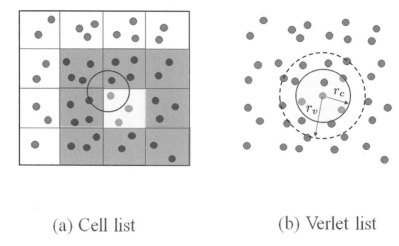

(a) Cell list (b) Verlet list

Figure 6.16 Illustration of two commonly used neighbor lists: (a) cell list and (b) Verlet list

a global search, and this is not only time consuming but also expensive in terms of computation time and storage. This is especially important when we calculate an atomistic system of fluids or amorphous polymers, in which there is no specific microstructure to constrain atom motions.

Since 1970s, various search methods have been developed. In most MD simulations, two neighbor list methods have been frequently used: (1) cell list and (2) Verlet list, as illustrated in Fig. 6.15.

Generally speaking, a cell list or a Verlet list is a data structure in MD simulations to efficiently maintain a list of all particles within a given cutoff distance of each other.

The construction of a cell list is an order $O(N^2)$ operation, which is one of efficient research algorithms. The main idea is that one first divides the simulation cells into many smaller cells. Because the size of the smaller cells is large enough, and each time the distance of an atom moves less than 0.2Å, so that one does not need always update the membership of the smaller cells. When one wants to update the neighbor list atom, one only needs to consider the atoms inside its own cell and the adjacent cells, i.e., the shaded cells in Fig. 6.16(a), which greatly reduces the amount of research tasks.

The main idea of the cell list is that the domain cell decomposition provides a buffer zone for any atom in a cell, which is the neighboring cells or the adjacent cells. Because of this "buffer zone," we do need to update the neighbor list by doing a global search, instead we just need to do a local search to update the neighbor list not every step but every several time steps or even more than 10 time steps, which greatly saves the CPU time. Similar to the idea of buffer zone, another commonly adopted approach is the so-called Verlet neighbor list. As shown in Fig. 6.17, a Verlet list of a given atom is a list of neighboring atoms in a sphere centered at the given atom. The radius of the sphere r_L is larger than the actual cutoff distance r_{cut}, and the sphere annulus, i.e., the region $r_{cut} \leq r \leq r_L$ is the buffer zone or "skin." Since $r_L > r_{cut}$, we do not need to

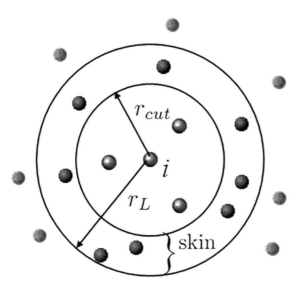

Figure 6.17 Illustration of the Verlet neighbor list and the "skin"

update the Verlet list every time step, the frequency to update the Verlet list is given by

$$\frac{N_L |\bar{v}| \Delta t}{2} = r_L - r_{cut} \rightarrow N_L = \frac{2(r_L - r_{cut})}{|\bar{v}| \Delta t},$$

where N_L is the number of time steps that we need to update a Verlet list and $\bar{v}$ is the average velocity of the atoms. For the detailed Verlet neighbor list update algorithm and analysis, the readers may further consult Chialvo and Debenedetti (1990).

6.8 Homework Problems

Problem 6.1 Consider the molecular dynamics equation,

$$m_i \ddot{\mathbf{r}}_i = \mathbf{F}_i, \quad i = 1, 2, \ldots, N,$$

where

$$\mathbf{F}_i = -\frac{\partial V}{\partial \mathbf{r}_i}.$$

Assume that $V = V(r_{ij})$ is a pair potential and $r_{ij} = |\mathbf{r}_j - \mathbf{r}_i|$.
Calculate

$$\mathbf{F}_i = -\frac{\partial V}{\partial \mathbf{r}_i} ?$$

where $\mathbf{r}_i = x_{i1} \mathbf{e}_1 + x_{i2} \mathbf{e}_2 + x_{i3} \mathbf{e}_3$ and

$$r_{ij} = |\mathbf{r}_j - \mathbf{r}_i| = \sqrt{(x_{j1} - x_{i1})^2 + x_{j2} - x_{i2})^2 + x_{j3} - x_{i3})^2}.$$

Problem 6.2 Consider the following Lennard-Jones potential,

$$V_{LJ}(r) = 4\epsilon \left[\left(\frac{\sigma}{r}\right)^{12} - \left(\frac{\sigma}{r}\right)^{6} \right],$$

where σ is a constant. Find the force field, i.e.,

$$\mathbf{F}_i(r) = ?$$

and find the equilibrium distance, r_0, between two atoms, at which the interaction force is zero. Plot $V_{IJ}(r)$ and $\mathbf{F}_i(r)$ in 1D case.

7 Molecular Dynamics Time Integration Techniques

An important computational procedure in MD computations is how to integrate trajectories of particles (atoms or molecules). This seems to be a numerical computation problem, but it is actually related to the underlying molecular physics. For instance, in crystalline solids, atoms are usually vibrating around their equilibrium positions. Both the frequency and the period of this lattice vibration, which can be nonlinear, will affect the material properties, i.e., they have intrinsic timescale that will also restrict the length of the integration time step.

On the other hand, if one wishes to simulate a nonequilibrium molecular event to study both stationary and time-dependent behaviors of the nonequilibrium molecular systems. The timescale of the nonequilibrium diffusion will come to play, which may depend on the spatial scale of the system. For instance, if the dimension of the system approaches to submicron, one may expect a meaningful simulation time up to micron second or even longer. In this case, we may need 10^{10} time steps to finish the simulation. Thus, it may be possible that any small numerical error may grow into a big amount error destroying any hope to extrapolate a sensible or useful physical result.

One of the keys to control the integration error and maintain physical fidelity of molecular modeling is to properly select and use right integration algorithms. Even though the time integration algorithms have been selected carefully and implemented in every MD source codes, it is still essential to understand them as a research developer or a computational specialist, because there are always new cases or special cases to be considered.

7.1 Basic Concept of Time Integration

In MD, the time integration is the procedure to find or to update the trajectory of each atom incrementally, i.e., integrating the equations of motion for each atom step by step,

$$M_i \frac{\partial^2 \mathbf{r}_i}{\partial t^2} = \mathbf{F}_i = -\frac{\partial V}{\partial \mathbf{r}_i}, \quad i = 1, 2, \ldots, N. \tag{7.1}$$

Now, we assume that the atomistic system is at a known configuration of $t = t_n$, and we would like to advance the atomistic system or molecular system to $t = t_{n+1}$ configuration. Considering a representative atom i, we have

$$d\mathbf{r}_i = \mathbf{v}_i(t)dt \rightarrow \int_{t_n}^{t_{n+1}} d\mathbf{r}_i = \int_{t_n}^{t_{n+1}} \mathbf{v}_i(t)dt \rightarrow \Delta \mathbf{r}_i \rightarrow \mathbf{v}_i(t^*)\Delta t, \ t^* \in (t_n, t_{n+1}).$$
(7.2)

The mean-value theorem of calculus tells us that there is a special $t^* \in (t_n, t_{n+1})$ that will make the above discrete formulation exact. However, we do not know $\mathbf{v}(t^*), t^* \in [t_n, t_{n+1}]$, so that if we choose $t^* = t_n$, the above expression becomes an approximation, i.e.,

$$\mathbf{r}_i(t_{n+1}) \approx \mathbf{r}_i(t_n) + \mathbf{v}_i(t_n)\Delta t. \tag{7.3}$$

This is because that

$$\mathbf{r}_i(t_{n+1}) = \mathbf{r}_i(t_n) + \mathbf{v}_i(t_n)\Delta t + \frac{1}{2}\frac{d\mathbf{v}_i}{dt}\bigg|_{t=t_n}\Delta^2 t + \cdots$$

and we call

$$\tau_n = \left\| \frac{1}{2}\frac{d\mathbf{v}_i}{dt}\bigg|_{t=t_n}\Delta^2 t + \cdots \right\| = O(\Delta^2 t)$$

as the local truncation error.

We then call the integration algorithm,

$$\mathbf{r}_i^{(n+1)} = \mathbf{r}_i^{(n)} + \mathbf{v}_i^{(n)}\Delta t, \tag{7.4}$$

$$\mathbf{v}_i^{(n+1)} = \mathbf{v}_i^{(n)} + \frac{1}{M_i}\mathbf{F}_i^{(n)}\Delta t, \tag{7.5}$$

as the forward Euler method.

Once we find $\mathbf{r}_i^{(n+1)}$, we can update the force field, i.e.,

$$\mathbf{F}_i^{(n+1)} = -\frac{\partial V}{\partial \mathbf{r}}\bigg|_{\mathbf{r}=\mathbf{r}^{n+1}},$$

we can then advance to $(\mathbf{v}_i^{(n+2)}, \mathbf{r}_i^{(n+2)})$ configuration by integrating Eqs. (7.1) and (7.2) as

$$\mathbf{v}_i^{(n+2)} = \mathbf{v}_i^{(n+1)} + \frac{1}{M_i}\mathbf{F}_i^{(n+1)}\Delta t, \tag{7.6}$$

$$\mathbf{r}_i^{(n+2)} = \mathbf{r}_i^{(n+1)} + \mathbf{v}_i^{(n+1)}\Delta t. \tag{7.7}$$

One may also choose $t^* = t_{n+1}$, and the integration algorithm becomes,

$$\mathbf{v}_i^{(n+1)} = \mathbf{v}_i^{(n)} + \frac{1}{M_i}\mathbf{F}_i^{(n+1)}(t_{n+1}, \mathbf{r}_i^{(n+1)})\Delta t, \tag{7.8}$$

$$\mathbf{r}_i^{(n+1)} = \mathbf{r}_i^{(n)} + \mathbf{v}_i^{(n+1)}\Delta t. \tag{7.9}$$

Equations (7.8) and (7.9) are called the backward Euler integration algorithm, and it is clear that the above integration algorithm is not explicit, because the unknown variable $\mathbf{r}_i^{(n+1)}$ is in the both sides of Eq. (7.8). Thus, it may require iterative nonlinear solver to find the numerical solution.

In both integration algorithms, $\mathbf{r}_i^{(n)} \neq \mathbf{r}_i(t_n)$ and $\mathbf{v}_i^{(n)} \neq \mathbf{v}_i(t_n)$. We can define the global truncation error as

$$e_n = \|\mathbf{r}_i(t_n) - \mathbf{r}_i^{(n)}\| = \left\| \mathbf{r}_i(t_n) - \left(\mathbf{r}_i^{(0)} + \mathbf{v}_i^{(0)}\Delta t + \mathbf{v}_i^{(1)}\Delta t + \cdots + \mathbf{v}_i^{(n-1)}\Delta t \right) \right\|. \tag{7.10}$$

A numerical algorithm is said to be convergent, if the global truncation approaches to zero, i.e.,

$$\lim_{\Delta t \to 0} \max_n \|e_n\| = 0.$$

For the forward Euler integration algorithm, by definition, the global error may be bounded as

$$e_{n+1} = \|\mathbf{r}_i(t_{n+1}) - \mathbf{r}_i^{(n+1)}\| \leq \|\mathbf{r}_i(t_n) - \mathbf{r}_i^{(n)} + \left(\mathbf{v}_i(t_n) - \mathbf{v}_i^{(n)}\right)\Delta t\| + \tau_n$$

$$\leq \|\mathbf{r}_i(t_n) - \mathbf{r}_i^n\| + \|\mathbf{v}_i(t_n) - \mathbf{v}_i^{(n)}\|\Delta t + \tau_n. \quad (7.11)$$

If we assume that there exits a constant L such that the following Lipschitz condition is satisfied,

$$\|\mathbf{v}_i(t_n) - \mathbf{v}_i^{(n)}\| \leq L\|\mathbf{r}_i(t_n) - \mathbf{r}_i^{(n)}\|.$$

The above inequality may be written as

$$e_n \leq e_{n-1} + L\Delta t e_n + \tau_n = (1 + L\Delta t)e_n + \tau_n, \; n = 0, 1, 2, \ldots, N-1$$

$$\leq (1 + L\Delta t)^n e_0 + \left((1 + L\Delta t)^{n-1} + (1 + L\Delta t)^{n-2} + \cdots + (1 + L\Delta t)^0\right)\tau, \quad (7.12)$$

where $\tau_m := \max\{\tau_i, 1 \leq i \leq n\}$.

Choosing $e_0 = 0$ and considering the identity

$$\sum_{k=0}^{n-1} r^k = \frac{1 - r^n}{1 - r},$$

we have

$$e_n \leq \frac{\tau_m}{L\Delta t}(\exp(L(t_n - t_0)) - 1) \sim O(\Delta t) \quad (7.13)$$

because $(1 + L\Delta t) \leq \exp(L\Delta t)$ and $\tau_m \sim O(\Delta^2 t)$.

7.2 Verlet Algorithms

7.2.1 Original Verlet Integration Algorithm

The original Verlet integration algorithm is the following position update formula:

$$\mathbf{r}_i(t_n + \Delta t) = 2\mathbf{r}_i(t_n) - \mathbf{r}_i(t_n - \Delta t) + \mathbf{a}_i(t_n)\Delta t^2. \quad (7.14)$$

We denote $\mathbf{r}_i(t_{n+1})$ as $\mathbf{r}_i(t_n + \Delta t)$ and $\mathbf{r}_i(t_{n-1})$ as $\mathbf{r}_i(t_n - \Delta t)$. This displacement or position update formula does not contain velocity field, i.e., we do not need the information on the velocities of atoms, but we need atom position configurations at both $t = t_{n-1}$ and $t = t_n$ in order to find the position at $t = t_{n+1}$. In fact, we can rewrite Eq. (7.14) as

$$\mathbf{r}_i(t_n + \Delta t) = \mathbf{r}_i(t_n) + \frac{\mathbf{r}_i(t_n) - \mathbf{r}_i(t_n - \Delta t)}{\Delta t}\Delta t + \mathbf{a}_i(t_n)\Delta t^2. \quad (7.15)$$

One can see that by storing two time steps of position vectors at $t = t_{n-1}$ and $t = t_n$ is equivalent to provide a velocity field at $t = t_n - \Delta/2$. Based on this interpretation, the integration algorithm does not seem to be fit, because it looks like the following form:

$$\mathbf{r}_i(t_n + \Delta t) = \mathbf{r}_i(t_n) + \text{``}\mathbf{v}_i(t_n - \Delta t/2)\Delta t\text{''} + \mathbf{a}_i(t_n)\Delta t^2.$$

However, we can rewrite Eq. (7.15) as

$$\frac{\mathbf{r}_i(t_n + \Delta t) - \mathbf{r}_i(t_n)}{\Delta t}\Delta t = \frac{\mathbf{r}_i(t_n) - \mathbf{r}_i(t_n - \Delta t)}{\Delta t}\Delta t + \mathbf{a}_i(t_n)\Delta t^2. \quad (7.16)$$

Everything seems to make sense now. This integration algorithm is essentially a leapfrog half step velocity update

$$\mathbf{v}_i(t_n + \Delta t/2) = \mathbf{v}_i(t_n - \Delta t/2) + \mathbf{a}_i(t_n)\Delta t.$$

As mentioned earlier, in principle, the original Verlet algorithm does not need the information for the velocity field in order to update displacement. However, in MD simulation, the velocity field information is also important, which is needed to calculate the system's energy, temperature, and stress among those statistical variables.

For this purpose, a so-called Störmer–Verlet formula is used for calculating velocity field, i.e.,

$$\mathbf{v}_i(t_n) = \frac{\mathbf{r}_i(t_n + \Delta t) - \mathbf{r}_i(t_n - \Delta t)}{2\Delta t}, \quad (7.17)$$

which is essentially the midpoint rule in finite difference.

In numerical analysis, the truncation error is referred to the error that rises from using the finite difference operator to replace a differential operator, which may be represented by an infinite sum of Taylor expansion. If we only use the first few terms of the Taylor expansion to approximate a Taylor series with infinitely many terms, all the terms that were truncated constitute the truncation error.

To find the truncation error for the original Verlet algorithm, we first take the third-order Taylor expansion,

$$\mathbf{r}_i(t_n + \Delta t) = \mathbf{r}_i(t_n) + \mathbf{v}_i(t_n)\Delta t + \frac{1}{2}\mathbf{a}_i(t_n)\Delta t^2 + \frac{1}{3!}\dot{\mathbf{a}}_i(t_n)\Delta t^3 + O(\Delta t^4), \quad (7.18)$$

where $O(X)$ is a symbol of order of magnitude. In passing, we note that

$$\lim_{X \to 0} \frac{O(X)}{X} \to constant \neq 0.$$

Whereas the order of magnitude symbol $o(X)$ indicates

$$\lim_{X \to 0} \frac{o(X)}{X} \to 0.$$

Taking a step back, we have the Taylor expansion

$$\mathbf{r}_i(t_n - \Delta t) = \mathbf{r}_i(t_n) - \mathbf{v}_i(t_n)\Delta t + \frac{1}{2}\mathbf{a}_i(t_n)\Delta t^2 - \frac{1}{3!}\dot{\mathbf{a}}_i(t_n)\Delta t^3 + O(\Delta t^4). \quad (7.19)$$

Adding (7.18) and (7.19), i.e., (7.18) + (7.19) the second and the fourth terms of the right-hand side will be canceled out, which yields

$$\mathbf{r}_i(t + \Delta t) + \mathbf{r}_i(t - \Delta t) = 2\mathbf{r}_i(t) + \mathbf{a}_i(t)\Delta t^2 + O(\Delta t^4). \quad (7.20)$$

This expression shows that the original Verlet has a fourth-order truncation error.

To find the truncation error for Stömer's velocity update, we substrate (7.19) from (7.18), i.e., (7.18)–(7.19), which yields

$$\mathbf{v}_i(t_n) = \frac{1}{2\Delta t}[\mathbf{r}_i(t_n + \Delta t) - \mathbf{r}_i(t_n - \Delta t)] + O(\Delta t^2). \quad (7.21)$$

This indicates that the Stömer's velocity update has a second-order truncation error.

Both estimates (7.20) and (7.21) are local, i.e., they are the estimates at $t = t_n + \Delta t$ with respect to time $t = t_n$. If we keep integrating, the local truncation error will accumulate, we would like to ask how this accumulation will affect the overall error estimate or the global error estimate.

To formulate the question in precise mathematical term, we would like to know

$$\text{Error}[\mathbf{r}_i(t_0 + n\Delta t)] = ?, \quad \Delta t = \frac{T}{n},$$

where T is the time duration of the integration, n is the number of time steps, and $\Delta t = T/n$ is the time step increment.

Since

$$\mathbf{r}_i(t_0 + \Delta t) = 2\mathbf{r}_i(t_0) - \mathbf{r}_i(t_0 - \Delta t) + \mathbf{a}_i(t_0)\Delta t^2 + O(\Delta t^4),$$

we have

$$\text{Error}[\mathbf{r}_i(t_0 + \Delta t)] \sim O(\Delta t^4). \quad (7.22)$$

Considering

$$\mathbf{r}_i(t_0 + 2\Delta t) = 2\mathbf{r}_i(t_0 + \Delta t) - \mathbf{r}_i(t_0) + \mathbf{a}_i(t_0 + \Delta t)\Delta t^2 + O(\Delta t^4), \quad (7.23)$$

we have

$$\text{Error}[\mathbf{r}_i(t_0 + 2\Delta t) = 2\text{Error}[\mathbf{r}_i(t_0 + \Delta t)] + O(\Delta t^4) = 3O(\Delta t^4). \quad (7.24)$$

Note that the error for the term $\mathbf{a}_i(t_0 + \Delta t)\Delta t^2$ is at least up to $O(\Delta t^6)$. Thus, it does not contribute to the error estimate in Eq. (7.24).

Similarly, one may show that

$$\text{Error}[\mathbf{r}_i(t_0 + 3\Delta t) = 2\text{Error}[\mathbf{r}_i(t_0 + 2\Delta t)] - \text{Error}[\mathbf{r}_i(t_0 + \Delta t)] + O(\Delta t^4) = 6O(\Delta t^4), \quad (7.25)$$

and

$$\text{Error}[\mathbf{r}_i(t_0 + 4\Delta t)] = 10O(\Delta t^4) \tag{7.26}$$
$$\text{Error}[\mathbf{r}_i(t_0 + 5\Delta t)] = 15O(\Delta t^4). \tag{7.27}$$

By induction, we can show that

$$\text{Error}[\mathbf{r}_i(t_0 + n\Delta t)] = \frac{n(n+1)}{2}O(\Delta t^4), \text{ where } n = \frac{T}{\Delta t}. \tag{7.28}$$

Hence, the leading term of the global integration error is

$$\text{Error}[\mathbf{r}_i(t_0 + T)] = O(\Delta t^2). \tag{7.29}$$

For the Stömer's velocity update, we know that its local truncation error is,

$$\mathbf{v}_i(t) = \frac{1}{2\Delta t}(\mathbf{r}_i(t + \Delta t) - \mathbf{r}_i(t - \Delta t)) + O(\Delta t^2).$$

Thus, we have

$$\mathbf{v}_i(t_0 + n\Delta t) = \frac{1}{2\Delta t}(\mathbf{r}_i(t_0 + (n+1)\Delta t) - \mathbf{r}_i(t_0 + (n-1)\Delta t)) + O(\Delta t^2).$$

On the other hand, based on the local error estimate of the displacement field, i.e., Eq. (7.28) we have

$$\text{Error}[\mathbf{r}_i(t + (n+1)\Delta t)] = \frac{(n+1)(n+2)}{2}O(\Delta t^4) \tag{7.30}$$
$$\text{Error}[\mathbf{r}_i(t + (n-1)\Delta t)] = \frac{(n)(n-1)}{2}O(\Delta t^4). \tag{7.31}$$

Since

$$\frac{(n+1)(n+2)}{2}O(\Delta t^4) - \frac{(n)(n-1)}{2}O(\Delta t^4) = (2n+2)O(\Delta t^4) \sim O(\Delta t^3),$$

we have

$$\text{Error}[\mathbf{v}_i(t + n\Delta t)] = \frac{1}{2\Delta t}(\text{Erorr}[\mathbf{r}_i(t + (n+1)\Delta t)]$$
$$- \text{Erorr}[\mathbf{r}_i(t + (n-1)\Delta t)]) + O(\Delta t^2)$$
$$\sim \frac{1}{2\Delta t}\left(2(n+1)O(\Delta t^4)\right) + O(\Delta t^2) = O(\Delta t^2). \tag{7.32}$$

Thus, for the Stömer velocity update, its global integration error is also at the second order, which is the same as its local error estimate.

In summary for the original Verlet, the integration error for both the displacement fields and the velocity fields are all second order, i.e.,

$$\text{Error}[\mathbf{r}_i(t + n\Delta t)] \sim \frac{n(n+1)}{2}O(\Delta t^4) \sim O(\Delta t^2) \tag{7.33}$$

$$\text{Error}[\mathbf{v}_i(t + n\Delta t)] \sim \frac{1}{2\Delta t}\left(2(n+1)O(\Delta t^4)\right) + O(\Delta t^2) \sim O(\Delta t^2). \tag{7.34}$$

Thus the original Verlet is a second-order integrator.

7.2.2 Velocity Verlet Algorithm

Velocity Verlet integration algorithm is one of the most popular and also accurate integration algorithms that are used in MD computations as well as in other fields of computational physics and engineering computations.

It has a half-step version, and an equivalent full step version. We first start to discuss the half-step version.

1. Half Step Version
The half-step version of the velocity Verlet algorithm consists of three steps as shown in the following equations:

$$\text{Step 1:} \quad \mathbf{p}_i^{(n+1/2)} = \mathbf{p}_i^{(n)} + \frac{\Delta t}{2}\mathbf{F}_i^{(n)} \quad (7.35)$$

$$\text{Step 2:} \quad \begin{cases} \mathbf{r}_i^{(n+1)} = \mathbf{r}_i^{(n)} + \frac{\Delta t}{m_i}\mathbf{p}_i^{(n+1/2)} \\ \mathbf{F}_i^{(n+1)} = -\frac{\partial V}{\partial \mathbf{r}}\Big|_{\mathbf{r}_i^{(n+1)}} \end{cases} \quad (7.36)$$

$$\text{Step 3:} \quad \mathbf{p}_i^{(n+1)} = \mathbf{p}_i^{(n+1/2)} + \frac{\Delta t}{2}\mathbf{F}_i^{(n+1)}. \quad (7.37)$$

The truncation error of the half-step velocity Verlet algorithm can be found as follows:

$$\text{Step 1:} \quad \mathbf{v}_i(t_n + \Delta t/2) = \mathbf{v}_i(t_n) + \mathbf{a}_i(t_n)\frac{\Delta t}{2} + O(\Delta t^2) \quad (7.38)$$

$$\text{Step 2:} \quad \mathbf{r}_i(t_n + \Delta t) = \mathbf{r}_i(t_n) + \mathbf{v}_i(t_n + \Delta t/2)\Delta t + O(\Delta t^2) \quad (7.39)$$

$$\text{Step 3:} \quad \mathbf{v}_i(t_n + \Delta t) = \mathbf{v}_i(t_n + \Delta t/2) + \mathbf{a}_i(t_n + \Delta t)\frac{\Delta t}{2} + O(\Delta t^2). \quad (7.40)$$

2. Full-Step Version
The full-step version is a two-step integration algorithm,

$$\text{Step 1:} \quad \mathbf{r}_i^{(n+1)} = \mathbf{r}_i^{(n)} + \mathbf{v}_i^{(n)}\Delta t + \frac{1}{2}\mathbf{a}_i^{(n)}\Delta t^2, \quad (7.41)$$

$$\text{Update force field:} \quad \mathbf{F}_i^{(n+1)} = -\frac{\partial V}{\partial \mathbf{r}}\Big|_{\mathbf{r}_i^{(n+1)}}, \quad \text{and} \quad (7.42)$$

$$\text{Step 2:} \quad \mathbf{v}_i^{(n+1)} = \mathbf{v}_i^{(n)} + \frac{\Delta t}{2}\left(\mathbf{a}_i^{(n)} + \mathbf{a}_i^{(n+1)}\right). \quad (7.43)$$

To find its local truncation error, we first consider the following Taylor expansion:

$$\mathbf{r}_i(t_n + \Delta t) = \mathbf{r}_i(t_n) + \mathbf{v}_i(t_n)\Delta t + \frac{1}{2}\mathbf{a}_i(t_n)\Delta t^2 + O(\Delta t^3). \quad (7.44)$$

Comparing Eq. (7.41) with Eq. (7.44), we can see clearly that the truncation error for the displacement field of velocity Verlet is the third of Δt, i.e., $O(\Delta t^3)$.

Second, we consider the following Taylor expansion:

$$\mathbf{v}_i(t_n + \Delta t/2) = \mathbf{v}_i(t_n) + \mathbf{a}_i(t_n)\frac{\Delta t}{2} + \dddot{\mathbf{v}}_i(t_n)\frac{\Delta t^2}{8} + O(\Delta t^3), \quad (7.45)$$

and
$$v_i(t_n + \Delta t/2) = v_i(t_n + \Delta t) - a_i(t_n + \Delta t)\frac{\Delta t}{2} + \ddot{v}_i(t_n + \Delta t)\frac{\Delta t^2}{8} + O(\Delta t^3). \quad (7.46)$$

From Eqs (7.45)–(7.46), we can find that

$$v_i(t_n + \Delta t) = v_i(t_n) + \frac{1}{2}(a_i(t_n) + a_i(t_n + \Delta t))\Delta t + (\ddot{v}_i(t_n) - \ddot{v}_i(t_n + \Delta t))\frac{\Delta t^2}{8} + O(\Delta t^3). \quad (7.47)$$

Using the Taylor expansion, we may show that the fourth term in the above equation has an error estimate of $O(\Delta t^3)$. This is because that

$$(\ddot{v}_i(t_n) - \ddot{v}_i(t_n + \Delta t)) = -\dddot{a}_i(t_n)\Delta t + O(\Delta t^2). \quad (7.48)$$

Thus, the truncation error for the velocity update of the velocity Verlet algorithm is also third order, i.e.,

$$v_i(t_n + \Delta t) = v_i(t_n) + \frac{1}{2}(a_i(t_n) + a_i(t_n + \Delta t))\Delta t + O(\Delta t^3). \quad (7.49)$$

This may be unexpected by some.

To find the truncation error for the displacement field, we first consider the Taylor expansion

$$r_i(t_0 + \Delta t) = r_i(t_0) + v_i(t_0) + \frac{1}{2}a_i(t_0)\Delta t^2 + O(\Delta t^3),$$

then, it clear that the truncation error the for the displacement field is

$$\text{Error}(r_i(t_0 + \Delta t)) = O(\Delta t^3). \quad (7.50)$$

To find the global error, i.e., $\text{Erorr}[r_i(t_0 + n\Delta t)]$, we use the Taylor expansion to find that

$$r_i(t + 2\Delta t) = r_i(t + \Delta t) + v_i(t + \Delta t)\Delta t + \frac{1}{2}a_i(t + \Delta t)\Delta t^2 + O(\Delta t^3) \quad (7.51)$$

and hence

$$\text{Error}[r_i(t + 2\Delta t)] = \text{Error}[r_i(t + \Delta t)] + O(\Delta t^3) = 2O(\Delta t^3). \quad (7.52)$$

This is because that

$$\text{Error}[v_i(t + \Delta t)\Delta t] \sim O(\Delta t^4) \text{ and } \text{Error}[a_i(t + \Delta t)\Delta t^2] \sim O(\Delta t^5).$$

Similarly, one may find that

$$\text{Error}[r_i(t + 3\Delta t) = 3O(\Delta t^3)$$

$$\vdots \quad \vdots$$

$$\text{Error}[r_i(t + n\Delta t) = nO(\Delta t^3),$$

which leads to

$$\text{Error}[r_i(t + T) \sim \frac{T}{\Delta t}O(\Delta t^3) \sim O(\Delta t^2).$$

Hence the integration error for the displacement is second order.

The global integration error for velocity field can be derived in similar fashion. First, we integrate one more step for velocity field, i.e.,

$$\mathbf{v}_i(t_0 + 2\Delta t) = \mathbf{v}_i(t_0 + \Delta t) + \frac{\Delta t}{2}(\mathbf{a}_i(t_0 + \Delta t) + \mathbf{a}_i(t_0 + 2\Delta t)) + O(\Delta t^3).$$

Thus,

$$Error[\mathbf{v}_i(t_0 + 2\Delta t)] = Error[\mathbf{v}_i(t_0 + \Delta t)] + O(\Delta t^3) = 2O(\Delta t^3),$$

and by continuing the process, one may find that

$$Error[\mathbf{v}_i(t + 3\Delta t)] = 3O(\Delta t^3),$$
$$= \cdots$$
$$Error[\mathbf{v}_i(t + n\Delta t)] = nO(\Delta t^3).$$

Because

$$n \sim \frac{T}{\Delta t}$$

we can conclude that

$$\rightarrow \quad Error[\mathbf{v}_i(t + n\Delta t)] \sim O(\Delta t^2);$$

therefore, the velocity Verlet is also a second-order time integrator.

7.2.3 Leapfrog Algorithm

There are three integration algorithms in the Verlet family of integrators. The third integration algorithm is the so-called leapfrog Verlet, which is the following two-step algorithm:

$$\mathbf{v}_i(t_n + \Delta t/2) = \mathbf{v}_i(t_n - \Delta t/2) + \mathbf{a}_i(t_n)\Delta t \tag{7.53}$$
$$\mathbf{r}_i(t_n + \Delta t) = \mathbf{r}_i(t_n) + \mathbf{v}_i(t_n + \Delta t/2)\Delta t. \tag{7.54}$$

One can see that the velocity update goes from the time step $t_n - \Delta t/2$ to $t_n + \Delta t/2$, whereas the displacement update goes from the time step t_n to $t_{n+1} = t_n + \Delta$. The value of displacement filed and velocity field are not obtained in the same time instance, but differ a half time step, as shown in Fig. 7.1. This situation is very much like a frog jump moving forward. Its front legs and rear legs are landing at different positions or maybe differing a half step in each jump. So is the name leapfrog comes from. Substituting the the value of the velocity field at $t_{n+1/2}$ into the displacement update formula, one may rewrite the two-step leapfrog algorithm as

$$\mathbf{v}_i(t_n + \Delta t/2) = \mathbf{v}_i(t_n - \Delta t/2) + \mathbf{a}_i(t_n)\Delta t \tag{7.55}$$
$$\mathbf{r}_i(t_n + \Delta t) = \mathbf{r}_i(t_n) + \mathbf{v}_i(t_n - \Delta t/2)\Delta t + \mathbf{a}_i(t_n)\Delta t^2,$$
$$\approx \mathbf{r}_i(t_n) + \mathbf{v}_i(t_n)\Delta t + \frac{1}{2}\mathbf{a}_i(t_n)\Delta t^2. \tag{7.56}$$

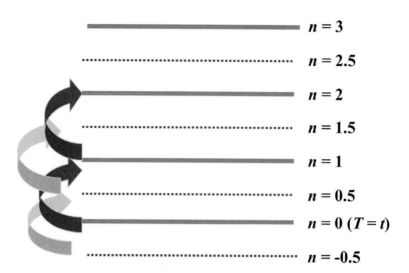

Figure 7.1 Illustration of leapfrog Verlet integration algorithm. The solid lines are the displacement field update time, and the dotted lines are the velocity field update time

In the last line, we used the approximation of $v_i(t_n - \Delta t/2) \approx \frac{1}{2}(v_i(t_n) + v_i(t_{n-1}))$ and $\frac{1}{2}v_i(t_{n-1}) + \frac{1}{2}a_i(t_n)\Delta t \approx \frac{1}{2}v_i(t_n)$. Thus, the leapfrog algorithm (7.53) and (7.54) may be viewed as a variant of Eqs. (7.41) and (7.43). Therefore, you may hear many people call the leapfrog algorithm as the Verlet leapfrog algorithm.

By the Taylor expansion, one can readily see that we have

$$\begin{aligned}
\mathbf{v}(t_n + \Delta t/2) &= \mathbf{v}_i(t_n - \Delta t/2) + \mathbf{a}_i(t_n - \Delta t/2)\Delta t + \frac{1}{2}\dot{\mathbf{a}}_i(t_n - \Delta t/2)\Delta t^2 + O(\Delta t^3) \\
&= \mathbf{v}_i(t_n - \Delta t/2) + \mathbf{a}_i(t_n)\Delta t - \frac{1}{2}\dot{\mathbf{a}}_i(t_n - \Delta t/2)\Delta t^2 \\
&\quad + \frac{1}{2}\dot{\mathbf{a}}_i(t_n - \Delta t/2)\Delta t^2 + O(\Delta t^3) \\
&= \mathbf{v}_i(t_n - \Delta t/2) + \mathbf{a}_i(t_n)\Delta t + O(\Delta t^3),
\end{aligned} \qquad (7.57)$$

and

$$\begin{aligned}
\mathbf{r}(t_n + \Delta t) &= \mathbf{r}_i(t_n) + \mathbf{v}_i(t_n)\Delta t + \frac{1}{2}\dot{\mathbf{a}}_i(t_n)\Delta t^2 + O(\Delta t^3) \\
&= \mathbf{r}_i(t_n) + \mathbf{v}_i(t_n + \Delta t/2)\Delta t - \frac{1}{2}\dot{\mathbf{a}}_i(t_n + \Delta t/2)\Delta t/2 \\
&\quad + \frac{1}{2}\dot{\mathbf{a}}_i(t_n - \Delta t/2)\Delta t^2 + O(\Delta t^3) \\
&= \mathbf{r}_i(t_n) + \mathbf{v}_i(t_n + \Delta t/2)\Delta t + O(\Delta t^3).
\end{aligned} \qquad (7.58)$$

Therefore, the local truncation error of the leapfrog algorithm is also second order for both the velocity field and the displacement field.

7.3 Predictor–Corrector Methods

To significantly increase integration accuracy, we often adopt the predictor–corrector method in MD simulations, which is a multiple substep method.

To explain the method, we use the forward Euler method as an example to integrate differential equation $\dot{y}(t) = f(y,t)$.

In the first substep of time step n, i.e., the predictor step, we employ the forward Euler method as the predictor, i.e.,

$$y_{n+1}^{(p)} = y_n + f(y_n, t_n)\Delta t, \tag{7.59}$$

and then the subsequent substep, i.e., the corrector step, we employ the middle point rule that utilizes the predictor result,

$$y_{n+1} = y_n + \frac{\Delta t}{2}\left(f(y_n, t_n) + f(y_{n+1}^{(p)}, t_{n+1})\right). \tag{7.60}$$

One may extend the corrector phase to multiple stages. For example, we consider the a general k-stage Runge–Kutta method to integrate $\dot{y}(t) = f(t,y)$, which is featured by a set of chosen parameters α_i, β_{ij} and γ_i for $i = 1, 2, \ldots k$ and $j < i$. Again, we still use the forward Euler to calculate slop, and the ith stage computes a slope s_i by evaluating $f(t,y)$ at a value of $t \in [t_n, t_{n+1}]$. The final value of y_{n+1} is obtained by adding to y_n and a linear combination of the previously computed slopes:

$$s_i = f\left(y_n + \Delta t \sum_{j=1}^{i-1} \beta_{ij} s_j\right). \tag{7.61}$$

The final value at $t = t_{n+1}$ will be explicitly expressed as

$$y_{n+1} = y_n + \Delta t \sum_{j=1}^{k} \gamma_i s_i. \tag{7.62}$$

For example, for the four-stage Runge–Kutta method, $\gamma_1 = 1/6, \gamma_2 = 2/6, \gamma_3 = 2/6,$ and $s_4 = 1/6$, and

$$y_{n+1} = y_n + \frac{\Delta t}{6}(s_1 + 2s_2 + 2s_3 + s_4),$$

in which

$$s_1 = f(t_n, y_n), \; s_2 = f\left(t_n + \Delta t/2, y_n + s_1 \frac{\Delta t}{2}\right), \; s_3$$
$$= f\left(t_n + \Delta t/2, y_n + s_2 \frac{\Delta t}{2}\right), \; s_4 = f(t_n + \Delta t, y_n + s_4 \Delta t).$$

7.3.1 Gear's Predictor–Corrector Algorithm

In MD simulations, we often employ the so-called Gear's predictor–corrector algorithm for some delicate time integrations involving simulations of phase transformation or other critical phenomena.

First, we can define,

$$\mathbf{r}_i^{(0)}(t_n) := \mathbf{r}_i(t_n), \quad \mathbf{r}_i^{(1)}(t_n) := \mathbf{v}_i(t_n)\Delta t, \quad \mathbf{r}_i^{(2)} := \mathbf{a}_i(t_n)\frac{\Delta t^2}{2!}, \quad \mathbf{r}_i^{(3)} := \dot{\mathbf{a}}_i(t_n)\frac{\Delta t^3}{3!}, \quad \ldots$$

Following the Taylor expansion, at $t = t_n$ we have

$$\mathbf{r}_i(t + \Delta t) = \mathbf{r}_i(t) + \mathbf{v}_i(t)\Delta t + \dot{\mathbf{v}}_i(t)\frac{\Delta t^2}{2!} + \dot{\mathbf{a}}_i(t)\frac{\Delta t^3}{3!} + O(\Delta t^4)$$

$$= \mathbf{r}_i^{(0)}(t) + \mathbf{r}_i^{(1)}(t) + \mathbf{r}_i^{(2)}(t) + \mathbf{r}_i^{(3)}(t) + O(\Delta t^4)$$

$$\mathbf{r}_i^{(1)}(t + \Delta t) = \mathbf{r}_i^{(1)}(t) + 2\mathbf{r}_i^{(2)}(t) + 3\mathbf{r}_i^{(3)}(t) + O(\Delta t^4)$$

$$\mathbf{r}_i^{(2)}(t + \Delta t) = \mathbf{r}_i^{(2)}(t) + 3\mathbf{r}_i^{(3)}(t) + O(\Delta t^4)$$

$$\mathbf{r}_i^{(3)}(t + \Delta t) = \mathbf{r}_i^{(3)}(t) + O(\Delta t^4).$$

In the predictor phase, we can first predict the values of $\mathbf{r}_i^{(j)}(t + \Delta t), j = 1, 2, 3, \ldots$ as

$$\mathbf{r}_i^{(P0)}(t + \Delta t) = \mathbf{r}_i^{(0)}(t) + \mathbf{r}_i^{(1)}(t) + \mathbf{r}_i^{(2)}(t) + \mathbf{r}_3 + O(\Delta t^4)$$

$$\mathbf{r}_i^{(P1)}(t + \Delta t) = \mathbf{r}_i^{(1)}(t) + 2\mathbf{r}_i^{(2)}(t) + 3\mathbf{r}_i^{(3)}(t) + O(\Delta t^4)$$

$$\mathbf{r}_i^{(P2)}(t + \Delta t) = \mathbf{r}_i^{(2)}(t) + 3\mathbf{r}_i^{(3)}(t) + O(\Delta t^4)$$

$$\mathbf{r}_i^{(P3)}(t + \Delta t) = \mathbf{r}_i^{(3)}(t) + O(\Delta t^4).$$

The above expressions can be written in a matrix with vector components,

$$\begin{bmatrix} \mathbf{r}_i^{(P0)}(t + \Delta t) \\ \mathbf{r}_i^{(P1)}(t + \Delta t) \\ \mathbf{r}_i^{(P2)}(t + \Delta t) \\ \mathbf{r}_i^{(P3)}(t + \Delta t) \end{bmatrix} = \begin{bmatrix} 1 & 1 & 1 & 1 \\ 0 & 1 & 2 & 3 \\ 0 & 0 & 1 & 3 \\ 0 & 0 & 0 & 1 \end{bmatrix} \begin{bmatrix} \mathbf{r}_i^{(0)}(t) \\ \mathbf{r}_i^{(1)}(t) \\ \mathbf{r}_i^{(2)}(t) \\ \mathbf{r}_i^{(3)}(t) \end{bmatrix}.$$

During the prediction phase, the force field, i.e., the acceleration field in the updated configuration, is not used. If we adjust the acceleration based on the predicted position,

$$\mathbf{a}_i = \mathbf{F}_i\left(\mathbf{r}_i^{(P0)}(t + \Delta t)\right)/m_i,$$

we can find the correction for acceleration

$$\mathbf{r}_i^{(C2)}(t + \Delta t) = \mathbf{a}(t + \Delta t)\frac{\Delta t^2}{2!}.$$

The amount error that needs to be corrected is

$$\mathbf{r}_i^{(C2)}(t + \Delta t) - \mathbf{r}^{(P2)}(t + \Delta t),$$

and the correction formula is

$$\mathbf{r}_i^{(Cj)}(t + \Delta t) = \mathbf{r}_i^{(Pj)}(t + \Delta t) + C_j\left(\mathbf{r}_i^{(C2)}(t + \Delta t) - \mathbf{r}_i^{(P2)}(t + \Delta t)\right), \quad j = 0, 1, 2, 3, \ldots,$$

where the constants, $C_j, j = 1, 2, \ldots$, can be derived based on stability analysis.

For the first three Gear's corrector–predictor algorithms, we list the coefficients $C_j, j = 0, 1, \ldots$ in Table 7.1. For details, readers may consult Gear (1966, 1971).

Table 7.1 Corrector coefficients for Gear's predictor–corrector algorithm

C_j	Third order	Fourth order	Fifth order
C_0	1/6	19/120	3/16
C_1	5/6	3/4	251/360
C_2	1	1	1
C_3	1/3	1/2	11/18
C_4	0	1/12	1/6
C_5	0	0	1/60

7.4 Symplectic Algorithm

In atomistic and molecular simulations, a molecular configuration at t, $\eta(t) = (\mathbf{q}^N, \mathbf{p}^N)$ is a set in the phase space. With an infinitesimal time increment, the system travels to a new state:

$$\eta(t + \Delta t) = \eta(t) + \dot{\eta}(t)\Delta t = \eta(t) + \mathbf{\Omega} \cdot \left.\frac{\partial H(\eta)}{\partial \eta}\right|_t \Delta t$$

which can be considered as an infinitesimal coordinate transformation in the phase space, i.e.,

$$\eta(t + \Delta t) = \xi(\eta(t)).$$

At a specific t, $\eta(t) = (\mathbf{q}, \mathbf{p})$ is a point in the phase space. With an infinitesimal time increment, the system travels to

$$\xi = \eta + \dot{\eta}\Delta t = \eta + \mathbf{\Omega} \cdot \frac{\partial H(\eta)}{\partial \eta}\Delta t, \tag{7.63}$$

which can be considered as an infinitesimal coordinate transformation in the phase space. The Jacobian matrix of the transformation is

$$\mathbf{M} = \frac{\partial \xi}{\partial \eta} = \mathbf{I} + \mathbf{\Omega} \cdot \frac{\partial^2 H(\eta)}{\partial \eta \partial \eta}\Delta t. \tag{7.64}$$

More precisely,

$$\mathbf{M} = \mathbf{I}^{6 \times 6} + \begin{bmatrix} 0 & \mathbf{I} \\ -\mathbf{I} & 0 \end{bmatrix} \begin{bmatrix} \frac{\partial^2 H}{\partial \mathbf{q} \partial \mathbf{q}} & \frac{\partial^2 H}{\partial \mathbf{q} \partial \mathbf{p}} \\ \frac{\partial^2 H}{\partial \mathbf{p} \partial \mathbf{q}} & \frac{\partial^2 H}{\partial \mathbf{p} \partial \mathbf{p}} \end{bmatrix} \Delta t$$

$$= \begin{bmatrix} \mathbf{I} + \frac{\partial^2 H}{\partial \mathbf{q} \partial \mathbf{q}}\Delta t & \frac{\partial^2 H}{\partial \mathbf{q} \partial \mathbf{p}}\Delta t \\ -\frac{\partial^2 H}{\partial \mathbf{p} \partial \mathbf{q}}\Delta t & \mathbf{I} - \frac{\partial^2 H}{\partial \mathbf{p} \partial \mathbf{p}}\Delta t \end{bmatrix}.$$

Since $d\xi = \mathbf{M}d\eta$, the determinant of the Jacobian is

$$det|\mathbf{M}| = \left|\frac{\partial \xi}{\partial \eta}\right| = 1 + \left(\frac{\partial^2 H}{\partial \mathbf{q}^2} : \frac{\partial^2 H}{\partial \mathbf{p}^2} - \left(\frac{\partial^2 H}{\partial \mathbf{q}\partial \mathbf{p}}\right)^2\right)\Delta t^2,$$
$$\approx 1 + O(\Delta t^2)$$

which means that as $\Delta t \to 0$, the volume of the phase space is invariant. If one translates $\Delta \xi$ over a finite period of time and $\Delta t = T/N$,

$$|\Delta \xi_N| \sim |\Delta \eta| + O(\Delta t).$$

The local volume of the phase space is invariant as $\Delta t \to 0$. This is the infinitesimal invariant.

In the following, we demonstrate that the velocity Verlet algorithm is a (absolutely) symplectic algorithm. Considering the form of three-step algorithm, we have

$$\mathbf{p}_i\left(t + \frac{\Delta t}{2}\right) = \mathbf{p}_i(t) + \frac{\Delta t}{2}\mathbf{F}_i(t) \tag{7.65}$$

$$\mathbf{r}_i(t + \Delta t) = \mathbf{r}_i(t) + \mathbf{p}_i\left(t + \frac{\Delta t}{2}\right)\Delta t/m_i \tag{7.66}$$

$$\mathbf{p}_i(t + \Delta t) = \mathbf{p}_i\left(t + \frac{\Delta t}{2}\right) + \frac{\Delta t}{2}\mathbf{F}_i(t + \Delta t). \tag{7.67}$$

The above algorithm may be viewed as a three-step variable transformation of $\eta = (\mathbf{r}, \mathbf{p})$, whose Jacobian is unity, i.e., it is volume preserving. To show this, we examine the algorithm in details:

- **Step 1**

First,

$$\xi_1(t) = \eta_1\left(t + \frac{\Delta t}{2}\right) \to \begin{cases} \mathbf{r}_i(t) &= \mathbf{r}_i(t) \\ \mathbf{p}_i(t + \Delta t/2) &= \mathbf{p}_i(t) + \frac{\Delta t}{2}\mathbf{F}_i(t) \end{cases} \tag{7.68}$$

we have

$$det M_1 = det \begin{vmatrix} 1 & 0 \\ \frac{\Delta t}{2}\frac{\partial \mathbf{F}}{\partial \mathbf{r}} & 1 \end{vmatrix} = 1.$$

- **Step 2**

Second,

$$\xi_2 = \eta\left(t + \frac{\Delta t}{2}\right) \to \begin{cases} \mathbf{r}_i(t + \Delta t) &= \mathbf{r}_i(t) + \frac{\Delta t}{m}\mathbf{p}_i(t + \Delta t/2) \\ \mathbf{p}_i(t + \Delta t/2) &= \mathbf{p}_i(t + \Delta t/2) \end{cases}, \tag{7.69}$$

we have

$$det M_2 = det \begin{vmatrix} 1 & \Delta t/m \\ 0 & 1 \end{vmatrix} = 1.$$

- **Step 3**

  In the third and he last step,

  $$\xi_3 = \eta(t + \Delta t) \rightarrow \begin{cases} \mathbf{r}_i(t + \Delta t) = \mathbf{r}_i(t + \Delta t) \\ \mathbf{p}_i(t + \Delta t) = \mathbf{p}_i(t + \frac{\Delta t}{2}) + \frac{\Delta t}{2}\mathbf{F}_i(t + \Delta t), \end{cases} \quad (7.70)$$

  we then have

  $$\det M_3 = \det \begin{vmatrix} 1 & 0 \\ \frac{\Delta t}{2}\mathbf{F}(t + \Delta t) & 1 \end{vmatrix} = 1.$$

  In summary, we have

  $$\det(\mathbf{M}) = (\det \mathbf{M}_1)(\det \mathbf{M}_2)(\det \mathbf{M}_3) = 1,$$

  which is not dependent on the size of the time increment. Therefore, the phase space volume is absolutely invariant.

7.5 Homework Problems

Problem 7.1 Find the truncation error for the velocity Verlet algorithm, and explain why the velocity Verlet algorithm is a second-order integrator.

8 Temperature Control in MD Simulations

In molecular dynamics (MD), the instantaneous temperature of a molecular system is related to the kinetic energy through each atom's momentum as follows:

$$\sum_{i=1}^{N} \frac{|\mathbf{p}_i|^2}{2m_i} = \frac{k_B T}{2}(3N - N_c) \tag{8.1}$$

where N is the total number of atoms, N_c is the number of constraints, and $3N - N_c = N_{df}$ is the total number of degrees of freedom. The ensemble average temperature is denoted as $<T>$ that is usually regarded as the macroscopic temperature of the system, which will be discussed later.

The ensemble average temperature is usually defined and acquired in an ensemble MD, in which the temperature is fixed. However, for an arbitrary ensemble equilibrium MD, for instance the microcanonical ensemble $<NVE>$ in which temperature is not fixed, or even a non-equilibrium MD, we can always use Eq. (8.1) to calculate the instantaneous temperature of a molecular system.

There are several MD techniques used in practice to control temperature by adding additional terms in the dynamic equations of MD, which we often call as thermostat. Among them, we may roughly divide them into three categories:

(i) Scaling velocities
  - Simple velocity scaling, e.g., isokinetic thermostat,
  - Berendsen thermostat;
(ii) Using stochastic forces or velocities
  - Andersen thermostat,
  - Langevin thermostat;
(iii) Using "extended Lagrangian" formalism
  - Nosé–Hoover thermostat,
  - Other thermostats.

8.1 Velocity Scaling

8.1.1 Simple Velocity Scaling

The simplest way to control temperature is to adjust velocity of each atom so that the temperature of the molecular system can be controlled at the desired value.

To do so, the velocities of all atoms are multiplied by a scaling factor λ at each time step, i.e.,

$$\lambda = \sqrt{\frac{T_0}{T_i}}, \tag{8.2}$$

where T_0 is the desired or targeted temperature and T_i is the instantaneous temperature at the ith time step.

The algorithm may be summarized as follows:

$$T(t) = \frac{1}{N} \sum_i \frac{m_i v_i^2}{2} /(k_B/2)$$

$$\Delta T = \frac{1}{N} \sum_i \frac{m_i (\lambda v_i)^2}{2} /(k_B/2) - \frac{1}{N} \sum_i \frac{m_i v_i^2}{2} /(k_B/2)$$

$$\Delta T = (\lambda^2 - 1) T(t)$$

$$\lambda = \sqrt{\frac{T_0}{T}} \quad \leftarrow \quad \text{Simple velocity scaling,}$$

where T_0 is the target temperature.

The following is a Fortran code script that shows how the simple velocity scaling is being implemented:

```
time: do step=1,Nsteps
   call Refold_Positions
   pos = pos + deltat*vel + 0.5d0*(deltat**2)*acc     ! r(t+dt)
   if (ConstantT .and. (temperature > 0) ) then      ! veloc rescale for const T
      call Compute_Temperature(ene_kin_aver,temperature) ! T(t)
      chi = sqrt( Trequested / temperature )
      vel = chi*vel + 0.5d0*deltat*acc                ! v(t+dt/2)
   else                                               ! regular constant E dynamics
      vel = vel + 0.5d0*deltat*acc                    ! v(t+dt/2)
   endif
   call Compute_Forces                                ! a(t+dt),ene_pot,virial
   vel = vel + 0.5d0*deltat*acc                       ! v(t+dt)
   call Compute_Temperature(ene_kin_aver,temperature) ! at t+dt, also ene_kin

   ene_pot_aver = sum( ene_pot ) / N
   ene_tot_aver = ene_kin_aver + ene_pot_aver
```

Even though it is a quite efficient algorithm to control temperature, the simple velocity scaling algorithm will severely influence the dynamics of the atomistic system, and it does not produce right MD thermodynamics ensemble that carries out the correct thermodynamics information. Hence its simulation results may not provide the correct thermodynamic properties that one wants to simulate. Thus, in serious MD modelings, it may not be used for the purpose to extrapolate thermodynamics information, while it is perfectly right to use it in the initialization phase to raise or to cool down the temperature of the system.

8.1.2 Berendsen Thermostat

The most popular and simplest thermostat may be the Berendsen thermostat, which couples the atomistic system to an external bath with the target temperature T_0, so that the rate of change in temperature is proportional to the difference in temperature, i.e.

$$\frac{dT(t)}{dt} = \frac{1}{\tau}(T_0 - T(t)), \tag{8.3}$$

where τ is the coupling parameter that determines the strength of the coupling between the system and the heat bath, or a relaxation parameter that determines how fast the system relaxes into the desired target temperature T_0.

A discrete form of Eq. (8.3) is

$$\Delta T \approx \frac{\delta t}{\tau}(T_0 - T(t)), \tag{8.4}$$

where δt is the time step size.

Substituting

$$\Delta T = (\lambda^2 - 1)T(t) \rightarrow \frac{\delta t}{\tau}(T_0 - T(t)),$$

one may find

$$\lambda^2 = 1 + \frac{\delta}{\tau}\left(\frac{T_0}{T(t)} - 1\right). \tag{8.5}$$

Often times, we use $T(t - \delta t/2)$ replacing $T(t)$ in Eq. (8.5) when the leapfrog or some other second-order central difference type of integration algorithms are used.

The effect of the Berendsen thermostat may be described by the following modified equation of motion,

$$m_i \frac{d\mathbf{v}_i}{dt} = \mathbf{F}_i + \frac{m_i}{\tau}\left(\frac{T_0}{T} - 1\right)\mathbf{v}_i, \tag{8.6}$$

which uses a weak damping coefficient $\gamma = \tau^{-1}$ to couple the atomistic system to an external heat bath of the target temperature T_0.

8.2 Stochastic Thermostat

8.2.1 Andersen Thermostat

The Andersen thermostat was the first thermostat proposed for MD simulations, and it permits one to use the canonical ensemble (NVT) in MD simulations. H. C. Andersen (1980) introduced a stochastic thermostat to control the temperature of a molecular dynamics system by having random collisions of molecules with an imaginary heat bath at the desired temperature. In MD simulation, a random particle is chosen and its velocity in the ith direction ($i = 1,2,3$) is reassigned randomly from a Maxwell–Boltzmann distribution at the desired temperature,

$$v_i = \mathcal{G}(\tilde{v}) = \left(\frac{m}{2\pi k_B T}\right)^{1/2} \exp\left[-\frac{m\tilde{v}^2}{2k_B T}\right], \quad (8.7)$$

where $\tilde{v}$ is an effective velocity in the average sense, i.e.,

$$\tilde{v} = \sqrt{\frac{\sum_i \mathbf{v}_i \cdot \mathbf{v}_i}{3N}}.$$

The number of collisions in a short time interval varies randomly, and it is based on the Poisson distribution, i.e.,

$$P(\Delta t) = \nu \exp(-\nu \Delta t) \approx -\nu \Delta t.$$

The probability of collision between the time interval $[t, t + \Delta t]$ is

$$P_{col} = P(\nu)\Delta t \approx \nu \Delta t + O(\Delta t^2).$$

When a random generated number, $0 \leq ranf \leq 1$ for a given atom, is less than $\nu \Delta$, i.e., $ranf(i) < \nu \Delta t$, we assign a new velocity to the atom as,

$$v_i = \mathcal{G}(\tilde{v}). \text{ and } \tilde{v} = \sqrt{\frac{\sum_i \mathbf{v}_i \cdot \mathbf{v}_i}{3N}},$$

where ν is the stochastic collision frequency, which is set as

$$\nu = \frac{2a\kappa V^{1/3}}{3k_B N} = \frac{2a\kappa}{3k_B \rho^{1/3} N^{2/3}},$$

where a is a dimensionless constant, κ is the thermal conductivity, V is the volume, k_B is the Boltzmann constant, and $\rho = N/V$ is the atom number density.

8.2.2 Langevin Thermostat

The Langevin thermostat is based on the stochastic Langevin dynamics, which we shall discuss in detail in a later chapter. Here, we briefly introduce it as a computational algorithm for controlling the temperature in molecular dynamics simulations.

In a way, the Langevin dynamics is similar to the implicit solvation method, in which the explicit solvent atoms are removed, however, their effects on solute particles are represented by a solvent viscosity or friction force, i.e.,

$$m_i \frac{d\mathbf{v}_i}{dt} = -\frac{\partial U(\mathbf{r}^N)}{\partial \mathbf{r}_i} + \mathbf{f}_{friction}. \quad (8.8)$$

In the Langevin dynamics, this friction force is described as

$$\mathbf{f}_{friction} = -\gamma_i m_i \mathbf{v}_i + \mathbf{R}_i, \quad (8.9)$$

where γ_i is the friction coefficient, and the term $\mathbf{R}_i$ is a random force due to stochastic collisions with the solvent, and it must complement with the viscous force in order to recover the proper canonical ensemble. The random force $\mathbf{R}_i$ obeys the Gaussian distribution with zero mean and a variance that depends on the viscous force coefficient, particle mass, temperature, and the time step:

$$\sigma_i^2 = 2\frac{m_i \gamma_i k_B T}{\Delta t},$$

where k_B is the Boltzmann constant.

8.3 Nosé–Hoover Thermostat

8.3.1 Preparation

In 1985, following the idea of extended Lagrangian of H.C. Andersen, S. Nosé introduced an additional "generalized" coordinate and its momentum coordinate forming the following extended Lagrangian,

$$\mathcal{L}_{ext} = \sum_{i=1}^{N} \frac{1}{2} m_i s^2 \dot{\mathbf{r}}_i^2 - U(\mathbf{r}^N) + \frac{Q\dot{s}^2}{2} - \frac{\ell}{\beta} \ln s,$$

where $\beta = (k_B T)^{-1}$. Note that how this expression reduces to the original Lagrangian if we set $s = 1$. The parameter ℓ will be chosen later on, and Q can be viewed as an effective mass associated with the coordinate s (a scale parameter of time).

The moments that conjugate with $\mathbf{r}_i$ and s are given by,

$$\mathbf{p}_i = \frac{\partial \mathcal{L}_{ext}}{\partial \dot{\mathbf{r}}_i} = m_i s^2 \dot{\mathbf{r}}_i$$

$$p_s = \frac{\partial \mathcal{L}_{ext}}{\partial \dot{s}} = Q\dot{s}.$$

Using these momenta, we can obtain the Hamiltonian for the extended system, which contains N-particles and the additional degree of freedom s, and p_s, i.e., there are $6N + 2$ degrees of freedom

$$H_{Nosé} = \sum_{i=1}^{N} \frac{\mathbf{p}_i \cdot \mathbf{p}_i}{2 m_i s^2} + \frac{p_s^2}{2Q} + U(\mathbf{q}^N) + \frac{\ell}{\beta} \ln s,$$

where $\beta = (k_B T_o)^{-1}$.

Based on the Nosé Hamiltonian, we can derive the equations of motion for virtual variables as:

$$\frac{d\mathbf{q}_i}{dt} = \frac{\partial H_{Nosé}}{\partial \mathbf{p}_i} = \frac{\mathbf{p}_i}{m_i s^2}$$

$$\frac{d\mathbf{p}_i}{dt} = -\frac{\partial H_{Nosé}}{\partial \mathbf{q}_i} = -\frac{\partial U(\mathbf{q})}{\partial \mathbf{q}_i}$$

$$\frac{ds}{dt} = \frac{\partial H_{Nosé}}{\partial p_s} = \frac{p_s}{Q}$$

$$\frac{dp_s}{dt} = -\frac{\partial H_{Nosé}}{\partial s} = \frac{1}{s}\left(\sum_{i=1}^{N} \frac{\mathbf{p}_i \cdot \mathbf{p}_i}{m_i s^2} - \ell k_B T_0\right). \tag{8.10}$$

In actual computations, we may use the so-called real variables:

$$\mathbf{p}' = \frac{\mathbf{p}}{s} \tag{8.11}$$

$$p'_s = p_s/s \tag{8.12}$$

$$\mathbf{q}' = \mathbf{q} \tag{8.13}$$

$$s' = s \tag{8.14}$$

$$\Delta t' = \Delta t/s.$$

Note that the real variables are denoted with a prime $'$, whereas the virtual variables are without prime.

Thus, with the change of variables, we have, e.g.,

$$\mathbf{p}_i := m_i s^2 \dot{\mathbf{r}}_i \quad \rightarrow \quad \frac{1}{2} m_i s^2 \dot{\mathbf{r}}_i^2 = \frac{1}{2} \frac{\mathbf{p}_i \cdot \mathbf{p}_i}{m_i s^2} = \frac{1}{2} \frac{\mathbf{p}'_i \cdot \mathbf{p}'_i}{m_i}$$

and

$$\dot{\mathbf{r}}_i = \frac{d\mathbf{r}_i}{dt} \quad \text{and} \quad \dot{\mathbf{r}}'_i = \frac{d\mathbf{r}_i}{dt'} = s \frac{d\mathbf{r}_i}{dt} = s \dot{\mathbf{r}}_i.$$

Thus, the kinetic energy in the Hamiltonian has the form

$$\frac{1}{2} m_i s^2 \dot{\mathbf{r}}_i^2 = \frac{\mathbf{p}'_i \cdot \mathbf{p}'_i}{2 m_i} = \frac{1}{2} m_i \dot{\mathbf{r}}'_i \cdot \dot{\mathbf{r}}'_i.$$

As a consequence, variable s can be interpreted as a scaling factor in the time step. Hence, the real time step fluctuates during the simulation! This is because $\Delta t = s \Delta t'$. We can then rewrite the Nosé Hamiltonian in terms of virtual variables,

$$H_{Nosé} = \sum_{i=1}^{N} \frac{\mathbf{p}'_i \cdot \mathbf{p}'_i}{2 m_i} + \frac{p_s^2}{2Q} + U(\mathbf{q}^N) + \frac{\ell}{\beta} \ln s,$$

where $\beta = (k_B T_o)^{-1}$.

Recall that the partition of function of a constant–NVE ensemble is

$$Z_{Nosé} = c \int d\mathbf{p} d\mathbf{q} dp_s ds \, \delta(H_{Nosé} - E).$$

By introducing an new independent variable, $\mathbf{p}' = \mathbf{p}/s$, and let

$$H(\mathbf{p}', \mathbf{q}) = \sum_{i=1}^{N} \frac{\mathbf{p}'_i \cdot \mathbf{p}'_i}{2 m_i} + U(\mathbf{q}).$$

Thus, we have

$$Z_{Nosé} = c \int dp_s ds d\mathbf{p}' d\mathbf{q} s^{3N} \delta \left(H(\mathbf{p}', \mathbf{q}) + \frac{p_s^2}{2Q} + \frac{\ell}{\beta} \ln s - E \right).$$

8.3.2 Main Results and Computational Algorithm

By introducing the Nosé Hamitonian, $H_{nosé}$, we can show that the dynamics generated in microcanonical ensemble, i.e., $<NVE>$ ensemble, with $H_{Nosé}$ is equivalent to the dynamics generated in a canonical ensemble, i.e., $<NVT>$ ensemble. That is for any random field variable, we have

$$<A>_{NVE,Nosé} = <A>_{NVT}. \tag{8.15}$$

Before we prove Eq. (8.15), we first consider an important property of the Dirac function,

$$\delta(h(s)) = \frac{\delta(s-s_0)}{h'(s)}, \quad s_0 \text{ is the root of } h(s), \text{i.e. } h(s_0) = 0,$$

where $h(s)$ is a continuous function of s.

To show this, we can prove alternatively that

$$\int f(s)\delta[h(s)]ds = \int f(s)\frac{\delta(s-s_0)}{h'(s_0)}ds$$

Since $h(s)$ is continuous around s_0, we may let

$$h(s) = h(s_0) + h'(s_0)(s-s_0) + \frac{1}{2!}h''(s_0)(s-s_0)^2 \ldots,$$

where $h(s_0) = 0$. Hence,

$$\int f(s)\delta(h(s))ds = \int f(s)\delta\left((s-s_0)h'(s_0)\right)ds = \int f(t/h'(s_0))\delta(t-t_0)\frac{dt}{h'(s_0)},$$

where $t = sh'(s_0)$ and $t_0 = s_0 h'(s_0)$.
This is because,

$$\int f(s)\delta(h(s))ds = \int f(s)\delta((s-s_0)h'(s_0))ds = \int f(s)\delta(t-t_0)\frac{dt}{h'(s_0)}$$
$$= \int f(t/h'(s_0))\delta(t-t_0)\frac{dt}{h'(s_0)}, \quad \text{with } t = sh'(s_0)$$
$$= f(s_0)/h'(s_0).$$

On the other hand,

$$\int f(s)\frac{\delta(s-s_0)}{h'(s_0)}ds = \frac{f(s_0)}{h'(s_0)}.$$

Therefore,

$$\delta(h(s)) = \frac{\delta(s-s_0)}{h'(s_0)}.$$

Now, we consider a special case that

$$h(s) = \frac{\ell}{\beta}\ln s + \mathcal{H}(\mathbf{p'},\mathbf{r}) + \frac{p_s^2}{2Q} - E = \frac{\ell}{\beta}\ln s + X,$$

which has a root at

$$s_0 = \exp\left[-\frac{\beta}{\ell}X\right], \text{ and } X = \mathcal{H}(\mathbf{p'},\mathbf{r}) + \frac{p_s^2}{2Q} - E, \; h'(s) = \frac{\ell}{\beta s}.$$

Hence,

$$\delta\left(H(\mathbf{p'},\mathbf{q}) + \frac{p_s^2}{2Q} + (3N+1)K_BT\ln s - E\right)$$

$$= \frac{\beta s}{\ell}\delta(s - s_0)$$

$$= \frac{\beta s}{\ell}\delta\left\{s - \exp\left[\frac{\beta}{\ell}\left(E - \left(H(\mathbf{p'},\mathbf{q}) + \frac{p_s^2}{2Q}\right)\right)\right]\right\}.$$

Therefore,

$$Z_{Nos\acute{e}} = c\int dp_s ds d\mathbf{p'} d\mathbf{q}\frac{\beta s^{3N+1}}{\ell}\delta\left(s - \exp\left[\frac{\beta}{\ell}\left(E - (H(\mathbf{p'},\mathbf{q}) + \frac{p_s^2}{2Q})\right)\right]\right).$$

Using another fundamental property of the delta function,

$$\int f(s)\delta(s-a)ds = f(a)$$

we can find that

$$Z_{Nos\acute{e}} = c\int dp_s d\mathbf{p'} d\mathbf{q} \left(\frac{\beta}{\ell}\right)\exp\left[\frac{\beta}{\ell}\left(E - (H(\mathbf{p'},\mathbf{q}) + \frac{p_s^2}{2Q})\right)\right]^{3N+1}$$

$$= c\exp\left[\frac{(3N+1)\beta E}{\ell}\right]\int dp_s \exp\left[-\frac{(3N+1)p_s^2\beta}{2Q\ell}\right]$$

$$\int d\mathbf{p'} d\mathbf{q}\exp\left[-\frac{(3N+1)\beta H(\mathbf{p'},\mathbf{q})}{\ell}\right]$$

$$= C_1\int d\mathbf{p'} d\mathbf{q}\exp\left[-\frac{(3N+1)\beta H(\mathbf{p'},\mathbf{q})}{\ell}\right]$$

$$\propto Z_{NVT} = c\int d\mathbf{p} d\mathbf{q}\exp\left[-\frac{H(\mathbf{p},\mathbf{q})}{K_BT}\right],$$

if we choose $\ell = (3N+1)$ and $\beta = (k_BT)^{-1}$.

Based on similar procedures, we can show that for a given field variable $A(\mathbf{p'},\mathbf{p})$,

$$c\int A(\mathbf{p'},\mathbf{p})dp_s d\mathbf{p'} d\mathbf{q}\left(\frac{\beta}{\ell}\right)\exp\left[\frac{\beta}{\ell}\left(E - (H(\mathbf{p'},\mathbf{q}) + \frac{p_s^2}{2Q})\right)\right]^{3N+1}$$

$$= C_1\int A(\mathbf{p'},\mathbf{p})d\mathbf{p'} d\mathbf{q}\exp\left[-\frac{H(\mathbf{p'},\mathbf{q})}{K_BT}\right].$$

Now we have proved that

$$\langle A(\mathbf{p'},\mathbf{q}) \rangle_{NVE,Nosé} = \frac{\int A(\mathbf{p'},\mathbf{q})\exp\left[-\frac{H(\mathbf{p'},\mathbf{q})}{K_B T}\right]d\mathbf{p'}d\mathbf{q}}{\int \exp\left[-\frac{H(\mathbf{p'},\mathbf{q})}{K_B T}\right]d\mathbf{p'}d\mathbf{q}} \equiv \langle A(\mathbf{p'},\mathbf{q}) \rangle_{NVT}.$$

Shuichi Nosé (Fig. 8.1) published this results in 1984 (Nosé 1984).

8.3.3 Hoover's Modifications

Hoover simplified Nosé's dynamics equations by introducing

$$\zeta = \frac{p_s}{Q} = \frac{sp'_s}{Q} = \dot{s} = \frac{d(\ln s)}{dt'},$$

which lead to the following variable transformations:

$$\frac{d\mathbf{q}_i}{dt} = \frac{\mathbf{p}_i}{m_i s^2} \qquad\qquad \frac{d\mathbf{q}'_i}{dt'} = \frac{\mathbf{p}'_i}{m_i}$$

$$\frac{d\mathbf{p}_i}{dt} = -\frac{\partial U(\mathbf{q})}{\partial \mathbf{q}_i} \qquad\Rightarrow\qquad \frac{d\mathbf{p}'_i}{dt'} = -\frac{\partial U}{\partial \mathbf{q}'_i} - \zeta \mathbf{p}'_i$$

$$\frac{ds}{dt} = \frac{p_s}{Q} \qquad\qquad \dot{s} = \zeta = \frac{d \ln s'}{dt'};$$

$$\frac{dp_s}{dt} = \left(\sum_{i=1}^{N} \frac{\mathbf{p}_i \cdot \mathbf{p}_i}{m_i s^2} - \ell k_B T_0\right)/s \qquad \frac{d\zeta}{dt'} = \frac{1}{Q}\left(\sum_{i=1}^{N} \frac{\mathbf{p}'_i \cdot \mathbf{p}'_i}{m_i} - \ell k_B T_0\right).$$

Considering the equation of motion of rigid particles in the Newton's second law, we have

$$m_i \frac{d^2\mathbf{q}'_i}{dt'^2} = -\frac{\partial U(\mathbf{q}^N)}{\partial \mathbf{q}_i} - \zeta \frac{d\mathbf{q}'_i}{dt'},$$

$$\frac{d\zeta}{dt'} = \frac{1}{Q}\left(\sum_i m_i \dot{\mathbf{q}}'_i \cdot \dot{\mathbf{q}}'_i - \ell k_B T_0\right),$$

where $\ell = 3N + 1$.

Moreover, if we define,

$$T_{ins} = \frac{1}{(3N+1)k_B}\sum_i m_i \dot{\mathbf{q}}'_i \cdot \dot{\mathbf{q}}'_i, \rightarrow \frac{d\zeta}{dt'} = \frac{(3N+1)k_B}{Q}(T - T_o).$$

In fact, from the Nosé Hamiltonian,

$$\mathcal{H}_{Nosé} = \sum_{i=1}^{N} \frac{\mathbf{p}_i^2}{2m_i s^2} + U(\mathbf{r}^N) + \frac{p_s^2}{2Q} + \frac{\ell}{\beta}\ln s,$$

we can derive the equations of motion for virtual variables as:

$$\frac{d\mathbf{r}_i}{dt} = \frac{\partial \mathcal{H}_{Nosé}}{\partial \mathbf{p}_i} = \frac{\mathbf{p}_i}{m_i s^2}; \quad \frac{d\mathbf{p}_i}{dt} = -\frac{\partial \mathcal{H}_{Nosé}}{\partial \mathbf{r}_i} = -\frac{\partial U(\mathbf{r}^N)}{\partial \mathbf{r}_i}$$

and

$$\frac{ds}{dt} = \frac{\partial \mathcal{H}_{Nosé}}{\partial p_s} = \frac{p_s}{Q}, \quad \text{and} \quad \frac{dp_s}{dt} = -\frac{\partial \mathcal{H}_{Nosé}}{\partial s} = \frac{1}{s}\left(\sum_i \frac{\mathbf{p}_i^2}{m_i s^2} - \frac{\ell}{\beta}\right).$$

We can also write the equations of motion in terms of real variables as

$$\frac{d\mathbf{r}'_i}{dt'} = \frac{\mathbf{p}'_i}{m_i}, \quad \frac{d\mathbf{p}'_i}{dt'} = -\frac{\partial U(\mathbf{r}^N)}{\partial \mathbf{r}'_i} - (s'p'_s/Q)\mathbf{p}'_i$$

and

$$\frac{1}{s'}\frac{ds'}{dt'} = s'p'_s/Q, \quad \text{and} \quad \frac{d}{dt'}(s'p'_s/Q) = \left(\sum_i \frac{\mathbf{p}'^2_i}{m_i} - \frac{\ell}{\beta}\right)/Q.$$

We can then express the Nosé Hamiltonian in terms of real variables,

$$\mathcal{H}'_{Nose}(\mathbf{r}'^N, \mathbf{p}'^N, s', p'_s) = \sum_{i=1}^{N} \frac{\mathbf{p}'^2_i}{2m_i} + U(\mathbf{r}'^N) + \frac{s'^2 p'^2_s}{2Q} + \frac{\ell}{\beta}\ln s'.$$

The equations of motion of real variables will render,

$$\frac{d\mathcal{H}'}{dt'} = \sum_i \left(\frac{\partial \mathcal{H}'}{\partial \mathbf{p}'_i}\frac{d\mathbf{p}'_i}{dt'} + \frac{\partial \mathcal{H}'}{\partial \mathbf{r}'_i}\frac{d\mathbf{r}'_i}{dt'} + \frac{\partial H'}{\partial p'_s}\frac{dp'_s}{dt'} + \frac{\partial \mathcal{H}'}{\partial s'}\frac{ds'}{dt'}\right) = 0.$$

Hence the Hamiltonian is conserved. However, we cannot derive the equations of motion for real variables from $\mathcal{H}'$, i.e.,

$$\frac{d\mathbf{p}'}{dt'} \neq -\frac{\partial \mathcal{H}'}{\partial \mathbf{r}'_i}, \quad \text{and} \quad \frac{dp'_s}{dt'} \neq -\frac{\partial \mathcal{H}'}{\partial s'}.$$

A velocity histogram of the Nosé–Hoover dynamics is shown in Fig. 8.2.

8.3.4 Nosé–Hoover Chain

In computation practice, the Nosé–Hoover dynamic equations may become very stiff. Therefore, instead of one thermostat, a chain of coupled thermostats may be used. For a system of N atoms with M thermostats, the equations of motion in terms of real variables are

$$\dot{\mathbf{q}}_i = \frac{\mathbf{p}_i}{m_i}, \quad i = 1, 2, \ldots, N \tag{8.16}$$

$$\dot{\mathbf{p}}_i = -\frac{\partial U(\mathbf{q}^N)}{\partial \mathbf{q}_i} - \frac{P_{\zeta_1}}{Q_1}\mathbf{p}_i \tag{8.17}$$

$$\dot{\zeta}_k = \frac{P_{\zeta_k}}{Q_k}, \quad k = 1, 2, \ldots, M \tag{8.18}$$

$$\dot{P}_{\zeta_1} = \left(\sum_{i=1}^{N} \frac{p_i p_i}{m_i} - \frac{\ell}{\beta}\right) - \frac{P_{\zeta_2}}{Q_2} P_{\zeta_1} \tag{8.19}$$

$$\dot{P}_{\zeta_k} = \left(\frac{P_{\zeta_{k-1}}^2}{Q_{k-1}} - \frac{1}{\beta}\right) - \frac{P_{\zeta_{k+1}}}{Q_{k+1}} P_{\zeta_k}, \quad k = 2, \ldots, M-1 \tag{8.20}$$

$$\dot{P}_{\zeta_M} = \left(\frac{P_{\zeta_{M-1}}^2}{Q_{M-1}} - \frac{1}{\beta}\right). \tag{8.21}$$

For the Nosé–Hoover chain, it can be shown that the following quantity is conserved:

$$H_{NHC} = \mathcal{H}(\mathbf{r}^N, \mathbf{p}^N) + \sum_{k=1}^{M} \frac{p_{\zeta_k}^2}{2Q_k} + \frac{\ell}{\beta}\zeta_1 + \sum_{k=2}^{M} \frac{1}{\beta}\zeta_k.$$

REMARK 8.1 **1. Why the Nosé–Hoover thermostat is time reversible?**
If one examines the equations of motion of the Nosé–Hoover MD,

$$m_i \frac{d^2 \mathbf{q}'_i}{dt'^2} = -\frac{\partial U(\mathbf{q}^N)}{\partial \mathbf{q}_i} - \zeta \frac{d\mathbf{q}'_i}{dt'},$$

$$\frac{d\zeta}{dt'} = \frac{1}{Q}\left(\sum_i m_i \mathbf{q}'_i \cdot \mathbf{q}'_i - \ell k_B T_o\right),$$

where $\ell = 3N + 1$ and

$$\zeta = \frac{p_s}{Q} = \frac{s p'_s}{Q} = \dot{s} = \frac{d(\ln s)}{dt'},$$

one may find that the above equations contain even expressions in t' and dt'. Thus, when t' change sign, the equations are invariant.

2. Q is a parameter or a virtual mass for s that has the dimension of *energy* $\times$ *(time)*^{2}. The choice of Q is crucial, and it controls how fast the system relaxes into a thermal equilibrium state, i.e., it may serve as a timescale of thermal fluctuations. When $Q \to \infty$, $d\zeta/dt' = 0$, if $\zeta(0) = 0$, the system degenerates to a microcanonical NVE ensemble.

8.3.5 Shuichi Nosé (1951–2005): In Memorium

After his doctoral work at Kyoto, Shuichi Nosé travelled to faroff Canada for three years as a postdoctoral fellow. He worked in relative isolation, listening to Dvorâk recordings, ice skating, and developing a new and fruitful approach to Gibbs statistical mechanics. I was excited by the new ideas he described in his two fundamental

Figure 8.1 Shuichi Nosé (June 17, 1951–August 17, 2005) was a Japanese chemist, who is best known for his contribution to canonical ensemble molecular dynamics, specifically for proposing the Nosé–Hoover thermostat (Photo courtesy of William G. Hoover)

papers of 1984. When I saw them, I immediately began to study and absorb the significance of this revolutionary approach to statistical mechanics. In the two papers, Nosé introduced and applied new algorithms for deterministic time-reversible simulations of isothermal systems. His new dynamics exactly reproduced Gibbs canonical ensemble. This was an intellectual feat! For 100 years dynamics and statistics were regarded as two distinct approaches. Nos'e united them. Today the applications of this development are well known, widespread, and far from exhausted.

That same year, 1984, I met Nosé for the first time, on a train platform in Paris – a real unplanned and unpredictable coincidence – my plane had been diverted to Orly from de Gaulle. While waiting at the airports train platform I noticed a Japanese with NOSE lettered on his suitcase. Was he the same Shuichi Nosé whose work I had been studying? Yes? he said, and we got acquainted. We were both in Paris to attend, several days later, a workshop on statistical mechanics and simulation organized by an early fan of parallel processing, Carl Moser. Our hotels were both within an easy walk of Notre Dame, so Nosé and I arranged to meet there, and had several hours discussions of his work on a bench in front of the cathedral. I learned enough from our Paris conversations to write a paper shortly thereafter (while visiting Philippe Choquard in Lausanne) stressing the importance of what is called the Nosé–Hoover version of his dynamical equations. Ever since 1984 I have worked hard to apply the new ideas to the understanding of both equilibrium and nonequilibrium systems.

I got to know Shuichi better in Japan, along with his charming wife Ibuki and (then) young son Atsushi (now a budding engineer). Shuichi had arranged for me to spend a sabbatical year with him at Keio University's Hiyoshi campus. My new wife Carol came too, and worked at Keio with Toshio Kawai and Taisuke Boku. Shortly after Carol and I were settled in Japan, we were invited to an elaborate dinner honoring Shuichi, sponsored by International Business Machines, which awarded him its Japan Science Prize in 1989. It was a great pleasure to see his parents enjoying their son's fame. We all watched a large-screen biographical video that IBM had made for this special occasion.

Nosé-san had arranged a very comfortable western-style apartment for us at Nestle House, within walking distance of the Keio campus. The walk passed by the preschool where we often saw Atsushi at play. At the University, Nosé-san arranged for all the amenities, along with plenty of computer time and numerous conference invitations. Nosé's 1984 work changed my own research direction profoundly, toward exploring the nonequilibrium applications of Nosé–Hoover mechanics. The resulting interpretation of nonequilibrium distribution functions as zero-measure fractals, rather than smooth distributions, provided a fascinating explanation of irreversibility. Nosé's approach made analyses of such systems much simpler. Nosé pursued a variety of topics, all related to computer simulation, after his revolutionary papers of 1984. His lasting legacy remains that early work together with the memories of those who knew him. I count myself fortunate to remember Shuichi. His early death is a shock for physics, and painful for his family and friends.

— By William Graham Hoover

8.4 How to Integrate Nosé–Hoover MD?

The integration of the Nosé–Hoover MD could get complicated, if one hopes to generate correct thermodynamics. In this section, we discuss several integration algorithms that are used to integrate the Nosé–Hoover MD.

The final equations of motions are:

$$m_i \frac{d^2 \mathbf{q}'_i}{dt'^2} = -\frac{\partial U(\mathbf{q}'^N)}{\partial \mathbf{q}_i} - \zeta \frac{d\mathbf{q}'_i}{dt'} = \mathbf{F}_i - \zeta \frac{d\mathbf{q}'_i}{dt'}$$

$$\frac{d\zeta}{dt'} = \frac{1}{Q}\left(\sum_i m_i \dot{\mathbf{q}}'_i \cdot \dot{\mathbf{q}}'_i - \ell k_B T_0\right),$$

where $\ell = 3N + 1$.

With the definition of instantaneous temperature,

$$T_{ins} = \frac{1}{3Nk_B}\sum_i m_i \dot{\mathbf{q}}'_i \cdot \dot{\mathbf{q}}'_i,$$

we have
$$\rightarrow \frac{d\zeta}{dt'} = \frac{3Nk_B}{Q}(T - T_0).$$

A simple integration algorithm for the Nosé–Hoover MD is to modify the following velocity Verlet algorithm:

$$\mathbf{q}_i(t + \Delta t) = \mathbf{q}_i(t)\Delta t + \frac{1}{2}\frac{\mathbf{f}_i(t)}{m_i}\Delta t^2 \quad (8.22)$$

$$\dot{\mathbf{q}}_i(t + \Delta t) = \dot{\mathbf{q}}_i(t) + \frac{1}{2m_i}\left(\mathbf{f}_i(t) + \mathbf{f}_i(t + \Delta t)\right)\Delta t. \quad (8.23)$$

For the Nosé–Hoover equations of motion, the velocity Verlet algorithm can be written as:

$$\mathbf{q}_i(t + \Delta t) = \mathbf{q}_i(t) + \dot{\mathbf{q}}_i(t)\Delta t + \frac{1}{2m_i}\left[\mathbf{f}_i(t) - \zeta(t)\dot{\mathbf{q}}_i(t)\right]\Delta t^2 \quad (8.24)$$

$$\zeta(t + \Delta t) = \zeta(t) + \frac{\Delta t}{Q}\left(\sum_{i=1}^{N}\frac{\mathbf{p}_i(t) \cdot \mathbf{p}_i(t)}{m_i} - 3Nk_BT\right) \quad (8.25)$$

$$\dot{\mathbf{q}}_i(t + \Delta t) = \dot{\mathbf{q}}_i(t) + \frac{1}{2m_i}\left[\mathbf{f}_i(t) + \mathbf{f}_i(t + \Delta t) - (\dot{\zeta}(t)\dot{\mathbf{q}}_i + \dot{\zeta}(t + \Delta t)\dot{\mathbf{q}}_i(t + \Delta t))\right]\Delta t. \quad (8.26)$$

In computer implementation, the above algorithm may be written in the following code form:

```
IF (Switch.EQ.1) THEN
! --- input: force and velocity at time t
! --- output: position and velocity at time  t+Delta
====solve equations of motion

delt2 = Delta*Delta/2.0d0
delth = Delta/2.0d0
sumv2 = 0.0d0

Do i = 1, NPART
     X(i) = X(i) + Delt*VX(i) + delta2*(Fx(i) ?PS*VX(i))
     Y(i) = Y(i) + Delt*VY(i) + delta2*(Fy(i) ?PS*VY(i))
     Z(i) = Z(i) + Delt*VZ(i) + delta2*(Fz(i)  ?PS*VZ(i))
     sumv2 = sumv2 +VX(i)**2+VY(i)**2 +VZ(i)**2
     VX(i) = VX(i)  +  delth*(Fx(i) ?PS*VX(i))
     VY(i) = VY(i)  +  delth*(Fy(i) ?PS*VY(i))
     VZ(i) = VZ(i)  +  delth*(Fz(i) ?PS*VZ(i))
ENDDO
     S = S + PS*Delt + (sumv2 -G*Temp)*delt2/Q
     PS =PS + (sumv2 -G*Temp)*delth/Q
ELSEIF (Switch. EQ. 2) THEN
```

8.4.1 Lie–Trotter Product Formula for Nosé–Hoover Dynamics

In a landmark paper, Martyna et al. (1996) applied the Trotter algorithm integrating the Nosé–Hoover MD. Considering

$$iL = iL_{NHC} + iL_p + iL_r,$$

they use a simple generalization of the Trotter formula for the Nosé–Hoover MD, and the evolution operator can be written as

$$\exp(iL\Delta t) = \exp\left(iL_{NHC}\frac{\Delta t}{2}\right)\exp\left(iL_p\frac{\Delta t}{2}\right)\exp(iL_r\Delta t)\exp\left(iL_p\frac{\Delta t}{2}\right)\exp\left(iL_{NHC}\frac{\Delta t}{2}\right) + O(\Delta t^3).$$

Here, the following form of the Trotter identity is used,

$$\exp(\mathbf{A}t + \mathbf{B}t + \mathbf{C}t) = \lim_{N\to\infty} \left(\exp(\mathbf{A}t/(2M))\exp(\mathbf{B}t/2M)\exp(\mathbf{C}t/M)\exp(\mathbf{B}t/2M)\exp(\mathbf{A}t/2M)\right)^M$$

or

$$\exp(iL_{NHC}t + iL_p t + iL_r t) =$$
$$\lim_{M\to\infty} \left(\exp(iL_{NHC}t/(2M))\exp(iL_p t/(2M))\exp(iL_r t/M)\exp(iL_p t/2M)\exp(iL_{NHC}t/(2M))\right)^M,$$

where

$$t = \Delta t M, \quad \Delta t = \frac{t}{M}.$$

In the following, in the spirit of the Trotter splitting algorithm, we present a multilevel Trotter algorithm that splits the differential operator system in Nosé–Hoover MD into the following three sets of equations:

$$\dot{\zeta} = \frac{1}{Q}\left(\sum_{i=1}^{N}\frac{\mathbf{p}_i\cdot\mathbf{p}_i}{m_i} - 3NK_BT\right), \quad m_i\ddot{\mathbf{q}}_i = \mathbf{F}_i - \zeta\dot{\mathbf{q}}_i$$

Step I: $\dot{\zeta} = \dfrac{1}{Q}\left(\displaystyle\sum_{i=1}^{N}\dfrac{\mathbf{p}_i\cdot\mathbf{p}_i}{m_i} - 3NK_BT\right), \quad m_i\ddot{\mathbf{q}}_i = -\zeta\dot{\mathbf{q}}_i, \quad \mathbf{A}\dfrac{t}{2N} = \mathbf{A}\dfrac{\Delta t}{2}$

Step II: $m_i\ddot{\mathbf{q}}_i = \mathbf{F}_i, \quad \mathbf{B}\dfrac{t}{N} = \mathbf{A}\Delta t$

Step III: $\dot{\zeta} = \dfrac{1}{Q}\left(\displaystyle\sum_{i=1}^{N}\dfrac{\mathbf{p}_i\cdot\mathbf{p}_i}{m_i} - 3NK_BT\right), \quad m_i\ddot{\mathbf{q}}_i = -\zeta\dot{\mathbf{q}}_i, \quad \mathbf{A}\dfrac{t}{2N} = \mathbf{A}\dfrac{\Delta t}{2}.$

Recall the Trotter identity again

$$\exp(iL_p t + iL_r t) = \lim_{N\to\infty} \left(\exp(iL_p t/(2N))\exp(iL_r t/N)\exp(iL_p t/2N)\right)^N.$$

In each of the above steps, we use the Trotter algorithm to split the integration into another set of substeps:

8 Temperature Control in MD Simulations

In Step I, we have

$$\dot{\zeta} = \frac{1}{Q}\left(\sum_{i=1}^{N} \frac{\mathbf{p}_i \cdot \mathbf{p}_i}{m_i} - 3NK_BT\right), \quad m_i\ddot{\mathbf{q}}_i = -\zeta\dot{\mathbf{q}}_i.$$

We further split them into **Step I.** Thermostat step (half step $\Delta t/2$)

I-1: $\quad \zeta(t + \Delta t/4) = \zeta(t) + \dfrac{\Delta t}{4}\dfrac{2(E_{kin} - \sigma)}{Q}$

I-2: $\quad \mathbf{v}^{(1)}(t + \Delta t/2) = \mathbf{v}(t)\exp\left(-\zeta(t + \Delta t/4)\dfrac{\Delta t}{2}\right)$

I-3: $\quad \zeta(t + \Delta t/2) = \zeta(t + \Delta t/4) + \dfrac{\Delta t}{4}\dfrac{2(E_{kin} - \sigma)}{Q}.$

Note that here the reduced unit is used.
In Step II, we have,

$$m_i\ddot{\mathbf{q}}_i = \mathbf{F}_i.$$

We further split it into **Step II.** Velocity Verlet step (full step Δt)

Get Force at t

II-1: $\quad \mathbf{v}_i^{(2)}(t + \Delta t/2) = \mathbf{v}_i^{(1)}(t) + \dfrac{\Delta t}{2}\dfrac{\mathbf{f}_i}{m_i}$

II-2: $\quad \mathbf{r}_i(t + \Delta t) = \mathbf{r}(t) + \Delta t\,\mathbf{v}_i^{(2)}(t + \Delta t/2)$

Get Force at t+Δt

II-3: $\quad \mathbf{v}^{(2)}(t + \Delta t) = \mathbf{v}^{(2)}(t + \Delta t/2) + \dfrac{\Delta t}{2}\dfrac{\mathbf{f}_i}{m_i}.$

Last, in Step III, we have

$$\dot{\zeta} = \frac{1}{Q}\left(\sum_{i=1}^{N} \frac{\mathbf{p}_i \cdot \mathbf{p}_i}{m_i} - 3NK_BT\right), \quad m_i\ddot{\mathbf{q}}_i = -\zeta\dot{\mathbf{q}}_i.$$

We split them into **Step III.** Second thermostat step (half step $\Delta t/2$),

III-1: $\quad \zeta(t + 3\Delta t/4) = \zeta(t + \Delta t/2) + \dfrac{\Delta t}{4}\dfrac{2(E_{kin} - \sigma)}{Q}$

III-2: $\quad \mathbf{v}(t + \Delta t) = \mathbf{v}^{(2)}(t + \Delta t)\exp\left(-\zeta(t + 3\Delta t/4)\dfrac{\Delta t}{2}\right)$

III-3: $\quad \zeta(t + \Delta t) = \zeta(t + 3\Delta t/4) + \dfrac{\Delta t}{4}\dfrac{2(E_{kin} - \sigma)}{Q}.$

REMARK 8.2 This elegant algorithm is also a symplectic algorithm, and it can preserve desirable quantities such as Hamiltonian and phase space volume (Poisson's bracket). It is a consistently third-order accuracy algorithm, and we call it Trotter (identity) inside Trotter (identity).

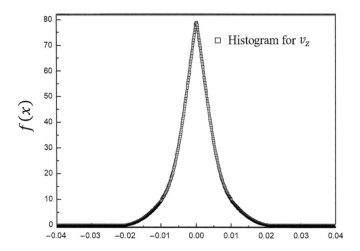

Figure 8.2 Velocity historgram for Nosé–Hoover dynamics

8.5 Other Thermostats

8.5.1 Evans–Hoover Thermostat: Isokinetic MD

Consider the following Gauss' least squares principle,

$$\operatorname*{argmin}_{\ddot{r}} C(\ddot{\mathbf{r}}_i) = \operatorname*{argmin}_{\ddot{r}} \left\{ \frac{1}{2} \sum_{j=1}^{N} m_j \left(\ddot{\mathbf{r}}_j - \frac{\mathbf{F}_j}{m_j} \right)^2 \right\}.$$

Then, the necessary condition for the minimum leads to

$$\frac{\partial C}{\partial \ddot{\mathbf{r}}_i} = 0, \quad \rightarrow \quad m_i \ddot{\mathbf{r}}_i - \mathbf{F}_i = 0.$$

Assume that the system is subjected the following constraint on its kinetic energy,

$$g(\dot{\mathbf{r}}_i, t) = \sum_{i=1}^{N} \frac{m_i \dot{\mathbf{r}}_i^2}{2} - \frac{3N k_B T}{2} = 0.$$

Find the constraint plane,

$$\frac{dg}{dt} = \sum_i m_i \dot{\mathbf{r}}_i \cdot \ddot{\mathbf{r}}_i = 0.$$

Then we consider the following constrained minimization problem with Lagrangian multiplier,

$$C = \frac{1}{2} \sum_{j=1}^{N} m_j \left(\ddot{\mathbf{r}}_j - \frac{\mathbf{F}_j}{m_j} \right)^2 + \lambda \left(\sum_{j=1}^{N} m_j \dot{\mathbf{r}}_j \cdot \ddot{\mathbf{r}}_j \right).$$

The stationary condition of minimum is

$$\frac{\partial C}{\partial \ddot{\mathbf{r}}_i} = 0, \quad \rightarrow \quad m_i \ddot{\mathbf{r}}_i = \mathbf{F}_i - \alpha \dot{\mathbf{r}}_i,$$

where $\alpha := \lambda m_i$. This is the equations of motion with the Gaussian isokinetic thermostat (Hoover et al. 1982 and Evans 1983) Evans 1983).
Substituting

$$m_i \ddot{\mathbf{r}}_i = \mathbf{F}_i - \lambda m_i \dot{\mathbf{r}}_i \text{ into } \sum_i m_i \dot{\mathbf{r}}_i \cdot \ddot{\mathbf{r}}_i = 0,$$

we can solve for λ,

$$\lambda = \frac{\sum_{i=1}^{N} \mathbf{F}_i \cdot \dot{\mathbf{r}}_i}{\sum_{i=1}^{N} m_i \dot{\mathbf{r}}_i^2} = \frac{\sum_{i=1}^{N} \mathbf{F}_i \cdot \dot{\mathbf{r}}_i}{3Nk_B T}.$$

8.5.2 What Is the Entropy of MD Systems with Thermostats?

Recalling the Liouville equation and the Liouville theorem, we have the balance of probability flow,

$$\frac{\partial}{\partial t} \int_\omega f d\Gamma = -\oint_{\partial \omega} f \dot{\Gamma} \cdot \mathbf{n} dS,$$

where $\Gamma = (\mathbf{q}, \mathbf{p})$. By the divergence theorem,

$$\int_\omega \frac{\partial f}{\partial t} d\Gamma = -\int_\omega \frac{\partial}{\partial \Gamma} \cdot \left(f \dot{\Gamma} \right) d\Gamma.$$

Since ω is arbitrary, we have the balance (conservation) of probability density (Liouville equation)

$$\frac{\partial f}{\partial t}\bigg|_\Gamma = iL[f] = -\left(f \frac{\partial}{\partial \Gamma} \cdot \dot{\Gamma} + \dot{\Gamma} \cdot \frac{\partial f}{\partial \Gamma} \right).$$

Let $\Gamma = (\mathbf{q}, \mathbf{p})$. Balance (conservation) of probability density (Liouville equation)

$$\frac{\partial f}{\partial t}\bigg|_\Gamma = iL[f] = -\left(f \frac{\partial}{\partial \Gamma} \cdot \dot{\Gamma} + \dot{\Gamma} \cdot \frac{\partial f}{\partial \Gamma} \right).$$

By chain rule, the total derivative

$$\frac{df}{dt} = \frac{\partial f}{\partial t} + \dot{\Gamma} \cdot \frac{\partial f}{\partial \Gamma} = -f \frac{\partial}{\partial \Gamma} \dot{\Gamma} =: -f \Lambda(\Gamma),$$

where $\Lambda(\Gamma)$ is called the *phase-space compression factor* and $\Lambda(\Gamma) = 0$ is called the *adiabatic incompressibility condition* (AIΓ).

For a Hamiltonian system, $\Lambda(\Gamma) = 0$,

$$\Lambda(\Gamma) = \sum_{i=1}^{N}\left(\frac{\partial}{\partial q_i}\cdot\dot{q}_i + \frac{\partial}{\partial p_i}\cdot\dot{p}_i\right) = \sum_{i=1}^{N}\left(\frac{\partial}{\partial q_i}\cdot\frac{\partial H}{\partial p_i} - \frac{\partial}{\partial p_i}\cdot\frac{\partial H}{\partial q_i}\right) \equiv 0.$$

Thus the existence of a Hamiltonian = the phase space is incompressible, i.e.,

$$\frac{df}{dt} = 0.$$

Considering that

$$S(t) = -k_B \int d\Gamma\, f(\Gamma) \log f(\Gamma), \quad \leftarrow \text{ (Gibbs entropy formula)} \quad (8.27)$$

and taking time derivative of entropy, we have

$$\dot{S} = -k_B \int d\Gamma [1 + \log f]\frac{\partial f}{\partial t} = k_B \int d\Gamma [1 + \log f]\frac{\partial}{\partial \Gamma}(f\dot{\Gamma}) \quad (8.28)$$

$$= -k_B \int d\Gamma\, f\dot{\Gamma} \cdot \frac{\partial}{\partial \Gamma}[1 + \log f] = -k_B \int d\Gamma\, \dot{\Gamma} \cdot \frac{\partial f}{\partial \Gamma}$$

$$= k_B \int d\Gamma\, f(t)\frac{\partial}{\partial \Gamma}\cdot\dot{\Gamma} = -3Nk_B <\zeta(t)>.$$

Recall

$$\dot{\mathbf{p}}_i = \mathbf{F}_i - \zeta\mathbf{p}_i, \text{ so } \frac{\partial}{\partial \Gamma}\dot{\Gamma} = -3N\zeta(t), \quad \leftarrow \Gamma = (\mathbf{p},\mathbf{q})$$

and

$$m_i\ddot{\mathbf{q}}_i = \mathbf{F}_i - \zeta m_i\dot{\mathbf{q}}_i.$$

Based on Hoover's definition,

$$\zeta = \frac{p_s}{Q} = \frac{sp'_s}{Q} = \frac{ds}{dt} \geq 0.$$

Then

$$\frac{dS}{dt} = -3Nk_B <\zeta(t)> \leq 0!$$

For Evans–Hoover MD, it yields the same result,

$$\frac{dS}{dt} = -3Nk_B <\alpha(t)> \leq 0!$$

Hoover argued that in the microscale we can have a MD system that generates negative entropy. However, if we consider the total system of the MD system + thermal bath the generation of the entropy may not be negative anymore. Therefore, using the thermostat technique in MD simulation does not violate the second law of thermodynamics. However, if one uses the thermostat technique to a macroscale dynamics system, proper care should be taken so that the second law of thermodynamics cannot be violated.

Figure 8.3 William Graham Hoover (1936–) is an American computational physicist (Photo courtesy of William G. Hoover)

8.5.3 William G. Hoover (1936–)

William G. Hoover (Fig. 8.3) was a chemistry professor at the University of California at Davis, who is best known for his contribution to computational statistical mechanics, in particular the Nosé–Hoover MD.

8.6 Homework Problems

Problem 8.1 Prove that for a given field distribution $A(\mathbf{q}, \mathbf{p}))$ the following equality holds:

$$<A(\mathbf{q},\mathbf{p}')>_{NVE,Nosé} = \frac{\int A(\mathbf{q},\mathbf{p}')\exp\left[-\frac{H(\mathbf{q},\mathbf{p}')}{k_B T}\right]d\mathbf{q}^N d\mathbf{p}'^N}{\int \exp\left[-\frac{H(\mathbf{q},\mathbf{p}')}{k_B T}\right]d\mathbf{q}^N d\mathbf{p}'^N} =:<A(\mathbf{q},\mathbf{p}')>_{NVT},$$

where $A(\mathbf{q}, \mathbf{p})$ is an arbitrary field in the phase space.

Problem 8.2 Consider the Nosé Hamiltonian,

$$H_{Nosé} = \sum_{i=1}^{N} \frac{\mathbf{p}_i \cdot \mathbf{p}_i}{2m_i s^2} + \frac{p_s^2}{2Q} + U(\mathbf{q}^N) + \frac{\ell}{\beta} \ln s,$$

where $\beta = (k_B T_o)^{-1}$.

(1) Derive the Nose molecular dynamics.

(2) Adopt the following relation between real variables and virtual variables,

$$\mathbf{p}' = \frac{\mathbf{p}}{s}$$
$$p'_s = p_s/s$$
$$\mathbf{q}' = \mathbf{q}$$
$$s' = s$$
$$\Delta t' = \Delta t/s,$$

Convert the Nosé molecular dynamics in terms of real variables. Note that the variables with "prime" are real variables, whereas the variables without prime are virtual variables.

(3) Consider Hoover's modification,

$$\zeta = \frac{p_s}{Q} = \frac{s p'_s}{Q} = \dot{s} = \frac{d(\ln s)}{dt'}.$$

Derive Nosé–Hoover molecular dynamics equations.

9 Andersen–Parrinello–Rahman Molecular Dynamics

The early MD simulation is a direct numerical simulation method, which is not connected to statistical physics on purpose, consciously or nonconsciously.

This state was changed in the early 1980s by a seminar work from Hans C. Andersen, who started to connect classical MD with statistical physics by building equilibrium ensemble MD.

In this chapter, we discuss the first equilibrium MD – the Parrinello–Rahman (PR) MD (PR-MD).

9.1 Andersen's NPH MD

In his 1980 seminal paper, Hans Andersen proposed the following way to simulate constant stress ensemble MD. Assuming that the MD cell is a square unit cube with each side having the length $\Omega^{1/3}$. Hence, the total volume of the cell is Ω. Then the three basis vectors of the MD cell form the following second-order tensor:

$$\mathbf{h} = \begin{bmatrix} \Omega^{1/3} & 0 & 0 \\ 0 & \Omega^{1/3} & 0 \\ 0 & 0 & \Omega^{1/3} \end{bmatrix}. \tag{9.1}$$

Recall that we can have the following isochoric deformation transformation,

$$\mathbf{F} = (J^{1/3}\mathbf{I})\bar{\mathbf{F}} = J^{1/3}\bar{\mathbf{F}} \quad \rightarrow \quad det(\bar{\mathbf{F}}) = 1,$$

where $J^{1/3} \sim \Omega^{1/3}$ in Eq. (9.1).

Unlike (NVE) MD or (NVT) MD, we allow the volume of the cell to fluctuate, so $\Omega = \Omega(t)$, and we assume that the pressure of the system is fixed.

The central ingredient of Andersen method is the following coordinate transformation (scaling):

$$\mathbf{r}_i(t) = \Omega^{1/3}(t)\mathbf{s}_i(t).$$

The central idea of Andersen method is to make the following coordinate transformation (scaling):

$$\mathbf{r}_i(t) = \Omega^{1/3}(t)\mathbf{I} \cdot \mathbf{s}_i(t).$$

By doing so, Andersen showed us that it leads to some magics.

First, Andersen proposed the following extended Lagrangian,

$$L_A = \frac{1}{2}\sum_{i=1}^{N} m_i |\Omega^{1/3}\dot{\mathbf{s}}_i|^2 - U(\Omega^{1/3}\mathbf{s}_i) + \frac{1}{2}M\dot{\Omega}^2 - P\Omega$$

and its corresponding Hamiltonian is

$$H_A = \frac{1}{2}\sum_{i=1}^{N} m_i |\Omega^{1/3}\dot{\mathbf{s}}_i|^2 + U(\Omega^{1/3}\mathbf{s}_i) + \frac{1}{2}M\dot{\Omega}^2 + P\Omega$$

where the term $P\Omega$ is in fact the external mechanical potential. Compare this with the definition of enthalpy,

The central idea of Andersen method is to make the following coordinate transformation (scaling),

$$\mathbf{r}_i(t) = \Omega^{1/3}(t)\mathbf{I} \cdot \mathbf{s}_i(t)$$

Then there are the magic appearing.

The Andersen Lagrangian leads to the following Lagrangian equations,

$$\frac{d}{dt}\left(\frac{\partial L_A}{\partial \dot{\mathbf{s}}_i}\right) - \frac{\partial L_A}{\partial \mathbf{s}_i} = 0$$

$$\frac{d}{dt}\left(\frac{\partial L_A}{\partial \dot{\Omega}}\right) - \frac{\partial L_A}{\partial \Omega} = 0.$$

It can then be expressed explicitly as,

$$\ddot{\mathbf{s}}_i = -\frac{2}{3}\Omega^{-1}\dot{\Omega}\dot{\mathbf{s}}_i - \frac{1}{m_i \Omega^{2/3}}\frac{\partial U(\Omega^{1/3}\mathbf{s}_i)}{\partial \mathbf{s}_i}$$

$$\ddot{\Omega} = \frac{1}{M}\left(\frac{1}{3\Omega}\sum_i m_i |\Omega^{1/3}\dot{\mathbf{s}}_i|^2 - \frac{\partial U(\Omega^{1/3}\mathbf{s}_i)}{\partial \Omega} - P\right).$$

If we let (there is a good reason for this):

$$P_{int} = \frac{1}{3\Omega}\sum_i m_i |\Omega^{1/3}\dot{\mathbf{s}}_i|^2 - \frac{\partial U(\Omega^{1/3}\mathbf{s}_i)}{\partial \Omega} \quad \text{and} \quad P_{ext} = P \rightarrow$$

$$\ddot{\Omega} = \frac{1}{M}(P_{int} - P_{ext}).$$

In Andersen's extended ensemble MD, the MD cell can only change its volume at the same rate in all three directions. Moreover, in Andersen MD, the ensemble only considered the fixed pressure, and it does not take into account shear stress and shear deformation, or the change of shape of the MD cell. Even for the volumetric part, the volume of the MD cell has to change isotropically, which cannot capture the volume as well as shape changes in most anisotropic lattices.

To develop a more general constant stress ensemble MD, Parrinello and Rahman proposed the following constant stress ensemble ($N\sigma H$) MD.

Figure 9.1 Hans C. Andersen (1941–) is a professor emeritus at Stanford University, who applies statistical mechanics to develop theoretical understanding of the structure and dynamics of liquids and new computer simulation methods to aid in these studies (Photo courtesy of Hans C. Andersen)

We shall see later that PR-MD is the departure point of contemporary multiscale simulation, and moreover, it is an extended ensemble MD that actually can solve some meaningful practical problems.

To follow the development history, we first present the original PR-MD formalism.

9.1.1 Hans C. Andersen (1941–)

Dr. Hans C. Andersen (Fig. 9.1) was David Mulvane Ehrsam and Edward Curtis Franklin Professor of Chemistry at Stanford University. Professor Andersen was one of the developers of what has come to be known as the Weeks–Chandler–Andersen theory of liquids, which is a way of understanding the structure, thermodynamics, and dynamics of simple dense liquids. His 1980 paper on the Andersen thermostat was among the first group of works studying the ensemble MD simulation.

Among many honors, he received the Theoretical Chemistry Award and Hildebrand Award in Theoretical and Experimental Chemistry of Liquids from the American Chemical Society. He has been elected a member of the National Academy of Sciences and a fellow of both the American Academy of Arts and Sciences and American Association for the Advancement of Science.

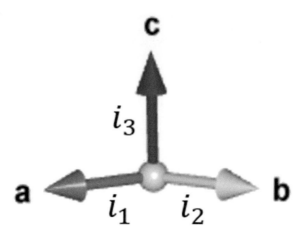

Figure 9.2 Molecular simulation cell edge vectors. They are not necessarily orthogonal

9.2 Parrinello-Rahman Formulation

Two years after the publication of Andersen's paper, in 1982, Parrinello and Rahman reformulated Andersen's extended Lagrangian for crystalline solids. Let **a**, **b**, and **c** be three position vectors representing the three sides of the MD cell as shown in Fig. 9.2.

We form a second-order tensor **h** by

$$\mathbf{h} = [\mathbf{a}|\mathbf{b}|\mathbf{c}] = \begin{bmatrix} a_1 & b_1 & c_1 \\ a_2 & b_2 & c_2 \\ a_3 & b_3 & c_3 \end{bmatrix}$$

or

$$\mathbf{h} = (a_i \delta_{j1} + b_i \delta_{j2} + c_i \delta_{j3}) \mathbf{e}_i \otimes \mathbf{i}_j, \quad i, j = 1, 2, 3.$$

We note that the volume of the MD cell is: $\Omega = |\mathbf{a} \cdot (\mathbf{b} \times \mathbf{c})|$.
In other words, if we let

$$h_{i1} = a_i, \ h_{i2} = b_i, \text{ and } h_{i3} = c_i, \text{ we have}$$

$$\mathbf{h} = h_{ij} \mathbf{e}_i \otimes \mathbf{i}_j, \text{ and } \Omega = \det\{\mathbf{h}\}$$

and $\mathbf{h}_1 = \mathbf{a} = h_{i1}\mathbf{e}_i, \mathbf{h}_2 = \mathbf{b} = h_{i2}\mathbf{e}_i$, and $\mathbf{h}_3 = \mathbf{c} = h_{i3}\mathbf{e}_i$.

In PR-MD, the volume of MD simulation cell is not fixed, i.e., it is not an NVE, NVT, or NVH ensemble MD simulation. Instead, we allow the volume of a MD cell to fluctuate, which means that the edge vectors of MD simulation cell are functions of time, i.e.,

$$\mathbf{a} = \mathbf{a}(t), \ \mathbf{b} = \mathbf{b}(t), \ \mathbf{c} = \mathbf{c}(t); \text{ and } \mathbf{h} = \mathbf{h}(t).$$

We define $\mathbf{h}_0 = \mathbf{h}(0)$ as the initial MD cell configuration tensor. Note that this initial time is the coarse-scale time, i.e., $\bar{t} = 0$.

To quantify the shape change of the MD cell, Parrinello and Rahman proposed the following "scaling" transformation (multiplicative decomposition) on atomic positions:

$$\mathbf{r}_I(t) = \mathbf{h}(t) \cdot \mathbf{s}_I(t), \text{ or } \mathbf{s}_I = \mathbf{h}^{-1} \cdot \mathbf{r}_I.$$

One may compare the above atom position decomposition with the original Andersen's decomposition that only takes into account the volume change of the MD cell, i.e.,

$$\text{(Cf Andersen's decomposition) } \mathbf{r}_I = \Omega^{1/3} \mathbf{I} \mathbf{s}_I.$$

At the initial time,

$$\mathbf{R}_I = \mathbf{h}_0 \cdot \mathbf{s}_I(t). \tag{9.2}$$

Note that the time variable in $\mathbf{h}(t)$ and $\mathbf{s}_I(t)$ may be different or in different timescales. This is because that the cell shape deformation and the local vibration of atoms are physical deformation events at different timescales. Thus, the "initial time" referred in the above is the initial coarse scale time, whereas for the vibration of atoms there is no such thing as the initial time, because atoms in crystals are always vibrating unless it is in the state of the absolute zero temperature. Therefore, the time variable t in $\mathbf{s}_I(t)$ in Eq. (9.2) may be regarded as the fine scale time.

In fact, if we write

$$\mathbf{s}_I(t) = \zeta_I^i(t)\mathbf{i}_i, \ i = 1, 2, 3; \ I = 1, 2, \ldots, N,$$

we can show that

$$\mathbf{r}_I(t) = \mathbf{h}(\zeta_I^j \mathbf{i}_j) = h_{ij} \mathbf{e}_i \zeta_I^j = \zeta_I^j(t) \mathbf{h}_j(t). \tag{9.3}$$

As an analog, we can borrow the language of continuum mechanics to illustrate atom motions in a crystal during PR-MD simulation. That is, first the vibration mode of the atoms may be viewed as the position in a fine scale time-dependent configuration, i.e., $\mathcal{B}_S(\mathbf{S}_i)$. Note that in contrast to the continuum mechanics, the parametric configuration here is really a parametric equilibrium state in which the image or parametric point of each atom vibrates all the time. We can then map each atom position into a physical referential configuration marked as the initial configuration of coarse scale time, i.e., $\mathbf{R}_i = \mathbf{h}_0 \mathbf{S}_i(t)$. Finally, we may find the instantaneous atom position by the linear map $\mathbf{r}_i = \mathbf{h}(t)\mathbf{h}_0^{-1}\mathbf{R}_i$, as shown in Fig. 9.3.

What is the meaning of *multiscale multiplicative decomposition* on atomic positions:

$$\mathbf{r}_I(t) = \mathbf{h}(t) \cdot \mathbf{s}_I(t)?$$

where $\mathbf{h}(t)$ is the time history of the stretch of the cell and $\mathbf{s}_I(t)$ is the particle fluctuations.

Comparing the additive multiscale decomposition,

$$\mathbf{r}_I(t) = \bar{\mathbf{r}}_I(t) + \mathbf{r}'_I(t)$$

Part II Statistical Molecular Dynamics

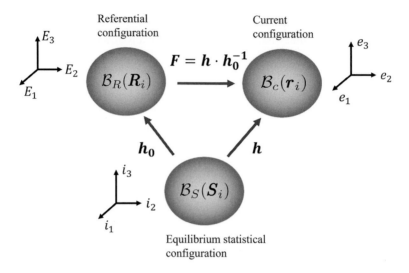

Figure 9.3 The Andersen–Parrinello–Rahman deformation map

with PR-MD, we have multiplicative decomposition,

$$\mathbf{r}_I(t) = \mathbf{h}(\bar{t})\mathbf{s}(t').$$

If we let

$$\mathbf{s}_I(t) = \zeta_I^i(t)\mathbf{i}_i, \ i = 1,2,3; \ I = 1,2,\ldots,N,$$

we have shown that

$$\mathbf{r}_I(t) = \mathbf{h}\zeta_I^j \mathbf{i}_i = \zeta_I^i(t)\mathbf{h}_i(t),$$

so that ζ_I^i may be treated as convective coordinates.

However, in PR-MD, $\{\mathbf{h}_i(t)\}$ is not related to a Lagrangian coordinate or configuration, but a coarse scale configuration. The particle fluctutation is characterized by the fine scale variable $\mathbf{s}_i(t')$.

In fact, it is more correct to write

$$\mathbf{r}_i(t) = \mathbf{h}(\bar{t}) \cdot \mathbf{s}_I(t'), \ \text{Cf} \ \mathbf{r}_I(t) = \bar{\mathbf{r}}_I(\bar{t}) + \mathbf{r}'_I(t').$$

Therefore, $\{\mathbf{h}_i(\bar{t})\}$ is an arbitrary Lagrangian–Eulerian frame.

In MD, it does no make sense to talk about reference configuration, because every atom oscillates around its equilibrium position, i.e., it makes no sense to talk about

$$\mathbf{r}_i(0) = \mathbf{R}_i = \mathbf{h}_0 \cdot \mathbf{s}_i(0),$$

but it makes a lot of sense to talk about,

$$\mathbf{r}_i(t') = \mathbf{R}_i(t') = \mathbf{h}_0 \cdot \mathbf{s}_i(t'), \ \text{at} \ \bar{t} = 0.$$

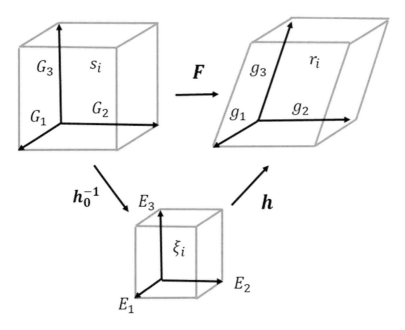

Figure 9.4 The Andersen–Parrinello–Rahman deformation map illustrated as a physical deformation event

Define the deformation gradient (coarse scale) as

$$\mathbf{F} := \frac{\partial \mathbf{x}(\bar{t})}{\partial \mathbf{X}}, \quad \text{where } \mathbf{X} = \mathbf{x}\Big|_{\bar{t}=0}.$$

We have

$$\mathbf{r}_i(t) = \mathbf{F}(\bar{t}) \cdot \mathbf{R}_i(t') = \mathbf{F} \cdot \mathbf{h}_0 \cdot \mathbf{s}_i(t') \to \mathbf{r}_i(t) = \mathbf{h}(\bar{t}) \cdot \mathbf{s}_i(t')$$

with

$$\mathbf{R}_i = \mathbf{h}_0 \cdot \mathbf{s}_i, \text{ and } \mathbf{h} = \mathbf{F} \cdot \mathbf{h}_0 \to \mathbf{F} = \mathbf{h} \cdot \mathbf{h}_0^{-1}.$$

The Andersen–Parrinello–Rahman deformation map may be illustrated in real physical deformation sequence of a unit cell as shown in Fig. 9.4.

REMARK 9.1
1. Causality assumption on fine scale time evolution
Sine $t' \ll \bar{t}$, we assume that fine scale time is only a temporal measure of statistical motion, and it is independent from the coarse scale time measure $\bar{t}$, and hence it has no causal consequence from the coarse scale, i.e.,

$$\mathbf{s}_i(t')\Big|_{\bar{t}_1} = \mathbf{s}_I(t')\Big|_{\bar{t}_2}, \quad \bar{t}_1 \neq \bar{t}_2.$$

2. We can define a reciprocal tensor for MD simulation cell edge vector frame,

$$\mathbf{\Pi} = \Omega(\mathbf{h}^T)^{-1} = 2\pi[\mathbf{b} \times \mathbf{c}, \mathbf{c} \times \mathbf{a}, \mathbf{a} \times \mathbf{b}] \to \mathbf{h} \cdot \mathbf{\Pi} = \Omega \mathbf{I}$$

and
$$\Pi = [\mathbf{H}_1|\mathbf{H}_2|\mathbf{H}_3].$$

Let
$$\mathbf{G}_i := \mathbf{H}_i(0) \rightarrow \Pi(0) = [\mathbf{G}_1|\mathbf{G}_2|\mathbf{G}_3].$$

Hence, we may write
$$\mathbf{h} = \mathbf{h}_i \otimes \mathbf{i}^i, \quad \mathbf{h}_0^{-1} = \mathbf{i}_j \otimes \mathbf{G}^j.$$

9.3 PR MD for NPH Ensemble

In 1982, Parrinello and Rahman first proposed the following extended Lagrangian:

$$\mathcal{L}_{PR} = \frac{1}{2} \sum_{i=1}^{N} m_i \dot{\mathbf{s}}_i \cdot \mathbf{G} \cdot \dot{\mathbf{s}}_i - \frac{1}{2} \sum_i \sum_{j \neq i} U(r_{ij}) + \frac{1}{2} W Tr(\dot{\mathbf{h}}^T \cdot \dot{\mathbf{h}}) - p\Omega, \quad (9.4)$$

where $\mathbf{G} = \mathbf{h}^T \cdot \mathbf{h}$. In Eq. (9.4), W is a parameter that is defined as

$$\frac{1}{2} W Tr(\dot{\mathbf{h}}^T \cdot \dot{\mathbf{h}}) = \frac{W}{2}(\dot{a}^2 + \dot{b}^2 + \dot{c}^2).$$

Then the Lagrangian equations can be derived as follows:

$$\frac{d}{dt}\left(\frac{\partial \mathcal{L}_{PR}}{\partial \dot{\mathbf{s}}_i}\right) - \frac{\partial \mathcal{L}_{PR}}{\partial \mathbf{s}_i} = 0 \quad (9.5)$$

$$\frac{d}{dt}\left(\frac{\partial \mathcal{L}_{PR}}{\partial \dot{\Omega}}\right) - \frac{\partial \mathcal{L}_{PR}}{\partial \Omega} = 0, \quad (9.6)$$

which lead to the following equation of motion for the PR-MD:

$$\ddot{\mathbf{s}}_i = -\sum_{j \neq i} \left(\frac{U'(r_{ij})}{m_i r_{ij}}\right)(\mathbf{s}_i - \mathbf{s}_j) - \mathbf{G}^{-1}\dot{\mathbf{G}}\dot{\mathbf{s}}_i, \quad (9.7)$$

$$W\ddot{\mathbf{h}} = -(\sigma_{virial} + p\mathbf{I}) \cdot \Pi, \quad (9.8)$$

where σ is the so-called virial stress,

$$\sigma_{virial} = \frac{1}{\Omega} \sum_I \left\{ -m_I \mathbf{v}_I \otimes \mathbf{v}_I + \frac{1}{2} \sum_{I \neq J} \left(\frac{U'(r_{IJ})}{r_{IJ}}\right) \mathbf{r}_{IJ} \otimes \mathbf{r}_{IJ} \right\}.$$

We shall discuss the general definition of the virial stress in MD simulation in the next chapter.

In fact, the PR-MD Lagrangian proposed by Parrinello and Rahman in Eq. (9.4) is the Lagrangian of the MD system. The total kinetic energy of the original PR-MD Lagrangian system is

$$\frac{1}{2}\sum_I m_I \dot{\mathbf{r}}_I \cdot \dot{\mathbf{r}}_I = \frac{1}{2}\sum_I m_I (\dot{\mathbf{h}}(\bar{t})\cdot \mathbf{s}_I + \mathbf{h}\cdot\dot{\mathbf{s}}_I(t')) \cdot (\dot{\mathbf{h}}(\bar{t})\cdot \mathbf{s}_I + \mathbf{h}\cdot\dot{\mathbf{s}}_I(t'))$$

$$= \frac{1}{2}\sum_I m_I\left(\mathbf{s}_I \dot{\mathbf{h}}^T \cdot \dot{\mathbf{h}}\cdot \mathbf{s}_I + \mathbf{s}_I\dot{\mathbf{h}}^T\cdot\mathbf{h}\cdot\dot{\mathbf{s}}_I + \dot{\mathbf{s}}_I\cdot\mathbf{h}^T\cdot\dot{\mathbf{h}}\cdot\mathbf{s}_I + \dot{\mathbf{s}}_I\cdot\mathbf{h}^T\cdot\mathbf{h}\cdot\dot{\mathbf{s}}_I\right),$$

among which, the first term is the coarse scale kinetic energy, and the last term is the fine scale kinetic energy. In PR-Lagrangian, we approximate them as

$$E_{kin} = \frac{1}{2}\sum_I m_I \dot{\mathbf{s}}_I \cdot \mathbf{G} \cdot \dot{\mathbf{s}}_I + \frac{1}{2}WTr(\dot{\mathbf{h}}^T\cdot\dot{\mathbf{h}}),$$

where $\mathbf{G} = \mathbf{h}^T\cdot\mathbf{h}$, and we neglect the two cross terms, which means that there is no correlation between the coarse scale dynamics and fine scale dynamics, and we shall discuss it in later chapters.

9.3.1 Derivations of Lagrangian Equations

To show how to derive the equations of motion of PR-MD, in the following, we show a step-by-step derivation:

1. $\mathbf{r}_{JI} = \mathbf{r}_I - \mathbf{r}_J = \mathbf{h}\cdot\mathbf{s}_I - \mathbf{h}\cdot\mathbf{s}_J = \mathbf{h}\cdot(\mathbf{s}_I - \mathbf{s}_J),$

hence

2. $\mathbf{r}_{JI}\cdot\mathbf{r}_{JI} = \mathbf{r}_{JI}^T\cdot\mathbf{r}_{JI} = (\mathbf{s}_I-\mathbf{s}_J)\cdot\mathbf{h}^T\cdot\mathbf{h}\cdot(\mathbf{s}_I - \mathbf{s}_J).$

Define a second-order tensor,

$$\mathbf{G} = \mathbf{h}^T\cdot\mathbf{h} \quad \rightarrow \quad r_{IJ}^2 = (\mathbf{s}_I - \mathbf{s}_J)\cdot\mathbf{G}\cdot(\mathbf{s}_I - \mathbf{s}_J).$$

One can show that

$$\frac{\partial r_{JI}}{\partial \mathbf{s}_I} = \mathbf{G}\cdot\frac{\mathbf{s}_I - \mathbf{s}_J}{r_{IJ}} \quad \rightarrow \quad \frac{\partial}{\partial \mathbf{s}_I}(\mathbf{r}_{JI}\cdot\mathbf{r}_{JI}) = 2\mathbf{r}_{JI}\frac{\partial \mathbf{r}_{JI}}{\partial \mathbf{s}_I} = 2\mathbf{G}\cdot(\mathbf{s}_I - \mathbf{s}_J).$$

Recalling the reciprocal tensor for the MD cell edge vector frame, we have the relation,

$$\mathbf{\Pi} = \Omega(\mathbf{h}^T)^{-1} = 2\pi[\mathbf{b}\times\mathbf{c}, \mathbf{c}\times\mathbf{a}, \mathbf{a}\times\mathbf{b}] \quad \rightarrow \quad \mathbf{h}\cdot\mathbf{\Pi} = \Omega\mathbf{I},$$

and PR MD Lagrangian can be expressed as

$$\mathcal{L}_{PR} = \frac{1}{2}\sum_{i=1}^N m_i \dot{\mathbf{s}}_i\cdot\mathbf{G}\cdot\dot{\mathbf{s}}_i - \frac{1}{2}\sum_i\sum_{j\neq i} U(r_{ij}) + \frac{1}{2}WTr(\dot{\mathbf{h}}^T\cdot\dot{\mathbf{h}}) - p\Omega.$$

Since $\mathbf{r}_{JI} = \mathbf{h}\cdot(\mathbf{s}_I - \mathbf{s}_J)$, we have

$$\frac{\partial \mathcal{L}_{PR}}{\partial \mathbf{s}_I} = -\sum_{J\neq I} U'(r_{IJ})\frac{\partial r_{IJ}}{\partial \mathbf{s}_I} = -\sum_{J\neq} U'(r_{IJ})\frac{\mathbf{G}\cdot(\mathbf{s}_I - \mathbf{s}_J)}{r_{IJ}} \quad \text{and}$$

$$\frac{\partial \mathcal{L}_{PR}}{\partial \dot{\mathbf{s}}_I} = m_I\mathbf{G}\cdot\dot{\mathbf{s}}_I, \quad \rightarrow \quad \frac{d}{dt}\frac{\partial \mathcal{L}_{PR}}{\partial \dot{\mathbf{s}}_I} = m_I\dot{\mathbf{G}}\cdot\dot{\mathbf{s}}_I + m_I\mathbf{G}\cdot\ddot{\mathbf{s}}_I.$$

Then the Euler–Lagrangian equations become

$$\frac{d}{dt}\frac{\partial \mathcal{L}_{PR}}{\partial \dot{\mathbf{s}}_i} - \frac{\partial \mathcal{L}_{PR}}{\partial \mathbf{s}_i} \rightarrow m_i \mathbf{G} \cdot \ddot{\mathbf{s}}_i = -\sum_{j \neq i} U'(r_{ij})\frac{\mathbf{G} \cdot (\mathbf{s}_i - \mathbf{s}_j)}{r_{ij}} - m_i \dot{\mathbf{G}} \cdot \dot{\mathbf{s}}_i,$$

which can be further simplified as

$$\ddot{\mathbf{s}}_i = -\sum_{j \neq i} \frac{U'(r_{ij})}{m_i r_{ij}}(\mathbf{s}_j - \mathbf{s}_j) - \mathbf{G}^{-1} \cdot \dot{\mathbf{G}} \cdot \dot{\mathbf{s}}_i \quad i = 1, 2, \ldots, N.$$

Furthermore, one can derive the equations of motion based on $\mathcal{L}_{PR}$ with respect to variable $\mathbf{h}$,

$$\left.\frac{\partial \mathcal{L}_{PR}}{\partial \mathbf{h}}\right|_{term1} = \sum_i m_i (\mathbf{h} \cdot \dot{\mathbf{s}}_i) \otimes \dot{\mathbf{s}}_i = \sum_i m_i (\mathbf{h} \cdot \dot{\mathbf{s}}_i) \otimes (\mathbf{h}^{-1} \cdot \mathbf{h} \cdot \dot{\mathbf{s}}_i)$$

$$= \sum_i m_i (\mathbf{h} \cdot \dot{\mathbf{s}}_i) \otimes (\mathbf{h} \cdot \dot{\mathbf{s}}_i) \cdot \mathbf{h}^{-T}$$

$$= \frac{1}{\Omega}\sum_i m_i (\mathbf{h} \cdot \dot{\mathbf{s}}_i) \otimes (\mathbf{h} \cdot \dot{\mathbf{s}}_i) \cdot \Pi \quad \leftarrow \Pi = \Omega \mathbf{h}^{-T}$$

$$= \frac{1}{\Omega}\sum_i m_i (\mathbf{v}_i \otimes \mathbf{v}_i) \cdot \Pi,$$

where $\mathbf{v}_i := \mathbf{h} \cdot \dot{\mathbf{s}}_i$ is not the total velocity.

Note that since

$$\frac{\partial \mathbf{G}}{\partial \mathbf{h}} = \frac{\partial}{\partial h_{mn}}(h_{ki}h_{kj})\mathbf{e}_i \otimes \mathbf{e}_j \otimes \mathbf{e}_m \otimes \mathbf{e}_n = (\delta_{in}h_{mj} + \delta_{jn}h_{mi})\mathbf{e}_i \otimes \mathbf{e}_j \otimes \mathbf{e}_m \otimes \mathbf{e}_n,$$

where $i, j, k, m, n = 1, 2, 3$, and one can show that

$$\dot{\mathbf{s}}_i \cdot \frac{\partial \mathbf{G}}{\partial \mathbf{h}} \cdot \dot{\mathbf{s}}_i = (\mathbf{h} \cdot \dot{\mathbf{s}}_i) \otimes \dot{\mathbf{s}}_i,$$

by using the indicial notation.

Moreover, we have

$$\left.\frac{\partial \mathcal{L}_{PR}}{\partial \mathbf{h}}\right|_{term2} = -\sum_{j \neq i} U'(r_{ij})\frac{\partial r_{ji}}{\partial r_{ji}} \cdot \frac{\partial r_{ji}}{\partial \mathbf{h}} \quad \leftarrow r_{ji} = \mathbf{h} \cdot (\mathbf{s}_i - \mathbf{s}_j)$$

$$= -\sum_{j \neq i} U'(r_{ij})\frac{\mathbf{r}_{ji}}{r_{ji}} \otimes (\mathbf{s}_i - \mathbf{s}_j), \quad \leftarrow \frac{\partial \mathbf{r}_{ji}}{\partial \mathbf{h}} = \mathbf{i} \otimes (\mathbf{s}_i - \mathbf{s}_j)$$

$$= -\sum_{j \neq i} U'(r_{ij})\frac{\mathbf{r}_{ji}}{r_{ji}} \otimes (\mathbf{h}^{-1} \cdot \mathbf{h} \cdot (\mathbf{s}_i - \mathbf{s}_j)),$$

$$= -\frac{1}{\Omega}\sum_{j \neq i} U'(r_{ij})\frac{\mathbf{r}_{ji}}{r_{ij}} \otimes (\mathbf{h} \cdot (\mathbf{s}_I - \mathbf{s}_j))\mathbf{h}^{-T}\Omega$$

$$= -\frac{1}{\Omega}\sum_{j \neq i} U'(r_{ij})\frac{\mathbf{r}_{ji}}{r_{ij}} \otimes (\mathbf{h} \cdot (\mathbf{s}_i - \mathbf{s}_j)) \cdot \Pi$$

$$= -\frac{1}{\Omega}\sum_{j \neq i} U'(r_{ij})\frac{\mathbf{r}_{ji} \otimes \mathbf{r}_{ji}}{r_{ij}} \cdot \Pi.$$

If we add the two terms that were just derived above together, we have

$$\frac{\partial \mathcal{L}_{PR}}{\partial \mathbf{h}}\bigg|_{term1} + \frac{\partial \mathcal{L}_{PR}}{\partial \mathbf{h}}\bigg|_{term2}$$

$$= -\sigma \cdot \Pi = -\frac{1}{\Omega} \sum_i \left\{ -m_i \mathbf{v}_i \otimes \mathbf{v}_i + \sum_{i \neq j} \left(\frac{U'(r_{ij})}{r_{ij}} \right) \mathbf{r}_{ji} \otimes \mathbf{r}_{ji} \right\} \cdot \Pi,$$

we then obtain the expression for the virial stress,

$$\sigma = \frac{1}{\Omega} \sum_i \left\{ -m_i \mathbf{v}_i \otimes \mathbf{v}_i + \sum_{i \neq j} \left(\frac{U'(r_{ij})}{r_{ij}} \right) \mathbf{r}_{ji} \otimes \mathbf{r}_{ji} \right\}$$

$$= \frac{1}{\Omega} \sum_i \left\{ -m_I \mathbf{v}_i \otimes \mathbf{v}_i + \sum_{i \neq j} \mathbf{f}_{ij} \otimes \mathbf{r}_{ij} \right\} \quad \leftarrow \quad \mathbf{f}_{ij} = U'(r_{ij}) \frac{\mathbf{r}_{ij}}{r_{ij}}.$$

One may compare it with the virial pressure defined in the Andersen MD,

$$P_{int} = \frac{1}{3\Omega} \sum_i m_i |\Omega^{1/3} \dot{\mathbf{s}}_i|^2 - \frac{\partial U(\Omega^{1/3} \mathbf{s}_i)}{\partial \Omega} \quad \leftarrow \quad P_{int} = -\frac{1}{3} \sigma_{ii}.$$

Now, we can calculate the third term of PR-MD Lagrangian,

$$\frac{\partial \mathcal{L}_{PR}}{\partial \mathbf{h}}\bigg|_{term3} = -p \frac{\partial \Omega}{\partial \mathbf{h}} = -p \frac{\partial \|\mathbf{h}\|}{\partial \mathbf{h}}.$$

In passing, we note that some simple formulas in tensor analysis are used here,

$$\frac{\partial \|\mathbf{h}\|}{\partial \mathbf{h}} = \|\mathbf{h}\| \mathbf{h}^{-T}, \text{ and } \frac{\partial}{\partial \mathbf{h}} tr\left(\mathbf{h}^T \cdot \mathbf{h}\right) = \frac{\partial}{\partial \mathbf{h}} \left(\mathbf{I}^{(2)} : (\mathbf{h}^T \cdot \mathbf{h})\right) = 2\mathbf{h}^T.$$

Finally, we have

$$\frac{\partial \mathcal{L}_{PR}}{\partial \mathbf{h}}\bigg|_{term3} = -p \frac{\partial \Omega}{\partial \mathbf{h}} = -p \|\mathbf{h}\| \mathbf{h}^{-T} = -p\Omega \mathbf{h}^{-T} = -p\Pi$$

and

$$\frac{\partial \mathcal{L}_{PR}}{\partial \mathbf{h}} = -(\sigma_{virial} + p\mathbf{I}) \cdot \Pi$$

On the other hand,

$$\frac{\partial \mathcal{L}_{PR}}{\partial \dot{\mathbf{h}}} = \frac{\partial}{\partial \dot{\mathbf{h}}} \left(\frac{1}{2} W Tr(\dot{\mathbf{h}}^T \cdot \dot{\mathbf{h}}) \right) = W \dot{\mathbf{h}}, \quad \leftarrow \quad \frac{\partial}{\partial \mathbf{A}} Tr(\mathbf{A}^T \cdot \mathbf{A}) = 2\mathbf{A}.$$

Hence,

$$\frac{d}{dt} \frac{\partial \mathcal{L}_{PR}}{\partial \dot{\mathbf{h}}} = W \ddot{\mathbf{h}}.$$

The Lagrangian equations become

$$\frac{d}{dt} \frac{\partial \mathcal{L}_{PR}}{\partial \dot{\mathbf{h}}} - \frac{\partial \mathcal{L}_{PR}}{\partial \mathbf{h}} = 0 \rightarrow W \ddot{\mathbf{h}} = -\left(\sigma_{virial} + p\mathbf{I}^{(2)} \right) \cdot \Pi.$$

Finally, we have derived the complete Lagrangian equations for the PR MD,

$$\ddot{\mathbf{s}}_i = -\sum_{j \neq i}\left(\frac{U'(r_{ij})}{m_i r_{ij}}\right)(\mathbf{s}_i - \mathbf{s}_j) - \mathbf{G}^{-1}\dot{\mathbf{G}}\dot{\mathbf{s}}_i, \quad \leftarrow \text{ Fine scale dynamics;}$$

$$W\ddot{\mathbf{h}} = -(\sigma + p\mathbf{i}) \cdot \Pi \quad \leftarrow \text{ Coarse scale dynamics.}$$

We can draw three conclusions based on the above derivations:

1. The macroscale driving force is the difference between the internal stress and external stress.
2. In the above isotropic PR-MD, it has an internal stress, but not an external stress, and it only has an external pressure, whereas in the Andersen MD, it has both internal and external stresses that are in the same form of pressure, and
3. This is the first success example of micro-to-macro transition for mechanical system modeling.

REMARK 9.2 1. As mentioned earlier, $\mathbf{v}_I = \mathbf{h} \cdot \dot{\mathbf{s}}_I$, is not the total velocity, but the fine scale velocity. The total velocity,

$$\dot{\mathbf{r}}_i(t) = \dot{\mathbf{h}}(\bar{t}) \cdot \mathbf{s}_i + \mathbf{h} \cdot \dot{\mathbf{s}}_i(t').$$

Comparing it with addictive decomposition, we can see that

$$\dot{\mathbf{r}}_i(t) = \dot{\bar{\mathbf{r}}}(t) + \dot{\mathbf{r}}'_i(t), \quad \bar{\mathbf{r}}(t) = \frac{1}{N}\sum_i \dot{\mathbf{r}}_i.$$

What are the differences between the two?

In multiplicative decomposition, the fine scale velocity is basically the atomistic oscillation velocity around the atom's own equilibrium position. Whereas in addictive decomposition, the fine scale velocity is the fluctuation around the velocity of the center of mass (mean velocity),

$$\sum_i \dot{\mathbf{r}}'_i = 0; \text{ and } \sum_i \bar{\mathbf{v}}(\bar{t})_i \cdot \mathbf{v}_i(t') = <\bar{\mathbf{v}}(\bar{t}) \cdot \mathbf{v}(t') \geq 0.$$

In literature, we call

$$\dot{\mathbf{r}}'_i = \dot{\mathbf{r}}_i - \dot{\bar{\mathbf{r}}}_i,$$

as the addictive peculiar velocity, and following the same manner, we may call

$$\mathbf{v}_I(t') = \mathbf{h} \cdot \dot{\mathbf{s}}_I(t'),$$

as the multiplicative peculiar velocity. Furthermore,

$$\sum_i \mathbf{v}_i = \sum_i \mathbf{h} \cdot \dot{\mathbf{s}}_i = \mathbf{h} \cdot \sum_i \dot{\mathbf{s}}_i = 0,$$

if we use the center of mass as the origin of the cell, we can show that

$$\sum_i m_i \mathbf{r}_i = m\mathbf{h} \cdot \sum_i \mathbf{s}_i = 0.$$

PR-MD is an equilibrium ensemble MD, i.e., an NPH MD in this case. Therefore, the Lagrangian equations are only for one cell of atoms with periodic boundary conditions. Can we extend PR-MD to nonequilibrium situations for general nanoscale simulations by using multiple cell approach, which will be discussed in later chapters.

9.3.2 Hamiltonian Approach

Using the PR-MD Lagrangian in Eq. (9.4), one may define the momenta $\mathbf{p}_i^s$ and $\mathbf{p}^h$ for $\mathbf{s}_i$ and $\mathbf{h}$ as follows:

$$\mathbf{p}_i^s := \frac{\partial \mathcal{L}_{PR}}{\partial \dot{\mathbf{s}}_i} = m_i \mathbf{G} \cdot \dot{\mathbf{s}}_i$$

$$\mathbf{p}^h := \frac{\partial \mathcal{L}_{PR}}{\partial \dot{\mathbf{h}}} = W\dot{\mathbf{h}} .$$

By using the Legendre transform, we can obtain the Hamiltonian for PR (NPH) MD,

$$\mathcal{H}_{PR} = \sum_i \mathbf{p}_i^s \cdot \dot{\mathbf{s}}_i + \mathbf{p}^h : \dot{\mathbf{h}} - \mathcal{L}_{PR}$$

$$= \frac{1}{2} \sum_i \frac{\mathbf{p}_i^s \cdot \mathbf{G}^{-1} \cdot \mathbf{p}_i^s}{m_i} + \frac{1}{2W} \mathbf{p}^h : \mathbf{p}^h + U(\mathbf{h} \cdot \mathbf{s}_i) + p\Omega$$

The Hamiltonian equations can be derived in straight forward manner as follows:

$$\dot{\mathbf{p}}_i^s = -\frac{\partial \mathcal{H}_{PR}}{\partial \mathbf{s}_i} = -\frac{\partial U(\mathbf{r}^N)}{\partial \mathbf{s}_i} = -\frac{\partial U(\mathbf{r}^N)}{\partial \mathbf{r}_i} \frac{\partial \mathbf{r}_i}{\partial \mathbf{s}_i} = -\frac{\partial U(\mathbf{r}^N)}{\partial \mathbf{r}_i} \cdot \mathbf{h};$$

$$\dot{\mathbf{s}}_i = \frac{\partial \mathcal{H}_{PR}}{\partial \mathbf{p}_i^s} = \frac{1}{m_i} \mathbf{G}^{-1} \cdot \mathbf{p}_i^s;$$

$$\dot{\mathbf{p}}^h = -\frac{\partial \mathcal{H}_{PR}}{\partial \mathbf{h}} = -(\sigma_{virial} + p\mathbf{i}) \cdot \mathbf{\Pi};$$

$$\dot{\mathbf{h}} = \frac{\partial \mathcal{H}_{PR}}{\partial \mathbf{p}_h} = \frac{\mathbf{p}^h}{W} .$$

9.3.3 Aneesur Rahman (1927–1987)

Aneesur Rahman (see Fig. 9.5) was a physicist in Argonne National Laboratory and is known worldwide as the father of MD, who pioneered using molecular dynamics to simulate physical systems.

His 1964 paper on liquid argon studied a system of 864 argon atoms on a CDC 3600 computer, using a Lennard-Jones potential, was among the first MD simulations. His MD algorithms still form the basis for many codes written today. Moreover, he worked on a wide variety of problems, such as the microcanonical ensemble approach to lattice gauge theory, which he invented with David J. E. Callaway, and the PR-MD with M. Parrinello.

In recognizing his contribution to the development of molecular dynamics and computational physics, the American Physical Society (APS) established the Aneesur

Figure 9.5 Aneesur Rahman (August 24, 1927–June 6, 1987) – father of molecular dynamics and an Argonne physicist (Photo courtesy of MarzMehter and Wikipedia.org)

Rahman Prize, which is annually awarded to the scientists who made outstanding achievements in computational physics research. First awarded in 1993, today the Aneesur Rahman Prize is the highest honor in the field of computational physics given by APS.

9.4 Physical Justification of PR MD

As Parrinello and Rahman commented in their 1981 paper, "... *Whether such a Lagrangian is derivable from first principles is a question for further study; its validity can be judged, as of now, by the equations of motion and the statistical ensembles that it generates. . . . .*"

To translate the above comments into mathematical language, we can rephrase the question as: can one derive the PR-MD Lagrangian

$$\mathcal{L}_{PR} = \frac{1}{2}\sum_{i=1}^{N} m_i \dot{\mathbf{s}}_i \cdot \mathbf{G} \cdot \dot{\mathbf{s}}_i - \frac{1}{2}\sum_{i}\sum_{j \neq i} U(r_{ij}) + \frac{1}{2}WTr(\dot{\mathbf{h}}^T \dot{\mathbf{h}}) - p\Omega \qquad (9.9)$$

from the original or first-principles Lagrangian,

$$\mathcal{L} = \sum_{i=1}^{N} m_i \dot{\mathbf{r}}_i \cdot \dot{\mathbf{r}}_i - \frac{1}{2}\sum_i \sum_{j \neq i} U(r_{ij}) - p\Omega? \qquad (9.10)$$

In particular, there is an arbitrary parameter W in Eq. (9.9), which has a unit of mass, and it determines the relaxation time of an equilibrium state from a nonequilibrium state due to the imbalance between the external pressure and the internal stress. In previous implementations, the choice of the parameter W is problematic and empirical.

We start with the original Lagrangian, or first-principles Lagrangian, as called by Parrinello and Rahman,

$$K = \frac{1}{2}\sum_i m_i \dot{\mathbf{r}}_i \cdot \dot{\mathbf{r}}_i = \frac{1}{2}\sum_i m_i (\dot{\mathbf{h}} \cdot \mathbf{s}_i + \mathbf{h} \cdot \dot{\mathbf{s}}_i) \cdot (\dot{\mathbf{h}} \cdot \mathbf{s}_i + \mathbf{h} \cdot \dot{\mathbf{s}}_i)$$

$$= \underbrace{\frac{1}{2}\dot{\mathbf{h}}^T \dot{\mathbf{h}} : \sum_i m_i \mathbf{s}_i \otimes \mathbf{s}_i}_{K_1} + \underbrace{\frac{1}{2}\sum_i m_i \dot{\mathbf{s}}_i \cdot \mathbf{G} \cdot \dot{\mathbf{s}}}_{K_2}$$

$$+ \underbrace{\frac{1}{2}\dot{\mathbf{h}}^T \mathbf{h} : \sum_i m_i \mathbf{s}_i \otimes \dot{\mathbf{s}}_i}_{K_3} + \underbrace{\frac{1}{2}\mathbf{h}^T \dot{\mathbf{h}} : \sum_i m_i \dot{\mathbf{s}}_i \otimes \mathbf{s}_i}_{K_4}$$

where $\mathbf{G} = \mathbf{h}^T \cdot \mathbf{h}$.

Parrinello and Rahman made the following approximation:

$$K_3 \approx 0, \text{ and } K_4 \approx 0.$$

To justify these approximations, we consider that the first term may be written as

$$\mathcal{K}_1 = \frac{1}{2}(\dot{\mathbf{h}}^T \cdot \dot{\mathbf{h}}) : \sum_i m_i \mathbf{S}_i \otimes \mathbf{S}_i.$$

We introduce the following statistical assumption,

$$\sum_i m_i \mathbf{S}_i \otimes \mathbf{S}_i = W\delta_{IJ}\mathbf{E}_I \otimes \mathbf{E}_J = \mathbf{J} = const.,$$

where $I, J = 1, 2, 3$, and we coined as the first PR closure condition.

If we assume that $J_{11} = J_{22} = J_{33} = W$ (spherical), we can have the following result used by Parrinello and Rahman:

$$\mathcal{K}_1 = \frac{1}{2}W tr(\dot{\mathbf{h}}^T \dot{\mathbf{h}}). \qquad (9.11)$$

In practice, one may make the following approximation:

$$\mathbf{J} = \sum_i m_i \mathbf{S}_i \otimes \mathbf{S}_i \approx \sum_i m_i \mathbf{S}_i(0) \otimes \mathbf{S}_i(0).$$

Thus, we state the first statistical assumption: $\mathbf{J}$ is a constant spherical tensor.

If $J_{11} = J_{22} = J_{33} = W$ (spherical), we have the following result from Parrinello and Rahman,

$$\mathcal{K}_1 = \frac{1}{2} W tr(\dot{\mathbf{h}}^T \dot{\mathbf{h}}). \tag{9.12}$$

In practice, one may make the following approximation:

$$\mathbf{J} = \sum_i m_i \mathbf{s}_i \otimes \mathbf{s}_i \approx \sum_i m_i \mathbf{s}_i(0) \otimes \mathbf{s}_i(0).$$

The physical meaning of this assumption is that the spatial space is statistically isotropic, and homogeneous. In other words, we regard $\{s_i\}$ as a set of random variables,

$$-1 \leq s_i \leq 1, \quad \sum_i s_i = 0.$$

Parrinello and Rahman made another choice, and they choose to let,

$$\mathcal{K}_{3\&4} = \frac{1}{2} \sum_i m_i \left(\mathbf{S}_i(\dot{\mathbf{h}}_\alpha^T \mathbf{h}_\alpha) \cdot \dot{\mathbf{S}}_i + \dot{\mathbf{S}}_i(\mathbf{h}_\alpha^T \dot{\mathbf{h}}_\alpha) \mathbf{S}_i \right) = 0,$$

which we call as the second PR closure.

In order to have the term $\mathcal{K}_{3\&4} = 0$, we choose the following statistical closures,

$$\left(\sum_i m_i \mathbf{s}_i \otimes \dot{\mathbf{s}}_i \right) = 0, \; (a) \text{ and } \left(\sum_i m_i \dot{\mathbf{s}}_i \otimes \mathbf{s}_i \right) = 0, (b).$$

In fact, the first PR closure,

$$\sum_i m_i \mathbf{s}_i \otimes \mathbf{s}_i = const.$$

implies that

$$\left(\sum_i m_i \mathbf{s}_i \otimes \dot{\mathbf{s}}_i \right) + \left(\sum_i m_i \dot{\mathbf{s}}_i \otimes \mathbf{s}_i \right) = 0.$$

Define a tensorial function:

$$\mathbf{AC}(\tau) = <\mathbf{S}_i(t) \otimes \mathbf{S}_i(t+\tau)> := \sum_i m_i \mathbf{S}_i(t) \otimes \mathbf{S}_i(t+\tau)$$

$$= \left(\sum_i m_i s_i(t) s_i(t+\tau) \right) \mathbf{E}_I \otimes \mathbf{E}_I$$

and a scalar function

$$f(\tau) = \sum_i m_i s_i(t) s_i(t+\tau),$$

which is an even function at $\tau = 0$, therefore,

$$\frac{d}{d\tau} f \bigg|_{\tau=0} = 0.$$

This leads to
$$\frac{d}{d\tau}\mathbf{AC}(\tau)\Big|_{\tau=0} = \sum_i m_i \mathbf{S}_i(t) \otimes \dot{\mathbf{S}}_i(t+\tau) = 0.$$

Similarly or equivalently
$$\sum_i m_i \dot{\mathbf{S}}_i(t) \otimes \mathbf{S}_i = 0.$$

The physical meaning of the above expression is that the random velocities and displacements have no correlation.
Note that this is a tensorial correlation function,
$$\mathbf{AC}(\tau) = <\mathbf{S}_i(t) \otimes \mathbf{S}_i(t+\tau)> := \sum_i m_i \mathbf{S}_i(t) \otimes \mathbf{S}_i(t+\tau)$$
$$= \left(\sum_i m_i s_i(t) s_i(t+\tau)\right) \mathbf{E}_I \otimes \mathbf{E}_I.$$

However, the standard correlation function should have the form,
$$AC(\tau) = \sum_i \mathbf{s}_i(t) \cdot \mathbf{s}_i(t+\tau).$$

Finally, we can write
$$\mathcal{L} = \frac{1}{2}\sum_{i=1}^N m_i \dot{\mathbf{s}}_i \cdot \mathbf{G} \cdot \dot{\mathbf{s}}_i - \frac{1}{2}\sum_i \sum_{j \neq i} U(r_{ij}) + \frac{1}{2}WTr(\dot{\mathbf{h}}^T \dot{\mathbf{h}}) - p\Omega$$

from
$$\mathcal{L} = \sum_{i=1}^N m_i \dot{\mathbf{r}}_i \cdot \dot{\mathbf{r}}_i - \frac{1}{2}\sum_i \sum_{j \neq i} U(r_{ij}) - p\Omega.$$

This is the complete justification (or proof) of PR Lagrangian.
Similar arguments may be made in the Andersen Lagrangian. Thus, in case of the Andersen Lagrangian, the scaling decomposition is $\mathbf{r}_i = \Omega^{1/3}\mathbf{s}_i$, and total kinetic energy is
$$K = \frac{1}{2}\sum_{i=1}^N m_i \dot{\mathbf{r}}_i^2 = \frac{1}{2}\sum_{i=1}^N m_i \left(\Omega^{1/3}\dot{\mathbf{s}}_i + \frac{1}{3}\Omega^{-2/3}\dot{\Omega}\mathbf{s}_i\right)^2.$$

Expanding the quadratic form in above expression, we have
$$K = \frac{1}{2}\sum_i m_i |\Omega^{1/3}\dot{\mathbf{s}}_i|^2 + \frac{2}{3}\Omega^{-1/3}\dot{\Omega}\sum_i m_i \mathbf{s}_i \cdot \dot{\mathbf{s}}_i + \frac{1}{9}\Omega^{-4/3}\dot{\Omega}^2 \sum_i m_i \mathbf{s}_i^2.$$

We assume that
$$\frac{2}{9}\Omega^{-4/3}\sum_i m_i \mathbf{s}_i^2 = const. = M \rightarrow \frac{1}{9}\Omega^{-4/3}\dot{\Omega}^2 \sum_i m_i \mathbf{s}_i^2 = \frac{1}{2}M\dot{\Omega}^2,$$

which makes sense because that the random variables must satisfy the condition $\sum_i m_i s_i^2 = 1$. Subsequently, we have

$$\sum_i m_i \mathbf{s}_i \cdot \dot{\mathbf{s}}_i = 0,$$

which leads to the following Andersen Lagrangian

$$\mathcal{L}_A = \frac{1}{2}\sum_{i=1}^{N} m_i|\Omega^{1/3}\dot{\mathbf{s}}_i|^2 + \frac{1}{2}M\dot{\Omega}^2 - U(\Omega^{1/3}\mathbf{s}_i). \qquad (9.13)$$

9.5 PR MD for (NσH) or (NτH) Ensemble

To consider more general external loading condition, Parrinello and Rahman constructed a more general (NσH) ensemble MD.

The main difference between (NPH) and (NσH) is the external potential. In (NPH) PR-MD

$$\mathcal{L}_{PR} = \frac{1}{2}\sum_{i=1}^{N} m_i \dot{\mathbf{s}}_i \cdot \mathbf{G} \cdot \dot{\mathbf{s}}_i - \frac{1}{2}\sum_i \sum_{j \neq i} U(r_{ij}) + \frac{1}{2}WTr(\dot{\mathbf{h}}^T \cdot \dot{\mathbf{h}}) - p\Omega \quad \text{and}$$

the external potential is $p\Omega$ or $p(\Omega - \Omega_0)$.

For a full external stress load, the strain energy induced by the environment should be,

$$V_{ext} = \mathbf{S} : \mathbf{E}\Omega_0,$$

where $\mathbf{S}$ is the second Piola–Kirchhoff stress tensor and $\mathbf{E}$ is the Green–Lagrangian strain tensor.

In nonlinear continuum mechanics, the Lagrangian strain is defined as

$$\mathbf{E} = \frac{1}{2}(\mathbf{F}^T \cdot \mathbf{F} - \mathbf{I}).$$

Since $\mathbf{F} = \mathbf{h} \cdot \mathbf{h}_0^{-1}$, we have

$$\mathbf{E} = \frac{1}{2}\left((\mathbf{h}_0^{-T} \cdot \mathbf{h}^T) \cdot (\mathbf{h} \cdot \mathbf{h}_0^{-1}) - \mathbf{I}\right) = \frac{1}{2}\left(\mathbf{h}_0^{-T} \mathbf{G} \cdot \mathbf{h}_0^{-1} - \mathbf{I}\right).$$

In passing, we recall Push forward operation in co-tangent space

$$\chi_*(\cdot^\flat) = \mathbf{F}^{-T}(\cdot^\flat)\mathbf{F}^{-1}.$$

Hence,

$$V_{ext} = (\mathbf{S} + p\mathbf{I}) : \mathbf{E}\Omega_0 - p\mathbf{I} : \mathbf{E}\Omega_0$$
$$= Tr[\mathbf{S} + p\mathbf{I}) \cdot \mathbf{E})]\Omega_0 - Tr[\mathbf{E}]p\Omega_0.$$

Here, we use the identity $\mathbf{A} : \mathbf{B} = Tr(\mathbf{A} \cdot \mathbf{B}^T)$.

To analyze the external work, we consider that $Tr[\mathbf{E}] = E_{11} + E_{22} + E_{33}$ is called the Lagrangian volumetric strain,

$$\mathrm{Tr}[\mathbf{E}] = E_{11} + E_{22} + E_{33} = \frac{\Omega - \Omega_0}{\Omega_0} = \frac{\Omega}{\Omega_0} - 1.$$

Hence,

$$V_{ext} = \mathrm{Tr}[(\mathbf{S} + p\mathbf{I}) \cdot \mathbf{E}]\Omega_0 - p\left(\frac{\Omega - \Omega_0}{\Omega_0}\right)\Omega_0$$

$$= \frac{1}{2}\mathrm{Tr}[(\mathbf{S} + p\mathbf{I}) \cdot (\mathbf{h}_0^{-T} \cdot \mathbf{G} \cdot \mathbf{h}_0^{-1} - \mathbf{I})]\Omega_0 - p(\Omega - \Omega_0).$$

Both terms,

$$\mathrm{Tr}(\mathbf{S} + p\mathbf{I})\Omega_0 \text{ and } p\Omega_0$$

are constant terms in Lagrangian equations, so that they will not affect final results of the equations of motion. We may then drop them out of V_{ext},

$$V_{ext} = -p\Omega + \frac{1}{2}\mathrm{Tr}[(\mathbf{S} + p\mathbf{I}) \cdot (\mathbf{h}_0^{-T} \cdot \mathbf{G} \cdot \mathbf{h}_0^{-1})]\Omega_0$$

$$= -p\Omega + \frac{1}{2}\mathrm{Tr}[(\mathbf{h}_0^{-1} \cdot (\mathbf{S} + p\mathbf{I}) \cdot \mathbf{h}_0^{-T}) \cdot \mathbf{G}]\Omega_0 \leftarrow \mathrm{Tr}[\mathbf{A} \cdot \mathbf{B}] = \mathrm{Tr}[\mathbf{B} \cdot \mathbf{A}].$$

Define the pullback stress measure,

$$\boldsymbol{\Sigma} := \mathbf{h}_0^{-1} \cdot (\mathbf{S} + p\mathbf{I}) \cdot \mathbf{h}_0^{-T} = \mathbf{S} + p.$$

In passing, we recall that the pullback operation in tangent space is

$$\chi_*(\cdot^\flat) = \mathbf{F}^{-1}(\cdot^\sharp)\mathbf{F}^{-T}.$$

Then the Lagrangian of ($N\sigma H$) ensemble PR-MD will be

$$\mathcal{L}_{PR} = \frac{1}{2}\sum_{i=1}^{N} m_i \dot{\mathbf{s}}_i \cdot \mathbf{G} \cdot \dot{\mathbf{s}}_i - \frac{1}{2}\sum_i \sum_{j \neq i} U(r_{ij}) + \frac{1}{2} W Tr(\dot{\mathbf{h}}^T \cdot \dot{\mathbf{h}}) - p\Omega + \frac{\Omega_0}{2}\boldsymbol{\Sigma} : \mathbf{G}.$$

The Hamiltonian for PR ($N\sigma H$) ensemble MD will be,

$$\mathcal{H}_{PR} = \sum_i \mathbf{p}_i^s \cdot \dot{\mathbf{s}}_i + \mathbf{p}^h : \dot{\mathbf{h}} - \mathcal{L}_{PR}$$

$$= \frac{1}{2}\sum_i \frac{\mathbf{p}_i^s \cdot \mathbf{G}^{-1} \cdot \mathbf{p}_i^s}{m_i} + \frac{1}{2W}\mathbf{p}^h : \mathbf{p}^h + U(\mathbf{h} \cdot \mathbf{s}_i) + p\Omega - \frac{\Omega_0}{2}\boldsymbol{\Sigma} : \mathbf{G}.$$

Since

$$\mathcal{L}_{PR}^{(N\sigma H)} = \mathcal{L}_{PR}^{(NPH)} + \frac{\Omega_0}{2}\boldsymbol{\Sigma} : \mathbf{G}, \text{ and } \mathcal{H}_{PR}^{(N\sigma H)} = \mathcal{H}_{PR}^{(HPH)} - \frac{\Omega_0}{2}\boldsymbol{\Sigma} : \mathbf{G},$$

and the added term is not involved with the variable $\mathbf{s}_i, \mathbf{p}_i^s$, the fine scale equations of motion will remain intact.

Since,
$$\frac{\partial}{\partial \mathbf{h}}\left(\frac{\Omega_0}{2}\Sigma : \mathbf{G}\right) = \Omega_0 \mathbf{h} \cdot \Sigma.$$

For the coarse scale equation of motion,
$$\frac{\partial \mathcal{L}_{PR}^{(N\sigma H)}}{\partial \mathbf{h}} = -(\sigma_{virial} + p\mathbf{I}) \cdot \Pi + \Omega_0 \cdot \mathbf{h} \cdot \Sigma,$$

and finally,
$$\frac{d}{dt}\frac{\partial \mathcal{L}_{PR}^{(N\sigma H)}}{\partial \dot{\mathbf{h}}} - \frac{\partial \mathcal{L}_{PR}^{(N\sigma H)}}{\partial \mathbf{h}} = 0 \;\rightarrow\; W\ddot{\mathbf{h}} = -(\sigma_{virial} + p\mathbf{I}) \cdot \Pi + \Omega_0 \cdot \mathbf{h} \cdot \Sigma \;\text{and}$$

$$\frac{d}{dt}\frac{\partial \mathcal{L}_{PR}^{(N\sigma H)}}{\partial \dot{\mathbf{h}}} - \frac{\partial \mathcal{L}_{PR}^{(N\sigma H)}}{\partial \mathbf{h}} = 0 \;\rightarrow\; W\ddot{\mathbf{h}} = -(\sigma_{virial} + p\mathbf{I}) \cdot \Pi + \Omega_0 \mathbf{h} \cdot \Sigma.$$

Consider
$$\Sigma := S + p\mathbf{I} \;\rightarrow\; \Omega_0 \mathbf{h} \cdot \Sigma = \Omega_0 \mathbf{h}\left(S_{ext} + p\mathbf{I}\right).$$

Moreover, the reciprocal tensor is defined as,
$$\Pi = \Omega \mathbf{h}^{-T},$$

and hence
$$(\sigma_{virial} + p\mathbf{I}) \cdot \Pi = \Omega \mathbf{h}\mathbf{h}^{-1}(\sigma_{virial} + p\mathbf{I})\mathbf{h}^{-T}$$
$$= \Omega_0 \mathbf{h}(S_{virial} + p\mathbf{I}).$$

Finally, we have
$$\frac{d}{dt}\frac{\partial \mathcal{L}_{PR}^{(N\sigma H)}}{\partial \dot{\mathbf{h}}} - \frac{\partial \mathcal{L}_{PR}^{(N\sigma H)}}{\partial \mathbf{h}} = 0 \;\rightarrow\; W\ddot{\mathbf{h}} = -\Omega_0 \mathbf{h} \cdot (S_{virial} - S_{ext}).$$

In above derivation, we have used identities,
$$S = JF^{-1}\sigma F^{-T}, \;\text{and}\; \frac{\Omega}{\Omega_0} = det(F) = J.$$

In summary, PR's extended Lagrangian may be written as
$$\mathcal{L}_{PR} = \frac{1}{2}\sum_{i=1}^{N} m_i \dot{\mathbf{s}}_i \cdot \mathbf{G} \cdot \dot{\mathbf{s}}_i - \frac{1}{2}\sum_{i}\sum_{j\neq i} U(r_{ij}) + \frac{1}{2}W\mathrm{Tr}(\dot{\mathbf{h}}^T\dot{\mathbf{h}}) - p\Omega + \frac{1}{2}\mathrm{Tr}(\Sigma \cdot \mathbf{G}),$$

where
$$\Sigma = \mathbf{h}_0^{-1} \cdot (S + p\mathbf{I}) \cdot \mathbf{h}_0^{-T}\Omega_0, \;\; \mathbf{G} = \mathbf{h}^T \cdot \mathbf{h},$$

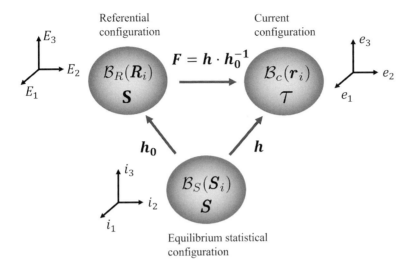

Figure 9.6 Stress measures in different continuum mechanics configurations: It is an (NτH) ensemble MD!

and the equations of motion of the PR-MD in anisotropic case can be expressed as,

$$\ddot{\mathbf{s}}_i = -\sum_{j \neq i} \left(\frac{U'(r_{ij})}{m_i r_{ij}} \right)(\mathbf{s}_i - \mathbf{s}_j) - \mathbf{G}^{-1}\dot{\mathbf{G}}\dot{\mathbf{s}}_i,$$

$$W\ddot{\mathbf{h}} = (\sigma_{virial} + p\mathbf{I})\mathbf{\Pi} - \Omega_0 \mathbf{h} \cdot \mathbf{\Sigma}, \quad \text{with } \mathbf{\Pi} = \Omega^{-1}\mathbf{h}^{-T},$$

and

$$\sigma_{virial} = \frac{1}{\Omega} \sum_{i=1}^{N} \left\{ m_i \mathbf{v}_i \otimes \mathbf{v}_i - \sum_{j \neq i} U'(r_{ij}) \frac{\mathbf{r}_{ij} \otimes \mathbf{r}_{ij}}{r_{ij}} \right\}.$$

This is the end of the derivation for PR-MD equations of motion.

REMARK 9.3 In Parrinello and Rahman's original paper, they called this ensemble MD as an NσH ensemble MD. Consider that

$$\mathbf{S} = \mathbf{h}_0^{-1}\mathbf{S}\mathbf{h}_0^{-T} \quad \text{and} \quad \mathbf{S} = \mathbf{F}^{-1}\tau^{\sharp}\mathbf{F}^{-T}.$$

If we require that

$$\mathbf{S} = \tau$$

during the cell deformation process, the anisotropic PR-MD is in fact an **NτH** ensemble rather than an **NσH** ensemble, as shown in Fig. 9.6.

An immediate success of PR-MD is its application to simulate crystal structure phase transformation (see Fig. 9.7).

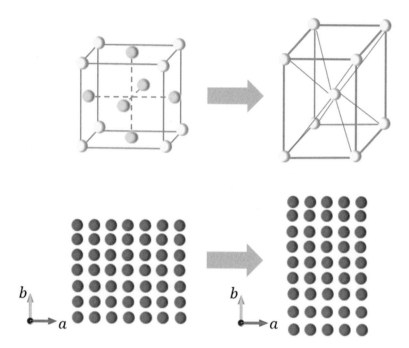

Figure 9.7 PR MD simulation of crystal microstructure change: a structure phase transformation of face-centered tetragonal (FCT) to a body-centered tetragonal (BCT).

9.6 Podio-Guidugli's Interpretation

The (Andersen–)PR (NσH) ensemble MD was reinterpreted by Paolo Podio-Guidugli (see *journal of elasticity* [2010], 145–153).

In the Podio-Guidugli version of PR-MD, he made the following changes in primary variables:

$$\mathbf{h} \to \mathbf{F}, \text{ and } \mathbf{s}_I \to \mathbf{R}_I,$$

in meaning as well as in notation.

The Podio-Guidugli's version of PR-MD Lagrangian is given as follows:

$$\mathcal{L}_{PG-PR} = \frac{1}{2}\sum_i m_i (\mathbf{F}^T \cdot \mathbf{F}) : (\dot{\mathbf{s}}_i \otimes \dot{\mathbf{s}}_i) + \frac{1}{2} W \|\dot{\mathbf{F}}\|^2 - U(\mathbf{F} \cdot \mathbf{s}_I) + \mathbf{S}_{ext} : \mathbf{C}\Omega_0, \quad (9.14)$$

where $\mathbf{C} = \mathbf{F}^T \cdot \mathbf{F}$ is the right Cauchy–Green tensor.

Note that in the original paper, the last term was $\Omega_0 \mathbf{S} \cdot \mathbf{F}$, which might be a typo, and we changed it to $\mathbf{S}_{ext} : \mathbf{C}\Omega_0$.

Based on Eq. (9.14), we can derive the equations of motion of the molecular system from the following Lagrangian equations:

$$\begin{cases} \dfrac{d}{dt}\left(\dfrac{\partial \mathcal{L}_{PG-PR}}{\partial \dot{\mathbf{s}}_i}\right) - \dfrac{\partial L_{PG-PR}}{\partial \mathbf{s}_i} = 0, \\ \dfrac{d}{dt}\left(\dfrac{\partial \mathcal{L}_{PG-PR}}{\partial \dot{\mathbf{F}}}\right) - \dfrac{\partial L_{PG-PR}}{\partial \mathbf{F}} = 0, \end{cases} \quad (9.15)$$

and then we obtain,

$$\ddot{\mathbf{s}}_i + \mathbf{C}^{-1} \cdot \dot{\mathbf{C}} \cdot \dot{\mathbf{s}}_i + \frac{1}{m_i} \sum_{i \neq j} \frac{1}{r_{ij}} \phi'(r_{ij})(\mathbf{s}_i - \mathbf{s}_j) = 0 \text{ and} \tag{9.16}$$

$$W\ddot{\mathbf{F}} = -\Omega_0 \mathbf{F} \cdot (\mathbf{S}_{virial} - \mathbf{S}_{ext}), \tag{9.17}$$

where

$$\mathbf{S}_{virial} = \frac{1}{\Omega_0} \sum_{i=1}^{N} \left\{ -m_i \dot{\mathbf{s}}_i \otimes \dot{\mathbf{s}}_i + \sum_{j \neq i} \phi'(r_{ij})(\mathbf{s}_i - \mathbf{s}_j) \otimes (\mathbf{s}_i - \mathbf{s}_j) \right\}. \tag{9.18}$$

The Hamiltonian for Podio-Guidugli's version of $(N\sigma H)$ ensemble MD is given as follows:

$$\mathcal{H}_{PG-PR} = \sum_i \mathbf{p}_i^s \cdot \dot{\mathbf{s}}_i + \mathbf{p}^F : \dot{\mathbf{F}} - \mathcal{L}_{PG-PR}$$

$$= \frac{1}{2} \sum_i \frac{\mathbf{p}_i^s \cdot \mathbf{G}^{-1} \cdot \mathbf{p}_i^s}{m_i} + \frac{1}{2W} \mathbf{p}^F : \mathbf{p}^F + U(\mathbf{F} \cdot \mathbf{s}_i) + p\Omega - \frac{\Omega_0}{2} \mathbf{S} : \mathbf{C},$$

where

$$\mathbf{p}_i^s = \frac{\partial \mathcal{L}_{PG-PR}}{\partial \dot{\mathbf{s}}_i} = m_i \mathbf{C} \cdot \dot{\mathbf{s}}_i, \text{ and } \mathbf{p}^F = \frac{\partial \mathcal{L}_{PG-PR}}{\partial \dot{\mathbf{F}}} = W\dot{\mathbf{F}}.$$

The main contribution of Podio-Guidugli's work is that he made some insightful interpretations or justification of the original PR-MD formulation. Podio-Guidugli argued that in order to prove that

$$\mathcal{K}_{3\&4} = \frac{1}{2} \sum_i m_i \left(\mathbf{s}_i (\dot{\mathbf{F}}^T \mathbf{F}) \cdot \dot{\mathbf{s}}_i + \dot{\mathbf{s}}_i (\mathbf{F}^T \dot{\mathbf{F}}) \mathbf{s}_i \right) = 0,$$

one must show that

$$\dot{\mathbf{F}}^T \mathbf{F} \sum_i m_i \mathbf{s}_i \otimes \dot{\mathbf{s}}_i = 0, \tag{9.19}$$

which hinges from the following condition:

$$\mathbf{W} \sum_i m_i \mathbf{s}_i \otimes \dot{\mathbf{s}}_i = 0, \tag{9.20}$$

where $\mathbf{W}$ is the spin tensor or antisymmetric part of $\dot{\mathbf{F}}^T \mathbf{F}$.

This is because that

$$\sum_i m_i \mathbf{s}_i \otimes \dot{\mathbf{s}}_i \tag{9.21}$$

is skew symmetric if the shape tensor of the simulation cell, where

$$\mathcal{S} = \sum_i m_i \mathbf{s}_i \otimes \mathbf{s}_i, \tag{9.22}$$

is a constant tensor.

This can be seen from the following argument.
Since the first PR closure may be expressed as,

$$\sum_i m_i s_i \otimes s_i = const.,$$

which implies that

$$\left(\sum_i m_i s_i \otimes \dot{s}_i\right) + \left(\sum_i m_i \dot{s}_i \otimes s_i\right) = 0. \tag{9.23}$$

It suggests that

$$\sum_i m_i s_i \otimes \dot{s}_i \text{ is skew-symmetric.}$$

Hence,

$$\dot{F}^T F \sum_i m_i s_i \otimes \dot{s}_i = L \cdot \sum_i m_i s_i \otimes \dot{s}_i = (D + W) \sum_i m_i s_i \otimes \dot{s}_i$$
$$= W \sum_i m_i s_i \otimes \dot{s}_i = 0. \tag{9.24}$$

In other words, if Eq. (9.24) holds, Eq. (9.19) holds.

9.7 Homework Problems

Problem 9.1 Consider the following extended (NσH) ensemble Lagrangian:

$$\mathcal{L}_{PG-PR} = \frac{1}{2} \sum_I m_I (F^T \cdot F) : (\dot{s}_I \otimes \dot{s}_I) + \frac{1}{2} W \|\dot{F}\|^2 - U(F \cdot s_I) - S : E\Omega_0,$$

where S is the second Piola–Kirchhoff stress tensor, $\|\dot{F}\|^2 := \dot{F} : \dot{F} = Trace(\dot{F} \cdot \dot{F})$, and $C = F^T \cdot F$ is the right Cauchy–Green tensor. Or, in another form,

$$\mathcal{L}_{PG-PR} = \frac{1}{2} \sum_I m_I (\dot{s}_I \cdot C \cdot \dot{s}_I) + \frac{1}{2} W \|\dot{F}\|^2 - U(F \cdot s_I) - S : E\Omega_0.$$

Derive the corresponding Euler–Lagrangian equations,

$$\begin{cases} \dfrac{d}{dt}\left(\dfrac{\partial \mathcal{L}_{PG-PR}}{\partial \dot{s}_I}\right) - \dfrac{\partial \mathcal{L}_{PG-PR}}{\partial s_I} = 0, \\ \dfrac{d}{dt}\left(\dfrac{\partial \mathcal{L}_{PG-PR}}{\partial \dot{F}}\right) - \dfrac{\partial \mathcal{L}_{PG-PR}}{\partial F} = 0. \end{cases}$$

Problem 9.2 In Problem 9.1, let

$$F = h(t) \cdot h_0^{-1}.$$

Derive the equation of motion for h?

10 Introduction to LAMMPS

In this chapter, we provide a short tutorial for a popular open source molecular dynamics (MD) simulation package, LAMMPS, which is an acronym for *Large-scale Atomic/Molecular Massively Parallel Simulator*. Figure 10.1 displays LAMMPS homepage: www.lammps.org/).

10.1 How to Download and Install LAMMPS

One can freely download LAMMPS source code from the following website: www.lammps.org/download.html. You can also download LAMMPS as a tarball from this page, using the "Download Now" button (see Figs. 10.2 and 10.3).

One may either download a tarball or pre-built executables. For instance, if you download a gzipped tar file for LAMMPS stable version (see Fig. 10.3), you can unpack it by typing the command

```
tar -xzvf lammps-stable.tar.gz
```

as instructed in the LAMMPS download page (Fig. 10.3) or by using any unzip software.

After downloading the source code, you may build LAMMPS executable code using "cmake" or "make" in Windows command prompt, for instance. On the other hand, we may also directly download the pre-built executable file from the LAMMPS website by selecting appropriate executable file for the desirable operating system platform (see Figs. 10.2 and 10.3).

Even if you do not want to build LAMMPS executable file by yourself, you may still want do download all the source code. For example, if you download the stable version of LAMMPS source code, all files are usually in the directory *lammps-stable*. When you open it, you may see different subdirectories, and there is an *examples* directory, which contains many useful example input script files or directories for demonstration purpose as the examples of how to use LAMMPS (see Fig. 10.4).

To install LAMMPS, the readers are referred to LAMMPS installation page for detailed installation information: https://docs.lammps.org/Install.html. Moreover, if you have Git installed on your computer, you can use Git access from https://github.com/lammps/lammps.

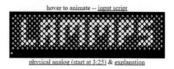

Figure 10.1 LAMMPS homepage (www.lammps.org/)

LAMMPS website

Download LAMMPS

You can download LAMMPS as a tarball from this page, using the "Download Now" button below.

There are several ways to get the LAMMPS software, either as a tarball, or from an active repository, or in executable form:

- Download a tarball (here or from GitHub)
- Git repository for LAMMPS
- SVN repository for LAMMPS
- Pre-built Linux executables
- Pre-built Mac executables
- Pre-built Windows executables

With source code, you have to build LAMMPS using "cmake" or "make". But you have more flexibility as to what features to include or exclude in the build. If you plan to modify or extend LAMMPS, then you need the source code.

The Install doc page lists what is included in the LAMMPS distribution.

Figure 10.2 LAMMPS download website (www.lammps.org/download.html)

In particular, for operating systems such as LINUX, OS_X, and Windows, there are downloadable pre-compiled LAMMPS executable files. For instance, if you would like to install Windows version LAMMPS in your computer, you can find the installer from the following link: https://docs.lammps.org/Install.html (see Fig. 10.5).

If you download and install a Windows version of precompiled LAMMPS executable file, its PATH environment in your computer will be automatically set by the installer, so that you can run the executable files from anywhere in your computer.

Download a tarball

Select the code you want, click the "**Download**" link, and your browser should download a gzipped tar file. Unpack it with the following command, and see the README file to get started.

```
tar -xzvf file.tar.gz
```

There were ~405,000 downloads of LAMMPS from Sept 2004 thru mid-year 2021. At that point we migrated our website to https://www.lammps.org and a different server, and stopped logging downloads.

LAMMPS **molecular dynamics package:**

LAMMPS Stable Release	29 Sep 2021	Latest stable release. 169 MB tarball with source and docs, GPLv2 license. Last updated 24 March 2022. Includes all bug fixes and new features described on this page, up to the date of the most recent stable release plus selected bugfixes from patch releases.	Download	Manual
LAMMPS Patch Release	4 May 2022	Latest patch release. 166 MB tarball with source and docs, GPLv2 license. Includes all bug fixes and new features described on this page.	Download	Manual

Legacy versions of LAMMPS:

LAMMPS 3 Mar 2020	last version of LAMMPS compatible with C++98, GPLv2 license, 164 MB, last updated 3 Mar 2020	Download	
LAMMPS 2001	older f90 version source tarball, GPLv2 license, 1.1 MB, last updated 17 Jan 2005	Download	Manual
LAMMPS 99	older f77 version source tarball, GPLv2 license, 844 kB	Download	Manual

All versions of LAMMPS are distributed under the terms of the GNU Public License Version 2. The earlier Fortran versions are no longer under active development.

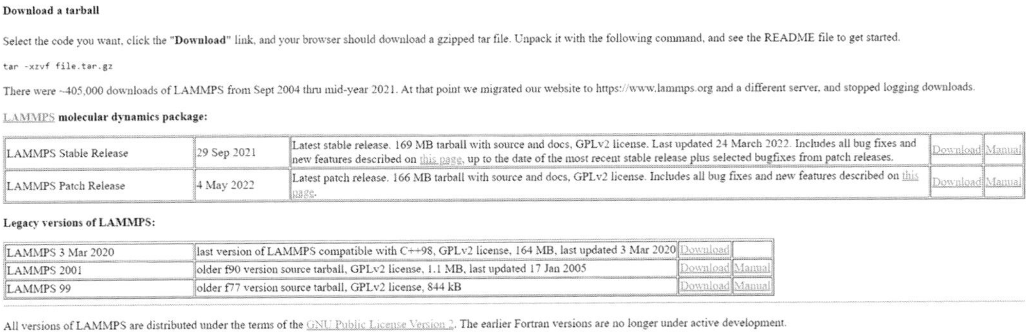

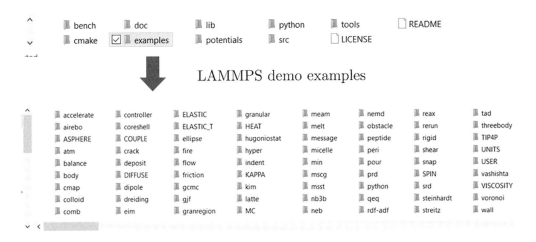

Figure 10.3 Download LAMMPS as a tarball (www.lammps.org/download.html)

Figure 10.4 LAMMPS source code directory and examples directory

Moreover the LAMMPS_POTENTIALS environment variable is preset to the folder that bundles many potential files together, so that they do not need to be copied into the respective working directory.

10.2 How to Run LAMMPS

LAMMPS is a text mode program, so that you must first start by clicking the Command Prompt entry in the Windows Start menu or just typing "cmd" in Windows Search Bar, to launch the Windows Power Shell.

There are three LAMMPS execution modes: **Serial Execution**, **Multi-threaded Parallel Execution** (OpenMP), and **Message Passing Interface (MPI) Parallel Execution**. The regular serial LAMMPS executable command is: *lmp_serial* or *lmp_mpi*.

Installing LAMMPS on Windows

There are installer packages for 32-bit and 64-bit versions of Windows available.

- Latest stable versions
- 32-bit Windows download area with all available installer versions
- 64-bit Windows download area with all available installer versions

The respective download directory will contain installer packages that are labeled with the date of the LAMMPS version and packages labeled as *latest*. It is usually recommended to download and install the latest package. The other packages are provided in case there is a problem with it. Download the installer executable suitable for your machine, execute it, and follow the instructions in the dialogs. Each version will install into a different directory, so it is possible to have multiple versions installed at the same time (however it is not recommended). Both kinds of packages contain:

- Either: a regular multi-threaded LAMMPS executable called lmp_serial. This should <u>always</u> work.
- Or: a multi-threaded LAMMPS executable that also supports parallel execution via MPI message passing. This executable is called lmp_mpi and requires installation of a suitable MPICH2 package to work.
- the LAMMPS manual in PDF format
- the LAMMPS developer guide in PDF format

Figure 10.5 Installing LAMMPS on Windows (from http://packages.lammps.org/windows.html)

10.2.1 Serial Execution

Assume that we have an input file called *in.file* that contains all the computation commands, material data, and model geometry and thermodynamic variables to run a particular MD simulation.

We can run this MD simulation via the "it -in" command line flag, or using the standard input command "<".

```
lmp_serial -in in.file
```

or

```
lmp_serial < in.file
```

10.2.2 Multi-Thread Parallel Execution

LAMMPS supports multi-threading via OpenMP (shared memory parallel), and by default one thread is enabled.

Most PCs and laptops have processors from 4 to 16 depending on their hardware specifications. If you would like to run more threads, for instance four, you can first set *OMP_NUM_THREADS=4*, via the *-pk omp 4* command line flag. After enabling the use of multiple threads, you also have to select styles in LAMMPS, which are multi-thread enabled and are usually identified by the /omp suffix. These can be selected explicitly with the suffix command or — most conveniently — using the *-sf omp* command line flag to the LAMMPS executable. For example,

```
set OMP_THREADS = 4
lmp_serial -pk omp 4 -sf omp -in in.file
```

or you can simply type in command line:

```
mpirun -np 4 lmp_mpi -in in.file
```

10.2.3 Message Passing Interface (MPI) Parallel Execution

The *lmp_mpi* executable supports parallel execution via MPI (message passing interface for distributed memory parallel). However, to use *lmp_mpi* you need to install MPICH2 from Argonne National Laboratory website: www.mpich.org/.

After the installation of the MPICH software, it needs to be integrated into the system. For this you need to start a Command Prompt in Administrator Mode (right click on the icon and select it), and then move into MPICH2 installation directory, and then further go to the subdirectory bin. After getting into the subdirectory bin, execute or type the command: *smpd.exe -install*.

MPI parallel executables have to be launched with *mpiexec -localonly # lmp_mpi* ⋯ or *mpiexec -np # lmp_mpi* ⋯ with "#" being the number of parallel processes to start. For the latter you may be asked for your password.

For example,

```
mpiexec localonly 4  lmp_mpi in  in.file
```

or

```
mpiexec np #  lmp_mpi in  in.file
```

10.3 Some Basic LAMMPS Commands

In this section, we discuss about how to write a LAMMPS input script file in order to carry out MD simulations.

LAMMPS input script is written by close to 200 commands with many variations, and in some way, it is a special purpose command language. To master all of the LAMMPS commands is not a trivial task, and it takes experience and practice. Here, we only introduce some basic LAMMPS commands to provide a foundation so that the readers can grow on it.

LAMMPS input script file has the following input script parsing rules:

- Each non-blank line is treated as a command.
- If the line ends with &, the command is assumed to continue on the next line.
- All characters following # are treated as a comment.
- A $ sign followed by characters indicates that this is the value of a variable (the characters), e.g., $temp or $temp.
- A line is broken into "words" separated by spaces OR tabs, e.g., letters, digits, underscores, or punctuation characters.

- The general format of a command is: first word is a command and successive words are ID (or its name), ID of objects that the command is applied, style, and its arguments.
- Text with spaces should be enclosed within double quotes, e.g., "dump modify" or "fix print."

Note that the LAMMPS line continuation character is "&," every command can be spread across many lines, though it is still a single command. For example, the following command lines

```
run 100000 every 1000 &
  "print 'Temp = $c'" &
  "print 'Press = $d'"
```

are equivalent to a single command line,

```
run 100000 every 1000  "print 'Temp = $c'"
"print 'Press = $d'"
```

For reference and documentation, LAMMPS has a detailed and complete online documentation, and you may download the LAMMPS command manual pdf file online: https://docs.lammps.org/Manual.pdf (see Fig. 10.6). In principle, you can always find the answer to any question that you may have on a specific command or operations.

There are close to 200 LAMMPS commands, and most of the commands have different combinations of options and usage specifications, which can have many different applications and modeling ramifications. However, for a specific modeling and simulation problem, you may not use all of them. As an introduction, we briefly introduce 20 basic types of LAMMPS commands in the following four groups.

(i) Initialization
(ii) Atom or particle definition
(iii) Computation or simulation setting
(ib) Execute a simulation

10.3.1 Initialization

The commands in this category are used to specify the characteristics of your LAMMPS simulation, i.e., units, processors, atom or particle style, boundary conditions, and so forth. For instance, what is the unit system of your simulation, or what types of particle models that you intend to use in your simulations, and the like. We now briefly discuss a few of them that are frequently used in LAMMPS simulations:

1. units (https://docs.lammps.org/units.html)

The syntax of the *units* command is:
units *style*

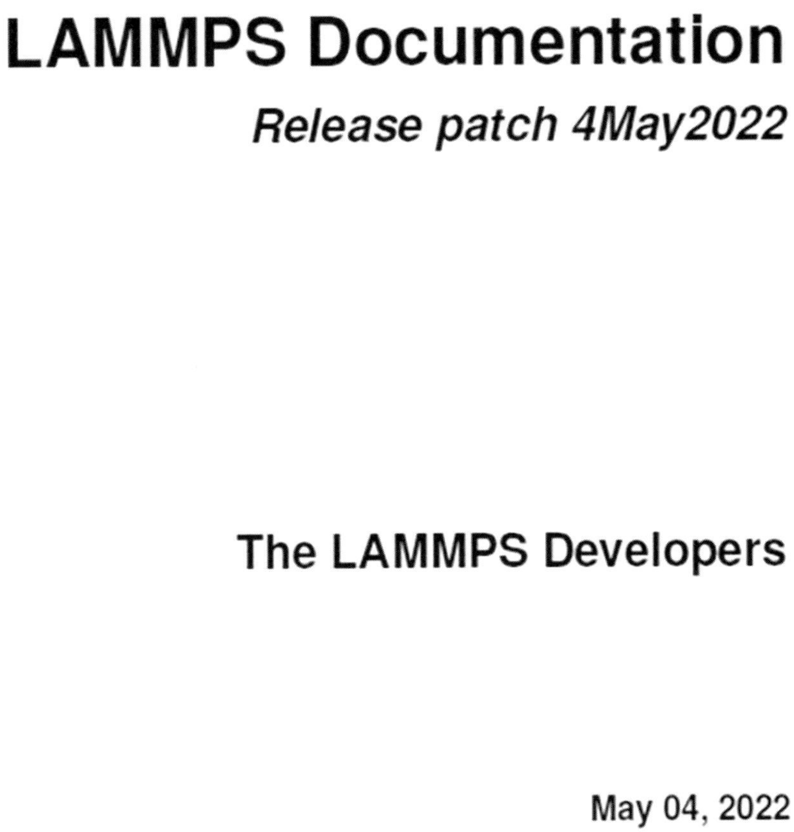

LAMMPS Documentation
Release patch 4May2022

The LAMMPS Developers

May 04, 2022

Figure 10.6 Screenshot for LAMMPS Manual (From https://docs.lammps.org/Manual.html)

where the argument *style* can be either *style* = *lj, real, metal, si, cgs, electron, micro*, or *nano*

For example, in the simulation of a metallic material, say, copper, you may start with the LAMMPS script with

```
units metal
```

When you do that, the units of your simulation system are as follows:

- mass = grams/mole
- distance = Angstroms
- time = picoseconds
- energy = eV
- velocity = Angstroms/picosecond
- force = eV/Angstrom
- torque = eV

- temperature = Kelvin
- pressure = bars
- dynamic viscosity = Poise
- charge = multiple of electron charge value 1.0 represents a proton
- dipole = charge*Angstroms
- electric field = volts/Angstrom
- density = gram/cm^{dim}

Or if you would like to model and simulate the interaction of many polymer chains, you may want to use the unit lj. When you use $units = lj$, the following reduced units will be used in LAMMPS calculations,

- mass = mass or m
- distance = σ, where $x^* = x/\sigma$
- time = τ, where $\tau^* = \tau\sqrt{\epsilon/(m\sigma^2)}$
- energy = ϵ, where $E^* = E/\epsilon$
- velocity = σ/τ, where $v^* = v(\tau/\sigma)$
- force = ϵ/σ, where $f^* = f(\sigma/\epsilon)$
- torque = ϵ, where $t^* = t/\epsilon$
- temperature = reduced LJ temperature, where $T^* = T(k_B/\epsilon)$
- pressure = reduced LJ pressure, where $p^* = p(\sigma^3/\epsilon)$
- dynamic viscosity = reduced LJ viscosity, where $\eta^* = \eta(\sigma^3/\epsilon\tau)$
- charge = reduced LJ charge, where $q^* = q/\sqrt{4\pi\epsilon_0\sigma\epsilon}$
- dipole = reduced LJ dipole, moment where $\mu^* = \mu/\sqrt{4\pi\epsilon_0\sigma^3\epsilon}$
- electric field = force/charge, where $E^* = E\sqrt{4\pi\epsilon_0\sigma}(\sigma/\epsilon)$
- density = mass/volume, where $\rho^* = \rho\sigma^{dim}$

For other unit arguments such as $real, si, cgm,$ and so forth the readers may find the detailed information at https://docs.lammps.org/units.html

2. boundary (https://docs.lammps.org/boundary.html)

The boundary command specifies the boundary conditions of the simulation. The syntax of the boundary command is

boundary *x y z*

This command sets the style of boundaries for the global simulation box in each dimension. The arguments of *x*, *y*, and *z* are the boundary conditions in *x*-direction, *y*-direction, and *z*-direction. They can be one or two of the following options: $x, y, z = p, s, f, m,$
in which
p is periodic boundary condition;
s is the shrink-wrapped (nonperiodic) boundary condition;
f is the fixed (nonperiodic) boundary condition;
m is the nonperiodic and shrink-wrapped boundary condition with a minimum value.

Note that if it is a single letter, it assigns the same style to both the lower and upper faces of the box. When two letters are assigned, the first style is assigned to the lower face and the second style to the upper face. The initial size of the simulation box is set by the *read_data, read_restart*, or *create_boxcommands*.

The so-called shrink-wrapped boundary condition means that the boundary condition can enforce all atoms staying in a fixed simulation domain and avoid any loss of atoms from your simulation domain. For example, you have a N number of atoms in a box, if the top of the box is open, some atoms may jump out of the box. By using shrink wrapping boundary condition, you can avoid any escape of atoms from the box, and the total number of atoms in the simulation domain remains constant. It is important to note that when I say the top of the box is open, which means that the dimension of the top is not defined, and as an atom goes out the height dimension will grow to keep with the outgoing atom so that it is still inside the box.

In some simulations, we may gradually add atoms into a box, and at the beginning the box may be empty. In this case, you may want to specify a minimum value of the box height, otherwise LAMMPS may think that box height is zero, and it shrinks to zero height resulting in the crash of the simulation. For example, you may find that the following types of boundary conditions are specified in a LAMMPS script file:

```
boundary p p p
boundary p fs p
boundary s f fm
```

3. atom_style (https://docs.lammps.org/atom_style.html)

This command defines which style of atoms or particles to use in a simulation. It is a command that determines what attributes are associated with the particles used in the simulation. This command must be used before a simulation is setup through either *read_data*, or *read_restart*, or *create_box* command.

The syntax of the command is:

atom_style *style args*

There are 25 values for variable *style*, and they are:
angle, atomic, body, bond, charge, dipole, dpd, edpd, electron, ellipsoid, full, line, mdpd, molecular, oxdna, peri , smd, sph, sphere, bpm/sphere, spin, tdpd, tri, template, hybrid.

In Table 10.1, we show the meanings of some most common atom_style(s).

The default style value for *atom_style* is *atomic*. In Table 10.1, for each type of atoms or particles, its attributes may need other commands to further specify or define, which will be discussed later.

There are four style values that have their own arguments: *body, tdpd, template*, and *hybrid*

Table 10.1 Examples of atom_style in LAMMPS

Style name	Particle attributes	Particle meaning and usage
angle	bonds and angles	bead-spring polymers with stiffness
atomic	only the default values	coarse-grain liquids, solids, metals
body	mass, inertia moments, quaternion, angular momentum	arbitrary bodies
bond	bonds	bead-spring polymers
charge	charge	atomic system with charges
dipole	charge and dipole moment	system with dipolar particles
dpd	internal temperature and internal energies	arbitrary bodies
tdpd	chemical concentration	tDPD particles
electron	charge and spin and eradius	electronic force field
ellipsoid	shape, quaternion, angular momentum	aspherical particles
meso	rho, e, cv	SPH particles
molecular	bonds, angles, dihedrals, impropers	uncharged molecules
peri	mass, volume	mesoscopic Peridynamic models
sphere	diameter, mass, angular velocity	(DEM) granular models
template	template index, template atom	small molecules with fixed topology

For the specification of these arguments and their meanings, the readers may consult https://docs.lammps.org/atom_style.html.

Here are a few examples:

```
atom_style atomic
atom_style bond
atom_style template myMols
atom_style tdpd 2
```

where the command *atom_style template myMols* means that this is a template style with the template ID=myMols, which is specified in a separate *molecular* command, and the command *atom_style tdpd 2* means that this is the tDPD particle, and there are two types of chemical species.

4. pair_style (https://docs.lammps.org/pair_style.html)

In LAMMPS, the command *pair_style* defines the pairwise potentials between a pair of two bonded atoms, and it sets the parameters for calculating pairwise interactions between two atoms or two particles, which could be coarse-grain particles or even peridynamic particles.

The syntax of the command is:

pair_style *style args*,

where the option variable *style* has over hundreds of values, and the following are main options and their meanings,

- none – turn off pairwise interactions
- hybrid – multiple styles of pairwise interactions
- dpd – dissipative particle dynamics (DPD)

- eam – embedded atom method (EAM)
- eam/alloy – alloy EAM
- lj/charmm/coul/long – CHARMM with long-range Coulomb
- lj/cut – cutoff Lennard-Jones potential with no Coulomb
- lj/gromacs – GROMACS-style Lennard-Jones potential
- morse – Morse potential
- reax/c – ReaxFF potential in C
- rebo – second-generation REBO potential of Brenner
- sw – Stillinger–Weber three-body potential
- tersoff – Tersoff three-body potential

For a complete list, readers can consult the online LAMMPS manual at page https://docs.lammps.org/pair_style.html.

Here are a few examples:

```
pair_style none
pair_style lj/cut 2.5
pair_style eam/alloy
pair_style hybrid lj/charmm/coul/long 10.0 eam
```

The first example, *pair_style none*, is the default value. In the second example, the pair potential is specified as the Lennard-Jones potential with a cutoff distance 2.5 length units. Note that the unit here is the LJ reduced unit, i.e., 2.5σ. The fourth example, *pair_style eam/alloy*, uses the *hybrid* style, where the atom pairs interact with each other with the pair potential, LJ-CHARMM with long-range Coulomb of the cutoff distance 10σ and the EAM potential.

It should be noted that there is another command in LAMMPS that is similar to the command *pair_style*, which is the command *bond_style*. Both of them specify interactions between a pair of atoms. However, the command *bond_style* specifies the bonded interaction between two atoms, for instance covalent bond force, not the dispersive force nor the Coulomb force, nor hydrogen force, and so forth.

5. Bond_style (https://docs.lammps.org/bond_style.html)
The syntax of the command is:

bond_style *style args*

Just like command *pair_style*, *bond_style* also has several style arguments:

- none – turn off bonded interactions
- zero – topology but no interactions
- hybrid – define multiple styles of bond interactions
- class2 – COMPASS (class 2) bond
- fene – FENE (finite-extensible non-linear elastic) bond
- harmonic – harmonic bond

in which the style *zero* means that the bond forces and energies will not be computed, but the geometry of the bond pairs is still accessible to other commands that need it.

The followings are some typical examples of the *bond_pair* command

```
bond_style harmonic
bond_style fene
bond_style hybrid harmonic fene
```

in which the style *harmonic* is one of the main bond potentials used in empirical MD.

10.3.2 Define Simulation Domain and Define Atoms and Particles

There are different ways to define atoms or particles in LAMMPS: (1) read the information from a data or restart file, which involves commands such as *read data, read restart* and (2) create atoms on a lattice, which involves commands such as *lattice, region, create box, create atoms*, and the like. Without considering input information from a data file, this group of commands can be divided into two categories: (1) defining simulation domain dimension, material microstructure (lattice), and (2) creating atomistic model.

In the following, we discuss several most used commands in this group.

1. lattice (https://docs.lammps.org/lattice.html)

The LAMMPS command *lattice* allows us to specify the lattice pattern in the simulation unit cell as well as to define the simulation unit cell itself. For instance, it may specify the simulation lattice as simple cubic, or BCC, or FCC, or HCP, and so forth as well as the lattice spacing. Furthermore, it may define the origin of the unit cell, its orientation, and side length.

The syntax of the command is:

lattice *style scale keyword values*

in which there are three option variables: *style, scale, keyword values*.

The variable *style* has nine values, i.e.,
$style\ value \in \{none, sc, bcc, fcc, hcp, diamond, sq, sq2, hex, custom\}$, in which the precise meaning of the nine values are as follows:

- none → no lattice structure;
- sc → simple cubic lattice;
- bcc → body center cubic lattice;
- fcc → face center cubic lattice;
- hcp → hexagonal close-packed lattice;
- diamond → diamond cubic lattice;
- sq → square lattice;
- hex → hexagonal lattice;
- custom → custom-made lattice

The variable *scale* specifies the lattice constants, and it uses either the reduced unit or real physical unit (Å). Using the following command, you may create a body-centered cubic (bcc) lattice having a lattice constant of 4.0Å,

```
lattice bcc 4.0
```

The variable *keyword values* has option values, and they are used to define the simulation unit cell within the lattice space:

- *origin values* $= x, y, z$ specifies the origin of the unit cell. If x, y, z are not all zero, it means that the lattice is shifted or translated when mapping it into the simulation box. When x, y, z values are fractional values ($0.0 <= x, y, z < 1.0$), it means that the lattice cell is shifted by a fraction of the lattice spacing in each dimension.
- *orient values* $= dim\ i\ j\ k$, in which $dim = x, y, z$ and (i, j, k) is the orientation vector that relates box size to lattice spacing. LAMMPS set the box size considering the lattice options. This means that the box size in each direction will be determined by the orientation vectors. The default value of orientation vector is $(100), (010), (001)$. If you don't use "orient" option in the lattice command, the box size will be set in proportion to the lattice parameter itself. However, if you use "orient" option, the box size will be set with respect to the orientation vectors.
- *spacing values* $= dx\ dy\ dz$, where dx, dy, dz = lattice spacings in the x, y, z box direction;
- a_1, a_2, a_3 *values* $= x\ y\ z$, where x, y, z = primitive vector components that define unit cell; and
- *basis values* $= x\ y\ z$, where x, y, z = fractional coordinates of a basis atom ($0 <= x, y, z < 1$).

For example, the following command

```
lattice sq 0.8 origin 0.0 0.5 0.0 orient x 1 1 0 orient y -1 1 0
```

specifies a two-dimensional (2D) square lattice with lattice constant 0.8Å, the unit cell origin is at $(0.0, 0.5, 0.0)$, the x-axis of the simulation box is along $(1,1,0)$ direction, and the y-axis of the simulation box is along the $(-1,1,0)$ direction.

Definition of Basis Atom in LAMMPS

Before moving to discussions of other commands, here we briefly introduce the concept of the basis atom in LAMMPS. In LAMMPS, the basis atom is defined through lattice style, such as SC, FCC, BCC, and Diamond cubic. However, for HCP lattice structures, the definition of basis atoms is not straightforward. This is because the definition of the LAMMPS' "basis atom" is about not how many atoms in a conventional unit cell, nor number of the lattice motif, i.e., number of atoms in a Wigner–Seitz cell.

In LAMMPS, it defines the basis atom for different lattice structures based on a cubic unit cell with edge length $= 1.0$. This means $\mathbf{a}_1 = (1, 0, 0), \mathbf{a}_2 = (0, 1, 0)$, and $\mathbf{a}_3 = (0, 0, 1)$.

When we count how many atoms in a cubic conventional unit cell, we usually use the following formula:

$$\text{\# of basis atoms} = \text{\# of interior atoms} + (\text{\# of corner atoms})/8$$
$$+ (\text{\# of face atoms})/2$$

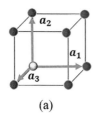 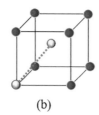 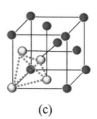 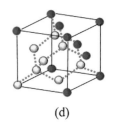

(a) (b) (c) (d)

Figure 10.7 Placements of the basis atoms in (a) simple cubic (SC) lattice, (b) body-centered cubic (BCC) lattice, (c) face-centered cubic (FCC) lattice, and (d) diamond cubic lattice.

Thus, based on this formula, SC lattice has one basis atom, BCC lattice has two basis atoms, and FCC lattice has four atoms. Since the diamond structure may be viewed as two FCC lattices overlapped together, so it has eight basis atoms as shown in Fig. 10.7.

One may explain this based on the lattice motif. As we know that the definition of the Bravais lattice is that there is only one atom in its lattice motif in the Wigner–Seitz cell of the lattice. In Appendix A, one may find that SC lattice, BCC lattice, and FCC lattice are Bravais lattices, and they only have one atom in the lattice motif, whereas the diamond cube lattice is not a Bravais lattice, and it has two atoms in a lattice motif. However, in LAMMPS, we would like to view things according to the definition of a simple cubic unit cell. We may view a BCC lattice as a simple cubic lattice with two atoms in a lattice motif; and a FCC may be viewed as a simple lattice with four atoms in a lattice motif, while the diamond cubic structure is a simple cubic lattice with eight atoms in its lattice motif, as indicated in Fig. 10.7.

In Fig. 10.7, we also show the placements of the basis atoms in SC, BCC, FCC, and diamond cubic lattice. A SC lattice has one basis atom at the high-left-bottom corner of the cube. A BCC lattice has two basis atoms, one at the corner and one at the center of the cube. An FCC lattice has four basis atoms, one at the corner and three at the cube face centers, and the basis atoms arrangement for a diamond cubic lattice is shown in Fig. 10.7(d).

For the hexagonal close packed (HCP) lattice, this becomes a complicated story. The LAMMPS-style HCP has a hexagonal lattice basis vectors $\mathbf{a}_1 = (1,0,0)$, $\mathbf{a}_2 = (1/2, \sqrt{(3)}/2, 0)$, and $\mathbf{a}_3 = (0, 0, \sqrt{(8/3)})$. However, the conventional unit cell for an HCP lattice is a hexagonal lattice with three interior atoms. If we calculate how many atoms in a hexagonal primitive unit cell, we may find that

$$\text{\# of atoms in unit cell} = 3 + 12 \ (corner\ atoms)/6 + 2 \ (face\ atoms)/2 = 6,$$

which is not the LAMMPS basis atom number. In fact, HCP is also a non-Bravais lattice, and it has two atoms in a lattice motif. This can be seen from the fact that each HCP hexagonal prim has three equivalent prim Wigner–Seitz cells. In each of the Wigner–Seitz cells, it has two atoms as shown in Fig. 10.8. This is because each Wigner–Seitz cell has $8 \ (corner\ atoms)/8 + 1 \ interior\ atom = 2 \ atoms$.

One can see that no matter in which way you calculate, the basis atoms in HCP is not the number given in LAMMPS manual. As mentioned earlier, LAMMPS defines

$\mathbf{a}_1 = (1,0,0)$
$\mathbf{a}_2 = (1/2, \sqrt{3}/2, 0)$
$\mathbf{a}_3 = (0, 0, \sqrt{8/3})$

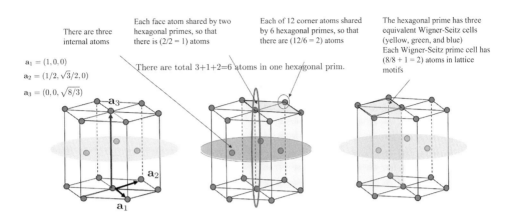

Figure 10.8 Hexagonal close packed (HCP) lattice and its Wigner–Seitz cell

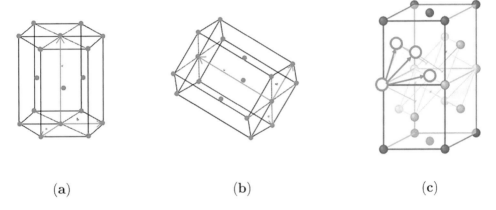

Figure 10.9 Equivalence between HCP and FCC lattices and placement of basis atoms in an hexagonal close packed (HCP) lattice: (a) An original HCP lattice; (b) The rotated HCP lattice, and (c) The corresponding FCC lattice.

the *number of basis atoms* as the number of atoms inside a *simple cubic lattice unit cell*. Thus, to calculate the basis atom number inside an HCP lattice, we have to count how many atoms are present in a simple cubic unit cell of HCP lattice. To do so, we rotate around the axis $\mathbf{a} = (\cos 30^o, \sin 30^o, 0)$ as shown in Fig. 10.9(b). After rotating, we can see that the rotated HCP hexagonal lattice is just an FCC lattice. Or one may say that FCC lattice has a hexagonal structure in a certain direction. Because of such correspondence between HCP lattice and FCC lattice, we can use it to calculate the basis atom for HCP lattices. Therefore, we say HCP lattice has four basis atoms, and their placement is shown in Fig. 10.9, in which two atoms are in the $z = 0$ plane and two atoms in the $z = 0.5$ plane.

LAMMPS has some 2D lattices of styles: *sq*, *sq2*, and *hex*. The style *sq* defines a square unit cell with edge length = 1.0. That is $\mathbf{a}_1 = (1, 0, 0)$ and $\mathbf{a}_2 = (0, 1, 0)$. The square lattice defined by the style *sq* has one basis atom that is placed at the lower-left corner of the square (see Fig. 10.10(a)). The *sq2* lattice is also a square lattice, but it has two basis atoms, one at the corner and one at the center of the square

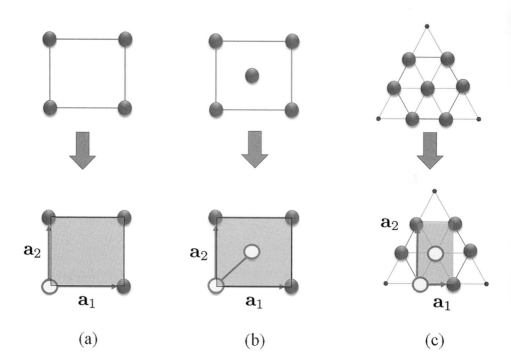

Figure 10.10 Two-dimensional lattice styles in LAMMPS: (a) *sq*, (b) *sq2*, and (c) *hex*

(see Fig. 10.10(b)). The style *hex* is also a 2D lattice, but the unit cell is rectangular, with $\mathbf{a}_1 = (1,0,0)$ and $\mathbf{a}_2 = (0, \sqrt{3}, 0)$. It has two basis atoms: one at the corner and another at the center of the rectangle (see Fig. 10.10(c)).

A lattice of style custom allows you to specify $\mathbf{a}_1, \mathbf{a}_2, \mathbf{a}_3$, and a list of basis atoms to put in the unit cell. By default, $\mathbf{a}_1$, $\mathbf{a}_2$, and $\mathbf{a}_3$ are three orthogonal unit vectors (edges of a unit cube). However, you can specify them to be of any length and nonorthogonal to each other, so that they describe a tilted parallelepiped. By using the basis keyword you add atoms, one at a time, to the unit cell. Its arguments are fractional coordinates ($0.0 <= x, y, z < 1.0$). The position vector x of a basis atom within the unit cell is thus a linear combination of the unit cells' three edge vectors, i.e., $\mathbf{x} = b_x \mathbf{a}_1 + b_y \mathbf{a}_2 + b_z \mathbf{a}_3$ where b_x, b_y, b_z are the three values specified for the basis keyword.

2. region (https://docs.lammps.org/region.html)

LAMMPS command *region* defines a geometric region of space or lattice space. Various other commands can use the defined regions for many different purposes. For example, the region can be filled with atoms via the *create_atoms* command. A simulation box can be defined by specifying the bounding box around a defined region via the *create_box* command. If we want to only simulate the atoms in a particular region, we can identify them as a group via the *group* command. Often times, we can use the surface of the region as a boundary wall via the fix wall/region command to specify boundary conditions of a MD simulation.

The syntax of the command is:

region ID *style args keyword arg*

The ID refers the region ID, because there may be several regions in a MD simulation. For each region that you create, you should first give it a name or ID. The option variable *style* has nine values: $style\ value \in \{delete, block, cone, cylinder, plane, sphere, union, intersect\}$.

Each of these nine values gives you a general idea what is the geometrical shape of the region. However, if you want to precisely define a region, you have to use the *args* option to define its size and dimension.

For instance, the style *block* indicates that the region is a rectangular-shaped body, and you can use the following arguments to specify its dimension:

block args = xlo xhi ylo yhi zlo zhi

where xlo, xhi, ylo, yhi, zlo, zhi bounds of block in all dimensions with distance units from the low coordinates to the high coordinates.

For example, we previously defined a BCC lattice by the command

```
lattice bcc   4.0
```

We can then define a region subsequently by using the following command:

```
region Box1  block  0.0, 10.0, 1.0, 9.0, 0.5, 10.2
```

where *Box1* is the region's ID or name, and the region is defined as a rectangular box in the space specified as,

$0\text{Å} < x < 10 \times 4\text{Å}; \quad 1.0\text{Å} < y < 9 \times 4\text{Å}; \text{ and } 0.5 \times 4\text{Å} < z < 10.2 \times 4\text{Å}.$

In general, the style arguments for *block* is expressed as

```
block args = xlo xhi ylo yhi zlo zhi
xlo,xhi,ylo,yhi,zlo,zhi = bounds of block in all dimensions (distance units)
```

Moreover, the *region* command has a *keyword/arg* variable that has five values: *side, units, move, rotate, open*, and each keyword value has its own arguments:

```
side value = in or out
  in = the region is inside the specified geometry
  out = the region is outside the specified geometry
units value = lattice or box
  lattice = the geometry is defined in lattice units
  box = the geometry is defined in simulation box units
move args = v_x v_y v_z
  v_x,v_y,v_z = equal-style variables for x,y,z displacement of region over time
rotate args = v_theta Px Py Pz Rx Ry Rz
  v_theta = equal-style variable for rotation of region over time (in radians)
  Px,Py,Pz = origin for axis of rotation (distance units)
  Rx,Ry,Rz = axis of rotation vector
open value = integer from 1-6 corresponding to face index (see below)
```

Now, we consider the following example:

```
region boxID block 0.1 10.1 0.1 10.1 0.1 10.1
```

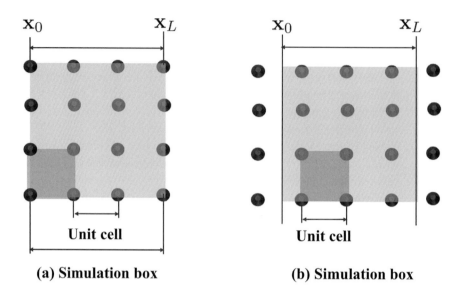

Figure 10.11 (a) A simulation box is perfectly aligned with the underlying lattice and (b) a simulation box has margin with the underlying lattice

If a simulation box that you created is perfectly aligned with the lattice point positions, there could be a problem with the atoms right at the boundary of the unit cell, when the periodic boundary condition is used. This situation can be illustrated by using a simple cubic lattice simulation cell as shown in Fig. 10.11(a). In this case, for the fixed vertical positions, the atom with coordinate X_0 and the atom with the coordinate X_L are the same atom, which can be simulated by using the specified periodic boundary condition. In principle, LAMMPS can figure out whether to put an atom at X_0 or X_L, and make sure that it does not add atoms at both positions. However, in practice, LAMMPS may be able to sort this out for cubic lattices, but it may not be able to figure this out for some complicated lattice types, particularly non-Bravais lattices or custom lattices. If LAMMPS does put two atoms in both positions, it will cause problems with your simulation, e.g., exploding simulations, extremely large forces, and the like. To avoid these problems, it is best to specify your simulation box in such a way so that there is a margin between your simulation box with the underlying lattice point positions as shown in Fig. 10.11(a). This is the reason why the region block command in this example specifies each axis range as 0.1 to 10.1, instead of 0.0 to 10.0.

Before we consider another example, we first the general expression for style *cylinder*, which is defined as follows:

```
cylinder args = dim c1 c2 radius lo hi
  dim = x or y or z = axis of cylinder
  c1,c2 = coords of cylinder axis in other 2 dimensions (distance units)
  radius = cylinder radius (distance units)
  c1,c2, and radius can be a variable (see below)
  lo,hi = bounds of cylinder in dim (distance units)
```

This definition looks almost self-explanatory, but not entirely. We may use the following example to explain how to use it:

```
region Cylinder1   cylinder z 2.0 3.0   5 -5.0 EDGE units box
```

Here, the region ID is Cylinder1, and the region is specified as a cylinder that has its axis along the z-direction, and its center is located at $x = 2.0$ and $y = 3.0$ with a radius of 5.0, while the cylinder's height extends in the z-direction from -5.0 to *its upper box boundary*, which is what the term "*EDGE*" means. In LAMMPS, *region* command's style *cylinder*, also for *block*, *cone*, and *prism* as well, the lo, hi values for cylinder or for block, cone, or prism style as well, can be specified as EDGE or INF, in which EDGE means they extend all the way to the global simulation box boundary, while INF means a large negative or positive number (1.0e20), so it should encompass the simulation box even if it changes size. If a region is defined before the simulation box has been created (via *create_box* or *read_data* commands), then an EDGE or INF parameter cannot be used.

One may see that in this example the distance unit is not the lattice distance unit, but specified as the box unit.

3. create_box (https://docs.lammps.org/create_box.html)

This command creates a simulation box based on the specified region. Thus, a region command must first be used to define a geometric domain. It also partitions the simulation box into a regular 3D grid of rectangular bricks, one per processor, based on the number of processors being used and the settings of the processors command. The partitioning can later be changed by the balance or fix balance commands.

The argument N is the number of atom types that will be used in the simulation. If the region is not of style prism, then LAMMPS encloses the region (block, sphere, etc.) with an axis-aligned orthogonal bounding box, which becomes the simulation domain.

For the command *region* style prime, one may find more information at https://docs.lammps.org/create_box.html.

The syntax of the command is:

create_box *N region-ID keyword value*

Unlike the command *region*, the command *create_box* does not has an ID or identifier. The option variables for this command are as follows:

- N = # of atom types to use in this simulation
- region-ID = ID of region to use as simulation domain
- zero or more keyword/value pairs may be appended

The keyword/value variable has eight values: *bond/types, angle/types, dihedral/types, improper/types, extra/bond/per/atom, extra/angle/per/atom, extra/dihedral/per/atom, extra/improper/per/atom*.

For example, we consider the following example:

```
create_box 1 Box1 bond/types 2 extra/bond/per/atom 1
```

In this example, we have only one type of atoms, and we have two types of bonds, and each atom can only have one bond.

4. create_atoms (htttps://docs.lammps.org/create_atoms.html)

Previously, we have defined and discussed LAMMPS commands *region* and *create_box*. The main function of these commands is to create a MD simulation box, but until now the box is empty. We still do not have any atoms in the simulation box. In LAMMPS, there are several ways to fill a simulation box with atoms or molecules. Among them, one of the simplest way to fill a simulation cell with atoms or particles is to use the command *create_atoms*.

The syntax of the command is:

create_atoms *type style args keyword values*

Unlike the command *region*, *create_atoms* command does not have an ID or identifier. The command basically has three option variables: *type, style/args*, and *keyword/values*.

The first option variable *type* is the atom type that you wish to create. The second option variable *style* has four values: *box, region, single, random*. Each style value has its own arguments, and they are listed as follows:

- box args = none
- region args = region-ID
  The purpose of using region-ID is to ensure that particles will only be created in that region.
- single args = x y z
  where (x, y, z) is the location or coordinates of the single particle under consideration (distance units).
- random args = N seed region-ID
  where N = number of random particles to create:
  seed = random # seed (the seed for pseudo random number generator (PRNG));
  region-ID $\rightarrow$ only creating atoms within in region.
  If one wants to create random particles for the entire simulation box use NULL.

The command *create_atoms* has several keyword/value pairs, and one can choose to use none or more depending on the requirements of the simulation.

These keywords are: {*mol, basis, ratio, subset, remap, var, set, rotate, units*}. For example,

- The keyword *mol* value = *template-ID seed*
  in which *template-ID* is the ID of the molecule template specified in a separate command *molecule*, and *seed* is the random PRNG random number seed (positive integer);
- The keyword *basis* values = M itype, where M indicates the number of basis atom and *type* is the number of atom types (1 to N).

For the rest of the keywords, the readers may consult https://docs.lammps.org/create_atoms.html for details.

Example 10.1

```
create_atoms 1 box
```

This command instructs to create the type 1 atoms and fill them in the entire simulation box of the lattice.

Example 10.2

```
create_atoms 3 region regsphere basis 2 3
```

This command instructs to create the type 3 atoms in the region *reqsphere*, and the second basis atom in lattice is assigned to be the type 3 atoms.

Example 10.3

```
create_atoms 3 single 0 0 5
```

This command instructs to create a single type 3 atom at the position or coordinates $(0, 0, 5)$.

The *create_atoms* command is one of the main LAMMPS commands, and it has several *styles* and several *keyword/values*, and it may take time and experience to master all of them.

10.3.3 Computation Settings

This is the most important group of commands, which can be further divided into four small groups:

(i) Specify force field parameters
- *pair_coeff*
- *bond_coeff*
- *angle coeff*

(ii) Specify simulation parameters.
- *group*
- *mass*
- *timestep*
- *velocity*
- *neighbor*
- *neighbor_modify*
- ...

(iii) Impose dynamic loading condition or dynamic boundary condition, specify time integration
- *fix*
- *compute*
- ...

(iv) Specify output variables and formats
- *thermo*
- *thermo_style*
- ...

I. Specify Force field parameters

We first discuss the commands that set force field

1. Pair_coeff (https://docs.lammps.org/pair_coeff.html)

This command specifies the pairwise force field coefficients for one or more pairs of atom types.

The syntax of the command is:

pair_coeff *I J args*

where I, J are the indices of the atom types and $args$ = coefficients for one or more pairs of atom types.

I and J can be specified in two ways: (1) I and J are numerical values and $I \leq J$, and (2) a wildcard asterisk can be used in place of or in conjunction with the I, J arguments to set the coefficients for multiple pairs of atom types. There are three wildcard asterisk forms "*", "*n", "n*", and "m * n". If N = the number of atom types, then an asterisk with no numeric values means all types from 1 to N. A leading asterisk *n means all types from 1 to n (inclusive). A trailing asterisk n* means all types from n to N (inclusive). A middle asterisk m * n means all types from m to n (inclusive). Note that only type pairs with $I \leq J$ are considered, and if asterisks imply type pairs where $J < I$, they are ignored.

For examples:

```
pair_coeff 1 2 1.0 1.0 2.5
pair_coeff 2 \* 1.0 1.0
pair_coeff 3\* 1\*2 1.0 1.0 2.5
pair_coeff \* \* 1.0 1.0
```

In the first example, the pair is between 1 and 2; in the second example, the pairs are between 2 to 2-N; in the third example, the pairs are between 3-N to 1-2, and in the last example the pairs are between 1-N to 1-N.

The number and meaning of the coefficients depends on the pair style. Pair coefficients can also be set in the data file read by the *read_data* command or in a restart file.

Different from *pair_coeff* command, LAMMPS has another command called bond_coeff.

2. bond_coeff (https://docs.lammps.org/bond_coeff.html) specifies the bond force field coefficients for one or more bond types.

The syntax of the command is:

bond_coeff *N args*

where *N* is the bond type or types and *args*= coefficients for one or more bond types.

Examples are

```
bond_coeff 5 80.0 1.2
bond_coeff * 30.0 1.5 1.0 1.0
bond_coeff 1*4 30.0 1.5 1.0 1.0
```

where the first line specifies the bond 5 coefficients, the second line specifies the coefficients of the bond from 1 to *N*, and the third line specifies the coefficients of the bond from No. 1 to No. 4. In the command *pair_coeff*, it specifies the coefficients of pair interactions between different types of atoms, whereas the command *bond_coeff* specifies the coefficients for the bond type *N*. Here, we do mention any atoms involved.

II. Specify Simulation Parameters

In the LAMMPS input script file, it provides execution commands to perform specific simulations for specific groups of atoms. Thus, one of the very useful command is:

II-1. group (https://docs.lammps.org/group.html)

It defines a collection of atoms as an atom group, and gives a group ID, so that we can use the group ID to refer this collection of atoms and assign them under various conditions or performing various simulations by using other commands such as *fix, compute, dump,* or *velocity*.

The syntax of the command is:

group *ID style args*

where *ID* is the group ID, and the option argument *style* has the following values: *delete clear empty region type id molecule variable include subtract union intersect dynamic static*. Each of the style has its own arguments, except the styles: *delete, clear, empty*.

For example,

```
group edge region regstrip
group water type 3 4
group sub id 10 25 50
group sub id 10 25 50 500:1000
```

The first *group* command defines all the atoms in the region with ID *regstrip* as the group *edge*. The second *group* command defines all the atoms of the atom type 3 and 4 as the group *water*. The third group command identifies three atoms with atom id: 10, 25, and 50 as the group *sub*, whereas the fourth *group* command identifies three atoms with atom id: 10, 25, 50 plus those atoms with atom id from 500 to 1000 together as the group *sub*. For more information, readers are referred to https://docs.lammps.org/group.html.

In LAMMPS, to set the mass for a given type of atoms, we use the command:

II-2. mass (https://docs.lammps.org/mass.html)

The syntax of the command is:

mass *I value*

where *I* is the atom type and *value* is the magnitude of mass based on its unit.
For examples,

```
mass 1 1.0
mass * 62.5
mass 2* 62.5
```

The first *mass* command assigns the type 1 atoms with mass=1.0; the second *mass* command assigns the atom types from 1 to N with mass = 62.5, and the third *mass* command assigns the atom types from 2 to N with mass = 62.5.

In LAMMPS, sometimes we need to assign a group atoms with initial velocities so that the group has a starting temperature, or to prescribe moving boundary conditions. To do so, we need another command:

II-3. velocity (https://docs.lammps.org/velocity.html)

The velocity command is used to set or change the velocities of a group in different prescribed ways (styles). In each style, there are required arguments and optional keyword/value parameters.
The syntax of the command is:

velocity *group-ID style args keyword value*

in which
  group-ID is the group of atoms that you would like to impose velocity.
  The value of *style* is one of these: *create, set, scale, ramp, zero*, which have their corresponding arguments:

- create args = temp seed temp = temperature value (temperature units) seed = random # seed (positive integer)
- set args = vx vy vz vx,vy,vz = velocity value or NULL (velocity units) any of vx,vy,vz can be a variable (see subsequet discussion)
- scale arg = temp temp = temperature value (temperature units)
- ramp args = vdim vlo vhi dim clo chi vdim = vx or vy or vz vlo,vhi = lower and upper velocity value (velocity units) dim = x or y or z clo,chi = lower and upper coordinate bound (distance units)
- zero arg = linear or angular linear = zero the linear momentum angular = zero the angular momentum

The *create* style generates an ensemble of velocities using a random number generator with the specified seed at the specified temperature. For example,

```
velocity all create 300.0 4928459
```

This means that we generate ensemble velocities by assigning random velocities to every atoms in the group all, such that the system temperature will be maintained at 300 K°. Since the velocity for each atom is randomly assigned by using the pseudo-random number generator (PRNG), the computer needs a seed number to start with. For the results repeatable, we usually assign a seed number, and here the seed number is 0.4938459.

The set style sets the velocities of all atoms in the group to the specified values. If any component is specified as NULL, then it is not set. Any of the v_x, v_y, v_z velocity components can be specified as an equal-style or atom-style variable. If the value is a variable, it should be specified as v_{name}, where name is the variable name. In this case, the variable will be evaluated, and its value used to determine the velocity component. Note that if a variable is used, the velocity it calculates must be in box units, not lattice units. For example, the command line

```
velocity border set NULL 4.0 v_vz sum yes units box
```

instructs the set the velocities for all atoms in the group *border* as follows:

$$v_x = 0, \ v_y = 4.0, \ v_z = v_{vz},$$

where v_z is a variable or variable name. The *sum* keyword is used here, which means that the new velocities of all atoms in the group *border* will be added to the previous velocities, because *sum = yes*. If *sum = no*, the previous velocity will be replaced by the current velocity.

The last keyword/value is: *unit box*, this is because $v_z = v_{vz}$ is a variable. The *units* keyword is used by style options *set* and *ramp*. If *units = box*, the velocities and coordinates specified in the velocity command are in the standard units described by the units command, e.g., real unit. If *units = lattice*, velocities are in units of lattice spacings per time, e.g., spacings/fmsec, and coordinates are in lattice spacings.

The *scale* style is a instruction to compute the current temperature of the group of atoms by rescaling the velocities to the specified temperature. For example, the command line

```
velocity flow scale 300.0
```

is to achieve the system temperature by scaling velocity or by using velocity scaling algorithm (see Chapter 8). For other style options and keyword values, readers can consult LAMMPS manual page https://docs.lammps.org/velocity.html.

II-4. neighbor (https://docs.lammps.org/neighbor.html)

The command *neighbor* sets parameters for building of pairwise neighbor lists. As mentioned in all atom pairs within a neighbor cutoff distance equal to the their force cutoff plus the skin distance are stored in the list (see Chapter 6).

The syntax of the command is:

neighbor *skin style*

in which *skin* is the extra distance beyond force cutoff (see Chapter 6). The *style* has three options: *bin, nsq, multi*.

The *bin* style creates a neighbor list by "binning" which is an operation that scales linearly with the number of atoms per processor, i.e., N/P, where N = total number of atoms and P = number of processors. The *nsq* style uses search algorithm that scales quadratically with N/P. It is almost always faster than the *nsq* style, which scales as $(N/P)^2$. However, for unsolvated small molecules in a nonperiodic box, the *nsq* option can sometimes be faster. The *multi* is a style that uses a modified binning algorithm, which is useful for systems with a wide range of cutoff distances, e.g., due to different size particles.

Here are two examples:

```
neighbor 0.3 bin
neighbor 2.0 nsq
```

In the first example, the skin is 0.3 distance unit thick, and the binning algorithm is used. In the second example, the skin is 2 distance unit thick, and the neighbor list is constructed with a full search without binning.

Typically, the larger the skin distance, the less often neighbor lists need to be built, but more pairs must be checked for possible force interactions every time step.

The command *neighbor* is usually used together with the command
II-5. neighbor_modify (https://docs.lammps.org/neigh_modify.html). This command sets parameters that updates the pairwise neighbor lists. Depending on what pair interactions and other commands are defined, a simulation may require one or more neighbor lists.

The syntax of this command is:

neighbor_modify *keyword values*

The command *neighbor_modify* has several keywords, i.e.,
{*delay, every, check, once, cluster, include, exclude, page, one, binsize*}.

The *every, delay, check*, and *once* options affect how often a neighbor list is being built as a simulation runs. The *delay* option means never build new lists until at least N steps after the previous build. The *every* option means build lists every M steps (after the delay has passed). If the *check* = *no*, the lists are built on the first step that satisfies the *delay* and *every* settings. If the *check* =*yes*, then the *every* and *delay* settings determine when a build may possibly be performed, but an actual build only occurs if some atom has moved more than half the skin distance (specified in the neighbor command) since the last build. The *exclude type* option turns off pairwise interactions between certain pairs of atoms, by not including them in the neighbor list. The *exclude group* option turns off the interaction if one atom is in the first group and the other in the second. Now we consider the following examples:

```
neigh_modify every 2 delay 10 check yes
neigh_modify exclude type 2 3
neigh_modify exclude group frozen frozen check no
```

The first command line instructs that every 2 time steps the neighbor list is updated, and then delay updating the neighbor list for 10 time steps. After that, check to see if some atom has moved half the skin distance or more, and if this is true, start building neighbor list again. In the second example, we turn off the pairwise interaction between atom types 2 and 3, while in the third example, we turn off the pairwise interaction in the group frozen, and we set *check=no*, which means that the neighbor lists are built on the first step according to keywords *delay* and *every* settings for updating.

II-6. variable (https://docs.lammps.org/variable.html)

The command *variable* assigns a name to a value or values to be used elsewhere in the LAMMPS input script for parameterization or used in output.

The syntax of the command is:

variable *name style args*

where the argument of the *name* is the name of the variable, and *style* is the *variable* style. The *variable* command has 15 styles or style name:
$style \in \{$ *delete, index, loop, world, universe, uloop, string, format, getenv, file, atomfile, python, internal,*
equal, vector, atom$\}$.

Most of the styles have their own arguments and the specific format to assign these arguments, which can be found in https://docs.lammps.org/variable.html.

The followings are a few examples:

```
\item variable x index run1 run2 run3 run4
\item variable j loop $n
\item variable T equal c_myTemp
```

In the first example, the variable name is x and the style name is *index*; it has been assigned with four strings: *run1 run2 run3 run4*.

In the second example, the variable name is j, and the style name is *loop*. The argument of the style *loop* is assigned with integer value of another variable n, i.e., n. In the third example, the name of the variable is T, and the style name is *equal*. Usually the argument of the style *equal* is an expression that contains numbers, thermo keywords defined in the command *thermo_style*, math operations, group functions, atom values, and vectors. In this case, it is another variable *c_myTemp* that has been assigned with a specific value.

(III) Imposing Dynamic Loading Condition and Specifying Time Integration

Now, we discuss one group of the most important commands in LAMMPS that instruct how to impose dynamic boundary conditions and specifying time integration. We start with the command:

III-1. fix (https://docs.lammps.org/fix.html). In LAMMPS, the command *fix* is used to call any operation to apply on a system during time stepping or minimization processes. Its most common usages are: updating of atom positions and velocities due to time integration, controlling temperature, applying constraint forces to atoms or particles, enforcing boundary conditions, computing diagnostics, among others. There are hundreds of fixes defined in LAMMPS, and it is probably the most important command in all LAMMPS commands.

The syntax of the command is:

fix ID group-ID style args

where

- ID = user-assigned name for the fix,
- group-ID = ID of the group of atoms to apply the fix to
- style = one of a long list of possible style names (see subsequent discussion)
- args = arguments used by a particular style

Fixes perform their operations at different stages of the computation or time. If two or more fixes operate at the same time, they will have to be specified in an order: which operation is first and which operation is next in the LAMMPS input script.

The ID of a *fix* can only contain alphanumeric characters and underscores. A *fix* command can be deleted by using the *unfix* command.

For example, we consider the following command line in a LAMMPS script file:

```
fix my_nvt all nvt temp 300.0 300.0 0.01
```

in which, *my_nvt* is the ID or the name of this *fix* command. The group-ID of this *fix* command is *all*, and the style of this *fix* command is *nvt*, which means that it is a NVT time integration via Nosé–Hoover thermostat.

For the *fix ID group-ID nvt* command, it has many keyword values, which may be found on LAMMPS manual page: https://docs.lammps.org/fix_nh.html.

In particular for the above example, the keyword for *fix ID group-ID nvt* command is: *temp*, and the syntax to specify the keyword value is:

temp values = Tstart Tstop Tdamp

where
Tstart,Tstop = external temperature at start/end of run
Tdamp = temperature damping parameter (time units)
Thus in this example, $Tstart = 300$ K, $Tstop = 300$ K, and the Nosé–Hoover thermostat damping parameter is 0.1.

III-2. Compute (https://docs.lammps.org/compute.html)
The LAMMPS command *Compute* is a computation diagnostic command in LAMMPS, which is used to call for calculation or extrapolation of thermodynamics variables/information or other system variables for a group of atoms, while the time integration or minimization process is progressing. For example, one may use *compute* to calculate system temperature, pressure, or the center of mass.

The syntax of the command is:

computer *ID group-ID style agrs*

where

- ID = user-assigned name for the computation
- group-ID = ID of the group of atoms to perform the computation on
- style = one of a list of possible style names (see on (see details in https://docs.lammps.org/compute.html)
- args = arguments used by a particular style

For example, we consider the following *compute* command:

```
compute thermo_press all pressure temp
```

in which, *thermo_press* is the ID of the *compute* command, *all* is the *group-ID*, and both *pressure* and *temp* are the styles that *pressure* means to calculate the total pressure and pressure tensor, while *temp* means to calculate the temperature of group of atoms.

(IV) Specify output variables and formats

This group of LAMMPS commands specifies the computation outputs.

IV-1. thermo (https://docs.lammps.org/thermo.html)

The LAMMPS command *thermo* specifies the detailed instruction on how to output thermodynamic information, e.g., temperature, energy, and pressure during time integration.

The syntax of the command is:

thermo *N*

Here, N is the time steps of an interval that the computer will consecutively output the prescribed thermodynamic information of the system every N time steps. If $N=0$, the computer will only print the thermodynamics information of the system at the beginning and at the end of the computation. For example,

```
thermo 100
```

This command instructs the thermodynamics output every 100 time steps.

Note that N can be also a variable.

```
variable  s equal logfreq(10,3,10)
thermo    v_s
```

This set of commands first define a variable $s = logfreq(10, 3, 10)$ and then N is the value of variable s. In LAMMPS, the logfreq(x,y,z) function uses the current time step to generate a new time step sequence, such that $x, y, z > 0$ and $y < z$ are required. The generated time steps are on a base-z logarithmic scale, starting with x, and the y value is how many of the $z - 1$ possible time steps within one logarithmic interval are generated. That is, the time steps follow the sequence $x, 2x, 3x, y*x, x*z, 2x*z, 3x*z, y*x*z, x*z^2, 2x*z^2, \ldots$. For any current time step, the next time step in the sequence is returned.

The command *thermo* is usually used together with the command

III-2. thermo_style (https://docs.lammps.org/thermo_style.html)

The LAMMPS command *thermo_style* has several usages.

In general, the command *thermo_style* sets the style and content for printing thermodynamic data to the screen and log file. The syntax of the command is:

thermo_style *style args*

in which it has three styles: {*one, multi, custom*}. Both *one* and *multi* have no arguments, or *one args = none* and *multi args = none*, whereas style *custom* has many keyword arguments, and readers can consult https://docs.lammps.org/thermo.html.

Example 10.4

```
thermo syle one
```

This command prints a one-line summary of thermodynamic information on the computer screen, which is the equivalent of

```
thermo_style custom step temp epair emol etotal press
```

However, this line contains only numeric values of: *step* (time step), *temp* (temperature), *epair* (pairwise energy (van der Waals pairwise + Coulombic pairwise + long-range kspace energies)), *emol* (molecular energy (bond energy + angle energy + dihedral energy + improper energy)), *etotal* (total energy (kinetic + potential energies)), and *press* (pressure)).

Example 10.5

```
thermo style multi
```

The style *multi* prints a multiple-line listing of thermodynamic information that is the equivalent of

```
thermo_style custom etotal ke temp pe ebond eangle edihed
eimp evdwl ecoul elong press
```

in which *ke* is kinetic energy and *pe* is the potential energy. The computer screen only shows numeric values and a string ID for each quantity.

Example 10.6

```
thermo_style custom step temp pe etotal press vol
```

This command prints *step* (time step), *temp* (temperature), *pe* (potential energy), *etotal* (total energy), *press* (pressure), and *vol* (volume).

III-3. dump (https://docs.lammps.org/dump.html)

The command *dump* outputs the current state or configuration of the atomistic system under simulation, and it takes a snapshot of atom quantities of one or more files every N time steps in one of several styles that the user prescribed.

The syntax of the command is:

dump *ID group-ID style N file args*

in which

- ID = user-assigned name for the dump
- group-ID = ID of the group of atoms to be dumped
- style ∈ {*atom, atom/gz, atom/mpiio,cfg,cfg/gz,cfg/mpiio,custom, custom/gz, custom/mpiio, dcd,h5md,image,local,local/gz,molfile, movie,netcdf, netcdf/mpiio,vtk,xtc,xyz,xyz/gz,xyz/mpiio*}
- N = dump everything this many time steps
- file = name of file to write or dump the system configuration/information to
- args = list of arguments for a particular style

The followings are two examples of *dump* command:

```
dump myDump all atom 100 dump.atom
dump hisDump all atom/mpiio 100 dump.atom.mpiio
```

In the first command whose name is *myDump*, we dump all the atoms in the group *all* every 100 time steps into an output file named *dump.atom*. For style *atom*, the coordinates of all atoms in the specified group are written to the output file, along with the atom ID and atom type.

In the second command whose name is also *hisDump*, we dump the atom configuration by using *atom/mpiio* style, which means that we write the information of a single dump file in parallel fashion via the standard MPI-IO library.

IV. Run or Minimize

The commands in this category are execution commands that are mainly used to instruct time integration, relaxation, or minimization.

The first command we discussed is:

IV-1. run (https://docs.lammps.org/run.html)

This command instructs the computer to run the molecular dynamics simulation for a specific number of time steps.

The syntax of the command is:

run *N keyword values.*

In this instruction format, N is the number of time steps that the MD shall carry out. If $N = 0$, only the thermodynamics of the system are computed and printed without taking a single time step. The additional integration features are specified in the *keyword* with given *values*. There are six keywords: *upto, start, stop, pre, post, every*. Their assigned value formats can be found in https://docs.lammps.org/run.html.

Here we look at a few examples:

```
run 10000
run 100000 upto
run 100 start 0 stop 1000
run 10000 every 1000 NULL
```

The first command line instructs a LAMMPS program to carry out time integration for 10,000 time steps. The second command line instructs a LAMMPS program to carry out time integration until 10,0000 time steps starting from the current time step, which is what the keyword *upto* means. For example, the current time step is $T_c = 10,000$, and we shall go on for another $100,000 - 10,000 = 90,000$ time steps before stop. The third command line instructs to carry out a time integration of 100 time steps between 0th time step and 1000th time step. The *start/stop* keywords are useful in the situation system when the system state variable, say temperature, keeps updating. For example, consider a *fix* command followed by five run commands,

```
fix        1 all nvt 200.0 300.0 1.0
run        1000 start 0 stop 5000
run        1000 start 0 stop 5000
run        1000 start 0 stop 5000
run        1000 start 0 stop 5000
run        1000 start 0 stop 5000
```

The NVT *fix* command ramps the target temperature from 200.0K° to 300.0K° during a run. If the run commands did not have the start/stop keywords (just "run 1000"), then the temperature would ramp from 200.0 to 300.0 during the 1000 steps of each run. With the start/stop keywords, the ramping takes place over the 10,000 steps of all five runs together. One may say that this may be achieved by the command lines,

```
fix        1 all nvt 200.0 300.0 1.0
run        5000
```

However, in the above *run* command of 5000 time steps, the temperature will not be updated in between.

The *every* keyword provides a means of breaking a LAMMPS run into a series of shorter runs. Thus, the fourth command line breaks a 10,000 time step run into a series ten 1000 time step runs. This is useful, if you would like to execute other commands in between every 1000 time steps. However, if the command "*NULL*" is used, there is nothing in between every 1000 time step runs.

Considering a nontrivial example: We can use the keyword *every* to succinctly write the commands

```
variable q equal x[100]
run 6000 every 2000 "print 'Coord = $q'"
```

which are the equivalent of the following command lines:

```
variable q equal x[100]
run 2000
print "Coord = $q"
run 2000
print "Coord = $q"
run 2000
print "Coord = $q"
```

which does three runs of 2000 steps and prints the *x*-coordinate of the 100th atom between runs. Note that the variable "$q" will be evaluated afresh each time the print command is executed.

In MD, we often need to perform an energy minimization of the system, so that we can relax the system into an equilibrium state. Such minimization or relaxation is accomplished by using various iterative methods to adjust atom coordinates or the system's configuration. In computations, these iteration methods will have convergence criteria so that the iterations are terminated when one of the stopping criteria is satisfied. In LAMMPS, this is achieved by the command:

minimize (https://docs.lammps.org/minimize.html)

The syntax of the command is:

IV-2. minimize *etol ftol maxiter maxeval*

where

- etol = stopping tolerance for energy;
- ftol = stopping tolerance for force;
- maxiter = max iterations of minimizer;
- maxeval = max number of force/energy evaluations.

The objective function being minimized is the total potential energy of the system as a function of the N atom coordinates:

$$E(\mathbf{r}_1, \mathbf{r}_2, \ldots, \mathbf{r}_N) = \sum_{i,j} E_{pair}(r_{ij}) + \sum_{i,j} E_{bond}(r_{ij}) + \sum_{i,j,k} E_{angle}(\mathbf{r}_i, \mathbf{r}_j, \mathbf{r}_k)$$
$$+ \sum_{i,j,k,\ell} E_{dihedral}(\mathbf{r}_i, \mathbf{r}_j, \mathbf{r}_k, \mathbf{r}_\ell) + \sum_{i,j,k,\ell} E_{improper}(\mathbf{r}_i, \mathbf{r}_j, \mathbf{r}_k, \mathbf{r}_\ell)$$
$$+ \sum_i E_{fix}(\mathbf{r}_i), \tag{10.1}$$

where $r_{ij} = |\mathbf{r}_j - \mathbf{r}_i|$. In Eq. (10.1), the first term is the sum of all nonbonded pairwise interactions including long-range van der Waals interaction and the Coulombic interactions, the second through fifth terms are bond, angle, dihedral, and improper interactions, respectively. The last term in Eq. (10.1), $E_{fix}(\mathbf{r}_i)$, is the energy contribution due to the prescribed constraints or applied forces to atoms.

In computations, the approximated critical point of the objective function is an approximated equilibrium configuration, which may or may not be a local minimizer or a global minimizer.

The starting point for the minimization is the current configuration of the atoms. The minimization procedure stops if any of several criteria are met,

- the change in energy between outer iterations is less than *etol*;
- the 2-norm (length) of the global force vector is less than the *ftol*;
- the line search fails because the step distance backtracks to 0.0;
- the number of outer iterations exceeds *maxiter*; and
- the number of total force evaluations exceeds *maxeval*.

We now consider the following examples:

```
minimize 1.0e-4 1.0e-6 100 1000
minimize 0.0 1.0e-8 1000 100000
```

The first command line instructs a minimization of energy critical tolerance value of 1.0e-4 and the force tolerance value of 1.0e-6. The maximum outer iteration number *maxiter* is set to be 100, and the maximum total force evaluation number *maxeval* is set to be 1000. The first command line instructs a minimization of $etol = 0.0$, $ftol = 1.0e\text{-}8$, $maxiter = 1000$, and $maxeval = 100,000$.

10.4 Case Study (I): Simulation of Three-Dimensional Nano-indentation

As an introduction, in the following, we explain a short LAMMPS script file to the readers, which is an example of three-dimensional (3D) nano-indentation simulation. By doing so, readers can have a starting point in learning how to use LAMMPS to conduct MD simulations. The input script file name is *in.indent3d*, which is displayed in its entirety in the following:

Example 10.7 A LAMMPS Command Script File for 3D Nano-indentation

```
# 3d indenter simulation of copper
clear
units metal
echo both
atom_style atomic
dimension 3
boundary p s p

region box block 0 35 0 26 0 35 units box
create_box 1 box
lattice fcc 3.61
region cu block 0 35 0 26 0 35 units box
create_atoms 1 region cu units box
timestep 0.002
```

```
pair_style eam/alloy
pair_coeff * * Cu_mishin1.eam.alloy Cu

# Energy Minimization
minimize 1.0e-6 1.0e-6 10000 10000

# rigid boundary
region 1 block 0 35 0 3.0 0 35 units box
group anvil1 region 1
region 2 block 0 3 0 26 0 35 units box
group anvil2 region 2
region 3 block 32 35 0 26 0 35 units box
group anvil3 region 3
group anvil union anvil1 anvil2 anvil3

group mobile subtract all anvil
log log_indent3d.dat

# initial velocities
compute new mobile temp
compute CSP all centro/atom fcc
velocity mobile create 300 482748 temp new
fix 1 mobile nvt temp 300.0 300.0 0.05
fix 2 anvil setforce 0.0 0.0 0.0

# assigning velocity to the indenter
# in y direction/loading direction
variable y equal "37.0-step*dt*0.03"
print "y is $y"

# indenter position and radius at onset of loading
fix 4 mobile indent 1000.0 sphere 17.5 v_y 17.5 10.0 units box

thermo 100
thermo_style custom step temp v_y f_4[1] f_4[2] f_4[3]
thermo_modify norm no
dump 1 all custom 1000 indent3d.lammpstrj id type x y z c_CSP

run 80000
```

To run the script file in Windows in your laptop in serial computation without using MPI parallel computation, you can simply type:

```
lmp_serial -in in.indent3d
```

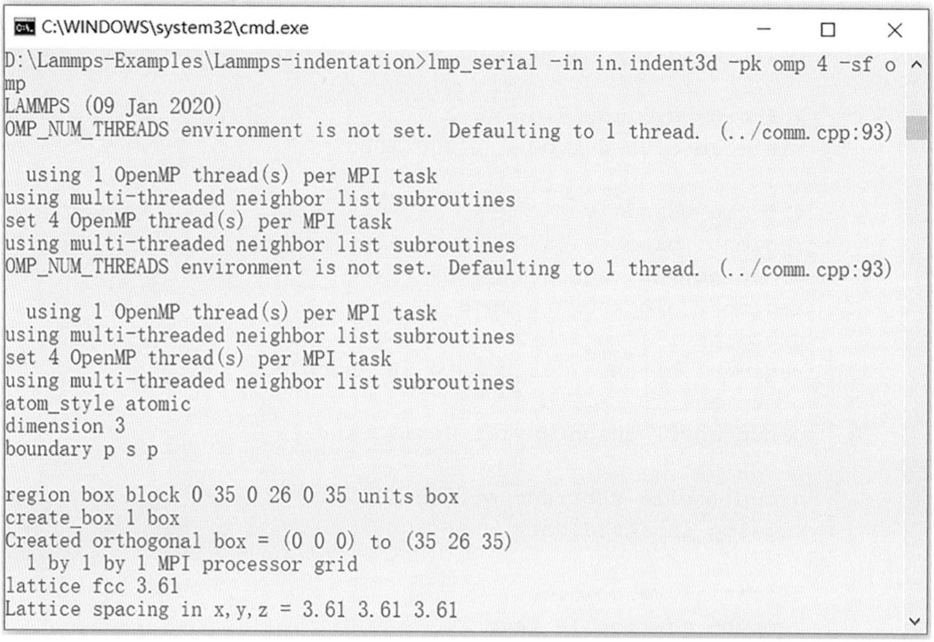

```
C:\WINDOWS\system32\cmd.exe                                    —    □    ×
D:\Lammps-Examples\Lammps-indentation>lmp_serial -in in.indent3d -pk omp 4 -sf o
mp
LAMMPS (09 Jan 2020)
OMP_NUM_THREADS environment is not set. Defaulting to 1 thread. (../comm.cpp:93)

  using 1 OpenMP thread(s) per MPI task
using multi-threaded neighbor list subroutines
set 4 OpenMP thread(s) per MPI task
using multi-threaded neighbor list subroutines
OMP_NUM_THREADS environment is not set. Defaulting to 1 thread. (../comm.cpp:93)

  using 1 OpenMP thread(s) per MPI task
using multi-threaded neighbor list subroutines
set 4 OpenMP thread(s) per MPI task
using multi-threaded neighbor list subroutines
atom_style atomic
dimension 3
boundary p s p

region box block 0 35 0 26 0 35 units box
create_box 1 box
Created orthogonal box = (0 0 0) to (35 26 35)
  1 by 1 by 1 MPI processor grid
lattice fcc 3.61
Lattice spacing in x,y,z = 3.61 3.61 3.61
```

Figure 10.12 Screenshot of how to run the LAMMPS input script file *in.indent3d* (the beginning).

which is using an OpenMP default setting of 1 thread. If you want to run the serial executable with the support of multi-threading parallelization from the styles in the USER-OMP packages. For instance, with four threads (most common personal laptop computers may have up to eight cores and sixteen threads), you can type this command at prompt,

```
lmp_serial -in in.indent3d -pk omp 4 -sf omp
```

as shown in Fig. 10.12 (in the beginning) and Fig. 10.13 (in the end). The computation results or the deformed configuration of the copper substrate is shown in Fig. 10.14.

Most of the commands in this script file are discussed in detail in this section, except a few cases. The *log* command in LAMMPS closes the current LAMMPS log file, opens a new file with the specified name, and begins logging information to it. The syntax of the command is **log** *file* where *file* is the new log file name. In the script input file *in.indent3d*, the new log file name is: *log_indent3d.dat*.
The command line:

```
compute CSP all centro.atom fcc
```

instructs a computation procedure denoted as *CSP* that calculates all groups' centrosymmetric parameter (CSP) for each atom for the entire FCC lattice (Cu is an FCC lattice).

```
C:\WINDOWS\system32\cmd.exe                                    —    □    ×
MPI task timing breakdown:
Section   |  min time   |  avg time   |  max time   |%varavg| %total
----------------------------------------------------------------------
Pair        554.04         554.04        554.04        0.0     97.33
Neigh       0.0338         0.0338        0.0338        0.0      0.01
Comm        2.172          2.172         2.172         0.0      0.38
Output      2.2162         2.2162        2.2162        0.0      0.39
Modify      9.0144         9.0144        9.0144        0.0      1.58
Other                      1.786                                0.31

Nlocal:    3000 ave 3000 max 3000 min
Histogram: 1 0 0 0 0 0 0 0 0 0
Nghost:    3036 ave 3036 max 3036 min
Histogram: 1 0 0 0 0 0 0 0 0 0
Neighs:    208128 ave 208128 max 208128 min
Histogram: 1 0 0 0 0 0 0 0 0 0
FullNghs:  416220 ave 416220 max 416220 min
Histogram: 1 0 0 0 0 0 0 0 0 0

Total # of neighbors = 416220
Ave neighs/atom = 138.74
Neighbor list builds = 5
Dangerous builds = 0

Total wall time: 0:19:15
```

Figure 10.13 Screenshot of how to run the LAMMPS input script file *in.indent3d* (the end).

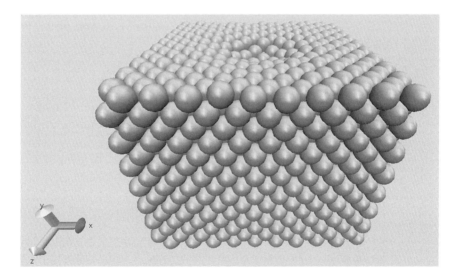

Figure 10.14 Deformed configuration of copper substrate under the rigid sphere indentor.

In solid-state physics, the CSP for each atom or particle was defined by Kelchner et al. (1998). It is a useful measure of the local lattice disorder around an atom and can be used to characterize whether an atom is part of a perfect lattice, a local defect, e.g., a dislocation or stacking fault, or located at a surface.

The CSP value of an atom having N nearest neighbors ($N = 12$ for FCC lattice, $N = 8$ for BCC lattices) is given by

$$p_{CSP} = \sum_{i=1}^{N/2} |\mathbf{R}_i + \mathbf{R}_{i+N/2}|^2, \qquad (10.2)$$

where $\mathbf{R}_i$ and $\mathbf{R}_{i+N/2}$ are vectors pointing from the central atom to a pair of opposite neighbors. For lattice sites in an ideal centrosymmetric crystal, the contributions of all neighbor pairs in this formula will cancel each other, and the resulting p_{CSP} value will hence be zero. Whereas the atomic sites within a defective crystal region will have a disturbed, non-centrosymmetric neighborhood, which results in the related CSP becoming non-zero or positive. Using an appropriate threshold, to filter out the small perturbations due to thermal fluctuations, the CSP value can be used as an order parameter to identify atoms that are part of crystal defects.

The most important command line of this script file is:

```
fix 4 mobile indent 1000.0 sphere 17.5 v_y 17.5 10.0 units box
```

The syntax of command *fix indent* is **fix** *ID group-ID* indent *K keyword values*. The ID for this particular *fix indent* command is 4, and it applies to the group *mobile*, i.e., all the atoms in region *cu*.

The indenter is spherical in shape, its center position or coordinate is $(x, y, z) = (17.5, v_y, 17.5)$, and its radius is 10 with the units of box. Note that v_y is the value of variable y that is defined as

$$y(t) = 37.0 - n\Delta t * 0.03,$$

where n is the time step number, Δt is the time step size, 37 is the initial height of the indenter, and 0.3 is a small parameter. As time increases, the indenter moves downward.

The command *fix indent* calculates the force acting on the group *mobile*, i.e., copper atom substrate. The force acting on each Cu atom is calculated based on the formula

$$F(r) = -K(r - R)^2,$$

where K is the specified force constant, and in this case $K = 1000.0$; r is the distance from the atom to the center of the indenter; and R is the radius of the indenter. The force is repulsive and $F(r) = 0, \forall r > R$.

The total force acting on the indentor is expressed in LAMMPS as $f_4[1], f_4[2], f_4[3]$. If we want to output the value of this force, we can use the command *thermo_style* $f_{ID}args$, which is the part of the *thermo_style* command, i.e.,

```
thermo_style custom step temp v_y f_4[1] f_4[2] f_4[3]
```

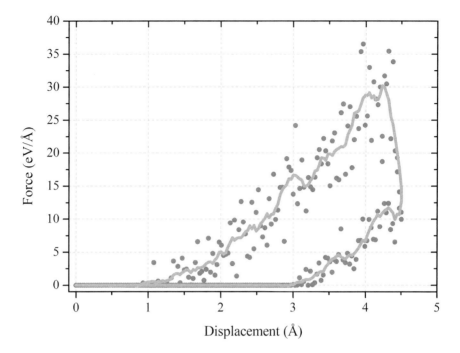

Figure 10.15 The displacement–force curve for a C-S-H model.

The load-deflection curve of the nano-indentation experiment can be found by plotting the relation between the indentation depth v_y and the vertical force that the indentor experienced, i.e., f_4 [2], which is shown in Fig. 10.15.

10.5 Case Study (II): MD Simulation of Mechanical Properties of Cement

In Chapter 4, we discussed how to use DFT simulating the mechanical properties of C-S-H crystal at nanoscale. However, the DFT C-S-H unit is very small with only a few nanometers in dimension, due to the limitation of computation resource and computing time.

In this case study, we present a study of using the ReaxFF reactive force-field MD simulation to model and simulate amorphous C-S-H structure at around 10 nanometer scale by using LAMMPS. The interested readers can download the LAMMPS script file from http://nanomechanics.berkeley.edu/introduction-to-computational-nanomechanics/.

The reactive force field (ReaxFF) interatomic potential is a powerful computational tool for developing and optimizing material properties. First-principles methods based on quantum mechanics, while offering valuable theoretical guidance at the electronic level, are often too computationally expensive for simulations of a large-scale C-S-H system. Alternatively, empirical interatomic potentials need significantly fewer computational resources, which enables simulations to better describe dynamic processes over longer time duration and on larger spatial scale. Such methods, however, typically require a predefined connectivity between atoms, excluding simulations of any chemical reactive events. The ReaxFF method was developed to help bridge the gap of quantum calculations and classical MD simulations. ReaxFF-based MD casts the empirical interatomic potential within a bond-order formalism, thus implicitly describing chemical bonding without expensive QM calculations, while enabling large-scale and relatively longer simulations.

The following LAMMPS script is an example of using ReaxFF force field to simulate mechanical property, i.e., stress–strain relation, of C-S-H material.

LAMMPS uses the following three commands:

pair_style reax/c
pair_style reax/kk
pair_style reax/omp

to conduct the ReaxFF force field-based MD simulations.
The syntax of the command is:
pair_style reax/c cfile keyword value
where

1. cfile = NULL or name of a control file
2. zero or more keyword/value pairs may be appended

```
#C-S-H/water input file

units real
#units metal
boundary p p p
atom_style full

read_data CSH.lammps05

pair_style    reax/c NULL
pair_coeff    * * ffield.reax.CaSiOH Ca Si O O O H

neighbor 2 bin
neigh_modify every 1 delay 0 check yes

velocity all create 300 5812775

fix 1 all npt temp 300 300 100 iso 0 0 100
fix 2 all qeq/reax 1 0.0 10.0 1e-6 reax/c
```

```
thermo 100

thermo_style custom step temp etotal pxx pyy pzz press
vol lx ly lz cella cellb cellc cellalpha cellbeta cellgamma

thermo_modify flush yes

run_style    verlet

timestep 0.25
run 5000

reset_timestep 0

unfix 1

fix 3 all nvt temp 300 300 100
#fix 4 all qeq/reax 1 0.0 10.0 1e-6 reax/c

dump 1 all xyz 1000 1.xyz

run 5000
write_data 11.data

reset_timestep 0

unfix 3

fix 5 all npt temp 300 300 100 x 0 0 100 z 0 0 100
fix 6 all deform 1 y erate 0.00004 remap v

compute deftemp all temp/deform
compute 4 all pressure deftemp

dump 2 all xyz 1000 2.xyz

run 50000

write_data 2.data
```

The interested readers can download the LAMMPS script and input file from http://nanomechanics.berkeley.edu/introduction-to-computationalnanomechanics/.

The simulations results can be visualized by using OVITO (see Fig. 10.16). In Fig. 10.17, we plot the simulated stress–strain relation for the C-S-H unit cell.

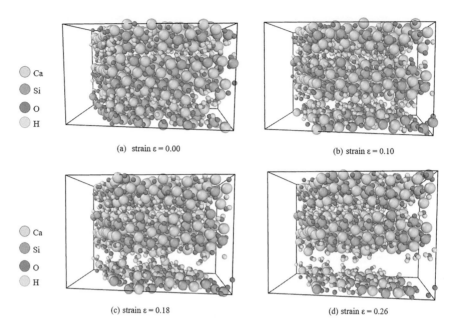

Figure 10.16 ReaxFF molecular dynamics simulation of the deformation process of a C-S-H unit cell under uniaxial tension test

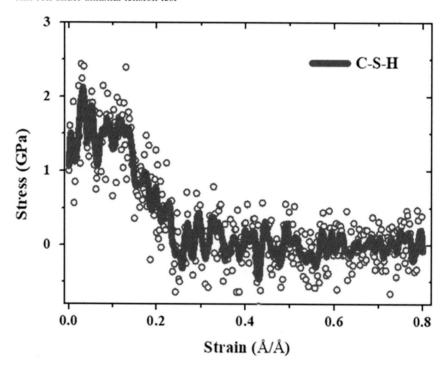

Figure 10.17 Molecular dynamics-simulated stress–strain relation for a C-S-H model

Figure 10.18 VMD homepage (www.ks.uiuc.edu/Research/vmd/)

10.6 MD Visualization Software: VMD and OVITO

In this section, we give a short tutorial for two popular MD visualization software: VMD and OVITO.

10.6.1 VMD

The full name of VMD is Visual Molecular Dynamics, which is a major reprocessing and post-processing computer software used for modeling, visualization, and analysis of atomistic and molecular systems, such as nanotubes, proteins, nucleic acids, and so forth, by using 3D graphics and built-in scripting. VMD is developed by the Theoretical and Computational Biophysics Group at the University of Illinois at Urbana-Champaign (see: Fig. 10.18). Among MD visualization programs, VMD is unique in its ability to efficiently operate on multi-gigabyte MD trajectories, and its interoperability with a large number of MD simulation packages, as well as its integration of structure and sequence information.

VMD supports computer operating systems LINUX, MacOSX, or Windows, is distributed free of charge, and includes source code. One can download VMD source code from the website: www.ks.uiuc.edu/Development/Download/download.cgi?PackageName=VMD after you have registered or logged in (see Fig. 10.19).

In the following, we introduce some basic features of VMD, and give a short tutorial, which can help new users quickly become familiar enough with VMD and its capabilities. The full range of the functionality and excellent tutorials of VMD can be found from the website: www.ks.uiuc.edu/Research/vmd/current/docs.html#tutorials.

Here, this short tutorial is divided into two parts: one covers the basics of molecular graphics representations and will introduce how to generate nice graphics; another one focuses on scripting in VMD.

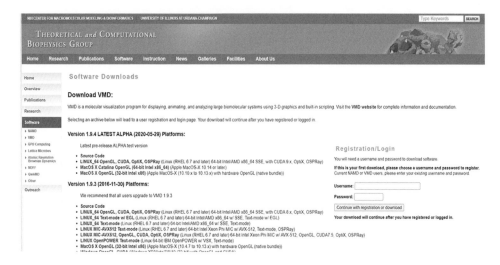

Figure 10.19 Installing VMD after registering or logging in

First, we shall give some basic information about the filetypes used in VMD. VMD specializes in the visualization of data from molecular simulations. There are several popular molecular data file formats that are widely used to describe molecular systems, such as PDB coordinate files, CHARMM-, NAMD-, and X-PLOR style PSF topology files, CHARMM-, NAMD-, and X-PLOR-style DCD trajectory files, and the like. Here, we will introduce three common molecular data file formats: PSF topology files, PDB coordinate files, and DCD trajectory files.

1. **PSF Topology File**

A PSF file, which is also called a protein structure file, contains all of the molecule-specific information needed to describe a molecular system using a particular force field CHARMM. The CHARMM force field is divided into a topology file and a parameter file. The former file is needed to generate the PSF file, and the latter file supplies specific numerical values for the generic CHARMM potential function. The PSF file defines the atom types, residues, covalent bonds, and selections of atoms used to determine force terms in MD or Monte Carlo simulations (such as angle and dihedral terms) but NOT coordinates. As a result, a PSF file must be loaded along with a PDB or DCD file, which defines the atomic coordinates. It contains six main sections: atoms, bonds, angles, dihedrals, impropers (dihedral force terms used to maintain planarity), and cross-terms. People can find more details in the website: https://www.ks.uiuc.edu/Training/Tutorials/namd/namd-tutorial-unix-html/node23.html.

For the illustration examples discussed in this chapter, the interested readers can download the example related PSF, PDB, and DCD files from: http://nanomechanics.berkeley.edu/introduction-to-computational-nanomechanics/.

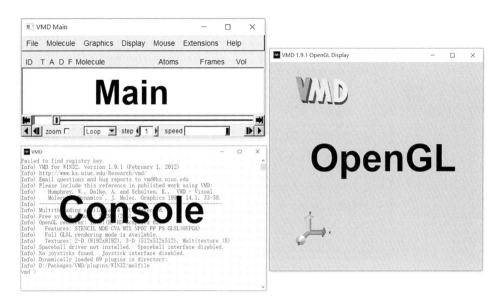

Figure 10.20 Three windows of VMD: the Main, OpenGL Display, and Console windows

2. **PDB Topology File**

A PDB file, which is a standard structure file, contains atomic coordinates and type alongside information about how they are organized into residues. PDB files may include multiple sets of coordinates for some or all atoms, except the coordinates of hydrogen atoms. VMD ignores everything in a PDB file except for the ATOM and HETATM records. When writing PDB files, the ATOM record type is used for all atoms in the system, including solvent and ions. People can find more details in the website: www.ks.uiuc.edu/Training/Tutorials/namd/namd-tutorial-unix-html/node22.html.

3. **DCD Trajectory File**

A DCD file is a binary file, which only contains coordinates. Since there is no information about atoms or their connectivity, the DCD file must be used in conjunction with a structure file. For example, a DCD file can be specified along with the PDB file, in which case the PDB file will be read as normal, and then coordinate sets are read from the DCD file until the end of the file is reached.

Next, we will give some examples to introduce how to use VMD in Windows system. After opening VMD, there are three windows (see Fig. 10.20): the Main, OpenGL Display, and Console windows. If you want to close the VMD, you can go to the Main window and choose File → Quit, or closing the Main window or Console window.

1. **Viewing Single Molecules**

In this example, we will load a single structure in the PDB format and learn how to view it from different angles and to alter the way that it is rendered on screen.

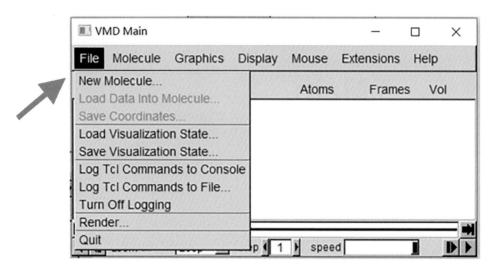

Figure 10.21 The Main window for loading the molecule

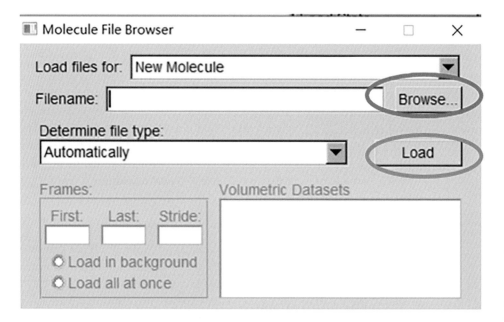

Figure 10.22 The Molecule File Browser window.

(1) Loading a Molecule

The first step is to load the molecule. A pdb file contains the atom coordinates.

(a) Select the *File → New Molecular* in the Main window of VMD (see Fig. 10.21).

(b) Another window, the Molecule File Browser, should now appear in your screen (see Fig. 10.22). Select the *Browse* button to find the .pdb file.

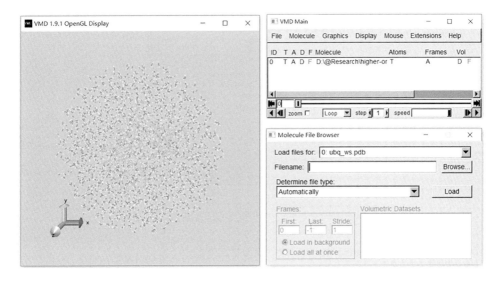

Figure 10.23 The Main window and the OpenGL Display window after loading the molecule

(c) Note that the molecule has not been loaded yet. After selecting the pdb file, you will be able return to the Molecule File Browser window. To actually load the file, you have to click the *Load* button (see Fig. 10.22). Then, the molecule should be loaded and displayed in the OpenGL Display window, as shown in Fig. 10.23. Note that once the molecule is loaded, the basic information including the name of the file and number of atoms will appear in the Main window. At that time, you may close the Molecule File Browser window at any time.

(2) **Displaying the Molecule**

To see the 3D structure of molecule, we will use the mouse in multiple modes. In the Main window of VMD, you can switch the mouse from the *Mouse* menu (see Fig. 10.24). We mainly focus on the first four options: Rotate Mode (R), Translate Mode (T), Scale Mode (S), and Center (C). Each option has a distinctive cursor that can help people identify which mode you are in (see Fig. 10.24).

(a) Rotate mode: In the OpenGL Display window, when the left mouse button is held, the molecule could be rotated around the vertical axis, or up and down around a horizontal one. When the right mouse button is held, the rotation will be performed around an axis perpendicular to the screen.

(b) Translate mode: You can move the molecule up, down, left, or right in the viewing plane while holding the left button down.

(c) Scale mode: When the left mouse is held, you can move the mouse left or right to zoom in or out.

(d) Center: The center option is not a direct method for altering the view point. It is used to specify a point about which rotations are done. Once you have selected a center point, you need to change to Rotate mode to perform the rotation.

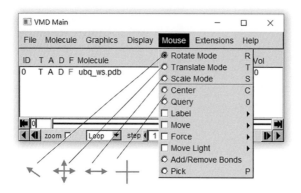

Figure 10.24 The Main window for displaying the molecule

If you want to reset the view, you can use menu option *Display* → *Reset View* or pressing "=" while selecting the OpenGL Display window.

(3) Changing the Drawing Styles of the Molecule

VMD can use a wide variety of drawing styles to display the molecule.

(a) Choose the *Graphics* → *Representations* from the menu bar of the Main window. A Graphical Representations window will then appear, as shown in Fig. 10.25. You can see the current graphical representation is being highlighted (see Fig. 10.25(a)), which shows the style of the representation (Lines), the coloring method (Name), and the selection of atoms to which this representation is applied. The current selection is for all atoms (see Fig. 10.25(b)).

(b) The coloring and drawing style representations can be changed by choosing different options from the drop-down menu, as shown in Fig. 10.25(c) and 10.25(d), respectively. Each Drawing Method has its own parameters. For instance, the thickness of the lines can be controlled by using the Thickness button, as shown in the Fig. 10.25(e).

Figure 10.26 shows different representations of the molecule, where Fig. 10.26(a) shows *Line* + *Name* representations, Fig. 10.26(b) shows *NewCartoon* + *Name* representations, and Fig. 10.26(c) shows *NewCartoon* + *ResType* representations.

2. Viewing Multiple Molecules

In this example, we shall discuss the case that multiple molecule files are used simultaneously. The interested readers can download alanin.psf, alanin.pdb, and alanin.dcd files from: http://nanomechanics.berkeley.edu/introduction-to-computational-nanomechanics/.

(1) Loading Multiple Molecules

(a) We first load the trajectory using the structure information in the PSF file. Open the Molecule File Browser from the *File* → *New Molecule*, and then select the alanin.psf file. Finally, click the *load* button to load the file. Note that, as shown in

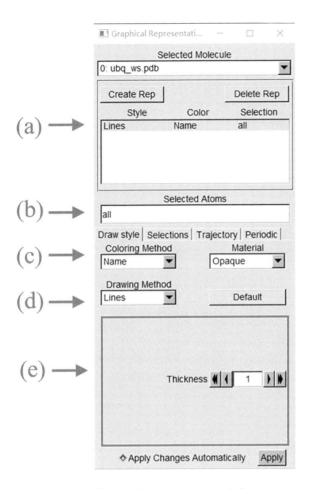

Figure 10.25 The Graphical Representations window

Fig. 10.27, there is no molecule in the OpenGL Display window, because the PSF file only contains the structural connectivity information without any coordinates. At the same time, the basic information including the name of the molecule and number of atoms will appear in the Main window, as shown in Fig. 10.28. But there is no frames.

(b) Next, we need to load coordinates into the molecule from a DCD file. Right click on the line for alanin.psf and select *Load Data into Molecule* from the menu. When the Molecule File Browser window appears, browse to the alanin.dcd file and load it. Now a structure can be visible in the OpenGL Display window and the alanin.psf file has 100 frames in the Main window, as shown in Fig. 10.29.

(c) Then, we need to load the second structure, that is the PDB file. Select *File → New Molecule* in the Main window. and then browse to the alanin.pdb file and load it from the Molecule File Browser window. The OpenGL Display window and Main

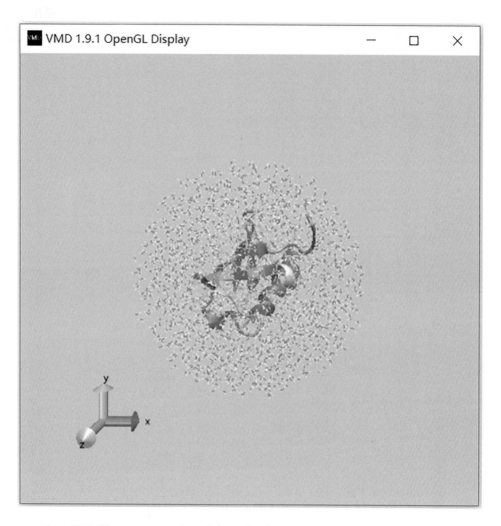

Figure 10.26 Three representations of the molecule

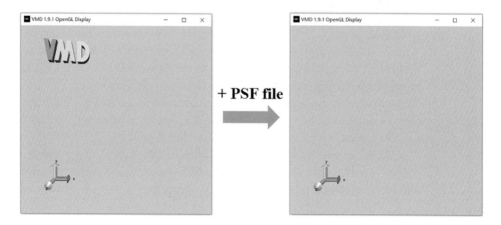

Figure 10.27 The OpenGL Display window after loading a PSF file

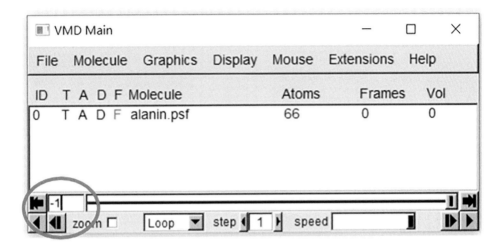

Figure 10.28 The Main window after loading a PSF file

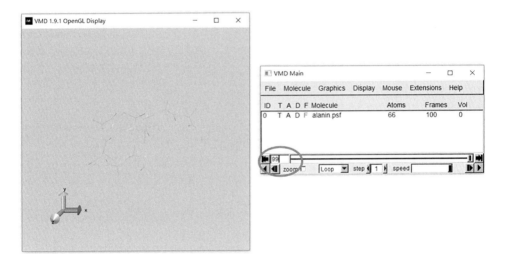

Figure 10.29 The OpenGL Display window and Main window after loading a PSF file and a DCD file

window will now look like Fig. 10.30. There are two entries for the two molecules in the Main window. Both have the same number of atoms but differ in the number of frames.

(2) **Changing the Drawing Styles of Multiple Molecules**

(a) Choose the *Graphics* → *Representations* from the menu bar of the Main window.

(b) Choose *Molecule* and *NewCartoon* for Colouring Method and Drawing Method, respectively.

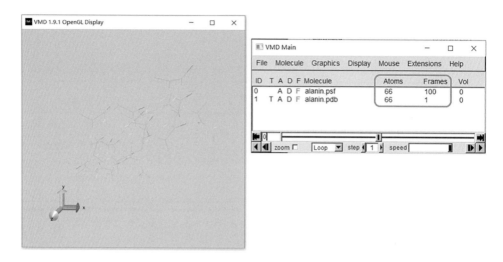

Figure 10.30 The OpenGL Display window and Main window after loading a PSF file and a PDB file

(c) Switch the molecule by clicking *the Selected Molecule* box and choosing "0: alanin.psf" from the menu list (see Fig. 10.31). Then alter the Graphical Representations to the same setting as the previous molecule. Finally, Fig. 10.32 shows the OpenGL Display window.

(3) **Setting Molecule Status**

In the Main window, as shown in Fig. 10.33, there are three letters to present different status settings for each molecule:

(a) Top (T): indicates the default molecule for all actions;

(b) Active (A): indicates if the trajectory of molecule will be updated in any animation;

(c) Drawn (D): indicates if the molecule is being displayed in the OpenGL Display window.

It may be noted that since there is only one molecule file that can have the "T" (top) status, if one wants to see the animation of atom motions and trajectories, one has to switch "T" status from the molecule file, alanin.psf, to the molecule file, alanin.pdb. To do so, one can double click the position of alanin.psf file in the T column as illustrated in Fig. 10.33.

(4) **Deleting Molecules**

A molecule could be deleted by right clicking on it in the Main window and then selecting *DeleteMolecule*, as shown in Fig. 10.34.

3. **Scripting in VMD**

In this example, we will discuss the basis of Tcl/Tk scripting language in VMD. To execute Tcl commands, you can use a convenient text console called bf Tk Console window. Select the *Extensions* → *TkConsole* menu item. A console window should appear (see Fig. 10.35). Then you can start entering Tcl/Tk commands in it.

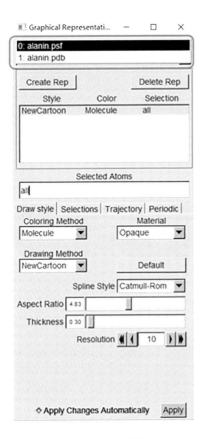

Figure 10.31 The Graphical Representations for multiple molecules

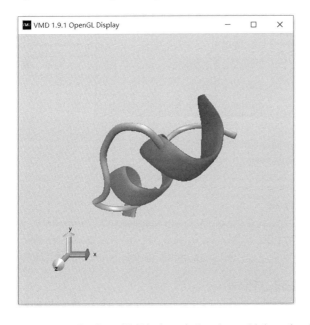

Figure 10.32 The OpenGL Display window for multiple molecules

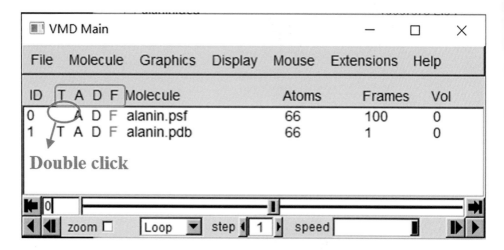

Figure 10.33 Setting molecule status

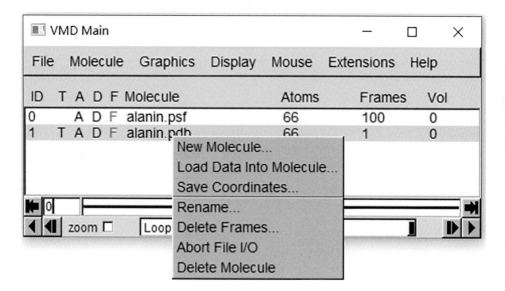

Figure 10.34 Deleting molecules

(1) Loading a Molecule

Navigate to the working directory using standard OS commands, and then type **mol new Ubiquitin.pdb**, as shown in Fig. 10.36. The Main window should show that a molecule is now loaded and it should be visible in the OpenGL Display window.

(2) Changing Molecule Properties

The molecule properties could be edited by using VMD's **atomselect** command. In the following, we shall give some examples.

```
76 VMD TkConsole                                    —   □   ×
File Console Edit Interp Prefs History Help
Main console display active (Tcl8.5.6 / Tk8.5.6)
(VMD) 1 %
```

Figure 10.35 The Tk Console window

```
76 VMD TkConsole                                    —   □   ×
File Console Edit Interp Prefs History Help
Main console display active (Tcl8.5.6 / Tk8.5.6)
(VMD) 1 % cd ..
>Main< (Packages) 2 % cd ..
>Main< () 3 % cd @research/higher-order\ quantum\ stress/Cambridge_Book/Examples
/Chapter10.6/VMD/
>Main< (VMD) 4 % dir
.:
1ubq.pdb              Ubiquitin.pdb         Ubiquitin.psf
VMD-basic.pdf         alanin.dcd            alanin.pdb
alanin.psf            equilibration.dcd     pdb5106.ent
proteins.tar.gz       pulling.dcd           ubiquitin1.psf
ubq.pgn               ubq_ws.pdb            vmd-tutorial-files.zip
vmd-tutorial.tar.gz
Ubiquitin.pdb Ubiquitin.psf
>Main< (VMD) 5 % mol new Ubiquitin.pdb
0
>Main< (VMD) 6 %
```

Figure 10.36 The Tk Console window after loading a molecule

```
>Main< (VMD) 5 % set crystal [atomselect top "all"]
atomselect0
>Main< (VMD) 6 %
```

Figure 10.37 The Tk Console window after typing *set crystal [atomselect top "all"]* command

(a) **Type Set Crystal [Atomselect Top "All"]**

This command creates a selection containing all the atoms in the molecule and assigning it to the variable crystal. Instead of a molecule ID (which is a number), we have used the shortcut **top** to refer to the top molecule. The result of atomselect is a function. Thus, **$crystal** is now a function that performs actions on the contents of the **all** selection. If the command is typed correctly, you can see a reply such as atomselect 0 in the Tk Console window (see Fig. 10.37).

(b) **Type $crystal num**

This returns the number of atoms in the molecule, which is same as that is shown in the Main window.

(c) **Making Some Measurement**

Here are some commands to measure the molecule.

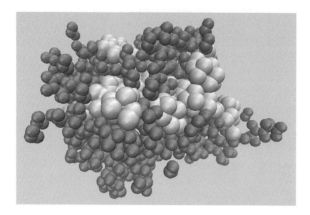

Figure 10.38 Ubiquitin in the VDW representation, colored according to the hydrophobicity of its residues

(I) **measure center $crystal**: get the coordinates for the geometric center of the molecule;

(II) **measure minmax $crystal**: get the coordinates for the minimum and maximum.

(d) **Moving the Molecule and Changing Its Coordinates**

Here are some commands to move the molecule on the screen and change its coordinates. Using these commands are different from rotating or translating the molecule with the mouse, which only changes the view and not the coordinates.

(I) **$crystal moveby {10 0 0}**: translate the molecule 10A along its x axis;

(II) **$crystal move [transaxis x 40 deg]**: rotate the molecule 40 degrees around its *x*-axis.

(III) **$crystal write new-Ubiquitin.pdb**: write the new coordinates to a pdb file in the current directory.

(e) **Type $ crystal set beta 0**

This command resets the **beta** field to be zero for all atoms. As you do this, you should observe that the atoms on your screen will suddenly change to a uniform color.

(f) **Type set sel [atomselect top "hydrophobic"]**

This command creates a selection that only contains all the hydrophobic residues.

(g) **Type $sel set beta 1**

This command will label all hydrophobic atoms by setting their beta values to 1. If the colors in the OpenGL Display do not get updated, click on the Apply button at the bottom of the Graphical Representations window.

(h) **Type $crystal set radius 1.0**

This command can be used to make all the atoms smaller and easier to see through.

(i) **Type $sel set radius 1.5**

This command can be used to make the atoms in the hydrophobic residues larger.

Finally, if you have followed the instructions correctly, the molecule should now look like the one in Fig. 10.38.

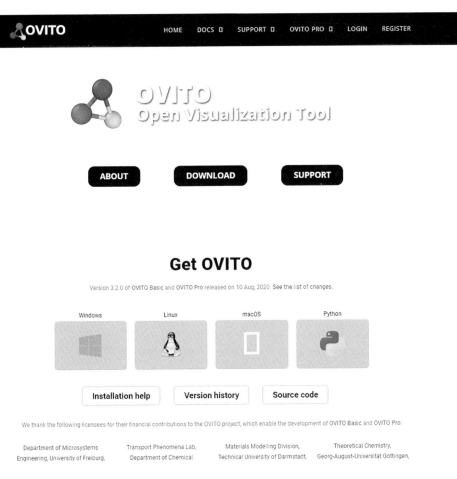

Figure 10.39 OVITO homepage www.ovito.org/

10.6.2 OVITO

The Open Visualization Tool (OVITO) is a scientific visualization and analysis software for atomistic and particle simulation data, typically generated in MD, atomistic Monte Carlo, and other particle-based simulations (see: Fig. 10.39). OVITO is developed by OVITO GmbH, which is freely available for all major platforms, including Windows, Linux, and macOS, under an open source license. For example, people can download OVITO freely for windows in the website: www.ovito.org/windows-downloads/. If you want to find more information about OVITO, please visit the website: www.ovito.org/. In the following, we shall give some basic information about OVITO and introduce how to use it.

1. **Importing Data**

The first step is to load a simulation file into OVITO by selecting *File* → *Load Files* from the menu or using the corresponding button in the toolbar. OVITO

Figure 10.40 Importing data in OVITO

can read the several file formats, such as LAMMPS dump, LAMMPS data, XYZ, PSOCAR / XDATCAR / CHGCAR, and CFG, etc. The imported database will be represented by the first item under the "Data source" section, as shown in Fig. 10.40.

2. **Viewport Windows**

OVITO's main window has four viewports, showing the loaded data from different perspectives, as shown in Fig. 10.41.

(a) **Navigation Functions**

The virtual camera of a viewport can be rotated or translational moved using the mouse.

(I) Left-click and drag to rotate the camera around the current orbit center, which is located in the center of the simulation box by default;

(II) Right-click or use the middle mouse button and drag to move the camera parallel to the projection plane;

10 Introduction to LAMMPS

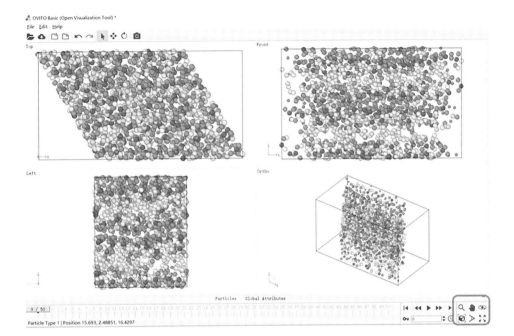

Figure 10.41 OVITO's main window

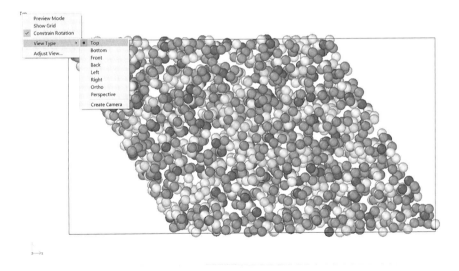

Figure 10.42 The viewport menu

(III) Use the mouse wheel to zoom in or out;
(IV) Double-click on an object to reposition the orbit center;
(V) Double-click in an empty region of a viewport to reset the orbit center to the center of the active dataset.

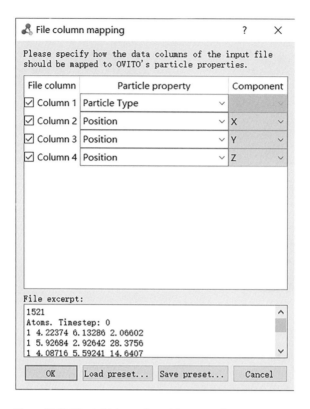

Figure 10.43 The initial set of particle properties

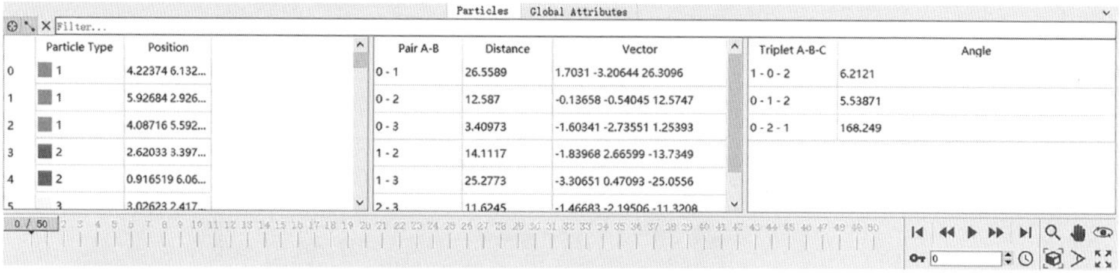

Figure 10.44 The OVITO's Data Inspector panel

(b) **Viewport Toolbar**

The viewport toolbar is located below the viewports, which contains buttons to activate various navigation input modes, as shown in the colored box in the Fig. 10.41.

(c) **Viewport Menu**

The viewport menu could be opened by clicking the text label in the upper left corner of a viewport, such as Top, Ortho, Left, and Perspective, as displayed in the Fig. 10.42. The viewport menu can switch to one of the standard viewing orientations, orthogonal, or perspective projection. The Preview Mode option activates a virtual frame that is displayed in the viewport to indicate the region that will be visible in

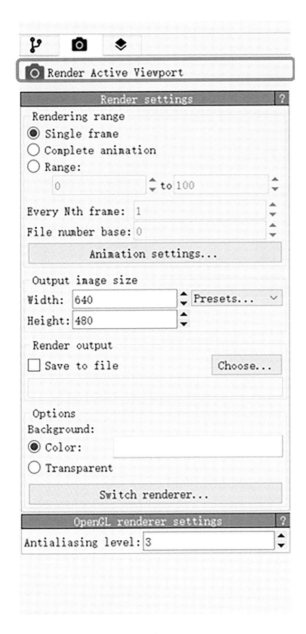

Figure 10.45 The Rendering tab

rendered images. A camera object could be inserted into the scene using the Create Camera function.

3. **Particle Properties**

Particle properties, such as position, chemical bond, and velocity, are important in the data model of OVITO. An initial set of properties is automatically created by

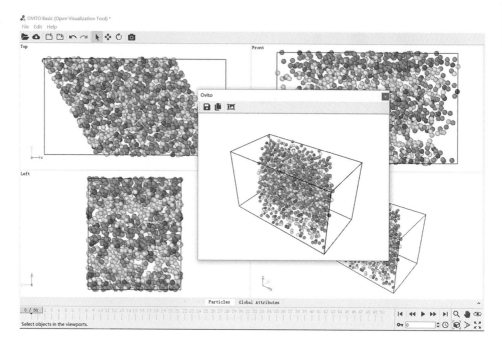

Figure 10.46 The generated image

OVITO whenever you open a simulation data file. For example, when you open a XYZ file, the particle properties including particle type and position are automatically mapped to corresponding particles (see Fig. 10.43). If you want to find out which properties are currently associated with the particles, you can open OVITO's Data Inspector that is a panel located below the viewport area (see Fig. 10.44). OVITO provides a rich set of functions for modifying the assigned properties of particles. For instance, Real(R,G,B) function controls the display color of particles or individual bonds, and Real(X,Y,Z) function represents the coordinates of particles. OVITO also provides a number of particle selection modifiers, which allow you to set the value of the Selection property. Of course, it is possible to export the particle property values to an output file with a variety of formats, such as XYZ format.

4. **Rendering**

If you want to produce images or a movie after loading a data file for visualization and analysis, you can go to **Rendering** tab in the command panel, as shown in Fig. 10.45. The Render Active Viewport button launches the image rendering process for the active viewport. Then a separate window showing the generated image will appear (see Fig. 10.46). The Render settings panel controls various settings, such as the resolution of the generated image, the background color, and the filename of generated images or movies.

OVITO can also render an animation of the loaded simulation trajectory. To render a video, choose the *Complete animation* option in the Render settings panel and set the frame rate for the output video, as shown in Fig. 10.47.

Figure 10.47 The render settings for rendering an animation

5. **Exporting Data**

The results of the current data pipeline in OVITO can be exported through *File* → *Export File* to a file with a variety of formats, such as LAMMPS dump, LAMMPS data, XYZ, POSCAR and so forth.

10.7 Homework Problems

Problem 10.1 **I.** Create a CNT with different chirality parameter (5, 10), (10, 10), (15, 0) and (20, 5) with length 5 nm.

II. Create a double layer CNT (a,a) with the distance between the two layers as 1nm. Set the inner CNT as (7, 7).

III. Move the CNT (10,10) to the center of the coordinate system, and rotate the CNT with 30° about y-axis. Output your results.

IV. Create a 5 nm × 10 nm of graphene sheet. Output your graphene sheet.

V. Create a POPE membrane that has a dimension of 100 unit × 100 unit with CHARMM36 force field. Output pdb file.

VI. Create a (7,7) CNT using nanotube builder, and save it to create a pdb file and a psf file for the (7,7) CNT.

VI. Using the file nanotube.tcl to create a pdb file and a psf file for a (7,7) CNT by using VMD.

VII. Create a solvent (water) box that contains the above CNT.

VIII. Add NaCl ions into the solvent box.

11 Monte Carlo Methods

In this chapter, we discuss the Monte Carlo method, and its applications in atomistic and molecular modeling and simulations.

11.1 Monte Carlo Sampling for Integrations

One of the main objectives of molecular dynamics (MD) simulations is to find the desired material properties at macroscale or mesoscale, or the thermodynamic properties of the material ensemble under study, through statistical ensemble averaging under given conditions, such as stresses, strains, temperature, specific heat, and diffusivity, and the like.

To accurately find the thermodynamic properties of a molecular ensemble, we first need to conduct lengthy molecular dynamics simulations to find the desired equilibrium state of the system. Second, once we have the simulation data, we need to perform ensemble average integration to find the macroscale material properties that we are looking for.

For both finding the final equilibrium state of the system and calculating the ensemble average integration, the so-called Monte Carlo method is a powerful tool. In this section, we start with the discussion on how to use the Monte Carlo method to efficiently calculate ensemble average integration, and then we shall provide a detailed introduction to the Metropolis–Hastings algorithm.

We first consider the problem of how to evaluate the following ensemble average integration:

$$<A> = \int d\mathbf{p}^N d\mathbf{q}^N A(\mathbf{q}^N, \mathbf{q}^N) \frac{\exp(-\beta H(\mathbf{q}^N, \mathbf{p}^N))}{Z},$$

which has $6N$ degrees of freedom, i.e., a $6N$-dimension of integration. When $N \gg 1$, this will be a very expensive integration. In practice, it is almost impossible to integrate this integral by numerical integration. Say, if the minimum quadrature points in one-dimensional is m, it then requires an order of m^{3N} or m^{6N} evaluations.

For the sake of simplicity, we first omit the variables $\mathbf{p}^N$ and only consider evaluating the following integration:

$$<A(\mathbf{q}^N)> = \int A(\mathbf{q}^N) p(\mathbf{q}^N) d\mathbf{q}^N, \qquad (11.1)$$

where $p(\mathbf{q}^N)$ is the probability density of the random variable configuration $\mathbf{q}^N$. For the canonical ensemble, we have,

$$p(\mathbf{q}^N) = \frac{\exp(-\beta E(\mathbf{q}^N))}{\int \exp(-\beta E(\mathbf{q}^N))d\mathbf{q}^N} = \frac{\exp(-\beta E(\mathbf{q}^N))}{Z}. \tag{11.2}$$

In numerical computations, we may use the average of a finite number of the random position configurations (sub-ensembles) to approximate the ensemble average that may be calculated by using quadrature integration, i.e.,

$$E[A] = <\mathbf{A}(\mathbf{q}^N)> \approx \frac{\sum_{i=1}^{N_q} A_i(\mathbf{q}^N)\exp(-\beta E_i(\mathbf{q}^N))\Delta\mathbf{q}^N}{\sum_{i=1}^{N_q} \exp(-\beta E_i(\mathbf{q}^N))\Delta\mathbf{q}^N}. \tag{11.3}$$

However, in order to obtain accurate results, the number of N_q has to be very large, and on the other hand, many of them may have little contribution to the ensemble average $<\mathbf{A}(\mathbf{q}^N)>$.

The so-called Monte Carlo method is basically a statistical sampling method that can find an accurate ensemble average without breaking the bank. In Eq. (11.3), $\mathbf{q}^N$ is a higher dimensional random variable, and it has the probability distribution shown in Eq. (11.2). We can, based on this distribution, randomly draw n random variables $\{\mathbf{q}_1^N, \mathbf{q}_2^N, \ldots, \mathbf{q}_n^N\}$ that form a sequence of independent and identically distributed (iid) random variable sequence. Then, the corresponding sequence $\{\mathbf{A}(\mathbf{q}_1^N), \mathbf{A}(\mathbf{q}_2^N), \ldots, \mathbf{A}(\mathbf{q}_n^N)\}$ is also an (iid) sequence, and we can calculate its sample mean as,

$$\hat{\mathbf{A}} = \frac{1}{n}\sum_{i=1}^{n} \mathbf{A}(\mathbf{q}_i^N).$$

Based on the central limit theorem (CLT), the sample mean will converge to the expectation of $\mathbf{A}$, i.e., $E[\mathbf{A}]$, which is the integral under evaluation, i.e., the ensemble average,

$$\lim_{n\to\infty} \hat{\mathbf{A}} = E[\mathbf{A}] = <\mathbf{A}>.$$

To evaluate the convergence, we may define an unbiased variance (not sample variance nor population variance) of $\mathbf{A}$ as

$$Var_n(\mathbf{A}) = \sigma_n^2 = \frac{1}{n-1}\sum_{i=1}^{n}(\mathbf{A}(\mathbf{q}_i^N) - E[\mathbf{A}])^2,$$

thus, we can introduce a biased sample variance as

$$Var(\hat{\mathbf{A}}) = \frac{1}{n^2}\sum_{i=1}^{n} Var_n(\mathbf{A}) = \frac{Var_n(\mathbf{A})}{n}.$$

Then, the standard error of the sample mean is

$$\sigma_{\hat{A}} = \frac{\sqrt{Var_n(\mathbf{A})}}{\sqrt{n}} = \frac{\sigma_n}{\sqrt{n}} \to 0, \text{ as } n \to 0.$$

This result is remarkable, because it is independent from the dimension of the phase space, which is $3N$ here or $6N$ in general for MD simulations, while the accuracy of conventional numerical integration algorithms usually deteriorates as the dimension of function integration space increases, which is often referred to as "the curse of dimensionality." It should be noted that the magic of the Monte Carlo method is that the (iid) sequences $\{\mathbf{q}_1^N, \mathbf{q}_2^N, \ldots, \mathbf{q}_n^N\}$ and $\{\mathbf{A}(\mathbf{q}_1^N), \mathbf{A}(\mathbf{q}_2^N), \ldots, \mathbf{A}(\mathbf{q}_n^N)\}$ are not the order sequences in conventional numerical integrations, and they are random variable sequences that are drawn based on their probability distributions.

In computations, it is easy to draw an (iid) sequence from a uniform random distribution. In numerical computations, we often use the so-called inverse transform technique to draw an independent random variable sequence for arbitrary given probability distribution, which is explained as follows.

Let U be a uniform random variable in the range $f_U(u) = 1, u \in [0, 1]$. If $x = F_X^{-1}(u) = g(u)$ or $u = F_X(x) = g^{-1}(x)$, X will be a random variable with the given probability distribution function (PDF) $f_X(x)$. This can be shown readily by using the transform method,

$$f_X(x) = F_X'(x) = F_U'(g^{-1}(x)) \left| \frac{dg^{-1}(x)}{dx} \right| = f_U(u) \left| \frac{du}{dx} \right| = f_X(x). \quad (11.4)$$

In other words, if $X = F^{-1}(U)$, X will be a random variable with cumulative distribution function $F_X(x) = F(x)$, because that

$$F_X(x) = P(X \leq x) = P(F^{-1}(U) \leq x) = P(U \leq F(x)) = F(x). \quad (11.5)$$

Therefore, if we have a random number generator to generate random numbers according to the uniform distribution, we can generate random variable with any given probability distribution. For example, in MATLAB, one can generate a uniform random number by just typing:

```
X= rand~,
```

which will generate a single uniformly distributed random number in the interval $(0, 1)$. If you type

```
X= rand (3, n),
```

it will generate an (iid) sequence of uniformly distributed random vectors in three-dimensional space $\mathbf{R}^3$, i.e. $\{\mathbf{X}_1, \mathbf{X}_2, \ldots, \mathbf{X}_i, \ldots \mathbf{X}_n\}$ where $\mathbf{X}_i \in \mathbf{R}^3$.

In practice, an easy way to generate a uniform distribution random number is using a so-called pseudorandom number generator (PRNG), also known as a deterministic random bit generator (DRBG). PRNGs are central in applications of the Monte Carlo method. Those topics are out of the scope of this book. For the sake of complete-

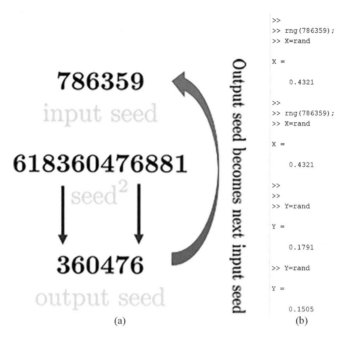

Figure 11.1 (a) Von Neumann's middle-square method to generate pseudorandom number and (b) MATLAB script of commands *rng* and *rand*

ness, here we briefly outline an early version of PRNG method, i.e., the **John von Neumann's middle-square method**. To generate a sequence of n-digit pseudorandom numbers, an n-digit starting value is created and squared, producing a $2n$-digit number. If the result has fewer than $2n$ digits, leading zeroes are added to compensate. The middle n-digits of the result would be the next number in the sequence, and returned as the result. This process is then repeated to generate more numbers as shown in Fig. 11.1. In MATLAB, one can specify the initial seed when one uses *rand* or other random number-generating command. By doing so, one can consistently have the same random number generated. Otherwise, each time you use the *rand* command, you will obtain a different number $x \in (0, 1)$ as shown in Fig. 11.1(b).

In summary, the idea of Monte Carlo techniques is to draw a set of (iid) random variables (samples) from a given distribution function $p(X)$ in order to approximate $p(X)$ with the empirical distribution

$$p(X) \approx p_e(X) = \frac{1}{n} \sum_{i=1}^{n} \delta(X - X_i) \to \int_{-\infty}^{\infty} f(x)p(x)dx \approx \int_{-\infty}^{\infty} f(x)p_e(x)dx$$

$$= \frac{1}{n} \sum_{i=1}^{n} f(x_i), \tag{11.6}$$

where $\sum_{i=1}^{n} \delta(X - X_i)$ is the sampling operator or the Dirac delta function and X_i here is a random variable, i.e., it is not true that $X_1 < X_2 < X_3 < \cdots < X_i < \cdots < X_n$.

There are two applications of the above Monte Carlo method: (1) we can use it to approximate a complicated integration by "summation average" or sampling and (2) if we use a random variable sequence, we can calculate its distribution function or probability mass function by calculating the sampling frequency,

$$p(X) = \frac{1}{n} \sum_{j=1}^{m} \frac{n_j}{n} \delta(X - X_j),$$

$$\sum_{i}^{m} n_j = n, \ j = 1, 2, \ldots, m \text{ and } x_1 < x_2 < \ldots x_j < \cdots < x_m.$$

11.1.1 Monte Carlo Sampling for Integration – Sample Mean Method

The first application of the Monte Carlo sampling method is to use it as a very efficient integration method to calculate the following integration:

$$I = \int_a^b f(x) dx. \tag{11.7}$$

We can rewrite Eq. (11.7) as,

$$I = (b - a) \int_{-\infty}^{\infty} f(x) p(x) dx, \tag{11.8}$$

where

$$p(x) = \begin{cases} \dfrac{1}{b-a}, & x \in (a,b) \\ 0, & x \notin (a,b) \end{cases} \tag{11.9}$$

is a piece-wise uniform PDF.

Thus, we can write

$$I = (b - a) \bar{f}, \tag{11.10}$$

where

$$\bar{f} = E[f(x)] = \int_{-\infty}^{\infty} f(x) p(x) dx. \tag{11.11}$$

Equation (11.10) is especially useful when the integration in Eq. (11.7) has no close form solution. In this case, we can approximate the expected value with a sample of n value sequence of random variables $\{x_1, x_2, \ldots x_i, \ldots x_n\}$ from the uniform probability distribution, by using

$$\hat{f} = \frac{1}{n} \sum_{i=1}^{n} f(x_i) \ \rightarrow \ \hat{f} \approx \bar{f} \ \rightarrow \ (b-a)\hat{f} \approx (b-a)\bar{f} = I. \tag{11.12}$$

Since $\{x_1, x_2, \ldots x_i, \ldots x_n\}$ is (iid), $\{f(x_1), f(x_2), \ldots f(x_i), \ldots f(x_n)\}$ is also an (iid) sequence of random variables, even though it may not have an uniform probability distribution. On the other hand, however, the sample mean,

$$\hat{f} = \frac{1}{n}\sum_{i=1}^{n} f(x_i), \tag{11.13}$$

will have the *t*-distribution, and it will become a normal distribution as $n \to \infty$ based on the **CLT**.

The CLT states that the random variable

$$\sqrt{n}\hat{z} = \frac{\hat{f} - \bar{f}}{\sigma}$$

follows the standard normal distribution $N(0, \sigma^2)$, where $\sigma = \sqrt{E[f^2] - (\bar{f}^2)}$ and $Var[f] = \sigma^2$.

That is,

$$\hat{f} \to \bar{f}$$

and since

$$\hat{z} = \frac{\hat{f} - \bar{f}}{\sigma/\sqrt{n}} \sim N(0, \sigma) \to Var[\hat{f}] = \frac{Var[f]}{n} = \frac{\sigma^2}{n}.$$

In fact, one can derive that

$$Var[\hat{f}] = E\left[(\hat{f} - \bar{f})^2\right] = \frac{1}{n^2}\left(\left(\sum_i f(x_i)\right)^2 - n\bar{f}^2\right)$$

$$= \frac{1}{n}\left(\frac{1}{n}\left(\sum_i f(x_i)\right)^2 - \bar{f}^2\right) = \frac{Var[f]}{n},$$

which leads to

$$\to std(\hat{f}) = \frac{std(f)}{\sqrt{n}}$$

so that the relative error of between $\hat{f}$ and $\bar{f}$ approaches to zero as $n \to \infty$.

Thus, if we define $\hat{I} = (b-a)\hat{f}$, we have

$$error = \sqrt{Var[\hat{I}]} = (b-a)\sqrt{Var[\hat{f}]} = (b-a)\sqrt{\frac{Var[f]}{n}} \to 0, \hat{I} \to I \text{ as } n \to 0.$$

We then have the convergence result of the sample mean method.

Example 11.1 Sample mean MC algorithm Use the sample mean MC algorithm to calculate the following integral:

$$\int_0^\pi \sin x\, dx = 2.$$

The following is a step-by-step procedure:

- Generate a sequence of n (PRNG) random variables: R_i;
- Compute $X_i = a + (b-a) \cdot R_i$;

- Compute $f(X_i), i = 1, 2, \ldots n$;
- Compute

$$I = \frac{(b-a)}{n} \sum_{i=1}^{n} f(x_i).$$

A MATLAB script file is listed as follows:

```
>xi=0.0;
>fi=0.0;
>a=0.0;
>b=pi;
>I=0.0;
>for i=1:1000
     xi= a + (b-a)*rand();
     fi= sin (xi);
     I = I + fi;
>end
>I=I* (b-a)/1000
>
>I = 2.0328
```

Note that even if $f(x)$ is not square integrable, the simple mean MC integration algorithm will still converge to its true value, but the error estimate may become unreliable.

REMARK 11.2 (Curse of dimensionality) In conventional numerical integration or quadrature integration, the integration error is usually at the order of $O(h^{k/d})$, where h is the discretization spacing, i.e., $h \sim L/N$; k is a method dependent parameter, e.g., $k = 2$ for the Simpson method; and d is the dimension of the integration domain. In terms of the number of quadrature points, the above error estimate may be written as $O(N^{-k/d})$, where N is the number of quadrature points in numerical integration. Obviously, as $N \to \infty$, the accuracy of the numerical integration will improve. However, as $d \to \infty$, the convergence rate will deteriorate rapidly. This phenomenon is called the *"curse of dimensionality."* On the other hand, if one uses the Monte Carlo method to carry out numerical integration, as shown previously, the error is always controlled at $O(N^{-1/2})$, and it will not be worse. This is the reason why the Monte Carlo integration method is very powerful in numerical integration of higher dimension space.

Example 11.3 The reject-and-accept method (hit-and-miss algorithm) In this example, we show how to use the so-called hit-and-miss algorithm to find the approximated value π. As shown in Fig. 11.2, a square box with each side length of $2r$ and a circle area of radius r is enclosed inside the square.

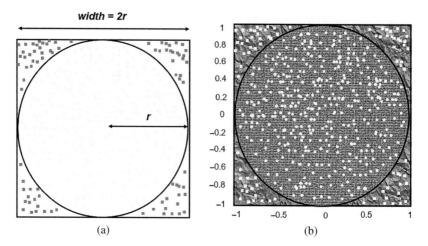

Figure 11.2 The hit-and-miss algorithm to calculate the value of π: (a) a square with side length of $2r$ and (b) the results of Monte Carlo method

Thus, the area of circle is $A_c = \pi r^2$ and the area of the square is $A_s = 4r^2$. Their ratio is

$$\frac{\text{area of square}}{\text{area of circle}} = \frac{4r^2}{\pi r^2} = \frac{4}{\pi} \rightarrow \pi = 4 \frac{\text{area of circle}}{\text{area of square}} = 4 \times \frac{\int_C ds}{\int_S ds}.$$

Based on the Monte Carlo method, we know that

$$A_s = \int_S ds = \frac{A_s}{N} \sum_{i=1}^{N} i = \sum_{i=1}^{N} \Delta s_i,$$

where $\Delta s_i = A_s/N$, whereas

$$A_c = \sum_{j=1}^{C} \Delta s_j, \quad x_j^2 + y_j^2 \leq r^2.$$

For uniform probability distribution, $\Delta s_i = \Delta s_j = \Delta s$, as $N \rightarrow \infty$. We then have

$$\pi = 4 \frac{A_c}{A_s} = 4 \times \frac{\sum_{j=1}^{C} j}{\sum_{i=1}^{N} i} = 4 \frac{C}{N} = \frac{\text{\# of dots inside the circle}}{\text{total number of dots}}, \quad N \rightarrow \infty. \quad (11.14)$$

We start by generating two sequences of N of PRN: $\{R_i\} \in [0, 1]$ and $\{R_j\} \in [0, 1]$. Assuming that $r = 2$, we can have

$$x_i = -1 + 2R_i, \text{ and } y_j = -1 + 2R_j$$

(i) Let C be the number of dots inside the circle;
(ii) Start from $C = 0$;
(iii) For $i = 1 : N$ if $x_i^2 + y_j^2 < 1$, $C = C + 1$;
(iv) Then π value is

$$\pi = 4\frac{C}{N}.$$

A MATLAB script is given as follows:

```
>N=100000;
>a=-1;
>r=1;
>x=a+(r-a)*rand(N,1);
>y=a+(r-a)*rand(N,1);
>radii=sqrt(x.^2+y.^2);
>i=radii<=r;
>hits=sum(i);
>misses=N-hits;
>Pi=4*(hits/N)
>plot(x(i),y(i),'.g');
>> hold;
Current plot held
>> plot(x(~i),y(~i),'.r')
```

The plot is shown in Fig. 11.2(b).

11.1.2 Importance Sampling

In many situations, particularly in MD simulations, we are interested in calculating the following assemble or sample integration,

$$\bar{f} = E[f] = \int_{-\infty}^{\infty} f(x)p(x)dx = \int_{-\infty}^{\infty} \left(f(x)\frac{p(x)}{q(x)}\right)q(x)dx = \bar{f}_q, \quad (11.15)$$

in which $p(x)$ is not the uniform distribution.

We still can use the Monte Carlo sample mean approximation to integrate it, i.e.,

$$\bar{f} = E[f] = \int_{-\infty}^{\infty} f(x)p(x)dx = \frac{1}{n}\sum_{i=1}^{n} f(x_i), \quad (11.16)$$

where both $\{x_i\}$ and $\{f(x_i)\}$ are (iid)s sampled from $p(x)$. However, some of the distribution $p(x)$ are too complex and involved with many parameters, or products of many components; or in some cases, we do not have the complete information about the probability distribution, and we only know the partial probability distribution. Thus, the sampling from the true probability distribution $p(x)$ can be either inefficient or we do not have enough information to do so.

On the other hand, if we can find a different distribution, say $q(x)$, that is concentrated in the area that hosts the importance part of $p(x)$, we may rewrite the integral,

$$\bar{f} = E[f] = \int_{-\infty}^{\infty} f(x)p(x)dx = \int_{-\infty}^{\infty} \left(f(x)\frac{p(x)}{q(x)} \right) q(x)dx = \bar{f}_q.$$

Thus, we may use the sample mean method to integrate it with respect to $q(x)$, i.e.

$$\hat{f}_q = \frac{1}{n} \sum_{i=1, x_i \sim q(x)}^{n} f(x_i)\frac{p(x_i)}{q(x_i)} \quad (11.17)$$

to replace the original sample mean,

$$\hat{f} = \frac{1}{n} \sum_{i=1, x_i \sim p(x)}^{n} f(x_i).$$

We can see that

$$E[\hat{f}_q] = \frac{1}{n} \sum_{i=1}^{n} \int_{-\infty}^{\infty} f(x_i)\frac{p(x_i)}{q(x_i)} q(x_i)dx_i = \frac{1}{n} \sum_{i=1}^{n} \int_{-\infty}^{\infty} f(x_i)p(x_i)dx_i = E[f] = \bar{f}.$$

It indicates that $\hat{f}_q$ is also an unbiased estimator. However, its variance,

$$Var[\bar{f}_q] = \int_{-\infty}^{\infty} \left(\frac{p(x)}{q(x)} f(x) - \bar{f} \right)^2 q(x)dx \approx Var[\hat{f}_q]$$

$$= \frac{1}{n} \sum_{i=1, x_i \sim q(x)}^{n} \left(\left(\frac{p(x_i)}{q(x_i)} f(x_i) \right)^2 - \bar{f}^2 \right),$$

strongly depends on the choice of $q(x)$.

In statistics, we often call $q(x)$ as the proposed distribution, or the bias potential in the umbrella sampling. Sometimes, we also call it the umbrella potential.

11.1.3 Umbrella Sampling

In canonical ensemble MD, we can use the simple mean method to calculate the ensemble average of a field variable $A(\mathbf{r})$ by using a finite number of configurations (sub-ensembles) to approximate the ensemble average,

$$< A(\mathbf{r}^N) > = \int A(\mathbf{r}^N) p(\mathbf{r}^N) d^N \mathbf{r} \approx \frac{1}{N_s} \sum_{i=1}^{N_s} A(\mathbf{r}_i^N), \text{ where } p(\mathbf{r}^N)$$

$$= \frac{\exp(-\beta E(\mathbf{r}^N))}{\int \exp(-\beta E(\mathbf{r}^N)) d^N \mathbf{r}}, \quad (11.18)$$

where $\{\mathbf{r}_i^N\}_{i=1}^{N_s}$ is an (iid) sequence and $A(\mathbf{r}_i^N)$ are solved in the canonical ensemble MD calculations, and each index i represents a microstate.

In the umbrella sampling, we replace the original probability distribution in the canonical ensemble system by

$$p(\mathbf{r}^N) \rightarrow q(\mathbf{r}^N) = p(\mathbf{r}^N)w(\mathbf{r}^N) = \frac{w(\mathbf{r}^N)\exp(-\beta E(\mathbf{r}^N))}{\int w(\mathbf{r}^n)\exp(-\beta E(\mathbf{r}^N))d^N\mathbf{r}}$$

$$= \frac{\exp(-\beta E(\mathbf{r}^N) + V(\mathbf{r}^N))}{\int \exp(-\beta E(\mathbf{r}^N) + V(\mathbf{r}^N))d^N\mathbf{r}}, \qquad (11.19)$$

where

$$V(\mathbf{r}^N) = -\frac{1}{\beta}\ln w(\mathbf{r}^N) \rightarrow w(\mathbf{r}^N) = \exp(-\beta V(\mathbf{r}^N)) \qquad (11.20)$$

is the biased potential.

It is straightforward to show that

$$<A>_p = \frac{\int (A(\mathbf{r}^N)/w(\mathbf{r}^N))w(\mathbf{r}^N)\exp(-\beta E(\mathbf{r}^N))d^N\mathbf{r}}{\int (1/w(\mathbf{r}^N))w(\mathbf{r}^N)\exp(-\beta E(\mathbf{r}^N))d^N\mathbf{r}}$$

$$= \frac{\int A(\mathbf{r}^N)/w(\mathbf{r}^N))\exp(-\beta(E(\mathbf{r}^N) + V(\mathbf{r}^N)))d^N\mathbf{r}}{\int (1/w(\mathbf{r}^N))\exp(-\beta(E(\mathbf{r}^N) + V(\mathbf{r}^N)))d^N\mathbf{r}}$$

$$= \frac{<A/w>_q}{<1/w>_q} \qquad (11.21)$$

Thus, the samplings for the nominator and denominator of Eq. (11.21) are simply

$$<A/w>_q = \frac{1}{N_s}\sum_{i=1}^{N_s}\frac{A(\mathbf{r}_i^N)}{w(\mathbf{r}_i^N)} \qquad (11.22)$$

and

$$<1/w>_q = \frac{1}{N_s}\sum_{i=1}^{N_s}\frac{1}{w(\mathbf{r}_i^N)}. \qquad (11.23)$$

It may be noted that there is a difference between the umbrella sampling and the importance sampling, i.e.,

$$<A>_{ub} = \frac{\frac{1}{N_s}\sum_{i=1}^{N_s}\frac{A(\mathbf{r}_i^N)}{w(\mathbf{r}_i^N)}}{\frac{1}{N_s}\sum_{i=1}^{N_s}\frac{1}{w(\mathbf{r}_i^N)}} \qquad (11.24)$$

and

$$<A>_{im} = \frac{\frac{1}{N_s}\sum_{i=1}^{N_s} A(\mathbf{r}_i^N)\frac{p(\mathbf{r}_i^N)}{w(\mathbf{r}_i^N)}}{\sum_{i=1}^{N_s}\frac{p(\mathbf{r}_i^N)}{w(\mathbf{r}_i^N)}}. \tag{11.25}$$

In fact, they are actually the same. The difference is just appearance, because the umbrella sampling is a special type of importance sampling. This is because in Eq. (11.24), both $A(\mathbf{r}_i^N)$ and $w(\mathbf{r}_i^N)$ for each atom are calculated based on the quantum mechanical force that is the spatial gradient of the energy $E(\mathbf{r}^N) + V(\mathbf{r}^N)$, whereas in Eq. (11.25), both $A(\mathbf{r}_i^N)$ and $w(\mathbf{r}_i^N)$ are calculated based on the quantum mechanical force that is only the spatial gradient of the energy $E(\mathbf{r}^N)$, provided both calculations are done in the canonical ensemble MD.

This method is especially useful for multiscale quantum mechanics/MD (QM/MM) simulations, because both $(A(\mathbf{r}_i^N)$ and $w(\mathbf{r}_i^N)$ are calculated based on the quantum mechanical force acting on each atom on the fly, which is the spatial gradient of the energy $E(\mathbf{r}^N) + V(\mathbf{r}^N)$, and we do not store the information for $E(\mathbf{r}_i^N)$ during QM/MM calculations.

In fact, the total energy of the quantum system is expensive to calculate, whereas $w(\mathbf{r}_i^N)$ is usually a simple function such as quadratic polynomial, i.e., a harmonic potential, it has the form

$$w(\mathbf{r}_i^N) = \sum_{j=1}^{N_q} \frac{K}{2}(\mathbf{r} - \mathbf{r}_j^i)^2 H(\mathbf{r} - \mathbf{r}_j^i), \tag{11.26}$$

where the index i is the microstate and the index j represents the spatial location, and

$$H(\mathbf{r} - \mathbf{r}_j^i) = \begin{cases} 1, & |\mathbf{r} - \mathbf{r}_j^i| \le r_c \\ 0, & |\mathbf{r} - \mathbf{r}_j^i| > r_c \end{cases}, \tag{11.27}$$

where r_c is the cutoff radius. Usually, $N_q \ll N_s$ and $\mathbf{r}_j^i, j = 1, 2, \ldots, N_q$ are placed on an area or a path where we wish to get the sampled values. Thus, it is much easier to calculate comparing with $E(\mathbf{r}_i^N)$.

Moreover, by definition, the Helmholtz free energy of the biased system may be written as

$$\begin{aligned} F_q &= -\frac{1}{\beta}\ln Z_q = -\frac{1}{\beta}\ln\left\{\int w(\mathbf{r}^N)\exp(-\beta E(\mathbf{r}^N))d^N\mathbf{r}\right\} \\ &= -\frac{1}{\beta}\ln\left\{\int \exp(-\beta[E(\mathbf{r}^N) + V(\mathbf{r}^N)])d^N\mathbf{r}\right\} \\ &= -\frac{1}{\beta}\ln\left\{\int \exp(-\beta E(\mathbf{r}^N))d^N\mathbf{r}\right\} + V(\mathbf{r}^N)) = F_p + V(\mathbf{r}^N), \tag{11.28} \end{aligned}$$

which leads to the relation,

$$F_p = F_q - V(\mathbf{r}^N), \tag{11.29}$$

where F_p is the original Helmholtz free energy.

11.2 Markov Chain Monte Carlo Method

A necessary condition to be able to use the Monte Carlo integration method is to have an (iid) sequence of random variables for a given probability distribution. Thus, the key of the Monte Carlo method is to generate a sequence of (iid) random variables based on a given distribution. However, this is very difficult to do. Previously, we show how to generate random variables by using a PRNG, which are not true (iid) sequences of random variables even for uniform probability distributions.

One of the most powerful methods for sampling from a posterior distribution is the Markov chain Monte Carlo (MCMC) method. The MCMC method provides a tool to generate an almost independent but identical distributed sequence in the sense that a random variable is only related with its adjacent random variable but not the rest of other random variables. Such sequence of random variables is called a Markov chain.

The samples generated by MCMC are not truly independent of each other, but the distribution of the sample will approach to the posterior distribution, as the sample size $n \to \infty$, that is we can use them to find the unbiased estimates for expected values. Moreover, MCMC methods are generally fast, and we do not need to calculate or even estimate of the partition function Z.

11.2.1 Markov Chain

Before discussing the MCMC method, we first outline a few properties of the Markov chain.

A Markov chain, which was named after a Russian mathematician, Andrey Markov, is a stochastic mathematical system or model that undergoes transitions from one state to another state in the phase space. It is a random process that is characterized as memoryless, i.e., the next state depends only on the current state but not on the further back events that preceded it. This specific kind of "memorylessness" is called the Markov property. Even though a Markov chain is not an (iid) sequence, the correlation between two arbitrary states are almost zero except the adjacent pairs, which is the best that we can have (so far) in computational materials.

To illustrate the concept of the Markov chain, we consider a finite state space $\mathcal{S}$:

$$\text{Discrete state space: } \mathbf{r}^N_{(i)} \in \mathcal{S} = \left\{ \mathbf{r}^N_{(1)}, \mathbf{r}^N_{(2)}, \ldots, \mathbf{r}^N_{(N_s)} \right\}.$$

A discrete Markov chain is a sequence of random variables $\mathbf{X}_1, \mathbf{X}_2, \ldots, \mathbf{X}_n$ that have the following Markov property:

$$P_r\left(\mathbf{X}_n = \mathbf{r}^N \middle| \mathbf{X}_1 = \mathbf{r}^N_{(1)}, \mathbf{X}_2 = \mathbf{r}^N_{(2)}, \ldots, \mathbf{X}_{n-1} = \mathbf{r}^N_{(n-1)}\right) = P_r\left(\mathbf{X}_n = \mathbf{r}^N \middle| \mathbf{X}_{n-1} = \mathbf{r}^N_{(n-1)}\right),$$

where $P_r(\mathbf{X}_n = \mathbf{r}^N | \mathbf{X}_1 = \mathbf{r}^N_{(1)}, \mathbf{X}_2 = \mathbf{r}^N_{(2)}, \ldots, \mathbf{X}_{n-1} = \mathbf{r}^N_{(n-1)})$ is the standard conditional probability, whose existence requires $P_r(\mathbf{r}^N | \mathbf{X}_1 = \mathbf{r}^N_{(1)}, \mathbf{X}_2 = \mathbf{r}^N_{(2)}, \ldots, \mathbf{X}_{n-1} = \mathbf{r}^N_{(n-1)}) > 0$. We usually call the conditional probability, $P_r(\mathbf{X}_n = \mathbf{r}^N_{(i)} | \mathbf{X}_{n-1} = \mathbf{r}^N_{(j)})$, as the transition probability between two states $\mathbf{r}^N_{(j)}$ and $\mathbf{r}^N_{(i)}$, and we may denote it as $p_{ij} = P_r(\mathbf{X}_n = \mathbf{r}^N_{(i)} | \mathbf{X}_{n-1} = \mathbf{r}^N_{(j)})$. For a finite state space, the transition probabilities form a matrix

$$\mathbf{P}_t = \{p_{ij}\}.$$

The general idea of the Markov Chain Monte Carlo (MCMC) method is to form a Markov chain, which may become a substitute of the (iid) sequence of random variables, such that it may represent a trajectory of the system states, by which we can explore the state space in a manner by spending more time in the most important regions, i.e., where $p(\mathbf{r}^N)$ is large.

How can we do that?

The meaning of this question is twofold: (1) How can we form a Markov chain in computations so that it will have a unique stationary distribution, and (2) If we are given a probability distribution of a phase space, or maybe only some information of a (posterior) probability distribution of the phase space, can we form a Markov chain as the sample sequence that has the desired posterior probability distribution?

In this section, we first answer the first question. Some Markov chains have **stationary distributions** or invariant distributions, meaning that the probability distribution in the Markov chain remains unchanged as the length of the chain or the sample number of the chain increases, or as the time progresses.

In modelings and simulations, we always approximate a continuous phase space as a discrete state space. For the sake of argument, we only consider a discrete state space, by the way, our universe is quantum mechanical, which implies that it is associated with a discrete state space. Therefore, we may represent the probability distribution in a discrete state space as a row vector $\mathbf{P}$, whose entries are the probabilities of each state. Thus, the sum of all entries of $\mathbf{P}$ adds to 1. As a definition of the stationary probability distribution, for a given transition matrix $\mathbf{P}_t$ in the state space, the so-called stationary probability distribution satisfies the condition,

$$\mathbf{P} = \mathbf{P}\mathbf{P}_t.$$

Thus, if we can find the corresponding Markov chain, we have successfully drawn samples from the desired distribution $\mathbf{P}$.

To illustrate the process, we consider two examples of state space with only three states with all its transition probabilities displayed in Fig. 11.3.

In specific, for the state space in Fig. 11.3(b), one can easily find the transition matrix for the state space expressed as follows,

$$\mathbf{P}_t = [p_{ij}] = \begin{bmatrix} 0.3 & 0.3 & 0.4 \\ 0.2 & 0.3 & 0.5 \\ 0.5 & 0.4 & 0.1 \end{bmatrix}.$$

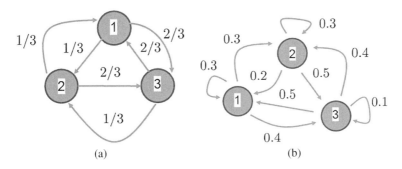

Figure 11.3 (a) A three-state irreducible Markov chain with all transition probabilities and (b) a three-state Markov chain with a uniform stationary probability distribution

Assume that the initial probability of the state space is

$$P_0 = [1, 0, 0].$$

Thus, we can keep making transition by multiplying the distribution vector with the transition matrix, i.e.,

$$\mathbf{P}_1 = \mathbf{P}_0 \mathbf{P}_t = [0.3000, 0.3000, 0.4000]$$
$$\mathbf{P}_2 = \mathbf{P}_1 \mathbf{P}_t = [0.3500, 0.3400, 0.3100]$$
$$\mathbf{P}_3 = \mathbf{P}_2 \mathbf{P}_t = [0.3280, 0.3310, 0.3410]$$
$$\mathbf{P}_4 = \mathbf{P}_3 \mathbf{P}_t = [0.3351, 0.3341, 0.3308]$$
$$\vdots$$
$$\mathbf{P}_8 = \mathbf{P}_7 \mathbf{P}_t = [0.3334, 0.3333, 0.3333]$$
$$\mathbf{P}_9 = \mathbf{P}_8 \mathbf{P}_t = [0.3333, 0.3333, 0.3333]. \tag{11.30}$$

From the calculations shown in Eq. (11.30), as iteration increases, it clearly shows that the Markov chain will reach to a uniform stationary probability distribution. Usually, this stationary distribution is what we would like to sample from.

Now, we still have two remaining questions:

(1) *Do all Markov chains can reach a stationary distribution?*
(2) *If a Markov chain has a stationary distribution, would this distribution be unique?*

The answer for the first question is NO. Only the *ergodic* Markov chains can have stationary distributions. Then what is an *ergodic* Markov chain?

DEFINITION 11.4 (Ergodic Markov chain) A state $\mathbf{r}_{(i)}^N$ is ergodic if it is *aperiodic* and positive *recurrent*. If all states in an *irreducible* Markov chain are ergodic, the Markov chain is ergodic.

To explain this definition, we need the following three concepts:

DEFINITION 11.5 (Reducibility) A Markov chain is irreducible if it is possible to start from any state to get to any state, or if all the states communicate with each other.

Both Markov chains are shown in Fig. 11.3 (a) and (b) are irreducible.

DEFINITION 11.6 (Recurrence)
- We say that a state, $r_{(i)}^N$, is transient if given that we start in a state, there is a likelihood or non-zero probability that we will never return to that state.
- On the other hand, a state, $r_{(i)}^N$, is recurrent if it is not transient.

As an example, the Markov chain in Fig. 11.4 has two communication classes: **Class 1** = $\{1,2,3\}$ and **Class 2** = $\{4,5,6\}$. Thus, this Markov chain is not irreducible. Moreover, among these two classes, Class 1 is transient, this is because if once we make a transition from state 1 to state 6, or from state 3 to state 4, we will never come back to Class 1 again. Whereas all the states in Class 2 are recurrent, because no matter which state you start from, you will have chances of coming back to all the states in Class 2 infinitely many times.

DEFINITION 11.7 (Aperiodicity) A state i has period k if any return to state i must occur in multiples of d time steps. Formally, the period of state i is defined as

$$d = \gcd\{n > 0 : \Pr(X_n = i \mid X_0 = i) > 0\},$$

where gcd denotes the greatest common divisor. When $d = 1$, the Markov chain is aperiodic.

Thus, if a Markov chain is aperiodic, when starting from a state i, we do not know (precisely) when we will return to the same state i after some transitions, and we return to the state i irregularly or aperiodically. For example, the Markov chain in Fig. 11.5 is periodic, and it has a periodicity $d = 3$. On the other hand, the Markov chain in Fig. 11.3(a) is aperiodic. For instance, if we consider the state 1: one departs from state 1 and returns back to it in $2, 3, 4, 5, 6, \ldots$. Hence, $d = 1$.

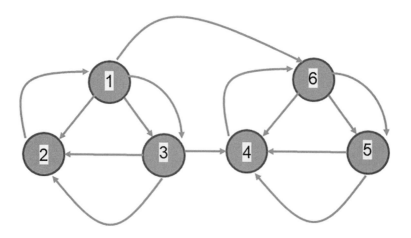

Figure 11.4 The Markov chain has two classes: **Class 1** = $\{1,2,3\}$ and **Class 2** = $\{4,5,6\}$

Therefore, one may say that in an aperiodic Markov chain, i.e., the state trajectories drawn from the transition does not have any periodic patterns, or one may say that an aperiodic Markov chain does not get trapped into any cycles.

In principle, a finite, aperiodic, and irreducible Markov chain $\{\mathbf{X}_n\}$ has a unique stationary probability distribution $P(\mathbf{r}^N)$. However, such Markov chain may not faithfully represent an equilibrium thermodynamic process. An important application of Monte Carol method to molecular simulations is its ability to mimic equilibrium ensemble MD with much large time steps. Therefore, we have to make sure that the Markov Chain Monte Carlo method used can generate desirable thermodynamics distribution, which requires the Monte Carlo simulations are conducted under specific thermodynamic equilibrium conditions. In order to do so, we are interested in a particular Markov chain: the reversible Markov chain.

DEFINITION 11.8 (Reversible Markov chain) *A Markov chain is reversible if for all pairs of states* $\mathbf{r}^N_{(i)}, \mathbf{r}^N_{(j)} \in \mathcal{S}^N$,

$$p(\mathbf{r}^N_{(i)}) \pi(\mathbf{r}^N_{(j)} | \mathbf{r}^N_{(i)}) = p(\mathbf{r}^N_{(j)}) \pi(\mathbf{r}^N_{(i)} | \mathbf{r}^N_{(j)}), \tag{11.31}$$

where $\pi(\bullet|\bullet)$ is the transition probability density. Equation (11.31) is known as the *detailed balance*.

For the reversible Markov chain, we state the following theorem without proof.

THEOREM 11.9 *An ergodic Markov chain in equilibrium state and satisfying the detailed balance condition Eq. (11.31) has a unique stationary distribution* $p(\mathbf{r}^N)$.

11.2.2 Metropolis–Hastings Method

To learn how to apply the Metropolis–Hastings method in MD simulations, we first give a popular receipt, i.e., a step-by-step instruction to implement it. Then, we discuss its justification and reasoning.

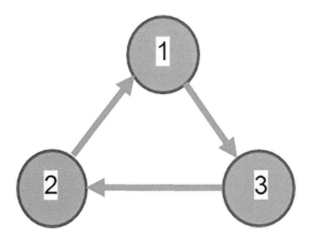

Figure 11.5 A periodic Markov chain with the periodicity $d = 3$

Metropolis–Hastings Algorithm for Molecular Dynamics

Step 0. Set the initial state for all the atoms $\mathbf{r}^N(0) = \{\mathbf{r}_i(0)\}_{i=1}^N$ and choose a starting atom position $\boldsymbol{\mu}_1 = \mathbf{r}_1(0)$ as the center of a cubic box with side length $\delta = 6\sigma$.

Choose the multivariate normal distribution $q = \mathcal{N}(\boldsymbol{\mu}, \sigma\mathbf{I})$ as the proposal distribution, and calculate the initial energy state $E(\mathbf{r}^N(0))$ and the posterior factor $p = \exp(-\beta E(\mathbf{r}^N(0)))$.

Repeat the following loop for $\ell = 1, 2, \ldots M$.

Repeat Step 1 to Step 3 for each atom $i = 1, 2, \ldots, N$ in each loop ℓ.

Step 1.
Let $\boldsymbol{\mu}_i = \mathbf{r}_i(\ell - 1)$; Use the proposal distribution $q = \mathcal{N}(\boldsymbol{\mu}_i, \sigma\mathbf{I})$ to draw a trial position for the ith atom $\mathbf{r}_{ci} = \mathbf{r}_i(\ell - 1) + \mathcal{N}(\boldsymbol{\mu}_i, \sigma\mathbf{I}) \to \mathbf{r}_c^N = \{\ldots, \mathbf{r}_{i-1}(\ell), \mathbf{r}_{ci}(\ell), \mathbf{r}_{i+1}(\ell - 1), \ldots\}$, and we then calculate trial energy state for the trial position, i.e., $E(\mathbf{r}_c^N)$;

Step 2.
Calculate acceptance probability

$$\rho = \frac{\exp(-\beta E(\mathbf{r}_c^N))|q(\mathbf{r}_i(\ell)|\mathbf{r}_{ci})|}{\exp(-\beta E(\mathbf{r}^N(\ell_{i-1})))|q(\mathbf{r}_{ci}|\mathbf{r}_i(\ell))|} = \frac{\exp(-\beta E(\mathbf{r}_c^N))}{\exp(-\beta E(\mathbf{r}^N(\ell_{i-1})))},$$

where $\mathbf{r}^N(\ell_{i-1}) = \{\ldots, \mathbf{r}_{i-1}(\ell), \mathbf{r}_i(\ell - 1), \mathbf{r}_{i+1}(\ell - 1), \ldots\}$ with $\mathbf{r}^N(\ell_1) = \mathbf{r}^N(\ell - 1)$.

Step 3.
If $\rho \geq 1$, accept the trial position: $\mathbf{r}_i(\ell) = \mathbf{r}_{ci}$ and $\mathbf{r}^N(\ell_i) = \mathbf{r}_c^N$.
If $\rho < 1$, draw a random variable $u \in (0, 1)$ of the uniform distribution.
If $\rho \geq u$, accept the trial position: $\mathbf{r}_i(\ell) = \mathbf{r}_{ci}$ and $\mathbf{r}^N(\ell_i) = \mathbf{r}_c^N$.
Otherwise ($\rho < u$), reject the trial position: $\mathbf{r}_i(\ell) = \mathbf{r}_i(\ell - 1)$ and $\mathbf{r}^N(\ell_i) = \mathbf{r}^N(\ell_{i-1})$.

Step 4.
Calculate the ensemble average

$$< A(\mathbf{r}^N) > = \frac{1}{MN} \sum_{j=1}^{MN} A_j(\mathbf{r}^N).$$

REMARK 11.10 1. Since the Gaussian distribution is symmetric, the proposal distribution term is canceled out when calculating the acceptance probability. This is the case for the original Metropolis algorithm.

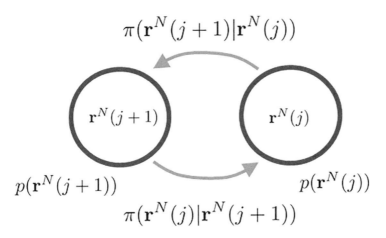

Figure 11.6 The detailed balance of transition probability between the state (i) and the state (i-1)

2. Because $\{\mathbf{r}_i(\ell)\}$ is an "almost iid," the same atomic configuration or microstates may occur or repeat during sampling process.

Now, we shall prove that the Markov chain generated by the Metropolis–Hastings algorithm satisfies the detailed balance condition, i.e.,

$$\frac{\pi(\mathbf{r}^N(j+1)|\mathbf{r}^N(j))}{\pi(\mathbf{r}^N(j)|\mathbf{r}^N(j+1))} = \frac{p(\mathbf{r}^N(j+1))}{p(\mathbf{r}^N(j))}, \tag{11.32}$$

as illustrated in Fig. 11.6. First, the Metropolis–Hastings acceptance probability is defined as

$$\rho[(j+1)|(j)] = \min\left(1, \frac{p(\mathbf{r}^N(j+1)q(\mathbf{r}^N(j)|\mathbf{r}^N(j+1))}{p(\mathbf{r}^N(j))q(\mathbf{r}^N(j+1)|\mathbf{r}^N(j))}\right) \tag{11.33}$$

or

$$\rho[(j)|(j+1)] = \min\left(1, \frac{p(\mathbf{r}^N_j)q(\mathbf{r}^N(j+1)|\mathbf{r}^N(j))}{p(\mathbf{r}^N(j+1))q(\mathbf{r}^N(j)|\mathbf{r}^N(j+1))}\right). \tag{11.34}$$

Combining Eqs. (11.33) and (11.34), we find that the ratio of two acceptance probability is

$$\frac{\rho[(j+1)|(j)]}{\rho[(j)|(j+1)]} = \frac{p(\mathbf{r}^N(j+1))q(\mathbf{r}^N(j)|\mathbf{r}^N(j+1))}{p(\mathbf{r}^N(j))q(\mathbf{r}^N(j+1)|\mathbf{r}^N(j))}. \tag{11.35}$$

The above equation may be rewritten as

$$\frac{\rho[(j+1)|(j)]q(\mathbf{r}^N(j+1)|\mathbf{r}^N(j))}{\rho[(j)|(j+1)]q(\mathbf{r}^N(j)|\mathbf{r}^N(j+1))} = \frac{p(\mathbf{r}^N(j+1))}{p(\mathbf{r}^N(j))}. \tag{11.36}$$

In the Metropolis–Hastings algorithm, we define and select the acceptance probability and the proposal probability such that the transition probability from the state $\mathbf{r}_{i-1}^N$ to the state $\mathbf{r}_i^N$ is:

$$\pi(\mathbf{r}^N(j+1)|\mathbf{r}^N(j)) := \rho[(j+1)|(j)]p_p(\mathbf{r}^N(j+1)|\mathbf{r}^N(j)). \qquad (11.37)$$

Substituting (11.37) into (11.36), we have the detailed balance equation as follows:

$$\frac{\pi(\mathbf{r}^N(j+1)|\mathbf{r}^N(j))}{\pi(\mathbf{r}^N(j)|\mathbf{r}^N(j+1))} = \frac{p(\mathbf{r}^N(j+1))}{p(\mathbf{r}^N(j))}. \qquad (11.38)$$

In the following, we show a segment of MATLAB script of a Monte Carlo NVT MD code for a given MCMC iteration ℓ from the above Steps 1–3:

```
sigma= 1/6;
for i = 1:natom
%
                % Suggest a trial move for each particle
                rtrial = r_old(:,i) + Delta* (normrnd(0,sigma, [3,1]);
%
                %periodic BC
                rtrial = PBC3D(rtrial,box);
%
                % Computing LJ potential energy associated with particle, i
                deltaE = LJ_EnergyChange(pos,rtrial,i,rcutoff,box);

                % Metropolis algorithm, check random number:
                if (rand < exp(-beta*deltaE))
                    % Accept atom displacement
                    r_new(:,i) = rtrial;
                    energy = energy + deltaE;
                else
                    r_new(:,i) = r_old(:,i);
                end
%
end
```

Note that $\mathbf{r}^N(\ell) = r_{new}(:,i)$ and $\mathbf{r}^N(\ell-1) = r_{old}(:,i)$. The reason for choice of $\sigma = 1/6$ is that $6\sigma = 1$, so that the normal distribution almost cover the entire side length of the cubic box for each dimension.

MCMC Example: Random Walk Proposal Distribution

In this example, the proposal distribution is generated by using a random walk or displacement trial move specification.

1. Consider a unit cell with oN atoms.
2. Random walk displacement trial:

Repeat the following steps $k = 1, 2, \ldots, N_s$ for atoms $i = 1, \ldots, N$ ($N_s = L \times N \gg N$):

Each atom is enclosed in a cubic box of dimension $\delta \times \delta \times \delta$, and it can randomly walk within the box in any directions with uniform distribution to reach a trial configuration as shown in Fig. 11.7.

3. For the canonical ensemble Monte Carlo simulations, we choose the posterior distribution as

$$p(\mathbf{r}_k^N) = \frac{1}{Z_N} \exp(-\beta E(\mathbf{r}_k^N)).$$

Once reaching a trial configuration, calculate the ratio of the probabilities for the new and old configurations, according to the Metropolis algorithm

$$\rho = \min\left(1, \frac{\exp(\beta E(\mathbf{r}_{trial}^N))}{\exp(-\beta E(\mathbf{r}_k^N))}\right).$$

4. When $\rho \geq 1$ accept the trial configuration: $\mathbf{r}_{k+1}^N = \mathbf{r}_{trial}^N$.
5. When $\rho \geq 1$, we draw a uniform random variable $u \in (0, 1)$ and if $\rho \geq u$ accept the trial configuration: $\mathbf{r}_{k+1}^N = \mathbf{r}_{trial}^N$.
6. If $\rho \leq u$ reject the trial configuration: $\mathbf{r}_{k+1}^N = \mathbf{r}_k^N$.
7. Calculate ensemble average for a given field $\mathbf{A}(\mathbf{r}^N)$,

$$<\mathbf{A}> = \frac{1}{N_s} \sum_{i=1}^{N_s} \mathbf{A}_i,$$

in which many microstates I are repeated.

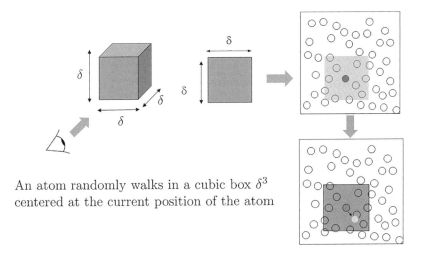

An atom randomly walks in a cubic box δ^3 centered at the current position of the atom

Figure 11.7 Random walk: Randomly displace a selected atom in a cubic box of dimension $\delta \times \delta \times \delta$ that is centered at the current position of the atom.

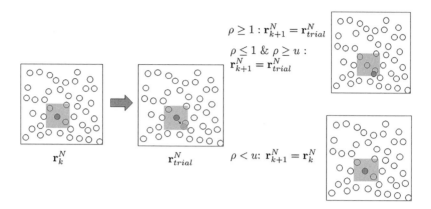

Figure 11.8 MCMC random walk algorithm

```
e=[1,1,1]';

for i = 1:natom
%
                % Suggest a trial move for each particle
                rtrial = pos(:,i) + deltaR*(rand(3,1)-0.5*e);
                rtrial = PBC3D(rtrial,box);
                % Computing energy associated with particle, i
                deltaE = LJ_EnergyChange(pos,rtrial,i,rcutoff,box);

                % Metropolis algorithm, check random number:
                if (rand < exp(-beta*deltaE))
                        % Accept atom displacement
                        pos(:,i) = rtrial;
                        energy = energy + deltaE;
                end
```

Note that in computations or in the computer code, the step

$$\rho \geq 1 : \mathbf{r}_{k+1}^N = \mathbf{r}_{trial}^N \text{ and } \rho \leq 1 \,\&\, \rho \geq u : \mathbf{r}_{k+1}^N = \mathbf{r}_{trial}^N$$

and the step

$$\rho \leq u : \mathbf{r}_{k+1}^N = \mathbf{r}_k^N$$

can be combined as a single step equivalently,

$$\rho \geq u : \mathbf{r}_k^N = \mathbf{r}_{trial}^N,$$

and thus $\mathbf{r}_k^N$ automatically becomes $\mathbf{r}_{k+1}^N$ without creating a new array to store $\mathbf{r}_{k+1}^N$. The MCMC random walk algorithm is summarized in Fig. 11.8.

The Metropolis–Hastings is very powerful and widely used. It reduces the problem of sampling of a difficult distribution $p(\mathbf{r}^N)$ to making proposals $p_p(\mathbf{r}_{i+1}^N | \mathbf{r}_i^N)$ and evaluating ratios of $p(\mathbf{r}_{i+1}^N)/p(\mathbf{r}_i^N)$.

MCMC Example Solution for Two-Dimensional Ising Model

Consider a finite size square lattice with $n_x \times n_y = n \times n$ nodes, i.e., the total number of nodes is: $N = n_x \times n_y = n \times n$. At each node, the magnetic dipole moment of the atomic spin is represented by the variable $\sigma_i = \pm 1, i = 1, 2, \ldots, N$.

If we only consider the nearest neighbor interaction (see Fig. 11.9), the total energy of the square lattice will be

$$E = \sum_i E_i = -\frac{J}{2} \sum_{i=1}^{N} \sum_{j(i)=1}^{4} \sigma_i \sigma_{j(i)}, \qquad (11.39)$$

where J is the intensity of the interaction of adjacent spins and $j(i) = 1, 2, 3, 4$ are nearest neighbors for a fixed node i. Note that the above energy expression implies that the appropriate periodic boundary condition in taken into account.

Thus, the set of total microstates is $\mathcal{M} = \{\sigma_1, \sigma_2, \ldots, \sigma_N\}_{\sigma_i = \pm}$. It has total number of 2^N microstates, i.e., $\{\mathcal{M}(k), k = 1, 2, 3, \ldots, 2^N\}$. For a given microstate $\mathcal{M}(n) = \{\sigma_1, \sigma_2, \ldots, \sigma_N\}$, we select the node i and flip its spin direction. The energy change will be

$$\Delta E = J\sigma_i \sum_{j(i)=1}^{4} \sigma_{j(i)},$$

- If $\Delta E \leq 0 \rightarrow \rho = \exp(-\beta \Delta E) \geq 1$, accept the flip: $\sigma_i(n+1) = \sigma_i(n)$; $\mathcal{M}(n+1) = \{\sigma_1(n), \ldots, \sigma_i(n+1), \ldots, \sigma_N(n)\}$;
- If $\Delta E \geq 0 \rightarrow \rho = \exp(-\beta \Delta E) \leq 1$, draw a uniform random number $u \in (0, 1)$:
  (i) If $\rho \geq u$, accept the flip: $\sigma_i(n+1) = \sigma_i(n) \rightarrow \mathcal{M}(n+1) = \{\sigma_1(n), \ldots, \sigma_i(n+1), \ldots, \sigma_N(n)\}$;
  (ii) If $\rho < u$, reject the flip: $\sigma_i(n+1) = \sigma_i(n) \rightarrow \mathcal{M}(n+1) = \mathcal{M}(n)\}$.

Example 11.11 In this example, we employ the Metropolis Monte Carlo method to calculate the magnetization of 2D Ising model.

We assume that the following conditions are satisfied and the following procedures are used:

(i) No external magnetic field is used in the simulation ($B = 0$);
(ii) Periodic boundary condition is applied;
(iii) Reduced unit (dimensionless unit) is used in the simulation;
(iv) Different lattice sizes are used in the simulation in order to observe the size effect (5×5, 10×10, 20×20, and 50×50);
(v) Quantities are reported per spin (normalized by $L \times L$);
(vi) Simulations are carried below and above the critical temperature, $T \in (1, 3.56) J/k_B$.

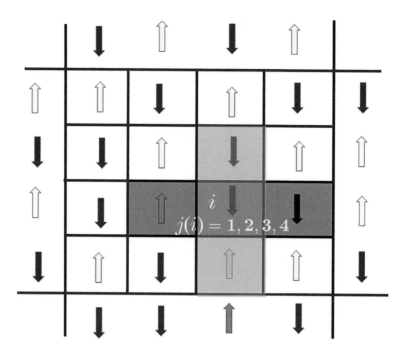

Figure 11.9 A two-dimensional Ising model on a square lattice box with periodic boundary condition

Interactions are all the same strength and are only between the spins on the sites that are the nearest neighbors on the lattice.

Repeat the MC steps for $(L \times L)$ times = 1 sweep of lattice is completed! Sample the energy and magnetization of the system at each sweep of the lattice

$$E/J = -\sum_{<i,j>} \sigma_i \sigma_j - B \sum_i \sigma_i.$$

Magnetization:

$$M = \sum_i \sigma_i.$$

Simulations are run up to 5,000 sweeps.

The average magnetization is calculated based on the following formula:

$$<M> = \frac{1}{N} \sum_{i=1}^{4000} M_i,$$

where the integer i is the sweep index, and in this example the total effect number of sweeps (independent trials) $N = 40,000$. The magnetization is calculated as follows:

$$\text{Magnetization} = <M> \pm \sigma_M$$

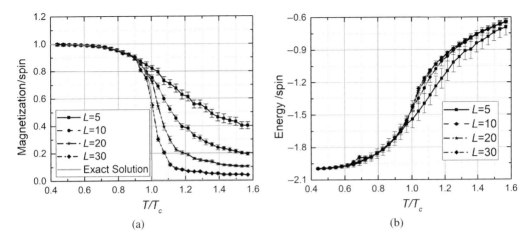

Figure 11.10 (a) Predicted magnetization (per spin) with different temperature values and different lattice sizes and (b) predicted energy (per spin) with different temperature values and different lattice sizes.

and

$$\sigma_M = \sqrt{\frac{1}{N-1}\left(<M>^2 - <M^2>\right)^2},$$

which is shown in Fig. 11.10(a). For the case of $B = 0$, the analytical solution is found by Onsager (1944)

$$<M> = \left(1 - \left[\sinh(2\beta J)\right]^{-4}\right)^{1/8},$$

where $\beta = 1/(k_B T)$.

Similarly, one can calculate the energy per spin as well. The results are shown in Fig. 11.10(b).

In Fig. 11.11(a), we displayed the calculated heat capacity based on the Monte Carlo simulation, and the heat capacity is calculated based on the formula:

$$C = \frac{\partial E}{\partial T} = \frac{1}{k_B T^2}\left(<E^2> - <E>^2\right).$$

We can also use the simulation results to calculate magnetic susceptibility based on the following formula,

$$\chi = \frac{1}{k_B T}\left(<M^2> - <M>^2\right),$$

and the results are presented in Fig. 11.11(b).

Last, we show the predicted magnetic phase field or phase separation in Fig. 11.12. At low temperatures, spins align with their neighboring spins (ferromagnetism). At high temperatures, aligned spin domains will become small and fluctuate (paramagnetism), and the temperature separating these two phases is called critical temperature, T_c, which is given as

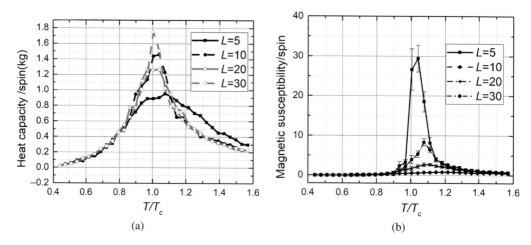

Figure 11.11 (a) Predicted heat capacity per spin with different temperature and different lattice sizes and (b) predicted susceptibility (per spin) with different temperature and differen lattice sizes

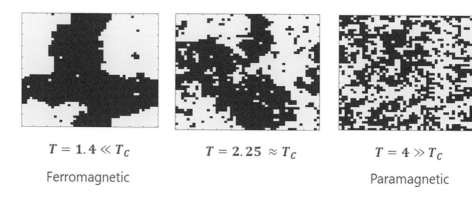

$T = 1.4 \ll T_C$ $T = 2.25 \approx T_C$ $T = 4 \gg T_C$

Ferromagnetic   Paramagnetic

Figure 11.12 Phase separation at different temperatures ($T_c = 2.2692 J/k_B$)

$$T_c = \frac{2J}{k_B \log(1 + \sqrt{2})} \approx 2.2692 \, J/k_B.$$

In the following, we list the main part of a MATLAB code for the above Metropolis Monte Carlo calculation:

```
%The Main MC Runs

N = 5000;    %Number of Sweeps (1 sweep = L^2 perturbations)
             %Number of MC Steps = N*L^2

for T = 1:0.08:3.56    %33 temperatures, 16 below and 16 above crit temp
    for L = [5,10,20,50] % 4 different lattice lengths

    % Initialization of the square lattice
    lattice = randi([0,1],L)*2-1;
```

```
    results = zeros(N,2);
       % Total MC Steps
       for m = 1:N
           % One Sweep
           for k=1:L^2
               % Pick a random site (center atom)
               i = randi([1,L]);
               j = randi([1,L]);
               center = lattice(i,j);
               % Nearest neighbors of the center atom (using periodic boundaries)
               up = lattice(mod(i-2,L)+1,j);
               down = lattice(mod(i,L)+1,j);
               left = lattice(i,mod(j-2,L)+1);
               right = lattice(i,mod(j,L)+1);
               % Energy Difference when a site is flipped
               dE = 2*center*(up+down+left+right);
               %Probability of acceptance
               prob = exp(-dE/T);
               if dE < 0 || (prob > rand)    %Accept the flip
                   lattice(i,j) = - lattice(i,j);
               end
           end
           % Total Energy
           E = 0;
           for i=1:L
               for j = 1:L
                   E = E-lattice(i,j)*(lattice(i,mod(j,L)+1)+lattice(mod(i,L)+1,j));
               end
           end

       % Total Magnetization

       M = sum(sum(lattice));

       % Store Energy and Magnetization after each Sweep
       results(m,1) = E;
       results(m,2) = M;
       end
    filename = ['MC_Results_L_' num2str(L) '_T_' num2str(T) '.xlsx'];
    xlswrite(filename,results)
    end
end
```

Previously, we have illustrated several computation examples of MCMC method and its applications. In general, in MCMC calculations, we do not need to know precisely the given probability distribution, but only the ratio $p(\mathbf{r}_{i+1}^N)/p(\mathbf{r}_i^N)$, so that we do not have to deal with partition function Z. In fact, in the language of statistics, if we wish to sample a posterior distribution, we only need some data or the likelihood function, which may not be the exact or complete probability distribution information.

In summary, the MCMC methods are a class of algorithms for sampling from a probability distribution constructed from a trajectory of a Markov chain that has the desired distribution as its equilibrium distribution. MCMC can create samples from a continuous random variable, with probability density proportional to a known function (up to a constant). These samples can be used to evaluate an integral over that variable, as its expected value or variance. In the following, we discuss its most common form that is extensively used in MD simulation.

11.3 Hamiltonian (Hybrid) Monte Carlo Method

As discussed in the previous sections, typical MCMC proposal distributions are constructed or implemented based on random walk type of algorithms, which have very slow pace to explore the phase space. Moreover, it does not consider kinetic energy. However, if one adds the kinetic energy into formulation, which is equivalent to adding a gradient bias, because of the quantum kinetic energy $K \sim -(\hbar^2/2m))\nabla^2$, and consequently, this will destroy detailed balance.

We can restore detailed balance by introducing fictitious state in the form of momentum variables. Now the state space is $\mathbf{r}^N$ (called position) and $\mathbf{p}^N$ (called momentum). We augment the energy function as well. We recover the distribution of $\mathbf{r}^N, p(\mathbf{r}^N)$ by forgetting the momentum coordinates.

Let $\mathbf{r}$ be position, $\mathbf{p}$ be the momentum, and H is the system's Hamiltonian. Then, the Hamiltonian equations of the system read as

$$\frac{d\mathbf{r}_i}{dt} = \frac{\partial H}{\partial \mathbf{p}_i}, \quad \text{and} \quad \frac{d\mathbf{p}_i}{dt} = -\frac{\partial H}{\partial \mathbf{r}_i},$$

where

$$H = \sum_{i=1}^{N} \frac{\mathbf{p}_i \cdot \mathbf{p}_i}{2m_i} + V(\mathbf{r}).$$

The probability for the "momentum" $\mathbf{p}$ is

$$p(\mathbf{p}) = \frac{1}{Z} \exp(-\beta K(\mathbf{p})), \quad \text{where } K(\mathbf{p}) = \frac{1}{2}\mathbf{p}^T \mathbf{M}^{-1} \mathbf{p}.$$

The probability for the position is

$$p(\mathbf{r}) = \frac{1}{Z} \exp(-\beta V(\mathbf{r}))$$

and

$$p(\mathbf{r},\mathbf{p}) = \frac{1}{Z} \exp(-\beta H(\mathbf{r},\mathbf{p})) = \frac{1}{Z} \exp(-\beta(K(\mathbf{p}) + V(\mathbf{r}))).$$

For a Hamiltonian system, even as the potential and kinetic energies oscillate, the Hamiltonian is a conserved quantity. The corresponding joint distribution is invariant to

$$p(\mathbf{r},\mathbf{p}) \propto \exp(-\beta(K(\mathbf{p}) + V(\mathbf{r}))),$$

which is not ergodic. The way to resolve this issue is randomizing the momenta, which is easy because $\mathbf{r}^N$ and $\mathbf{p}^N$ are independent, and their distributions are Gaussian.

Therefore, Hamiltonian Monte Carlo (HMC) method consists of two MCMC steps:

1. Randomize the momenta.
2. Simulate the dynamics, starting with these momenta.

In the following, we shall see a detailed formulation of Hamiltonian Monte Carlo.

Hamiltonian Monte Carlo Algorithm

Step 1. The initial momenta are drawn from a (zero-mean) Gaussian distribution at inverse temperature $\beta = (k_B T)^{-1}$,

$$p(\mathbf{p}) = \frac{1}{Z_0} \exp(-\beta K(\mathbf{p})), \text{ where } K(\mathbf{p}) = \frac{1}{2}\mathbf{M}^{-1}\mathbf{p}^T\mathbf{p},$$

and $Z_0 = (2\pi)^{-d/2} det(\mathbf{M}/\beta)^{-1/2}$.

Step 2. Integrate the Hamiltonian by using the leapfrog algorithm with time duration $L \times (\Delta t)$,

$$\mathbf{v}_i(t + \Delta t/2) = \mathbf{v}_i - \frac{\partial V}{\partial \mathbf{x}}\bigg|_{\mathbf{x}(t)} \frac{\Delta t}{2} \quad (11.40)$$

$$\mathbf{x}_i(t + \Delta t) = \mathbf{x}_i(\Delta t) + \mathbf{v}_i(t + \Delta t/2)\Delta t \quad (11.41)$$

$$\mathbf{v}_i(t + \Delta t) = \mathbf{v}_i(t + \Delta t/2) - \frac{\partial V}{\partial \mathbf{x}}\bigg|_{\mathbf{x}(t+\Delta t)} \frac{\Delta t}{2}. \quad (11.42)$$

Step 3. Accept or reject the position by using the Metropolis–Hastings acceptance probability

$$\rho = \min\left[1, \exp(-\beta(H(\mathbf{p}',\mathbf{x}') - H(\mathbf{p},\mathbf{x}))\right].$$

In summary, each iteration of HMC proceeds in two steps:

- Sample momentum $\mathbf{p}^N$ from a zero-mean normal distribution.
- Do L times.
- Propose a new state $(\mathbf{r}^N, \mathbf{p}^N)$ using Hamiltonian dynamics.
- Accept or reject the new state using the M-H test based on the energy difference ΔH.

Note that momentum is only preserved through the steps in the inner loop over L. In the following, we present a MATLAB code for one particle HMC sampling example with bivariate Gaussian distribution.

```
HYBRID MONTE CARLO SAMPLING -- BIVARIATE NORMAL
rand('seed',12345);
randn('seed',12345);
```

```
% STEP SIZE
delta = 0.3;
nSamples = 1000;
L = 20;

% DEFINE POTENTIAL ENERGY FUNCTION
U = inline('transpose(x)*inv([1,.8;.8,1])*x','x');

% DEFINE GRADIENT OF POTENTIAL ENERGY
dU = inline('transpose(x)*inv([1,.8;.8,1])','x');

% DEFINE KINETIC ENERGY FUNCTION
K = inline('sum((transpose(p)*p))/2','p');

% INITIAL STATE
x = zeros(2,nSamples);
x0 = [0;6];
x(:,1) = x0;

t = 1;
while t < nSamples
    t = t + 1;

    % SAMPLE RANDOM MOMENTUM
    p0 = randn(2,1);
%% SIMULATE HAMILTONIAN DYNAMICS
    % FIRST 1/2 STEP OF MOMENTUM
    pStar = p0 - delta/2*dU(x(:,t-1))';

    % FIRST FULL STEP FOR POSITION/SAMPLE
    xStar = x(:,t-1) + delta*pStar;

    % FULL STEPS
    for jL = 1:L-1
        % MOMENTUM
        pStar = pStar - delta*dU(xStar)';
        % POSITION/SAMPLE
        xStar = xStar + delta*pStar;
    end

    % LAST HALP STEP
    pStar = pStar - delta/2*dU(xStar)';
    pStar = - pStar
```

```
    % The Negative Momentum here may
    % make the proposal distribution symmetric.
    % HOWEVER WE THROW THIS AWAY FOR NEXT
    % SAMPLE, SO IT DOESN'T MATTER

    % EVALUATE ENERGIES AT
    % START AND END OF TRAJECTORY
    U0 = U(x(:,t-1));
    UStar = U(xStar);

    K0 = K(p0);
    KStar = K(pStar);

% ACCEPTANCE/REJECTION CRITERION
    alpha = min(1,exp((U0 + K0) - (UStar + KStar)));

    u = rand;
    if u < alpha
        x(:,t) = xStar;
    else
        x(:,t) = x(:,t-1);
    end
end
```

Note that the MATLAB command $randn(2,1)$ return a 2 by 1 of the standard normal random vector. In HMC, one may use the following script to draw random momenta:

```
d=2
mu=[0,0]';
M=[2,0.0;0.0,2];
sigma=M/beta;
p = mvnrand (mu, sigma, d);
```

We have one important question on the legitimacy and justification of HMC to generate a Markov chain with a stationary probability distribution.

How Can We Achieve the Detailed Balance in HMC?
In each integration cycle of HMC, the displacement integration is deterministic, and we can only draw samples of **p**. Thus, if we use the symplectic integrator in time integration, the total probability increment in the phase space should be conserved:

$$p_t(\mathbf{r}'|\mathbf{r},\ell)d\mathbf{r}' = p(\mathbf{p})d\mathbf{p} \quad \to \quad p_t(\mathbf{r}'|\mathbf{r}) = p(\mathbf{p})\frac{\partial \mathbf{p}}{\partial \mathbf{r}'}.$$

Note that the argument ℓ is the transition probability $p_t(\cdot | \cdot ,\ell)$ indicates that this transition probability is in loop ℓ.

On the other hand, if the symplectic integration is used, Hamiltonian or total energy of the system is conserved, i.e.,

$$H(\mathbf{r},\mathbf{p}) = H(\mathbf{r}',\mathbf{p}').$$

Moreover, the volume of phase space is conserved,

$$d\mathbf{r}d\mathbf{p} = d\mathbf{r}'d\mathbf{p}' \rightarrow \frac{\partial \mathbf{p}}{\partial \mathbf{r}'} = \frac{\partial \mathbf{p}'}{\partial \mathbf{r}}.$$

From the above results, we have

$$p(\mathbf{r})p_t(\mathbf{r}'|\mathbf{r},L) = p(\mathbf{r})p(\mathbf{p})\frac{\partial \mathbf{p}}{\partial \mathbf{r}'} = \exp(-H(\mathbf{r},\mathbf{p}))\frac{\partial \mathbf{p}}{\partial \mathbf{r}'}$$

$$= \exp(-H(\mathbf{r}',\mathbf{p}'))\frac{\partial \mathbf{p}}{\partial \mathbf{r}'}$$

$$= p(\mathbf{r}')p(\mathbf{p}')\frac{\partial \mathbf{p}'}{\partial \mathbf{r}}$$

$$= p(\mathbf{r}')p_t(\mathbf{r}|\mathbf{r}',\ell).$$

REMARK 11.12 The take-home message is: HMC only works if we use a symplectic time integration algorithm.

12 Langevin Equations and Dissipative Particle Dynamics

In this chapter, we discuss the theory and computation of Langevin equation and a related coarse-grained (CG) model – the dissipative particle dynamics (DPD).

12.1 Langevin Equation

In molecular simulations, a common tool to describe a stochastic process is to use the so-called Langevin equation, which was proposed by Paul Langevin (Fig. 12.1) to guarantee that the corresponding dynamics equations can generate a canonical ensemble. The formal form of Langevin equation can be expressed as follows:

$$m\frac{d\mathbf{v}_i}{dt} = \mathbf{F}_i - \alpha \mathbf{v}_i + \delta \mathbf{R}_i, \quad i = 1, 2, \ldots, N, \tag{12.1}$$

where the symbol δ denotes fast fluctuations, and

(i) $\mathbf{F}_i$ – interatomic or intermolecular force;
(ii) $\mathbf{F}_{di} = -\alpha \mathbf{v}_i$ – dissipative (drag) force;
(iii) $\delta \mathbf{R}_i$ – fast fluctuation force.

In the already-discussed three forces, we have studied intermolecular force before. In the following, we shall focus on the dissipative force and the fluctuation (random) force, and thus we may rewrite

$$m\frac{d\mathbf{v}_i}{dt} = -\alpha \mathbf{v}_i + \delta \mathbf{R}_i, \tag{12.2}$$

where the dissipative force $\alpha \mathbf{v}_i$ may be thought as the friction of an effective background medium (fluid or solvent), i.e., the drag force. By imitating the friction coefficient of a sphere of radius a in a Newtonian fluid, the friction coefficient of the Langevin particles are set in the following form,

$$\alpha = 6\pi \eta a, \tag{12.3}$$

where η is viscosity of the background "fluid" and a is the radius of the particle. On the other hand, the fluctuation force $\delta \mathbf{R}_i$ is a random force, which is assumed to have the following properties:

Figure 12.1 Paul Langevin (January 23, 1872–December 19, 1946) was a French physicist who developed Langevin dynamics and the Langevin equation (Photo courtesy of Henri Manuel and Wikipedia.org)

1. White noise $<\delta \mathbf{R}_i(t)> = 0$;
2. Uncorrelated with prior velocity: $<\mathbf{v}_i(0) \cdot \delta \mathbf{R}_j(t)> = 0$;
3. Markov process: $<\delta \mathbf{R}_i(t) \cdot \delta \mathbf{R}_j(\tau)> = C\delta(t-\tau)\delta_{ij}$;
4. White noise Gaussian distribution: $p(\mathbf{R}) = \dfrac{1}{\sqrt{2\pi}\sigma} \exp\left(-\dfrac{\mathbf{R}}{2\sigma^2}\right)$,

where

$$<f> := \int_V f(\mathbf{x})p(\mathbf{x})dV$$

is the ensemble average and $p(\mathbf{x})$ is the probability distribution. Thus condition (4) implies condition (1), because the mean of the white noise Gaussian distribution is zero, i.e., $\mu = 0$. We note in passing that the Markov process is a special memoryless stochastic process, such that the presence is dependent on the immediate past, but the future is independent from the past. Moreover, the third condition also states that the random forces acting on different particles are not correlated.

The reason why Langevin dynamics is able to generate a canonical ensemble is that one may view the effective background environment or "solvent" as the thermal reservoir. The interaction between the particles and thermal reservoir is controlled by both the "drag force" and the random force, which can be adjusted by the friction coefficient α or the amplitude of the random force σ. However, the viscous force alone is not able to generate a canonical ensemble system. To illustrate this point, we first let the random force being zero, and then in the one-dimensional (1D) case, the Langevin equation without fluctuation has the form,

$$m\frac{dv}{dt} = -\alpha v(t). \tag{12.4}$$

The solution of the above equation is,

$$v(t) = v(0)\exp(-2\alpha t/m), \tag{12.5}$$

and hence

$$<v^2> = <v(0)^2 \exp(-\alpha t/m)> \to 0, \text{ as } t \to \infty. \tag{12.6}$$

If we recall that in an NVT ensemble molecular dynamics (MD), we have

$$<\frac{m}{2}v^2> = \lim_{t\to\infty}\frac{1}{t}\int_{t_0}^{t_0+t}\frac{m}{2}v^2(\tau)d\tau = \frac{m}{2}<v^2(t)>_{NVT} = \frac{1}{2}k_BT, \tag{12.7}$$

which leads to

$$<v^2> = k_BT/m \neq 0.$$

Now one can see that without the random force, the viscous force alone cannot generate a valid canonical ensemble.

On the other hand, not every pair of drag force and random force can be put together to generate a canonical ensemble. In order to produce or simulate a canonical ensemble, the dissipative force and the random force have to satisfy certain relationship. First, we let $\delta \mathbf{R}_i = \sigma \boldsymbol{\theta}_i(t)$, where

$$<\theta_i^2> = 1,$$

and $\boldsymbol{\theta}_i(t)$ are dimensionless random variable vectors with the standard Gaussian distribution. Thus, the Langevin equation can be put into the form,

$$m_i\ddot{\mathbf{r}}_i = -\alpha\dot{\mathbf{r}}_i + \sigma\boldsymbol{\theta}_i(t), \tag{12.8}$$

where the friction coefficient and the random force coefficient are connected by the requirement

$$<\delta\mathbf{R}_i(t)\delta\mathbf{R}_j(\tau)> = \delta_{ij}\delta(t-\tau)2dk_BT\alpha, \tag{12.9}$$

where T is the system's temperature and d is the space dimension.
Equation (12.9) leads to

$$\sigma^2<\boldsymbol{\theta}_i(t)\boldsymbol{\theta}_j(\tau)> = 2d\alpha k_BT\delta_{ij}\delta(t-\tau). \tag{12.10}$$

Since by design, $\boldsymbol{\theta}_i$ have the standard Gaussian distribution

$$<\boldsymbol{\theta}_i(t)\boldsymbol{\theta}_j(\tau)> = \delta_{ij}\delta(t-\tau)$$

and hence
$$\sigma^2 = 2d\alpha k_B T, \tag{12.11}$$
where d is the space dimension.

12.1.1 Statistical Mechanics of the Langevin Equation

In this section, we discuss the statistical mechanics characters of Langevin dynamics. For simplicity, we only examine the 1D Langevin equation without intermolecular force.

First, we consider the following single particle Langevin equation with a random force $\delta R(t)$,
$$m\frac{dv}{dt} = -\alpha v(t) + \delta R(t). \tag{12.12}$$

As mentioned earlier, the random force is a white noise of the Markov process (uncorrelated), i.e.,
$$<\delta R(t)> = 0, \quad \text{and} \quad <\delta R(t)\delta R(\tau)> = \sigma^2 \delta(t-\tau). \tag{12.13}$$

The particular solution of Eq. (12.12) is
$$v(t) = v(0)\exp(-\alpha t/m) + \int_0^t d\tau \exp(-\alpha(t-\tau)/m)\delta R(\tau) \tag{12.14}$$

or
$$v(t) = v(0)\exp(-\xi t) + \int_0^t d\tau \exp(-\xi(t-\tau))\delta R(\tau), \tag{12.15}$$

where $\xi = \alpha/m$.

We can then calculate the kinetic energy of the system as follows:
$$<v(t) \cdot v(t)> = <v(0) \cdot v(0)> \exp(-2\xi t)$$
$$+ \frac{2}{m}\int_0^t d\tau \exp(-\xi(2t-\tau)) <v(0)\delta R(\tau)>$$
$$+ \frac{1}{m^2}\int_0^t d\tau' \int_0^t d\tau \exp(-\xi(2t-\tau-\tau')) <\delta R(\tau) \cdot \delta R(\tau')>.$$

Since that the initial velocity field is not correlated with the random force field, we have
$$<v(0)\delta R(\tau)> = 0, \quad \forall \tau > 0.$$

Moreover, by definition we have
$$<\delta R(\tau) \cdot \delta R(\tau')> = \sigma^2 \delta(\tau-\tau'). \tag{12.16}$$

Hence, we can obtain the result of the following integration:
$$\int_0^t d\tau' \int_0^t d\tau \exp(-\xi(2t-\tau-\tau')) <\delta R(\tau)\delta R(\tau')> = \frac{\sigma^2}{2\xi}(1-\exp(-2\xi t)). \tag{12.17}$$

Calculating the kinetic energy of the system, we find that

$$<v(t)v(t)> = \frac{k_BT}{m} = <v(0)v(0)> \exp(-2\xi t) + \frac{\sigma^2}{2\xi m^2}(1-\exp(-2\xi t)]. \quad (12.18)$$

Since

$$m <v(t)v(t)> = k_BT \quad \rightarrow \quad <v(t)v(t)> = \frac{k_BT}{m},$$

we let $t \rightarrow \infty$, and we then identify that

$$\sigma^2 = 2k_BTm\xi = 2k_BT\alpha, \quad (12.19)$$

which is exactly the $\sigma - \alpha$ relation shown in Eq. (12.11) for $d = 1$. We then recover the requirement or relation between the random force and the viscous force as

$$<\delta R(t)\delta R(\tau)> = 2k_BT\alpha\delta(t-\tau). \quad (12.20)$$

Furthermore, Eq. (12.20) allows us to calculate the friction coefficient of the drag force from random force as

$$\alpha = \frac{1}{2k_BT}\int_{-\infty}^{\infty} <R(t)R(\tau)> dt. \quad (12.21)$$

Equation (12.21) is essentially the relation between dissipation and fluctuation, and it is a special case of the dissipation–fluctuation theorem of statistical thermodynamics. Even though, this is not a proof of fluctuation–dissipation theorem, but Langevin equation is constructed in such way that it reflects the fluctuation–dissipation theorem.

The above results can be easily generated into three-dimensional (3D) cases. For instance, Eq. (12.18) can be readily extended into the 3D case as,

$$<\mathbf{v}(t) \cdot \mathbf{v}(t)> = 3\frac{k_BT}{m} = <\mathbf{v}(0) \cdot \mathbf{v}(0)> \exp(-2\xi t) + \frac{\sigma^2}{2\xi m^2}(1-\exp(-2\xi t)], \quad (12.22)$$

and let $t \rightarrow \infty$, we recover the temperature–kinematic energy relation in the canonical NVT ensemble,

$$<(\mathbf{v}(t))^2> = 3\frac{k_BT}{m} \quad \rightarrow \quad E_{kin} = \frac{3}{2}k_BT.$$

Moreover, we can readily extend Eq. (12.20) into the 3D case as

$$<\mathbf{R}_i(t) \cdot \mathbf{R}_j(\tau)> = 6\delta_{ij}k_BT\alpha\delta(t-\tau). \quad (12.23)$$

Note that $i, j = 1, 2, \ldots, N$ and $\delta_{ij} = 1 (i = j)$ $\delta_{ij} = 0, i \neq j$, and $\delta_{ii} \neq 3$. We can then find that

$$\sigma^2 = 6k_BTm\xi = 6k_BT\alpha. \quad (12.24)$$

REMARK 12.1 We can compare the fluctuation–dissipation formula,

$$\alpha = \frac{1}{6k_BT}\int_{-\infty}^{\infty} <\mathbf{R}(0) \cdot \mathbf{R}(t)> dt, \quad (12.25)$$

with the Green–Kubo relation,

$$D = \frac{1}{3}\int_0^\infty <\mathbf{v}(0)\cdot\mathbf{v}(t)> dt = \frac{1}{6}\int_{-\infty}^\infty <\mathbf{v}(0)\cdot\mathbf{v}(t)> dt, \quad (12.26)$$

where in the case of Langevin equation, $D = \mu k_B T$, i.e.,

$$\mu = \frac{1}{6k_B T}\int_{-\infty}^\infty <\mathbf{v}(0)\cdot\mathbf{v}(t)> dt.$$

Now, we can understand statistical mechanics implication of Langevin equation.

12.1.2 Langevin Dynamics as a Canonical Ensemble MD

In the previous section, we almost proved that the Langevin dynamics is equivalently a canonical ensemble MD. However, we have not shown that the particle velocity probability distribution is a normal or Gaussian distribution. In this section, we provide a simple heuristic derivation to show that the velocity probability distribution of an 1D single Langevin particle indeed obeys the normal distribution.

As we mentioned previously, for a free Brownian particle, the Langevin equation is,

$$m\frac{dv}{dt} = -\alpha v + \sigma\theta(t). \quad (12.27)$$

Based on the Liouville theorem, we can find that the probability of the velocity of the above Langevin equation will obey the following Fokker–Planck equation, i.e.,

$$\frac{\partial}{\partial t}P(v,t) = \frac{\partial}{\partial v}\left(\frac{\alpha}{m}vP(v,t)\right) + \frac{\alpha k_B T}{m^2}\frac{\partial^2}{\partial v^2}P(v,t), \quad (12.28)$$

where $P(v,t)$ is the probability to find the Brownian particles at the interval $[v, v + dv]$.

For simplicity, we now show that the probability of the Langevin particle velocity obeys the Fokker–Planck equation case for $m = 1$. In this case, Langevin equation has the form,

$$\dot{v} = -\alpha v + R(t) \quad \rightarrow \quad v^L(t + \Delta t) = v(t) + (-\alpha v(t) + R(t))\Delta t,$$

if we approximate the time derivative as a forward Euler difference operator, i.e.,

$$\frac{dv}{dt} \approx \frac{v(t+\Delta t) - v(t)}{\Delta t}.$$

Our task is to find the probability of the particle velocity at the interval $[v(t), v(t + \Delta t)]$. Based on the illustration in Fig. 12.2, one can find that

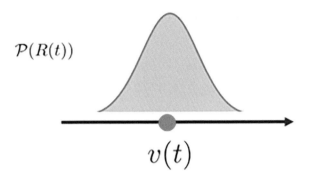

Figure 12.2 Probability distribution for the random force acting on a Langevin particle

$$P(v_{t+\Delta t}, t + \Delta t) = \int\int P(v_t,t)\mathcal{P}(R)\delta(v_{t+\Delta t} - v^L_{t+\Delta t})dv_t dR$$
$$= \int\int P(v_t,t)\mathcal{P}(R)\delta(v_{new} - (v_t + (-\alpha v_t + R)\Delta t))dv_t dR.$$
(12.29)

Using the following identity

$$\int f(y)\delta(x - \lambda y)dy = \frac{1}{\lambda}f(x/\lambda)$$

and letting $\lambda = (1 - \alpha\Delta t)$ and $x = v_{t+\Delta t} - R\Delta t$, we can simplify Eq. (12.29) as

$$P(v_{t+\Delta t}, t + \Delta t) = \frac{1}{\lambda}\int P\left(\frac{1}{\lambda}(v_{t+\Delta t} - R(t))\Delta t), t\right)\mathcal{P}(R)dR.$$

Considering that $(1 - x)^{-1} \approx 1 + x + \cdots$, we have

$$\frac{1}{\lambda} = (1 - \alpha\Delta t)^{-1} \approx 1 + \alpha\Delta t.$$

Thus, we can write

$$\frac{1}{\lambda}P\left(\frac{1}{\lambda}(v_{t+\Delta t} - R(t))\Delta t), t\right) \approx -P(v_{t+\Delta t} - (-\alpha v_{t+\Delta t} + R(t))\Delta t, t)(1 + \alpha\Delta t).$$

Now, we denote $v_{t+\Delta t} = v$, and we use Taylor expansion to expand $P(v + (\alpha v - R)\Delta t)(1 + \alpha\Delta t))$ to the second-order term, i.e.,

$$P(v + (\alpha v - R)\Delta t)(1 + \alpha\Delta t)) \approx P(v,t)(1 + \alpha\Delta t) + \frac{\partial P}{\partial v}(+\alpha v(t) - R(t))\Delta t$$
$$+ \frac{1}{2}\frac{\partial^2 P}{\partial v^2}(\alpha v(t) - R(t))\Delta t)^2 + O((\Delta t)^3).$$

Then, we can obtain the following integral equation,

$$P(v, t + \Delta t) = \int \left\{ P(v,t)(1 + \alpha \Delta t] \right.$$
$$\left. + \frac{\partial P}{\partial v}(+\alpha v(t) - R(t))\Delta t) + \frac{1}{2}\frac{\partial^2 P}{\partial v^2}(\alpha v(t) - R(t))\Delta t)^2 \right\} \mathcal{P}(R) dR. \quad (12.30)$$

Since

$$\int \mathcal{P}(R) dR = 1 \text{ and } \int \mathcal{P}(R) R dR = 0,$$

we can further simplify Eq. (12.30) as

$$P(v, t + \Delta t) = P(v,t)(1 + \alpha \Delta t)$$
$$+ \frac{\partial P}{\partial v}\alpha v(t)\Delta t + \int \frac{1}{2}\frac{\partial^2 P}{\partial v^2}(\alpha v(t) - R(t))\Delta t)^2 \mathcal{P}(R) dR al P(R) dR. \quad (12.31)$$

Since the variance of probability density of a time-dependent random force variable is defined as

$$\int R^2(t) \mathcal{P}(R) dR = <R(t)R(\tau)> \quad (12.32)$$

as $\tau = t + \Delta t$ and $\Delta t \to 0$. Thus,

$$<R(t)R(\tau)> = 2k_B T \alpha \delta(t - \tau) \approx 2k_B T \alpha \frac{1}{\Delta t}. \quad (12.33)$$

It should be noted that Eq. (12.33) reveals an important feature of the stochastic or time-dependent random force. That is, if $\mathbf{R}(t) = \sigma \theta(t)$ and $\theta(t)$ is a time-dependent random variable of standard normal distribution, then $\sqrt{<\mathbf{R}^2>} \sim \sigma(\Delta t)^{-1/2}$, where σ is the "static standard deviation" of the random force. This indicates that the standard deviation of the stochastic random force is its static standard deviation scaled by a factor of $(\Delta t)^{-1/2}$.

Therefore, based on this results, a first-order approximation of the last term in Eq. (12.31) will be

$$k_B T \alpha \Delta t \frac{\partial^2 P}{\partial v^2}.$$

To this end, we can rewrite Eq. (12.31) as follows:

$$\frac{P(v, t + \Delta t) - P(v,t)}{\Delta t} = \alpha P(v,t) + \frac{\partial P}{\partial v}\alpha v(t) + k_B T \alpha \frac{\partial^2 P}{\partial v^2}, \quad (12.34)$$

which results the canonical form of the Fokker–Planck equation as shown in Eq. (12.28). We also note that the Fokker–Planck equation, which is named after Adriaan Fokker and Max Planck (see Fokker (1914); Planck (1917)), is the governing stochastic partial differential equation that is specifically used to describe the time-dependent probability density function for the velocity of a Langevin particle under the influence of drag forces and random forces in Brownian motion. For a more general treatment of the Fokker–Planck equation, readers may consult Kadanoff (2000).

Now, we proceed to solve Eq. (12.28). By considering the initial condition: $v = v_0$, $t = 0$, it is straightforward to show that the solution of Eq. (12.28) has a standard form,

$$P(v,0) = \delta(v - v_0). \tag{12.35}$$

Thus, the solution of Eq. (12.28) can be found as

$$P(v,t) = \left(\frac{\beta m}{2\pi(1 - \exp(-2t/\tau_B))}\right)^{1/2} \exp\left(-\frac{\beta m}{2} \frac{(v - v_0 \exp(-t/\tau_B))^2}{[1 - \exp(-2t/\tau_B)]}\right).$$

When $t \to \infty$,

$$P(v,t) = \left(\frac{\beta m}{2\pi}\right)^{1/2} \exp(-\beta m v^2 / 2), \tag{12.36}$$

which has exactly the same form as the Maxwell velocity distribution, namely,

$$g(v_x) = \left(\frac{\beta m}{2\pi}\right)^{1/2} \exp\left(-\frac{1}{2}\beta m v_x^2\right).$$

12.2 LAMMPS Examples for Langevin Dynamics

In this section, we look into two LAMMPS script input files that use Langevin dynamics to simulate particle motions and polyethylene polymers under given temperature conditions.

Example 12.2 In this example, we use a LAMMPS script file named *in.Langevin* (you may rename it for other name) to simulate particle random motion at the given temperature. The purpose of this study is to examine whether or not the velocity probability distribution generated by the Langevin dynamics is truly a Maxwell distribution as claimed in the previous section.

To start, you can use the default command-line interpreter of Microsoft Windows by typing *cmd* entering the Command Prompt mode. Once you get in to the directory that the script file *in.Langevin* is located, type the following command:

lmp_serial -in in.Langevin

The code will run automatically as shown from the screenshot in Fig. 12.3.

The whole calculation only takes a few minutes in a laptop. The LAMMPS script instructs the computer program outputting each Langevin particle's velocity square and energy into a file named *id_E-Vel.txt*, and the script is given as follows,

```
# in.Langevin
#
# Test for Langevin

# Variable definition

variable x equal 10
```

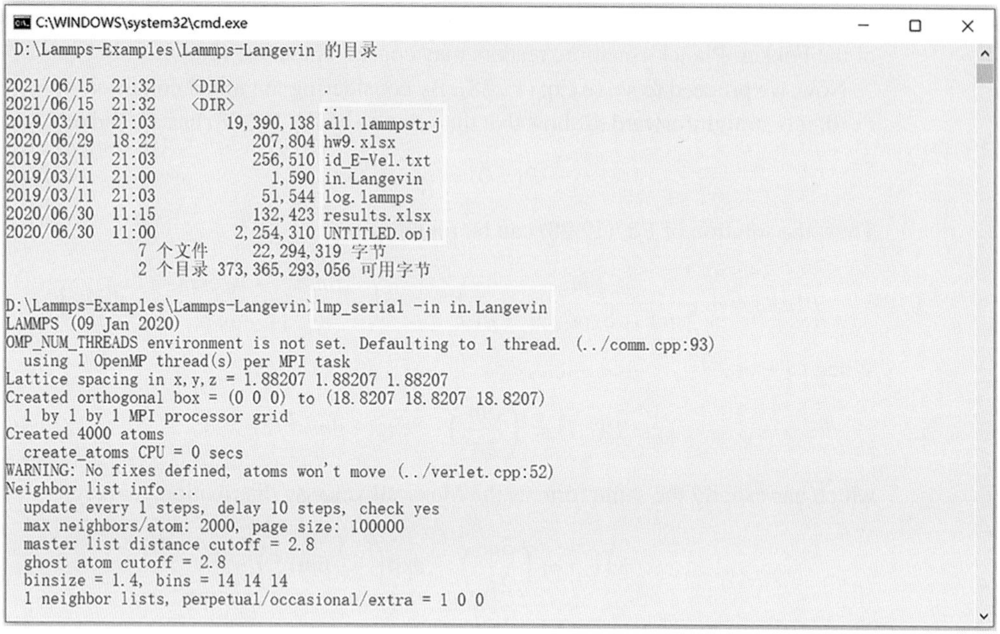

```
C:\WINDOWS\system32\cmd.exe                                    —    □    ×
D:\Lammps-Examples\Lammps-Langevin 的目录

2021/06/15  21:32    <DIR>          .
2021/06/15  21:32    <DIR>          ..
2019/03/11  21:03           19,390,138 all.lammpstrj
2020/06/29  18:22              207,804 hw9.xlsx
2019/03/11  21:03              256,510 id_E-Vel.txt
2019/03/11  21:00                1,590 in.Langevin
2019/03/11  21:03               51,544 log.lammps
2020/06/30  11:15              132,423 results.xlsx
2020/06/30  11:00            2,254,310 UNTITLED.opj
               7 个文件        22,294,319 字节
               2 个目录 373,365,293,056 可用字节

D:\Lammps-Examples\Lammps-Langevin>lmp_serial -in in.Langevin
LAMMPS (09 Jan 2020)
OMP_NUM_THREADS environment is not set. Defaulting to 1 thread. (../comm.cpp:93)
  using 1 OpenMP thread(s) per MPI task
Lattice spacing in x,y,z = 1.88207 1.88207 1.88207
Created orthogonal box = (0 0 0) to (18.8207 18.8207 18.8207)
  1 by 1 by 1 MPI processor grid
Created 4000 atoms
  create_atoms CPU = 0 secs
WARNING: No fixes defined, atoms won't move (../verlet.cpp:52)
Neighbor list info ...
  update every 1 steps, delay 10 steps, check yes
  max neighbors/atom: 2000, page size: 100000
  master list distance cutoff = 2.8
  ghost atom cutoff = 2.8
  binsize = 1.4, bins = 14 14 14
  1 neighbor lists, perpetual/occasional/extra = 1 0 0
```

Figure 12.3 Screenshot: running LAMMPS script file *in.Langevin*

```
variable y equal 10
variable z equal 10

variable lc  equal 0.6
variable    T   equal 2.0
variable rc  equal 2.5

# Setup unit and atom style

units lj
atom_style atomic

# Create the box and generate atoms
#
lattice    fcc ${lc}
region    box block 0 $x 0 $y 0 $z
create_box    1 box
create_atoms  1 box
mass    1 1.0

# LJ potential and parameters
pair_style   lj/cut ${rc}
pair_coeff   1 1 1.0 1.0

# Create the velocity according to the initial temperature

velocity    all create $T 12345
run    0
velocity    all scale $T
```

```
# Set up for neighbor list
neighbor   0.3 bin
neigh_modify delay 0 every 1

# Compute the interesting parameters

compute    1    all   pe/atom
compute    2    all   ke/atom
compute    3    all   temp

variable   energy atom    c_1+c_2
variable   velsqr atom    vx*vx+vy*vy+vz*vz
variable   Tdiff   equal    (c_3-v_T)*(c_3-v_T)

# Thermo output and dump output

thermo_style      custom step  temp v_Tdiff
thermo            10

dump          1    all     atom    100    all.lammpstrj
dump_modify   1    format line    "%7d %3d %8.5f %8.5f %8.5f" scale yes

# Equilibration run

fix   1 all nve
fix      2 all langevin ${T} ${T} 1.0   34224

run         10000

# Data collection in the equilibrium state, v_energy is pe+ke for each atom,
# v_velsqr is sqrt(vx*vx+vy*vy+vz*vz)
# id_E-Vel.txt is the file name.

dump           2    all     custom  2000  id_E-Vel.txt id v_energy v_velsqr
dump_modify    2    format line   "%7d %10.5f %10.5f"

run        2000
```

In this script file, the only command that relates to Langevin dynamics is the following one:

```
fix      2 all langevin ${T} ${T} 1.0   34224
```

In LAMMPS, we use a **fix langevin command** to run Langevin dynamics. The syntax of the command is :

  fix ID group-ID langevin Tstart Tstop damp seed keyword values ⋯ ,

where

- ID = user-assigned name for the fix;
- group ID = ID of the group of atoms to apply the fix to;
- langevin = style name of this fix command;
- Tstart,Tstop = desired temperature at start/end of run (temperature units), where Tstart can be a variable (see below);
- damp = damping parameter (time units);
- seed = random number seed to use for white noise (positive integer);

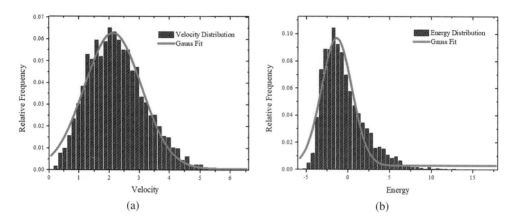

Figure 12.4 Statistical characters of a Langevin dynamics simulation: (a) the energy distribution and (b) the velocity distribution

- zero or more keyword/value pairs may be appended;
- keyword = angmom or omega or scale or tally or zero.

Based on these information, one can find that the particular Langevin fix ID is 2, and the fix command is applied to all the atoms, the start and the end temperature are variables, and in this case they are taking the value of variable T, which is defined as

```
variable T   equal 2.0
```

The damp coefficient in the script file is set at $\gamma = 1$, and the random seed number is set at 34,224. The LAMMPS command

```
dump      2    all    custom   2000   id_E-Vel.txt id v_energy v_velsqr
```

outputs energy and velocity square for each particle every 2,000 time steps into the file *id_E-Vel.text*. Using the output file, one can plot the energy and velocity distribution to see that indeed the Langevin dynamics generates a canonical ensemble statistical system (see Fig. 12.4).

For detailed information, interested readers can consult LAMMPS manual at https://docs.lammps.org/fix_langevin.html.

Example 12.3 In the second example, we use a LAMMPS script input file to model and simulate random motion of polyethylene polymer chains. The script file requires an input data file, which can be downloaded from: http://nanomechanics.berkeley.edu/introduction-to-computational-nanomechanics/.

```
# pe.lammps
#
# Langevin dynamics modeling of PE
#
#set up basic simulation stuff
units real
```

```
# Energy measured in Kcal/mole
# Distances in
# mass in g/mol
# time in fs
#

variable temp   equal 300     #K

atom_style molecular
boundary          p p p

#neighbor settings
neighbor 30.0 bin
atom_modify sort 1000 100
neigh_modify every 1 delay 0 check yes

#read input configuration with 5 polymers
read_data poly.input

# 1.66 monomers per step, and 28.05g/mol for a PE monomer
variable   massvalue equal   28.05*1.66
mass     *         ${massvalue}

#No pair interaction
pair_style         zero 30
pair_coeff         * *

#harmonic bonds    19.1 pN/ converted into units of Kcal/mol
variable k equal   0.275   # Kcal/mol
variable klammps   equal ${k}/2
bond_style harmonic
#   U(r)=K(r-r0)^2
#   K in Kcal/mol   r0/
bond_coeff * ${klammps}       0.0

#save images
shell "mkdir img"
shell "rm img/*"
dump   img all   image  1000 img/t*.jpg   type type adiam 0.1 bond type 0.8 zoom 1.6 view 0 0
dump_modify img backcolor white  boxcolor black
#save video

dump  video all   movie  100 movie.avi type type adiam 0.1 bond type 0.8 zoom 1.6 view 0 0
dump_modify video backcolor white  boxcolor black

fix integrator all nve
fix dynamics    all langevin ${temp} ${temp} 1000 252352

# specify timestep
timestep    10     #fs
thermo     100

run 100000
```

The key command of this script input file is

```
fix dynamics    all langevin ${temp} ${temp} 1000 252352
```

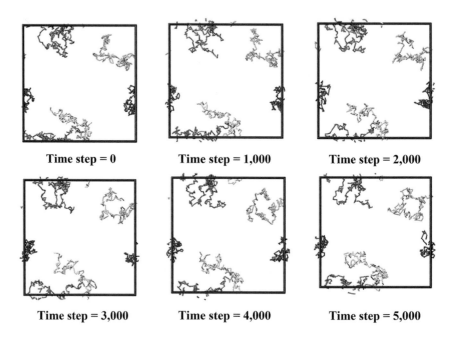

Figure 12.5 Dissipative particle dynamics simulation polyethylene polymer configurations at different time steps

As we explained previously, the word *dynamics* is the name of this fix command, and this command is applied to all atoms. The start and ending temperatures are the variable *temp equal 30*.

The following command

```
dump   img  all   image   1000 img/t*.jpg   type type adiam 0.1 bond type 0.8 zoom 1.6 view 0 0
```

specifies that in every 1,000 time steps the computed polyethylene polymer configuration will be output in an image file of .jpg format. In Fig. 12.5, we display the first six polymer configurations.

12.3 Dissipative Particle Dynamics

Dissipative particle dynamics (DPD) is a special type CG model of the Langevin MD that has been extensively used to model solvent molecules, polymers, and granule particulate matters or discrete amorphous materials (Hoogerbrugge and Koelman (1992); Koelman and Hoogerbrugge (1993)).

The basic concept of coarse-graining is similar to that of homogenization or model reduction. However, when we use the term "coarse-grained model," we usually mean that it is a homogenization model or reduced-order model (ROM) with

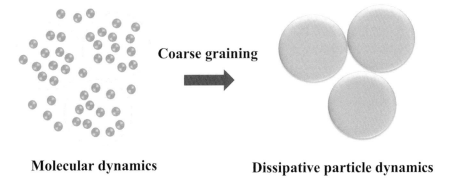

Figure 12.6 Illustration of coarse-graining process of the dissipative particle dynamics

correct statistical thermodynamics, instead of mechanical averaging or smoothing. In Fig. 12.6, we show a schematic illustration of coarse-graining process.

It is important to note that the critical property of a CG description should be true to the original system, meaning that it is a reduction or homogenization of the actual microscopic details. When we prescribe a CG description, we do not want to introduce any outside information. We do not want to add anything that is not from the microscale details. This is usually a challenge because a coarse-grain description may correspond to many molecular systems at microscale. The requirement of the coarse-grained system has the same thermodynamics properties in transient is one unique feature that distinguishes coarse-graining from other types of homogenization or reduced-order modelings.

12.3.1 Theory and Formulation of DPD

In the early 1990s, Hoogerbrigge and Koelman developed the DPD as a finite-sized CG model for fine-scale Langevin particles (see Hoogerbrugge and Koelman (1992); Koelman and Hoogerbrugge (1993)). A few years later, Español and Warren (1995) modified the early DPD formulation to make it consistent with statistical thermodynamics. It was not until 2007, Kinjo and Hyodo (2007) rigorously derived DPD formulation by using the Mori–Zwanzig formalism of nonequilibrium statistical thermodynamics. It is now a very useful modeling tool for simulation of a wide variety of complex hydrodynamic phenomena at mesoscale, including cell membrane dynamics, polymer dynamics, and mesoscale non-Newtonian flows.

The formulation of the DPD is very similar to that of the Langevin equation, and it read as

$$\frac{d\mathbf{P}_i}{dt} = \mathbf{F}_i = \sum_{j \neq i} \left(\mathbf{F}_{ij}^C + \mathbf{F}_{ij}^D + \delta \mathbf{F}_{ij}^R (\Delta t)^{-1/2} \right), \quad i = 1, 2, \ldots, N, \quad (12.37)$$

where $\mathbf{P}_i$ are the linear momentum of of the ith PDP particle, $\mathbf{F}_{ij}^C$ are the conservative internal forces, $\mathbf{F}_{ij}^D$ are the drag forces, and $\delta \mathbf{F}_{ij}^R$ are the random forces.

Different from the Langevin dynamics, the above forces have some common forms (not limited to) that are extensively used in practice, and we briefly discuss them in the following:

(i) Conservative force

$$\mathbf{F}_{ij}^C = \begin{cases} a_{ij}(b_i - r_{ij})\frac{\mathbf{r}_{ij}}{r_{ij}}, & \forall r_{ij} < b_i; \\ 0, & \forall r_{ij} \geq b_i, \end{cases} \quad (12.38)$$

where $r_{ij} = |\mathbf{r}_{ij}| = |\mathbf{r}_j - \mathbf{r}_i|$, a_{ij} is a maximum repulsion between particle i and j, and b_i is the radius of the particle i;

(ii) Drag force

$$\mathbf{F}_{ij}^D = -\gamma w^D(r) \left(\frac{\mathbf{r}_{ij} \cdot \mathbf{v}_{ij}}{r_{ij}} \right) \frac{\mathbf{r}_{ij}}{r_{ij}},$$

with $\mathbf{r}_{ij} = \mathbf{r}_i - \mathbf{r}_j$; and $\mathbf{v}_{ij} = \mathbf{v}_i - \mathbf{v}_j$, (12.39)

where $w^D(r)$ is a r-dependent weight function; and

(iii) Random force

$$\delta \mathbf{F}_{ij}^R = \sigma w^R(r) \frac{\theta_{ij} \cdot \mathbf{r}_{ij}}{r_{ij}} \frac{\mathbf{r}_{ij}}{r_{ij}}, \text{ with } <\theta_{ij}(t)> = 0,$$

and $<\theta_{ij}(t) \cdot \theta_{k\ell}(\tau)> = (\delta_{ik}\delta_{j\ell} + \delta_{i\ell}\delta_{jk})\delta(t - \tau),$ (12.40)

where $w^R(r)$ is also a r-dependent weight function.

REMARK 12.4 In above expressions, $\mathbf{n}_{ij} = \mathbf{r}_{ij}/r_{ij}$ is the direction of the interaction force, while $\mathbf{v}_{ij} \cdot \mathbf{n}_{ij}$ and $\theta_{ij} \cdot \mathbf{n}_{ij}$ are projections of velocities and the time-dependent standard normal random vectors onto the direction of $\mathbf{n}_{ij}$. Thus, corresponding to Langevin dynamics, we have

$$\alpha \rightarrow \gamma w^D(r); \quad \sigma \rightarrow \sigma w^R(r).$$

Since in Langevin dynamics the fluctuation–dissipation relation yields $\sigma^2 = 2\alpha k_B T$, we then have

$$\left(\sigma w^R(r)\right)^2 = 2\gamma w^D(r) k_B T.$$

Equating the coefficients in both constant part and inhomogeneous part, we find that

$$w^D(r) = (w^R(r))^2, \quad \sigma^2 = 2\gamma k_B T.$$

We often use,

$$w^D(r) = [w^R(r)]^2 = \begin{cases} (b - r)^2, & \forall r < b, \\ 0, & \forall r \geq b. \end{cases}$$

REMARK 12.5 In Eq. (12.37), the random force is written as $\mathbf{R}_{ij} = \delta \mathbf{F}_{ij}^R (\Delta t)^{-1/2}$. We have discussed this issue in the section of Langevin equation, and now we re-examine the issue through a different angle.

If we let
$$R_{ij} = \delta F_i^R = \sigma \theta_i,$$
without the factor of $(\Delta t)^{-1/2}$, the total force in a time interval t will be,
$$F_i^R = \int_0^t \delta F_i^R(\tau) d\tau,$$
and hence
$$< (F_i^R)^2 > = \left\langle \left(\sigma \int_0^t \theta_i(\tau) d\tau \right)^2 \right\rangle = \left\langle \sigma^2 \left(\sum_i^N \theta_i^2 \right) \left(\frac{t}{N} \right)^2 \right\rangle \le \sigma^2 t^2/N = t\sigma^2 \Delta t \quad (12.41)$$
because $< \theta_i \cdot \theta_j > = 0$, $i \ne j$, and $\Delta t = \frac{t}{N}$. Subsequently, we have
$$< (F_i^R)^2 > \to 0, \text{ as } \Delta t \to 0.$$
This does not make any sense. Hence, we must let
$$R_i = \delta F_i^R (\Delta t)^{-1/2} = \sigma \theta_i / \sqrt{\Delta t} \quad \to \quad < (R_i)^2 > \to t\sigma^2.$$
Therefore, in PDP the random force must be written as
$$R_{ij} = \sigma w^R(r_{ij}) \theta_{ij} (\Delta t)^{-1/2} \frac{\mathbf{r}_{ij}}{r_{ij}}.$$

In fact, this is a very important aspect of stochastic random variable. To explain this, we let the potential energy $V = 0$, and we may write Langevin equation as,
$$dp = -\xi p(t) dt + \sigma dW, \text{ where } dW = \delta F^R dt.$$
Define
$$\mu(v) = -\xi p(t).$$
We can then write the Langevin equation as follows:
$$dp = \mu(t) dt + \sigma dW,$$
where p is a random variable or a random variable of Ito process.

This is the generic form of the stochastic differential equation. For a general random field X_t in an Ito process, the canonical form of stochastic differential equation is defined as
$$dX_t = \mu(X_t) dt + \sigma(X_t) dW_t. \quad (12.42)$$

In fact, this is related to the celebrated Ito Lemma in stochastic differential equation. Consider a stochastic Ito drift–diffusion process,
$$dX_t = \mu(X_t) dt + \sigma(X_t) dW_t.$$

The Ito lemma is referred to the following differential relation for a random process $V(X_t, t)$,

$$dV(X_t,t) = \left(\frac{\partial V}{\partial t} + \mu\frac{\partial V}{\partial X_t} + \frac{1}{2}\sigma^2\frac{\partial^2 V}{\partial X_t^2}\right)dt + \sigma\frac{\partial V}{\partial X_t}dW_t.$$

Note that by using the chain rule, one has

$$dV(X_t,t) = \left(\frac{\partial V}{\partial t} + \mu\frac{\partial V}{\partial X_t}\right)dt + \mu\sigma\frac{\partial^2 V}{\partial X_t^2}dtdW_t +$$
$$+ \frac{1}{2}\sigma^2\frac{\partial^2 V}{\partial X_t^2}dW_t^2 + \sigma\frac{\partial V}{\partial X_t}dW_t + O((dt)^2).$$

As we discussed previously,

$$dW_t = \delta F^R dt \sim \alpha\theta dt/(\Delta t)^{1/2} \rightarrow (dW_t)^2 \sim dt.$$

We then arrive at the conclusion of the Ito lemma,

$$dV(X_t,t) = \left(\frac{\partial V}{\partial t} + \mu\frac{\partial V}{\partial X_t} + \frac{1}{2}\sigma^2\frac{\partial^2 V}{\partial X_t^2}\right)dt + \sigma\frac{\partial V}{\partial X_t}dW_t + O(dt^{3/2}). \quad (12.43)$$

12.3.2 LAMMPS Example of DPD Simulation

In this section, we present an example of LAMMPS script to demonstrate how to run a DPD simulation.

Example 12.6 In the first example, one can run the LAMMPS script file to simulate spinodal decomposition of two groups of particles, the red particles and the green particles, by using the DPD.

Spinodal decomposition is a thermodynamic phase transition or phase separation process in which the two-phase mixture separates each other, becoming two distinct phases, when there is no nucleation barrier to the decomposition.

```
#
# A total number of 24000 particles in 20^3 sigma^3 volume
# corresponds to a standard density of 3 particles per sigma^3.
#

#percentage of particles of type 1
variable npart1 equal 12000

#percentage of particles of type 2
variable npart2 equal 12000

#temperature in reduced units
variable temp equal 1.0

#set up basic simulation stuff
```

12 Langevin Equations and Dissipative Particle Dynamics

```
#dimension 2
units lj
atom_style molecular
boundary        p p p
comm_modify     vel yes

#neighbor settings
neighbor 0.5 bin
neigh_modify every 1 delay 0 check yes

#add additional particles to box
region box block -10 10 -10 10 -10 10

#2 is number of types of atoms in the simulation
create_box 2 box

#Put type 1 and 2 atoms randomly into the box
create_atoms 1 random ${npart1} 358723 box
create_atoms 2 random ${npart2} 125233 box

group apart type 1
group bpart type 2

#all particles has mass 1
mass    *       1

#Dissipative-particle-dynamics

#   Standard interactions are:
#
#   25 = type 1-1 and 2-2 particles likes each other
#   80 = type 1-2 particles dislikes each other
#
#   Here the standard choice of interactions are due
#   to Groot and Warren, J. Chem. Phys 107, 1997 (4423)
#
#     a_11=a_22 = 75 kT / rho = 25 kT at rho=3
#     to reproduce the compressibility of water with
#     the pure phases.
#
#     The cross term is a_12 = 75 kT/rho + Delta a
#     where Delta a is empirically related to
#     regular solution chi parameter via chi=0.286 Delta a
#
#     Hence a_12=80kT corresponds to Delta a=55kT
#     or chi=15.73. Remember the critical chi is 2
#     hence the mixture will phaseseparate when.
#     a_12>32 and stay mixed otherwise
#

pair_style      dpd  ${temp} 1.0 3845739
pair_coeff      1 1 25   4.5
pair_coeff      2 2 25   4.5
pair_coeff      1 2 80   4.5

#make pictures
shell "mkdir img"
shell "rm img/*"
dump        img all  image  100 img/t*.jpg    type type adiam 1.0 zoom 1.3
```

```
dump_modify img backcolor white   boxcolor black
dump_modify img pad 6

#save video
dump           video all   movie   10 movie.avi       type type adiam 1.0 zoom 1.3
dump_modify video backcolor white   boxcolor black

dump           video2 apart   movie   10 movie_apart.avi   type type adiam 1.0 zoom 1.3
dump_modify video2 backcolor white   boxcolor black

dump           video3 bpart   movie   10 movie_bpart.avi   type type adiam 1.0 zoom 1.3
dump_modify video3 backcolor white   boxcolor black

fix integrator all nve

# specify timestep
timestep 0.02
thermo    100

run 10000
```

In the above LAMMPS script file, the only relevant commands to DPD simulation are the following commands:

```
pair_style          dpd  ${temp} 1.0 3845739
pair_coeff          1 1  25    4.5
pair_coeff          2 2  25    4.5
pair_coeff          1 2  80    4.5
```

In LAMMPS, the *pair_style dpd* computes a force field for DPD based on the formulation in Groot and Warren (1997). For style *dpd*, the force on atom I due to atom J is given as a sum of the following three terms:

$$\mathbf{F}_i = \sum_j (F^C + F^D + F^R)\mathbf{n}_{ij}, \quad r_{ij} < r_c$$

$$F^C = Aw(r)$$
$$F^D = -\gamma w^2(r)(\mathbf{n}_{ij} \cdot \mathbf{v}_{ij});$$
$$F^R = \sigma w(r)\alpha(\Delta t)^{-1/2};$$
$$w(r) = 1 - r/r_c;$$
$$\mathbf{n}_{ij} = \frac{\mathbf{r}_{ij}}{r_{ij}}, \quad \mathbf{r}_{ij} = \mathbf{r}_j - \mathbf{r}_i \text{ and } r_{ij} = |\mathbf{r}_{ij}|, \qquad (12.44)$$

where α is a Gaussian random variable with zero mean and unit variance, Δt is the time increment size, and $w(r)$ is a weighting factor that varies between 0 and 1, r_c is the cutoff distance, σ is set equal to $\sqrt{2k_B T \gamma}$, where k_B is the Boltzmann constant, and T is the temperature parameter in the *pair_style* command.

In LAMMPS, companion with the command *pair_style dpd*, the command *pair_coeff* defines the following DPD force field parameters:

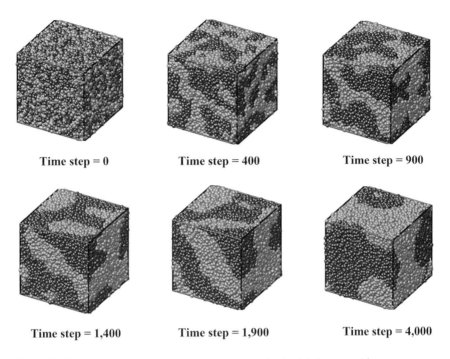

Figure 12.7 Dissipative particle dynamics simulation of spinodal decomposition

- A (force unit)
- γ (force/velocity unit)
- cutoff (distance unit)

for each pair of atoms types. Note that the last coefficient (cutoff) is optional. If not specified, the global DPD cutoff is used. Note that σ is set to as $\sigma = \sqrt{2Tk_B\gamma}$, where T is the temperature set by the *pair_style* command above, so that it does not need to be specified.

In this simulation, we have two types of particles: the dark atoms and the light atoms, and their force fields are set by the above statements. The simulation results are shown in Fig. 12.7, and as one may find that as time increases the two phases are gradually separate from each others. Note that in the LAMMPS output files, the dark atoms are in red color, and the light atoms are in green color.

12.3.3 Case Study: DPD Simulation of Lipid Bilayer Membrane

As discussed previously, DPD is a CG mesoscale particle dynamics that groups a few atoms into one mesoscale particle (or bead) by averaging some unessential degrees of freedom of those atoms. If this process can be done in a thermodynamically consistent manner, we may be able to capture the most fundamental physical and chemical properties of the system while simplifying the complexity of the full atom model. DPD is, in particular, a very efficient tool for simulation of mesoscale systems of polymers and biomembranes In the following example, we present a step-by-step instruction and procedure to show how to build a DPD lipid bilayer membrane model by using open source software and how to use LAMMPS to simulate functions of a lipid bilayer membrane.

Example 12.7 The biological membrane is made up of lipid bilayer with hydrophobic tails and hydrophilic heads.

To start building the DPD lipid bilayer membrane model, we need the following two data input files for lipid and water molecules. We label the first data input file as *lipid.xyz*, which provides the mesoscale microstructure of a lipid:

```
11
C 0.0 0.0 0.0
C 0.5 0.0 0.0
C 1.0 0.0 0.0
C 1.5 0.0 0.0
C 0.0 0.6 0.0
C 0.5 0.6 0.0
C 1.0 0.6 0.0
C 1.5 0.6 0.0
H 2.0 0.0 0.0
H 2.0 0.5 0.0
H 2.5 0.0 0.0
```

The second input data file is labeled as *water.xyz*, which provides a mesoscale water molecule particle (condensed from one oxygen atom and two hydrogen atoms) position:

```
W 0.0   0.0   0.0
```

Then, we employ an open source software, **PACKMOL** (Initial configurations for Molecular Dynamics Simulations by packing optimization), to create the initial configuration file of the lipid bilayer, which we label it as *bilayer.xyz*. PACKMOL creates an initial point for MD and related particle dynamics simulations by packing molecules in defined regions of space. The packing guarantees that short-range repulsive interactions do not disrupt the simulations

To download and install PACKMOL, one can visit the official website of PACKMOL (see Fig. 12.8; http://m3g.iqm.unicamp.br/packmol/home.shtml)

One may find a detailed description of PACKMOL in Martínez et al. (2009). To use PACKMOL to build the lipid bilayer model, one needs a PACKMOL script file, *bilayer-packmol.in*, which is provided as follows:

```
tolerance 0.5
filetype xyz
output bilayer.xyz
```

PACKMOL

Initial configurations for Molecular Dynamics Simulations by packing optimization

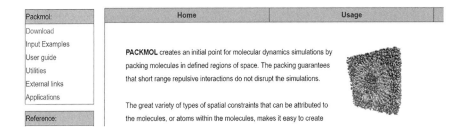

Figure 12.8 Website of PACKMOL: Initial Configurations for Molecular Dynamics Simulations by Packing Optimization

```
structure lipid.xyz
   number 250
   inside box 0.  0.  10.  19.428  19.428  13.0
   atoms  1
      below plane  0.  0.  1.  10.5
   end atoms
   atoms 11
      over plane  0.  0.  1.  12.5
   end atoms
 end structure

structure lipid.xyz
   number 250
   inside box 0.  0.  7.0  19.428  19.428  13.0
   atoms  1  5
      over plane  0.  0.  1.  9.5
   end atoms
   atoms 11
      over plane  0.  0.  1.  7.5
   end atoms
 end structure

 structure water.xyz
    number 8250
    inside box  0.  0.  12.5  19.428  19.428  19.428
 end structure
```

```
structure water.xyz
    number 8250
    inside box  0.  0.  12.5  19.428  19.428  7.5
end structure
```

In Fig. 12.9, we show how each part of the lipids and water molecules correspond to the segments of the PACKMOL script file.

After downloading *packmol.exe* from PACKMOL website, one can run the packmol script file *bilayer-packmol.in* to generate the initial configuration file (in (.xyz) format) for the lipid bilayer model as the operation shown in Fig. 12.10.

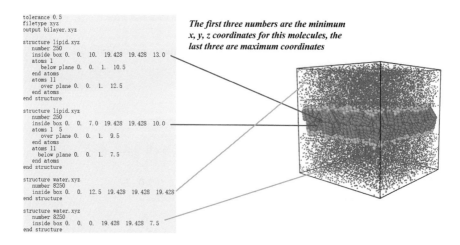

Figure 12.9 How does each part of PACKMOL script file, bilayer-packmol.in, correspond to the lipid bilayer microstructure

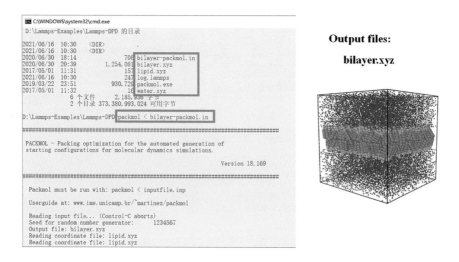

Figure 12.10 How to use PACKMOL to generate LAMMPS input data file, *bilayer.xyz*

12 Langevin Equations and Dissipative Particle Dynamics

With the configuration file *lipid.xyz* ready, we can use the visualization MD software, VMD (*https://www.ks.uiuc.edu/Research/vmd/*), to generate the LAMMPS data input file. To do this, one needs to use Tk Console in VMD. The procedure is outlined as follows:

First open VMD, and then click VMD–Extension–Tk Console. Once in the Tk Console mode, one can run the following Tk console file, *Tkconsole_script*, by typing the following command at command prompt:

```
% source   Tkconsole_script
```

Then the computer will take the file *bilayer.xyz* as the input file, and generate *bilayer.psf* file and *bilayer.data* file, which is the LAMMPS input data file.

The following is Tk console script file named *Tkconsole_script*:

```
mol new bilayer.xyz

pbc set { 19.428 19.428 19.428 }
topo clearbonds
for {set i 0} {$i < 500} {incr i} {
    topo addbond [expr $i*11]   [expr $i*11+1]
    topo addbond [expr $i*11+1] [expr $i*11+2]
    topo addbond [expr $i*11+2] [expr $i*11+3]
    topo addbond [expr $i*11+4] [expr $i*11+5]
    topo addbond [expr $i*11+5] [expr $i*11+6]
    topo addbond [expr $i*11+6] [expr $i*11+7]
    topo addbond [expr $i*11+3] [expr $i*11+8]
    topo addbond [expr $i*11+7] [expr $i*11+9]
    topo addbond [expr $i*11+8] [expr $i*11+9]
    topo addbond [expr $i*11+8] [expr $i*11+10]
    topo addangle [expr $i*11]   [expr $i*11+1] [expr $i*11+2]
    topo addangle [expr $i*11+1] [expr $i*11+2] [expr $i*11+3]
    topo addangle [expr $i*11+2] [expr $i*11+3] [expr $i*11+8]
    topo addangle [expr $i*11+4] [expr $i*11+5] [expr $i*11+6]
    topo addangle [expr $i*11+5] [expr $i*11+6] [expr $i*11+7]
    topo addangle [expr $i*11+6] [expr $i*11+7] [expr $i*11+9]
}

animate write psf bilayer.psf
topo writelammpsdata bilayer.data
```

Once we created the file, *bilayer.data*, we can start to simulate lipid membrane by using LAMMPS. We name the LAMMPS script file for lipid bilayer simulation as *in.DPD-bilayer*, which is displayed subsequently:

```
#
#   DPD lipid bilayer LAMMPS code: in.DPD-bilayer
#
units     lj
```

```
atom_style    full
comm_modify   vel    yes
pair_style    dpd    1.0    1.0    34387

read_data bilayer.data

# define masses and interaction coefficients
pair_coeff    1    1    10    4.5
pair_coeff    2    2    30    4.5
pair_coeff    3    3    25    4.5
pair_coeff    1    2    35    4.5
pair_coeff    1    3    75    4.5
pair_coeff    2    3    30    4.5

bond_style    harmonic
bond_coeff    1    128.0    0.5

angle_style   cosine/delta
angle_coeff   1    7.5    180.0

# create initial velocities
velocity    all    create    1.0    4928459    dist    gaussian

# change neighbor list parameters to avoid dangerous builds
neighbor      2.0    bin
neigh_modify  delay  3

#specify simulation parameters
timestep    0.04
thermo      10

#first equilibrate the initial condition
fix    1    all    nve
run    500

dump    traj    all xtc    50 traj.xtc
dump_modify    traj    unwrap yes

dump    traj_xyz    all xyz    50    traj.xyz

# production run
run 50000
```

In Fig. 12.11, we show the top half of the LAMMPS script file, in which the key commands are explained.

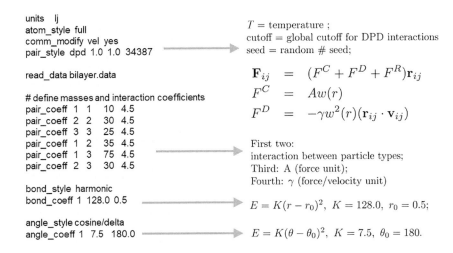

Figure 12.11 A snippet of the LAMMPS DPD code with comments and notes

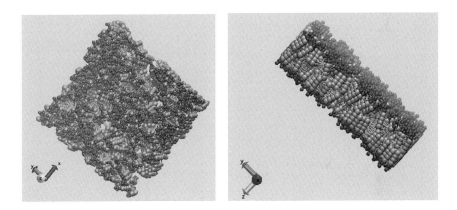

Figure 12.12 DPD model of a lipid bilayer in water environment

The simulation results are shown in Fig. 12.12. One can visualize the simulation movie made by Dr. Kaiyue Wang, and it was posted in YouTube with the following link: *https://youtu.be/A7M0cCPPBFQ*

12.4 Homework Problems

Problem 12.1 In the following, we focus on a single particle Langevin equation with only the dissipative force and the fluctuation (random) force,

$$m\frac{dv}{dt} = -\alpha v + R(t).$$

The fluctuation force R is a random force, which is assumed to satisfy the following conditions:

1. White noise $<R(t)> = 0$;
2. Uncorrelated with prior velocity: $<v(0) \cdot R(t)> = 0$;
3. Markov process: $<R(t) \cdot R(\tau)> = \sigma^2 \delta(t-\tau)$;
4. White noise Gaussion: $P(R) = \dfrac{1}{\sqrt{2\pi}\sigma} \exp\left(-\dfrac{R(t)}{2\sigma^2}\right)$.

Show that

1. $v(t) = v(0)\exp(-\xi t) + \displaystyle\int_0^t d\tau \exp(-\xi(t-\tau))R(\tau)$,

where $\xi = \alpha/m$, and

2. $\sigma^2 = 2k_B T m \xi = 2k_B T \alpha$,

as $t \to \infty$.

13 Nonequilibrium Molecular Dynamics

The natural world has intrinsic multiscale structures or we may say that the nature is multiscale in nature. In general, most of physical phenomena observed at macroscale are nonequilibrium physical processes, while most of well-developed physical models (not phenomenological) that we have are equilibrium thermodynamics or equilibrium statistical mechanics model. More precisely speaking, most of the macroscale or phenomenological physical models that have been developed are used to describe nonequilibrium processes, while most (not all of them) well-established microscale physical models are about the idealized equilibrium process. From this perspective, a key of the multiscale modeling is how we can start from an equilibrium microscale theory to predict the properties of a nonequilibrium system at macroscale.

On the other hand, in recent decades, both nonequilibrium statistical mechanics as well as nonequilibrium molecular dynamics (MD) have made great progress. To fully discuss the topic of nonequilibrium MD is out of the scope of this book. Interested readers may consult the special literatures on this subject, such as the monograph by Evans and Morriss (2007) and by Todd and Daivis (2017).

In this chapter, we introduce some well-established nonequilibrium MD methods as the first step for readers jumping into more sophisticated methods.

13.1 Green–Kubo Relation

At macroscale, we often encounter various transport phenomena. For instance, we often encounter heat conduction process in many thermomechanical engineering processes, in which the heat flux is proportional to temperature gradient (the Fourier law),

$$\mathbf{q} = -\lambda \nabla T, \tag{13.1}$$

where $\mathbf{q}$ is the heat flux, T is the temperature, and λ is the heat conductivity coefficient, i.e., the transport coefficient in this case.

In electrical engineering or electromagnets, the electric current is proportional to the gradient of the electric potential, i.e.,

$$\mathbf{J} = -\sigma \nabla \phi, \tag{13.2}$$

where **J** is the current density, ϕ is the electrical potential, and σ is the electrical conductivity. Another common nonequilibrium process is the Newtonian fluid flow, in which the shear stress is proportional to the shear strain rate, i.e.,

$$\sigma_{xy} = \eta \dot{\gamma}_{xy} = \eta(\dot{\gamma})\left(\frac{\partial \dot{u}}{\partial y} + \frac{\partial \dot{v}}{\partial x}\right), \tag{13.3}$$

where σ_{xy} is a shear stress component, $\frac{\partial \dot{u}}{\partial y} + \frac{\partial \dot{v}}{\partial x}$ is shear strain rate, $\eta(\dot{\gamma})$ is the strain rate, and η is the transport coefficient in this case – the shear viscosity.

In many engineering applications, we often encounter linear constitutive relations in transport processes, which may be expressed as the following general form:

$$\bar{\mathbf{J}} = \mathbf{L}\Big|_{F_e=0} \cdot \mathbf{F}_e, \tag{13.4}$$

where $\bar{\mathbf{J}}$ is called the thermodynamical flux, $\mathbf{F}_e$ is applied field, and **L** is linear (macroscale) transport coefficient. In some cases, the applied field is tensorial field, we have

$$\bar{\mathbf{J}} = \mathbf{L}\Big|_{F_e=0} : \mathbf{F}_e. \tag{13.5}$$

In actual applications, we need to accurately determine or predict the transport coefficient of a given thermomechanical process. In other words, we need to determine the constitutive relation of a non-equilibrium process. The so-called Green–Kubo relation is such a method that uses the equilibrium MD to determine the transport coefficient of a nonequilibrium process.

In other words, we can find the transport coefficient by using MD simulation via the **Green–Kubo relation:**

$$\frac{\mathbf{L}(\mathbf{F}_e = 0)}{V} = \beta \int_0^\infty < \mathbf{J}(0) \otimes \mathbf{J}(\tau) >_{F_e=0} d\tau, \quad \beta^{-1} = k_B T, \tag{13.6}$$

where V is the volume of the system, $\mathbf{F}_e$ is the negative potential gradient, and **J** is the thermodynamics flux. The Green–Kubo relation is a profound consequence of the *fluctuation–dissipation theorem* (FDT).

Note that the integrand $< \mathbf{J}(0) \otimes \mathbf{J}(\tau) >$ in the above integral is called the **autocorrelation function**. Therefore, before we prove the Green–Kubo relation, we first discuss what is the correlation function.

13.1.1 Correlation Function

The term "correlation" refers to a process for establishing whether or not relationships exist between two variables.

One way to find out this is to plot the two sets of data of respective variables onto a planar plot, which is usually called "scatterplot." Figure 13.1 shows several sets of scatterplots.

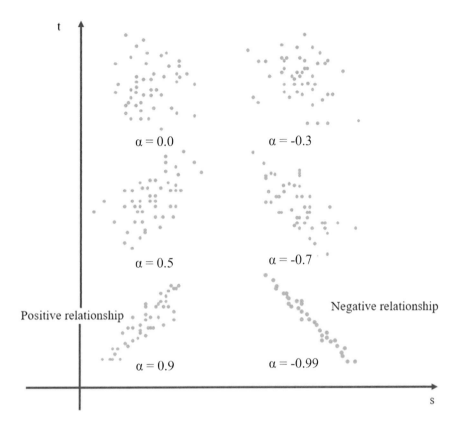

Figure 13.1 Scattered data with different correlations, and α is the slope of the linear regression line

In statistical mechanics, the correlation function is a measure of the order in a system, as characterized by a mathematical correlation function, and describes how microscopic variables at different positions are correlated.

A Nävie definition of the correlation of a set of random variable data $\{A_i\}_{i=1}^{N}$ and $\{B_i\}_{i=1}^{N}$ is expressed as

$$C_{AB} = \frac{1}{N} \sum_{i=1}^{N} A_i B_i \equiv \ <A_i B_i>. \tag{13.7}$$

One may want to normalize the correlation defined in the above equation as,

$$C_{AB} = \frac{\frac{1}{N}\sum_{i=1}^{N} A_i B_i}{\sqrt{\left(\frac{1}{N}\sum_{i=1}^{N} A_i^2\right)\left(\frac{1}{N}\sum_{i=1}^{N} B_i^2\right)}}, \tag{13.8}$$

which is called as the Pearson product moment correlation coefficient, in honor of its developer Karl Pearson.

If $\{A_i\}$ and $\{B_i\}$ are not white noise random variables, or their means are not zero, i.e.,

$$< A_i > \neq 0 \text{ and } < B_i > \neq 0.$$

One may want to calculate the correlation only on the fluctuation part,

$$C_{AB} = \frac{\frac{1}{N}\sum_{i=1}^{N}(A_i - < A >)(B_i - < B >)}{\sqrt{\left(\frac{1}{N}\sum_{i=1}^{N}(A_i - < A >)^2\right)\left(\frac{1}{N}\sum_{i=1}^{N}(B_i - < B >)^2\right)}}$$

$$= \frac{\sum_{i=1}^{N}A_i B_i - \frac{1}{N}\left(\sum_{i=1}^{N}A_i\right)\left(\sum_{i=1}^{N}B_i\right)}{\sqrt{\left(\frac{1}{N}\sum_{i=1}^{N}A_i^2 - \left(\frac{1}{N}\sum_{i=1}^{N}A_i\right)^2\right)\left(\frac{1}{N}\sum_{i=1}^{N}B_i^2 - \left(\frac{1}{N}\sum_{i=1}^{N}B_i\right)^2\right)}}.$$

In MD, variable data sets often are time-dependent, and the *time correlation coefficient*, which is also called *correlation function*, is defined as,

1. $C_{AB} = < A(0)B(t) > = \frac{1}{N}\sum_{i=1}^{N}A_i(0)B_i(t)$ and

or

2. $C_{AB}(t) = < A(t')B(t'+t) > = \lim_{\tau \to \infty}\frac{1}{\tau}\int_{0}^{\tau}A(t')B(t'+t)dt'$

$$= \frac{1}{MN}\sum_{j=1}^{M}\sum_{i=1}^{N}A_i(t_j)B_i(t_j + t),$$

where i is the index of particles and j is the index of different origins or initial conditions. Thus, the first average is over multiple origins or initial conditions. Hence, the above equation may be interpreted as

$$\begin{bmatrix} C(t = \Delta t) \\ C(t = 2\Delta t) \\ \vdots \\ C(t = n\Delta t) \end{bmatrix} = \frac{1}{M}\sum_{j=1}^{M}\begin{bmatrix} \frac{1}{N}\sum_{i=1}^{N}\left(\mathbf{v}_i(t=t_j)\cdot\mathbf{v}_i(t=t_j+\Delta t)\right) \\ \frac{1}{N}\sum_{i=1}^{N}\left(\mathbf{v}_i(t=t_j)\cdot\mathbf{v}_i(t=t_j+2\Delta t)\right) \\ \vdots \\ \frac{1}{N}\sum_{i=1}^{N}\left(\mathbf{v}_i(t=t_j)\cdot\mathbf{v}_i(t=t_j+n\Delta t)\right) \end{bmatrix}. \quad (13.9)$$

If we replace the time average with assemble average, we can express the time correlation as the ensemble average in the phase space, i.e.,

$$C_{AB}(t) = <\mathbf{A}(0)\mathbf{B}(t)> = \int d\mathbf{q}^N d\mathbf{p}^N f(\mathbf{q}^N,\mathbf{p}^N)\mathbf{A}(0,\mathbf{q}^N,\mathbf{p}^N)\mathbf{B}(t,\mathbf{q}^N,\mathbf{p}^N),$$

where $f(\mathbf{q}^N,\mathbf{p}^N)$ is the distribution function for the equilibrium state, providing supplements for those who are not familiar with the topic.

Before we proceed further, we first sort out various terminologies in correlation analysis and statistical analysis. The most general definition of the correlation function is,

$$C(t) = <A(t')B(t'')>, \quad \text{and } t'' = t' + t, \text{ here } t' \text{ is initial value.}$$

Usually, we choose the origin $t' = 0$ so $C(t) = <A(0)B(t)>$.

However, $C(t)$ does not depend on the choice of the time origin.

The difference between convolution and correlation is,

1. Convolution: $A(t) * B(t) = \int_{-\infty}^{\infty} A(\tau)B(t-\tau)d\tau;$

2. Correlation: $A(t) \otimes B(t) = \int_{0}^{\infty} A(\tau)B(\tau+t)d\tau.$

Simple Properties of Time Correlation Function

In the following, we list and discuss some simple properties of the time correlation function, which we may use in the rest of the book.

1.
$$\text{When } t \to 0, \ C_{AB}(0) = <A(t')B(t')> = <AB>,$$

it is a static correlation.

2.
$$\text{When } t \to \infty, \ C_{AB} \to <A><B>, \text{ and } C_{AB} \to 0$$

if $A(t)$ and $B(t)$ are random variables. In this case, $A(t)$ and $B(t)$ are no longer correlated, because they are not contemporaries.

3. The time correlation only depends on time separation, but not the absolute value of time, so

$$C(t) = <A(t')B(t'')>, \quad t = t'' - t'.$$

Then, we let

$$t' = 0, \ t'' = t; \ \to \ C(t) = <A(0)B(t)>$$

and

$$\text{let } t' = -t, \ t'' = 0; \ \to \ C(t) = <A(-t)B(0)>.$$

4. Therefore, if we choose, $t' = -t$ and $t'' = 0$, for autocorrelation,

$$C(t) = <A(-t)A(0)> = <A(0)A(-t)> \to t' = 0, \ \to \ C(t) = C(-t).$$

This means that the autocorrelation is an even function, so that

$$\frac{d}{dt}C(t)\bigg|_{t=0} = 0,$$

which is another proof that $C(0) = <A(0)^2>$ is a static correlation (see: Property 1). Following Chandler's notation (Chandler 1987), we can show that

$$\delta A(t) = A(t) - <A>, \quad \delta A(t) = \delta A(t; \mathbf{q}^N, \mathbf{p}^N) = \delta A(\mathbf{q}^N(t), \mathbf{p}^N(t)).$$

$$C_d(t) = <\delta A(0)\delta A(t)> = <(A(0) - <A>)(A(t) - <A>)>$$
$$= <A(0)(A(t) - <A(t)>)> = <A(0)A(t)> - <A>^2.$$

As mentioned earlier,

$$\text{when } t \to 0, \ C_d(0) = <(\delta A(0))^2>, \text{ and}$$
$$\text{when } t \to \infty, \ C_d(t) \to <\delta A(t')><\delta A(t'')> = 0.$$

13.1.2 Green–Kubo Formula

Before we prove the Green–Kubo formula, we first express the classical Green–Kubo formula as follows,

$$\lim_{t \to \infty} \frac{<[A(t) - A(0)]^2>}{2t} = \int_0^\infty d\tau <\dot{A}(0)\dot{A}(\tau)>,$$

which links the mean square displacement of a dynamic variable to its velocity correlation function. Furthermore, we can also link

$$\frac{<[A(t) - A(0)]^2>}{2t} \sim \gamma$$

to the coefficient of the related transport process, and hence,

$$\gamma \sim \int_0^\infty d\tau <\dot{A}(0)\dot{A}(\tau)>.$$

The contemporary version of the Green–Kubo formula may be stated as follows. If you have a macroscale linear constitutive relation,

$$\mathbf{J} = \mathbf{L}\bigg|_{\mathbf{F}_e=0} : \mathbf{F}_e, \tag{13.10}$$

where the linear transport coefficient is given as

$$\mathbf{L}\bigg|_{\mathbf{F}_e=0} =: \mathbf{L}(0), \tag{13.11}$$

then the general form of the Green–Kubo relation may be written as,

$$\mathbf{L}(0) = \beta V \int_0^\infty d\tau <\mathbf{J}(0) \otimes \mathbf{J}(\tau)>\Big|_{F_e=0}, \quad (13.12)$$

which can be proved by using the fluctuation and dissipation theorem. In this book, instead of providing a general proof, which can be found in many statistical mechanics books, we shall prove the Green–Kubo formulation for several individual cases, to connect it with actual applications.

Before proceeding further, we first derive the classical Green–Kubo formula.

13.1.3 Derivation of the Green–Kubo Formula

We first consider a random dynamic variable $A(t)$ and its time derivative, and it is straightforward that

$$A(t) - A(0) = \int_0^t \dot{A}(t')dt'$$

then,

$$[A(t) - A(0)]^2 = \left\{\int_0^t \dot{A}(t')dt'\right\}^2 = \left(\int_0^t \dot{A}(t')dt'\right)\left(\int_0^t \dot{A}(t'')dt''\right)$$
$$= \int_0^t dt'' \int_0^t dt' \dot{A}(t')\dot{A}(t'')$$

and

$$msd = <[A(t) - A(0)]^2> = \int_0^t dt'' \int_0^t dt' <\dot{A}(t')\dot{A}(t'')>.$$

By symmetry, we may only need to integrate on half of the square domain,

$$msd = 2\int_0^t dt'' \int_0^{t''} dt' <\dot{A}(t')\dot{A}(t'')>.$$

Since the correlation function is not affected by the shifting in time origin, i.e.,

$$<\dot{A}(t')\dot{A}(t'')> = <\dot{A}(0)\dot{A}(t''-t')>$$

as shown in Fig. 13.2, the change of variables, $\tau = t'' - t', d\tau = -dt'$, yields

$$msd = 2\int_0^t dt'' \int_0^{t''} d\tau <\dot{A}(0)\dot{A}(\tau)>, \quad (13.13)$$

where the minus sign in the front of $d\tau$ is absorbed by switching the upper and lower limits of the integration.

Changing the order of the integration for the double integral, we have

$$msd = 2\int_0^t d\tau \int_\tau^t dt'' <\dot{A}(\tau)\dot{A}(0)> = 2t\int_0^t d\tau <\dot{A}(0)\dot{A}(\tau)>\left(1 - \frac{\tau}{t}\right).$$

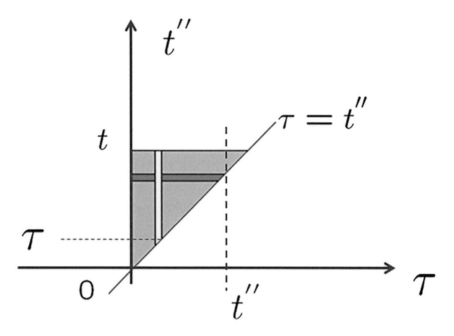

Figure 13.2 Integration domain and change of variable

Then,
$$\frac{<[A(t)-A(0)]^2>}{2t} = \int_0^t d\tau <\dot{A}(0)\dot{A}(\tau)> (1-\tau/t).$$

Taking the long time limit, we find that,
$$\lim_{t\to\infty} \frac{<[A(t)-A(0)]^2>}{2t} = \int_0^\infty d\tau <\dot{A}(0)\dot{A}(\tau)>. \tag{13.14}$$

This expression is the so-called Green–Kubo formula, which links the mean square displacement of a dynamic variable to its velocity correlation function.

Alternatively, the Green–Kubo formula may also be expressed as

$$\frac{d}{dt} <[A(t)-A(0)]^2> = 2\int_0^\infty d\tau <\dot{A}(0)\dot{A}(\tau)>. \tag{13.15}$$

At this point, we have not linked the autocorrelation to the transport coefficient yet, i.e.,

$$\mathbf{L}(0) = \int_0^\infty <\dot{\mathbf{A}}(0) \otimes \dot{\mathbf{A}}(t)> dt.$$

A general proof of the above relation by using the fluctuation–dissipation theorem will be given later. Before we present the general proof, we first discuss some special cases of the Green–Kubo relation.

13.1.4 Diffusion and the Einstein Relation

We now consider the common diffusion process that is governed by Fick's law

$$\mathbf{j} = -D\nabla c, \tag{13.16}$$

where D is the diffusion coefficient, and c is the concentration of a single species.

This is a macroscale-level description of diffusion that is by the gradient of concentration. On the other hand, the macroscale diffusion is a statistical average of the Brownian motion of many microscale particles. We need to find a micro–macro link that can be used to compute the diffusion coefficient D, which is a macroscale or statistical parameter.

Based on Fick's law, the conservation of species flux is,

$$\frac{\partial c(r,t)}{\partial t} + \nabla \cdot \mathbf{j}(r,t) = 0.$$

Combining with the first equation, we have,

$$\frac{\partial c(r,t)}{\partial t} - D\nabla^2 c(r,t) = 0.$$

Here r is radial coordinate in a spherical coordinates.

We assume that at the beginning all the particles are concentrated at the center, i.e.,

$$c(r,0) = \delta(r), \tag{13.17}$$

where $\delta(r)$ is the Dirac's delta function.

One can solve the equation, and find the solution

$$c(r,t) = \frac{1}{(4\pi Dt)^{d/2}} \exp\left(-\frac{r^2}{4Dt}\right). \tag{13.18}$$

For fixed t, Eq. (13.18) is a Gaussian distribution, and here d is the dimension of the space, and the following normalization condition must be satisfied,

$$\lim_{t \to \infty} \int_{V(t)} c(r,t) dV = 1. \tag{13.19}$$

Thus, one may view the species concentration distribution is a probability distribution of the microscale random particle positions. From this perspective, the second moment of the species distribution is equivalent to the mean square of particle displacements, i.e.,

$$< [\mathbf{r}(t) - \mathbf{r}(0)]^2 > = <r^2> = \int_V c(r,t) r^2 dV. \tag{13.20}$$

If we have the computer simulation data of microscale particle motions, this can be obtained by,

$$< \Delta r(t)^2 > = \frac{1}{N} \sum_{i=1}^{N} \Delta r_i^2.$$

For 3D cases, by definition we have,

$$<r^2> = \int_V c(r,t)r^2 dV. \tag{13.21}$$

Taking time derivative of the above expression,

$$\frac{\partial}{\partial t}<r^2> = D\int_V r^2 \nabla^2 c(r,t) dV$$

$$= D\int_V dV\left(\nabla \cdot (r^2 \nabla c) - \nabla r^2 \cdot \nabla c\right)$$

$$= D\oint_{\partial V} r^2 \nabla c \cdot \mathbf{n} dS - 2D\int_V \nabla r \cdot \nabla c dV,$$

$$= 0 - 2D\int_V \left(\nabla \cdot (\mathbf{r}c(r,t)) - (\nabla \cdot \mathbf{r})c(r,t)\right) dV$$

$$= -2D\int_{\partial V} \mathbf{r} \cdot \mathbf{n}c(r,t) dS + 2D\int_V (\nabla \cdot \mathbf{r})c(r,t) dV$$

$$= 6D\int_V c(r,t) dV.$$

In the above derivation, we assume that at boundary of space V the species concentration vanishes; and we used the identity that $\nabla \cdot \mathbf{r} = 3$.

When $t \to \infty$, we have

$$6D \lim_{t\to\infty} \int_{V(t)} c(r,t) dV = 6D = \lim_{t\to\infty} \frac{d}{dt} <[\Delta \mathbf{r}]^2>. \tag{13.22}$$

From Eqs. (13.19) and (13.22), we can then derive the Einstein (1905) relation,

$$D = \lim_{t\to\infty} \frac{1}{6}\frac{d}{dt} <[\mathbf{r}(t) - \mathbf{r}(0)]^2>. \tag{13.23}$$

This relation can be used to find a host of transport coefficients in diffusion processes such as heat conduction and mass diffusions.

By using the Einstein relation, we can link the transport coefficient to an ensemble average of particle position fluctuation for a class of diffusion processes.

In numerical computer simulation, we use the formula,

$$D = \frac{1}{6N} \lim_{t\to\infty} \frac{d}{dt} \sum_{i=1}^{N} [\mathbf{r}_i(t) - \mathbf{r}_i(0)]^2. \tag{13.24}$$

By using the Green–Kubo formula, we can have

$$\lim_{t\to\infty} \frac{<[A(t) - A(0)]^2>}{2t} = \int_0^\infty d\tau <\dot{A}(0)\dot{A}(\tau)>,$$

then
$$\frac{d}{dt} <[A(t) - A(0)]^2> = 2 \int_0^\infty d\tau <\dot{A}(0)\dot{A}(\tau)>.$$

Considering the fact that $A(t) - A(0) \to \mathbf{r}(t) - \mathbf{r}(0)$, we can then derive

$$\frac{d}{dt}\left(\Delta r^2(t)\right) := \frac{d}{dt} <[\mathbf{r}_i(t) - \mathbf{r}_i(0)]^2> = 2 \int_0^\infty d\tau <\mathbf{v}_i(0) \cdot \mathbf{v}_i(\tau)>.$$

Recalling
$$6D = \lim_{t \to \infty} \frac{d}{dt} <[\Delta \mathbf{r}_i]^2>,$$

we can find that
$$D = \frac{1}{3} \int_0^\infty dt <\mathbf{v}_i(0) \cdot \mathbf{v}_i(t)>. \quad (13.25)$$

This is the **Green–Kubo** relation that links the diffusivity with the autocorrelation of the particle velocity field of the diffusion process.

In the following example, we discuss how to use the Green–Kubo relation to calculate the macroscale viscosity of the Newtonian fluid by using the virial stress formulation of MD simulations.

13.1.5 Green–Kubo Relation for the Viscosity Coefficient

We first consider the following macroscale constitutive relation of the Newtonian fluid,

$$\bar{\sigma} = \eta : (\nabla \otimes \mathbf{v}). \quad (13.26)$$

Comparing with the Green–Kubo relation

$$\mathbf{L}(0) = \beta V \int_0^\infty d\tau <\mathbf{J}(0) \otimes \mathbf{J}(\tau)> \Big|_{F_e=0},$$

we can make the following variable substitutions, $\mathbf{L}(0) \to \eta$ and $\mathbf{J} \to \sigma/V$. Then the Green–Kubo relation for this particular case becomes

$$\eta = \frac{V}{k_B T} \int_0^\infty <\frac{\sigma(0)}{V} \otimes \frac{\sigma(t)}{V}> dt$$

$$= \frac{\rho}{3k_B T} \frac{1}{NM} \sum_{i=1}^N \sum_{j=1}^M \sigma(t_j) \otimes \sigma(t_j + t), \quad (13.27)$$

where $\rho = 1/V$. For the detailed discussion of the analytical expression of the Green–Kubo relation for viscosity, interested readers may consult the monograph of Evans and Morriss (2007).

The following LAMMPS script is excerpted from LAMMPS documentation Section **8.4.6 Calculate viscosity**: *https://lammps.sandia.gov/doc/Howto_viscosity.html*. It uses the Green–Kubo formula to calculate the average viscosity coefficient for the liquid Argon by using the Green–Kubo formula.

```
# Sample LAMMPS input script for viscosity of liquid Ar

units       real
variable    T equal 86.4956
variable    V equal vol
variable    dt equal 4.0
variable    p equal 400        # correlation length
variable    s equal 5          # sample interval
variable    d equal $p*$s      # dump interval

# convert from LAMMPS real units to SI

variable    kB equal 1.3806504e-23    # [J/K] Boltzmann
variable    atm2Pa equal 101325.0
variable    A2m equal 1.0e-10
variable    fs2s equal 1.0e-15
variable    convert equal ${atm2Pa}*$atm2Pa*${fs2s}*${A2m}*${A2m}*${A2m}

# setup problem

dimension    3
boundary     p p p
lattice      fcc 5.376 orient x 1 0 0 orient y 0 1 0 orient z 0 0 1
region       box block 0 4 0 4 0 4
create_box   1 box
create_atoms 1 box
mass         1 39.948
pair_style   lj/cut 13.0
pair_coeff   * * 0.2381 3.405
timestep     ${dt}
thermo       $d

# equilibration and thermalization

velocity    all create $T 102486 mom yes rot yes dist gaussian
fix         NVT all nvt temp $T $T 10 drag 0.2
run         8000

# viscosity calculation, switch to NVE if desired

#unfix      NVT
#fix        NVE all nve

\textcolor{blue}{
reset_timestep 0
variable    pxy equal pxy
variable    pxz equal pxz
variable    pyz equal pyz
fix         SS all ave/correlate $s $p $d &
            v_pxy v_pxz v_pyz type auto file S0St.dat ave running
variable    scale equal ${convert}/(${kB}*$T)*$V*$s*${dt}
variable    v11 equal trap(f_SS[3])*${scale}
variable    v22 equal trap(f_SS[4])*${scale}
variable    v33 equal trap(f_SS[5])*${scale}
thermo_style custom step temp press v_pxy v_pxz v_pyz v_v11 v_v22 v_v33
run         100000
variable    v equal (v_v11+v_v22+v_v33)/3.0
}
variable    ndens equal count(all)/vol
print       "average viscosity: $v [Pa.s] @ $T K, ${ndens} /A^3"
```

The key part of the script of how to calculate the correlation function of stress tenor as formulated in Eq. (13.27) is given as follows,

```
fix         SS all ave/correlate $s $p $d &
            v_pxy v_pxz v_pyz type auto file S0St.dat ave running
```

the values of variables s, p, and d are

- Nevery = use input values every this many time steps
- Nrepeat = # of correlation time windows to accumulate
- Nfreq = calculate time window averages every this many time steps

The following sentences are also excerpted form LAMMPS manual, which provide a clear explanation and a detailed demonstration on the meanings of parameters $Nevery, Nrepeat, and\ Nfreq$:

The Nevery, Nrepeat, and Nfreq arguments specify on what timesteps the input values will be used to calculate correlation data. The input values are sampled every Nevery timesteps. The correlation data for the preceding samples is computed on timesteps that are a multiple of Nfreq. Consider a set of samples from some initial time up to an output timestep. The initial time could be the beginning of the simulation or the last output time; see the ave keyword for options. For the set of samples, the correlation value Cij is calculated as:

$$C_{ij}(delta) = ave(V_i(t) * V_j(t + delta))$$

which is the correlation value between input values V_i and V_j, separated by time delta. Note that the second value Vj in the pair is always the one sampled at the later time. The symbol $ave()$ represents an average over every pair of samples in the set that are separated by time delta. The maximum delta used is of size (Nrepeat-1)*Nevery. Thus the correlation between a pair of input values yields Nrepeat correlation datums:

$$C_{ij}(0), C_{ij}(Nevery), C_{ij}(2 * Nevery), \ldots, C_{ij}((Nrepeat - 1) * Nevery)$$

For example, if Nevery=5, Nrepeat=6, and Nfreq=100, then values on time steps 0, 5, 10, 15, ?100 will be used to compute the final averages on timestep 100. Six averages will be computed:

$C_{ij}(0), C_{ij}(5), C_{ij}(10), C_{ij}(15), C_{ij}(20)$, and $C_{ij}(25)$. $C_{ij}(10)$ on the time step 100 will be the average of 19 samples, namely $V_i(0) * V_j(10), V_i(5) * V_j(15), V_i(10) * V_j(20), V_i(15) * V_j(25), \ldots, V_i(85) * V_j(95), V_i(90) * V_j(100)$.

Nfreq must be a multiple of Nevery; Nevery and Nrepeat must be non-zero. Also, if the ave keyword is set to one which is the default, then $Nfreq >= (Nrepeat - 1) * Nevery$ is required.

13.1.6 Green–Kubo Relation for Thermal Conductivity

Both MD and first-principles-based calculations are very useful methods the thermal conductivity in applications. In this section, we provide the formula as well as the derivation of the Green–Kubo relation for thermal conductivity, which mainly follows from the Evans and Morris' (2007) book *Statistical Mechanics of Non-equilibrium Liquids*.

We first consider the general Green–Kubo relation,

$$L(0) = \beta V \int_0^\infty d\tau <\mathbf{J}(0) \cdot \mathbf{J}(\tau)> \bigg|_{F_e=0}. \tag{13.28}$$

In the above equation, we let $L(0) = \lambda = T$ and $\mathbf{J} \to \mathbf{J}/V$. Then we have

$$\lambda = \frac{\beta V}{T} \int_0^\infty d\tau < \frac{\mathbf{J}(0)}{V} \cdot \frac{\mathbf{J}(\tau)}{V} >\bigg|_{F_e=0}, \tag{13.29}$$

which leads to

$$\lambda = \lim_{\tau \to \infty} \frac{1}{V k_B T^2} \int_0^\tau d\tau <\mathbf{J}(0) \cdot \mathbf{J}(\tau)> \bigg|_{F_e=0}, \tag{13.30}$$

where

$$\mathbf{J} = -\lambda \nabla T. \tag{13.31}$$

Then the Green–Kubo thermal conductivity formula is given as

$$\frac{1}{\lambda} = \lim_{\tau \to \infty} \frac{1}{V k_B T^2} \int_0^\tau <\nabla T(0) \cdot \nabla T(t)> dt. \tag{13.32}$$

The following proof of the Green–Kubo formula for the heat conductivity is based on the presentation from Evans and Morriss (2007).

During heat conduction, the continuity equation of internal energy gives,

$$\rho \frac{dU}{dt}\bigg|_V = \frac{dQ}{dt} = -\nabla \cdot \mathbf{J}_Q, \tag{13.33}$$

where $\mathbf{J}_Q$ is the heat flux, which is based on the Fourier's law,

$$\mathbf{J}_Q = -\lambda \nabla T.$$

These lead to heat diffusion equation,

$$\rho \frac{dU}{dt} = \lambda \nabla^2 T \quad \to \quad \rho \frac{d\Delta U}{dt} = \lambda \nabla^2 \Delta T,$$

for unsteady-state nonequilibrium heat conduction.

Based on the equilibrium thermodynamics,

$$\frac{1}{V}\frac{\partial E}{\partial T}\bigg|_V = \frac{\partial Q}{\partial T}\bigg|_V = \frac{\partial (\rho U)}{\partial T}\bigg|_V = \rho C_V, \quad C_V \text{ is the specific heat.}$$

This implies that for small temperature fluctuation,

$$\Delta T = \frac{\Delta(\rho U)}{\rho C_V}.$$

Therefore,

$$\rho \frac{d\Delta U}{dt} = \rho \Delta \dot{U} = \lambda \nabla^2 \Delta T = \lambda \nabla^2 \left(\frac{\Delta(\rho U)}{\rho C_V}\right) \quad \to \quad \rho \Delta \dot{U} = \frac{\lambda}{\rho C_V}\nabla^2 \rho \Delta U.$$

Let $D_T = \lambda/(\rho C_V)$ as the thermal diffusivity. We consider the following form of Fourier series solution of the internal energy,

$$\Delta U(\mathbf{k},t) = \exp(i\mathbf{k}\cdot\mathbf{r})u(t) \rightarrow \nabla^2 U(\mathbf{k},t) = -k^2 U(\mathbf{k},t).$$

Hence,
$$\rho\Delta\dot{U}(\mathbf{k},t) = -k^2 D_T(\mathbf{k},t)\rho\Delta U(\mathbf{k},t).$$

A generalization of $\rho\Delta\dot{U}(\mathbf{k},t) = -k^2 D_T(k,t)\rho\Delta U(\mathbf{k},t)$ is

$$\rho\frac{dU(\mathbf{k},t)}{dt} = -k^2\int_0^t d\tau\, D_T(k,t-\tau)\rho U(\mathbf{k},\tau), \qquad (13.34)$$

where $k = |\mathbf{k}|$.

Multiplying $\rho U(-\mathbf{k},0) = \rho U^*(\mathbf{k},t)$ through Eq. (13.34) yields,

$$\frac{d}{dt}C(k,t) = -k^2\int_0^t d\tau\, D_T(k,t-\tau)C(k,\tau). \qquad (13.35)$$

Taking the Fourier transform of Eq. (13.35) in time yields,

$$i\omega\tilde{C}(k,\omega) - C(k,0) = -k^2\tilde{D}_T(k,\omega)\tilde{C}(k,\omega), \qquad (13.36)$$

and hence we have

$$\tilde{C}(k,\omega) = \frac{C(k,0)}{i\omega + k^2 D_T(k,\omega)}. \qquad (13.37)$$

On the other hand,
$$\frac{d^2}{dt^2}C(k,t) = \frac{d}{dt}<\rho\frac{d\Delta U(\mathbf{k},t)}{dt}(\rho U(-\mathbf{k},0))>$$
$$= \frac{d}{dt}<D_T(-k^2)\rho\Delta U(\mathbf{k},t)(\rho U(-\mathbf{k},0))>$$
$$= <(-k^2 D_T\rho\Delta U(\mathbf{k},t)((-k^2 D_T\rho U(-\mathbf{k},0))>$$
$$= <\rho\dot{U}(\mathbf{k},t)\rho\dot{U}(-k,0)> = -\phi(k,t).$$

Note that $U^*(\mathbf{k},t) = U(-\mathbf{k},-t)$, so that $\rho k^2 D_T U(-\mathbf{k},0) = \rho\dot{U}(-\mathbf{k},0)$.

Recalling the equations,
$$\rho\dot{U}(\mathbf{k},t) = -\nabla\mathbf{J}_Q(\mathbf{k},t) = -\mathbf{k}\cdot\mathbf{J}_Q(\mathbf{k},t), \qquad (13.38)$$

and
$$\rho\dot{U}(-\mathbf{k},0) = -\nabla\mathbf{J}_Q(-\mathbf{k},0) = \mathbf{k}\cdot\mathbf{J}_Q(-\mathbf{k},0), \qquad (13.39)$$

we then have
$$\phi(k,t) = k^2 <\mathbf{J}_Q(\mathbf{k},t)\cdot\mathbf{J}_Q(-\mathbf{k},0)> = k^2 N_Q(k,t). \qquad (13.40)$$

We know that
$$\tilde{\phi}(k,\omega) = -\omega^2\tilde{C}(k,\omega) - i\omega C(k,0) = (i\omega)(i\omega\tilde{C}(k,\omega) - C(k,0))$$

and
$$\tilde{C}(k,\omega) = (i\omega)^{-1}\tilde{\phi}(k,\omega) + C(k,0).$$

Then we have

$$\tilde{D}_T(k,\omega) = \frac{i\omega\tilde{C}(k,\omega) - C(k,0)}{-k^2\tilde{C}(k,\omega)},$$

and hence

$$\tilde{D}_T(k,\omega) = \frac{\tilde{\phi}(k,\omega)}{k^2((i\omega)\tilde{\phi}(k,\omega) + C(k,0))} = \frac{N_Q(k,\omega)}{((i\omega)\tilde{\phi}(k,\omega) + C(k,0))}.$$

As $k \to 0$,

$$\tilde{D}_T(0,\omega) = N_Q(0,\omega)/C(0,0),$$

where $C(0,0) = <(\rho\Delta U(0,0))^2> = <(\Delta E)^2>$.

13.2 Example: LAMMPS Simulation of Thermal Conductivity

In this section, we present an example of simulation of heat conductivity of solid Argon by using the Green–Kubo formula that is implemented in LAMMPS. An outstanding feature of the Green–Kubo formula is that it can use the equilibrium MD to calculate the transport coefficient of a nonequilibrium process. Even though the accuracy of such calculation is still debatable, it nevertheless provides a tool to evaluate the material properties under the nonequilibrium state.

In the example, we use both NVT and NPT ensemble MD to simulate the heat conductivity coefficient from 10K° to 80K°. We use the Lennard-Jones potential for the solid Argon with parameter $\epsilon = 0.2381$ Kcal/mol, $\sigma = 3.405 \times 10^{-10}$ m. The thermal conductivity is calculated by using the Green–Kubo formula

$$\lambda = \frac{1}{3Vk_BT^2} \int_0^\tau <\mathbf{J}(0) \cdot \mathbf{J}(t)> dt, \qquad (13.41)$$

where V is the volume, T the temperature, and the angular brackets denote the ensemble average, or, in the case of a MD simulation, the average over time. The microscopic heat current is given by

$$\mathbf{J}(t) = \sum_{i=1}^N \mathbf{v}_i \epsilon_i + \frac{1}{2} \sum_{i,j, i \neq j} \mathbf{r}_{ij}(\mathbf{F}_{ij} \cdot \mathbf{v}_j), \qquad (13.42)$$

where $\mathbf{v}_i$ is the velocity of particle i and $\mathbf{F}_{ij}$ is the force on the atom i due to its neighbor atom j from the pair potential. The site energy ϵ_i is given by

$$\epsilon_i = \frac{1}{2}m_i|\mathbf{v}_i|^2 + \frac{1}{2}\sum_{i<j}(\mathbf{F}_{ij} \cdot \mathbf{v}_i)\mathbf{r}_{ij}, \qquad (13.43)$$

where m_i is the mass of the atom i.

A detailed discussion on the simulation method and results can be found in Tretiakov and Scandolo (2004).

In LAMMPS (*https://lammps.sandia.gov/doc/compute_heat_flux.html*), the syntax of **compute heat/flux command** is as follows:

Syntax
compute ID group-ID heat/flux ke-ID pe-ID stress-ID
where

- ID, group-ID are documented in compute command
- heat/flux = style name of this compute command
- ke-ID = ID of a compute that calculates per-atom kinetic energy
- pe-ID = ID of a compute that calculates per-atom potential energy
- stress-ID = ID of a compute that calculates per-atom stress

The LAMMPS script to use the Green–Kubo formula to calculate the thermal conductivity of Argon is called *in.conductivity*, which is listed subsequently:

```
# Set the number of processors to use
package omp 3
# Sample LAMMPS input script for thermal conductivity of solid Ar
units       real
variable    T equal 70
variable    V equal vol
variable    dt equal 4.0
variable    p equal 200      # correlation length
variable    s equal 10       # sample interval
variable    d equal $p*$s    # dump interval
# convert from LAMMPS real units to SI
variable    kB equal 1.3806504e-23    # [J/K] Boltzmann
variable    kCal2J equal 4186.0/6.02214e23
variable    A2m equal 1.0e-10
variable    fs2s equal 1.0e-15
variable    convert equal ${kCal2J}*${kCal2J}/${fs2s}/${A2m}
# setup problem
dimension   3
boundary    p p p
lattice     fcc 5.376 orient x 1 0 0 orient y 0 1 0 orient z 0 0 1
region      box block 0 4 0 4 0 4
create_box  1 box
create_atoms 1 box
mass        1 39.948
pair_style  lj/cut 13.0
pair_coeff  * * 0.2381 3.405
timestep    ${dt}
thermo      $d

# equilibration and thermalization
velocity    all create $T 102486 mom yes rot yes dist gaussian
fix         NVT all nvt temp $T $T 10 drag 0.2
run         8000
# thermal conductivity calculation, switch to NVE if desired
#unfix      NVT
#fix        NVE all nve
reset_timestep 0
compute     myKE all ke/atom
compute     myPE all pe/atom
compute     myStress all stress/atom NULL virial
compute     flux all heat/flux myKE myPE myStress
variable    Jx equal c_flux[1]/vol
variable    Jy equal c_flux[2]/vol
variable    Jz equal c_flux[3]/vol
fix         JJ all ave/correlate $s $p $d &
            c_flux[1] c_flux[2] c_flux[3] type auto file J0Jt.dat ave running
```

```
variable      scale equal ${convert}/${kB}/$T/$T/$V*$s*${dt}
variable      k11 equal trap(f_JJ[3])*${scale}
variable      k22 equal trap(f_JJ[4])*${scale}
variable      k33 equal trap(f_JJ[5])*${scale}
thermo_style custom step temp v_Jx v_Jy v_Jz v_k11 v_k22 v_k33
#output movie for visualization
#dump m2 all movie $d movie.avi type type size 1280 720
#output xyz file for VMD
dump t1 all xyz $d ar.xyz
run           100000
variable      k equal (v_k11+v_k22+v_k33)/3.0
variable      ndens equal count(all)/vol
print         "average conductivity: $k[W/mK] @ $T K, ${ndens} /A^3"
```

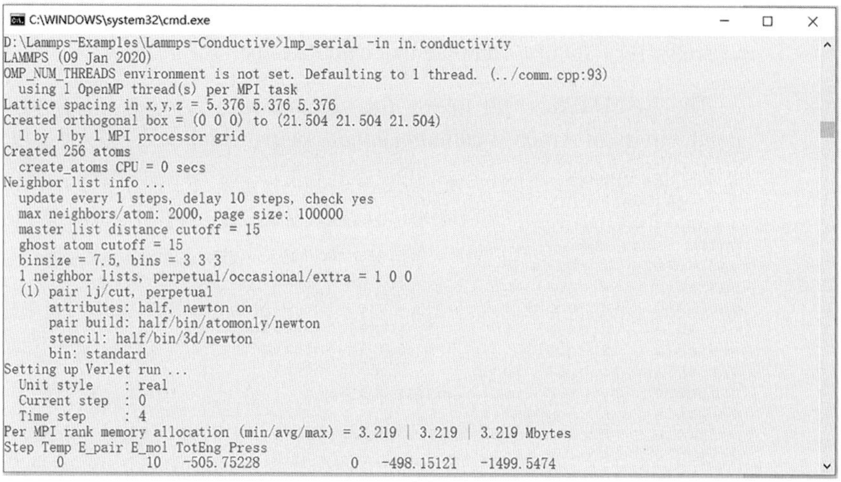

```
C:\WINDOWS\system32\cmd.exe
D:\Lammps-Examples\Lammps-Conductive>lmp_serial -in in.conductivity
LAMMPS (09 Jan 2020)
OMP_NUM_THREADS environment is not set. Defaulting to 1 thread. (../comm.cpp:93)
  using 1 OpenMP thread(s) per MPI task
Lattice spacing in x,y,z = 5.376 5.376 5.376
Created orthogonal box = (0 0 0) to (21.504 21.504 21.504)
  1 by 1 by 1 MPI processor grid
Created 256 atoms
  create_atoms CPU = 0 secs
Neighbor list info ...
  update every 1 steps, delay 10 steps, check yes
  max neighbors/atom: 2000, page size: 100000
  master list distance cutoff = 15
  ghost atom cutoff = 15
  binsize = 7.5, bins = 3 3 3
  1 neighbor lists, perpetual/occasional/extra = 1 0 0
  (1) pair lj/cut, perpetual
      attributes: half, newton on
      pair build: half/bin/atomonly/newton
      stencil: half/bin/3d/newton
      bin: standard
Setting up Verlet run ...
  Unit style    : real
  Current step  : 0
  Time step     : 4
Per MPI rank memory allocation (min/avg/max) = 3.219 | 3.219 | 3.219 Mbytes
Step Temp E_pair E_mol TotEng Press
       0           10   -505.75228            0   -498.15121   -1499.5474
```

(a)

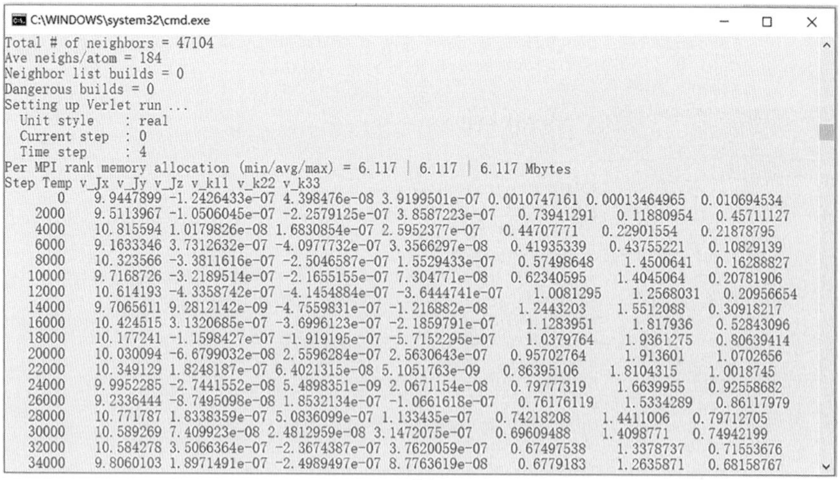

```
C:\WINDOWS\system32\cmd.exe
Total # of neighbors = 47104
Ave neighs/atom = 184
Neighbor list builds = 0
Dangerous builds = 0
Setting up Verlet run ...
  Unit style    : real
  Current step  : 0
  Time step     : 4
Per MPI rank memory allocation (min/avg/max) = 6.117 | 6.117 | 6.117 Mbytes
Step Temp v_Jx v_Jy v_Jz v_k11 v_k22 v_k33
       0    9.9447899 -1.2426433e-07 4.398476e-08 3.9199501e-07 0.0010747161 0.00013464965 0.010694534
    2000    9.5113967 -1.0506045e-07 -2.2579125e-07 3.8587223e-07 0.73941291 0.11880954 0.45711127
    4000    10.815594 1.0179826e-08 1.6830854e-07 2.5952377e-07 0.44707771 0.22901554 0.21878795
    6000    9.1633346 3.7312632e-07 -4.0977732e-07 3.3566297e-08 0.41935339 0.43755221 0.10829139
    8000    10.323566 -3.3811616e-07 -2.5046587e-07 1.5529433e-07 0.57498648 1.4500641 0.16288827
   10000    9.7168726 -3.2189514e-07 -2.1655155e-07 7.304771e-08 0.62340595 1.4045064 0.20781906
   12000    10.614193 -4.3358742e-07 -4.1454884e-07 -3.6444741e-07 1.0081295 1.2568031 0.20956654
   14000    9.7065611 9.2812142e-09 -4.7559831e-07 -1.216882e-08 1.2443203 1.5512088 0.30918217
   16000    10.424515 3.1320685e-07 -3.6996123e-07 -2.1859791e-07 1.1283951 1.817936 0.52843096
   18000    10.177241 -1.1598427e-07 -1.919195e-07 -5.7152295e-07 1.0379764 1.9361275 0.80639414
   20000    10.030094 -6.6799032e-08 2.5596284e-07 2.5630643e-07 0.95702764 1.913601 1.0702656
   22000    10.349129 1.8248187e-07 6.4021315e-08 5.1051763e-09 0.86395106 1.8104315 1.0018745
   24000    9.9952285 -2.7441552e-08 5.4898351e-09 2.0671154e-08 0.79777319 1.6639955 0.92558682
   26000    9.2336444 -8.7495098e-08 1.8532134e-07 -1.0661618e-07 0.76176119 1.5334289 0.86117979
   28000    10.771787 1.8338359e-07 5.0836099e-07 1.133435e-07 0.74218208 1.4411006 0.79712705
   30000    10.589269 7.409923e-08 2.4812959e-08 3.1472075e-07 0.69609488 1.4098771 0.74942199
   32000    10.584278 3.5066364e-07 -2.3674387e-07 3.7620059e-07 0.67497538 1.3378737 0.71553676
   34000    9.8060103 1.8971491e-07 -2.4989497e-07 8.7763619e-08 0.6779183 1.2635871 0.68158767
```

(b)

Figure 13.3 (a) The screenshot when using LAMMPS calculating the solid Argon's conductivity, and (b) the screen output of thermal flux and thermal conductivity values

One can run the LAMMPS script by simply typing:

`lmp_serial <in.conductivity`

In Fig 13.3, we show the screenshot of the running process at the command prompt.
In the LAMMPS script, the key command is:

`compute     flux all heat/flux myKE myPE myStress`

In case of two-body LJ interactions, the heat flux is defined as

$$\mathbf{J} = \frac{1}{V}\left[\sum_i \mathbf{v}_i \epsilon_i + \sum_{i<j} (\mathbf{F}_{ij} \cdot \mathbf{v}_j)\mathbf{r}_{ij}\right]$$

$$= \frac{1}{V}\left[\sum_i \mathbf{v}_i \epsilon_i + \frac{1}{2}\sum_{i<j} (\mathbf{F}_{ij} \cdot (\mathbf{v}_i + \mathbf{v}_j))\mathbf{r}_{ij}\right], \quad (13.44)$$

and they are consistent with Eqs. (13.42)–(13.43). The LAMMPS simulation results are presented in Fig. 13.4.

When a system is away from equilibrium due to external disturbance, and when the external disturbance is off, how does the system approach (come back) to a new equilibrium state? Specifically, we would like to ask the following two questions:

Question I: What is the system's response to the disturbance?
Question II: What is the macroscale response of a system due to fine-scale fluctuation?

The fluctuation and dissipation theorem (FDT) for the systems in nonequilibrium steady state provides the precise answers to these two questions. In the following, we first present the classical version of the fluctuation and dissipation theorem, and then provide a proof of the contemporary version of the FDT (Kubo (1966); Chandler (1987)).

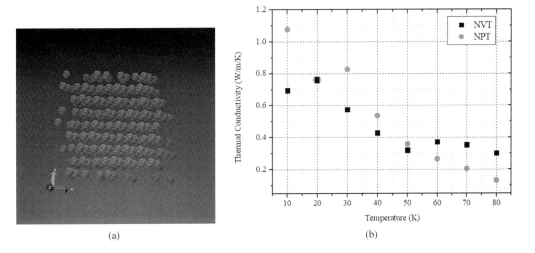

Figure 13.4 (a) Solid Argon lattice and (b) the thermal conductivity under different temperature

13.3 Fluctuation–Dissipation Theorem (FDT)

As discussed in the previous chapter, for the Langevin equation,

$$m_i \ddot{x} = -\nabla_i U - \alpha \dot{x}_i + R(t),$$

the dissipative or the drag force coefficient is related to the autocorrelation of the random force fluctuation, i.e.,

$$\alpha = \frac{1}{2k_B T} \int_{-\infty}^{\infty} <R(0) \cdot R(t)> dt. \tag{13.45}$$

Note that here the random force is not the flux, thus this is not exactly the Green–Kubo relation.

We may write the dissipation force in the following form of response function, i.e.,

$$\alpha \dot{x} = \int_{-\infty}^{t} \chi(\tau) \dot{x}(t - \tau) d\tau, \tag{13.46}$$

where $\chi(t) = 2\alpha \delta(t - 0)$ is the response function. Taking the Fourier transform for both sides of the above equation yields

$$\hat{\chi} = 2\alpha \quad \rightarrow \quad \alpha = \frac{1}{2}\hat{\chi}.$$

We may also rewrite Eq. (13.45) as

$$\alpha \, sgn(t) = \frac{1}{2k_B T} \int_{-\infty}^{\infty} <R(\tau)R(t+\tau)> d\tau. \tag{13.47}$$

Applying Fourier transform to both sides of Eq. (13.47) and recalling that $\mathcal{F}[sgn(t)] = 2/(i\omega)$, we have

$$\alpha \frac{2}{i\omega} = \frac{1}{2k_B T} \hat{\mathcal{R}}_{rr}(\omega) = \frac{1}{2k_B T} S_r(\omega), \tag{13.48}$$

where $i = \sqrt{-1}$; $S_{rr}(\omega)$ is the so-called power spectrum, i.e.,

$$S_{rr}(\omega) := <\hat{R}(\omega)\hat{R}^*(\omega)>, \tag{13.49}$$

and $\hat{\mathcal{R}}_{rr} = \mathcal{F}[\mathcal{R}_{rr}]$ is the Fourier transform of the following autocorrelation function,

$$\mathcal{R}_{rr} := \int_{-\infty}^{\infty} <R(\tau)R(t+\tau)> d\tau. \tag{13.50}$$

Equation (13.48) leads to a profound relation between the power spectrum and the response function in the Fourier space,

$$S_{rr}(\omega) = \frac{2k_B T}{\omega} Im(\hat{\chi}), \tag{13.51}$$

which is the classical form of the FDT.

As mentioned earlier, this is not a coincidence, but the classical formulation of the so-called fluctuation–dissipation theorem for the thermodynamic systems under nonequilibrium steady state.

The FDT in statistical physics provides a quantitative detailed balance between the energy dissipation into heat (e.g., drag force, friction) and the thermal fluctuations. Because FDT is the theoretical foundation of nonequilibrium molecular dynamics, in this section, we provide an introduction as well as the detailed derivations of FDT, catering those readers who do not have formal training in statistical physics while interested in using MD as a computational tool to solve scientific and engineering problems.

13.3.1 Fluctuation–Dissipation Theorem (FDT)

In this section, we shall give a formal proof of the classical fluctuation-dissipation theorem for a scalar dynamic variable $A(t) = A(x(t), p(t))$ in a general setting (not restricted in the case of the Langevin dynamics). The only assumption we made is that $A(t)$ is a real and symmetric time-dependent function.

Let $A(t)$ be an observable of a dynamical system with the Hamiltonian $H_0(x)$ subject to thermal fluctuations. The observable $A(t)$ will fluctuate around its mean value $<A(0)>$ with fluctuations that can be characterized by the power spectrum

$$S_{AA}(\omega) = <\hat{A}(\omega)\hat{A}^*(\omega)> \tag{13.52}$$

where $\hat{A}(\omega)$ and $\hat{A}^*(\omega)$ are Fourier transforms, i.e.,

$$\hat{A} = \int_{-\infty}^{\infty} A(t)\exp(i\omega t)dt, \text{ and } \hat{A}^* = \int_{-\infty}^{\infty} A^*(t)\exp(i\omega t)dt \tag{13.53}$$

and $A^*(t)$ is the complex of conjugate of $A(t)$.

The power spectrum of a dynamic time series $A(t)$ is an important variable, because based on the Wiener–Khinchin theorem (see Wiener (1964)) the inverse Fourier transform of the power spectrum of $S_{AA}(\omega)$ equals to the time autocorrelation function of $A(t)$, i.e.,

$$R_{AA}(\tau) = \int_{-\infty}^{\infty} A(t)A(t-\tau)dt = \frac{1}{2\pi}\int_{-\infty}^{\infty} \hat{A}(\omega)\hat{A}^*(\omega)d\omega = \mathcal{F}^{-1}(S_{AA}(\omega)). \tag{13.54}$$

Suppose that there exists a time-varying, spatially constant force field $f(t)$ that alters the original Hamiltonian to $H(A) = H_0(A) - f_0 A(0)$. The response of the observable $A(t)$ to a time-dependent field $f(t) = f_0 H(t)$ may be characterized to first order by the susceptibility or linear response function $\chi(t)$ of the system,

$$<A(t)> = <A(0)> + \int_{-\infty}^{t} \chi(t-\tau)f(\tau)d\tau = <A(0)> + f_0 \int_{-\infty}^{t} \chi(t-\tau)H(t)d\tau. \tag{13.55}$$

Consider

$$f(t) = f_0 H(-t) = \begin{cases} 1, & t > 0 \\ 0, & t \leq 0, \end{cases} \quad (13.56)$$

where $H(t)$ is the Heaviside function. Thus, the expectation of $A(t)$ should be

$$<A(t)>_0 = \frac{1}{Z_0} \int_{-\infty}^{\infty}\int_{-\infty}^{\infty} dxdp\, A(t)\exp(-\beta H_0(x,p)), \quad (13.57)$$

where $H_0(x, p)$ is the Hamiltonian function at the equilibrium state, and

$$Z_0 = \int_{-\infty}^{\infty}\int_{-\infty}^{\infty} \exp(-\beta H_0(x(t), p(t)))dxdp. \quad (13.58)$$

For the nonequilibrium state, we assume that it is a small fluctuation from the equilibrium state, so the Hamiltonian function becomes

$$H(A(t)) = H_0(x, p) - A(0)f,$$

where f is a perturbation filed, and

$$<A(t)> = \frac{1}{Z} \int_{-\infty}^{\infty}\int_{-\infty}^{\infty} dxdp\, A(t)\exp(-\beta H(A(t)))$$

$$= \frac{1}{Z} \int_{-\infty}^{\infty}\int_{-\infty}^{\infty} dxdp\, A(t)\exp(-\beta (H_0(x(t), p(t)) - A(0)f)),$$

where

$$Z = \int_{-\infty}^{\infty}\int_{-\infty}^{\infty} \exp(-\beta H(A(t)))dxdp$$

$$= \int_{-\infty}^{\infty}\int_{-\infty}^{\infty} \exp(-\beta (H_0(x(t), p(t)) - A(0)f)dxdp. \quad (13.59)$$

Now, we study the asymptotic behaviors of $<A(t)>$. Considering $H(A) = H_0(x, p) + \Delta H$, we have

$$<A(t)> = \frac{1}{Z} \int_{-\infty}^{\infty}\int_{-\infty}^{\infty} dxdp\, A(t)\exp(-\beta H(A(t)))$$

$$= \int_{-\infty}^{\infty}\int_{-\infty}^{\infty} dxdp\, A(t)\exp(-\beta (H_0(x(t), p(t)) + \Delta H))$$

$$\cdot \left(\int_{-\infty}^{\infty}\int_{-\infty}^{\infty} dxdp\, \exp(-\beta (H_0(x(t), p(t)) + \Delta H))\right)^{-1}. \quad (13.60)$$

We can then calculate the variational Taylor expansion of $<A>$, i.e.,

$$<A(t)> = <A(t)>_0 + \delta <A(t)> + O(f^2). \quad (13.61)$$

Considering $\Delta H = -f_0 A(0)$ and taking Gateaux derivative, we can obtain

$$\delta <A(t)> = \frac{d}{d\alpha} <A(t)>(\alpha f)\Big|_{\alpha=0}$$
$$= \frac{\partial <A(t)>}{\partial f_0}\Big|_{f_0=0} f_0 = -\beta\big(<\Delta \mathcal{H} A>_0 - <A(t)>_0<\Delta \mathcal{H}>_0\big)$$
$$= \beta f_0\big(<A(0)A(t)>_0 - <A(t)>_0<A(0)>_0\big).$$
$$= \beta f_0\big(<\delta A(0)\delta A(t)>_0\big). \tag{13.62}$$

Now, we define

$$\mathcal{A}(t) := <\delta A(0)\delta A(t)>_0. \tag{13.63}$$

We can then rewrite Eq. (13.62) as

$$<A> = <A>_0 + \beta f_0 \mathcal{A}. \tag{13.64}$$

By using Eqs. (13.55) and (13.64), one can find that

$$f_0 \int_0^\infty d\tau \chi(\tau) H(tau - t) = \beta f_0 R_{AA}(t). \tag{13.65}$$

Taking time derivatives on both sides of Eq. (13.65) yields,

$$-\chi(t) = \beta \frac{dR_{AA}}{dt} H(t). \tag{13.66}$$

The Fourier transform of both sides of Eq. (13.66) leads to

$$-\hat{\chi}(\omega) = i\omega\beta\hat{R}_{AA} - \beta R_{AA}(0) \rightarrow 2\text{IM}[\hat{\chi}(\omega)] = \omega\beta\hat{R}_{AA}, \tag{13.67}$$

where we assume that $R_{AA}(t)$ is real and symmetric.

Based on the Wiener–Khinchin theorem, i.e., Eqs. (13.54) and (13.67), we derived the mathematical expression of the classical FDT for a general dynamic variable,

$$\boxed{S_{AA}(\omega) = \frac{2k_B T}{\omega}\text{Im}[\hat{\chi}(\omega)].} \tag{13.68}$$

13.3.2 Onsager's Regression Hypothesis and FDT

In nonequilibrium statistical physics, the linear response theory forms a theoretical basis for evaluation of transport properties by using MD simulation. We are now looking at FDT through a slightly contemporary perspective, but we still consider the steady-static linear response. Now we examine how average of a mechanical property $A(t) := A(\mathbf{q}^N(t), \mathbf{p}^N(t))$ changes in the presence of a static external perturbation f_0. The following presentation is largely rephrased from Chandler (1987), thus the notations are slightly different from the previous section, even though the proof is essentially the same.

Consider the evolution of time-dependent variable,

$$A(\mathbf{p}^N(t), \mathbf{q}^N(t)) = A(T_t(\mathbf{p}^N(0), \mathbf{q}^N(0))) =: A(t). \tag{13.69}$$

By introducing the short-handed notation, we let

$$T_r[\exp(-\beta\mathcal{H})]A(\mathbf{q}^N, \mathbf{p}^N)] := \int_\Gamma d\mathbf{q}^N d\mathbf{p}^N \exp(-\beta\mathcal{H})A(t). \tag{13.70}$$

We first define,

$$<A(t)> = \frac{Tr[A(t)e^{-\beta\mathcal{H}_0}]}{Tr[e^{-\beta\mathcal{H}_0}]}, \quad \text{and} \quad \bar{A}(t) = \frac{Tr[A(t)e^{-\beta\mathcal{H}_0+\Delta\mathcal{H}}]}{Tr[e^{-\beta\mathcal{H}_0+\Delta\mathcal{H}}]}, \tag{13.71}$$

where $T_r[\cdot] = \int_\Gamma [\cdot] d\mathbf{q}^N d\mathbf{p}^N$.

The original equilibrium expectation is defined as,

$$<A(t)> := \frac{T_r[\exp(-\beta\mathcal{H}_0)A(t)]}{T_r[\exp(-\beta\mathcal{H}_0)]} = \frac{T_r[\exp(-\beta\mathcal{H}_0)A(0)]}{T_r[\exp(-\beta\mathcal{H}_0)]} = <A(0)> =: <A>. \tag{13.72}$$

This is because that the properties at equilibrium state do not depend on time. We further define that

$$\delta A(t) = A(t) - <A>,$$
$$\delta A(0) = A(0) - <A>,$$
$$\delta \bar{A}(t) = <A(t)> - <A>,$$
$$\delta \bar{A}(0) = <A(0)> - <A>,$$
$$C(t) = <\delta A(0)\delta A(t)> = <A(0)A(t) - <A><A>>$$
$$= <A(0)A(t)> - <A><A>;$$
$$C(0) = <\delta A(0)\delta A(0)> = <A(0)A(0)> - <A><A>.$$

By definition, we know that

$$\lim_{t \to 0} C(t) = <(\delta A(0))^2> \quad \text{is finite}, \tag{13.73}$$

and when $t \to \infty$, $\delta A(t)$ will be uncorrelated with $\delta A(0)$, i.e.,

$$\lim_{t \to \infty} C(t) = <\delta A(0)\delta A(t)> \to 0.$$

After applying perturbation to the Hamiltonian,

$$\mathcal{H} = \mathcal{H}_0 + \Delta\mathcal{H}, \quad \text{where} \quad \Delta\mathcal{H} = -fA;$$

note that the field value A here may be understood as $A(0)$ and,

$$f = -\frac{\partial F}{\partial A(0)},$$

where F is the free energy,

$$F = -k_B T \ln Z, \text{ and } Z = \int d\mathbf{q}^N d\mathbf{p}^N \exp(-\beta(\mathcal{H} - fA(0))).$$

In fact, it can be shown that

$$\frac{\partial F}{\partial A(0)} = -k_B T \frac{1}{Z} \frac{\partial}{\partial A(0)} \int d\mathbf{q}^N d\mathbf{p}^N \exp(-\beta(\mathcal{H} - fA(0))) = -f. \quad (13.74)$$

The nonequilibrium expectation of $A(t)$ may be calculated as if it is in an "equilibrium state":

$$\bar{A}(t) := \frac{\int_\Gamma d\mathbf{q}^N d\mathbf{p}^N e^{-\beta(\mathcal{H}_0 + \Delta\mathcal{H})} A(\mathbf{q}^N, \mathbf{p}^N, t)}{\int_\Gamma d\mathbf{q}^N d\mathbf{p}^N e^{-\beta(\mathcal{H}_0 + \Delta\mathcal{H})}}, \quad (13.75)$$

which describes the "*macroscopic relaxation*" of the observable value of $A(t)$, which evolves toward its equilibrium value $<A>$, while the system is initially in a state that is not far away from the equilibrium, and gradually approaches equilibrium with a thermal reservoir.

REMARK 13.1 Note that

- $<A(t)> = <A(0)>$ is denoted as $<A>$, which is the equilibrium expectation, and it is independent of time;
- $f(\mathbf{q}^N, t)$ is an applied field that couples with $A(\mathbf{q}^N, \mathbf{p}^N)$. In multiscale analysis, it can be a coarse-scale field.

The perturbation $f(\mathbf{q}^N, t)$ is applied to the system, and it has brought the original equilibrium state to a nonequilibrium but an autonomous state at $t = 0$, which may relax to a new equilibrium state, whose distribution function is the distribution function of the nonequilibrium state at $t = 0$, i.e.,

$$F(\mathbf{q}^N, \mathbf{p}^N) \propto \exp(-\beta(\mathcal{H} + \Delta\mathcal{H})),$$

and

$$\bar{A}(0) = Tr[\exp(-\beta(\mathcal{H} + \Delta\mathcal{H})) A(0)] / Tr[\exp(-\beta(\mathcal{H} + \Delta\mathcal{H}))].$$

As time evolves, $A(\mathbf{q}^N, \mathbf{p}^N, t)$ changes according to

$$\bar{A}(t) = Tr[\exp(-\beta(\mathcal{H} + \Delta\mathcal{H})) A(t)] / Tr[\exp(-\beta(\mathcal{H} + \Delta\mathcal{H}))].$$

Now we study the asymptotic behaviors of $\bar{A}(t)$. Consider

$$\bar{A}(t) = Tr[\exp(-\beta(\mathcal{H} + \Delta\mathcal{H})) A(t)] / Tr[\exp(-\beta(\mathcal{H} + \Delta\mathcal{H}))],$$

we can calculate the Taylor variational expansion of A,

$$\bar{A}(t) = <A(t)>_0 + \delta \bar{A}(t) + O((\Delta\mathcal{H})^2).$$

Considering the fact that $\Delta \mathcal{H} = -fA$ and taking Gateaux derivative, we have the following result:

$$\delta \bar{A}(t) = \frac{d}{d\alpha} \bar{A}(t)(\alpha f(t)) \Big|_{\alpha=0}$$

$$= \frac{\partial \bar{A}(t)}{\partial f} \Big|_{f=0} f = -\beta \big(<\Delta \mathcal{H} A>_0 - <A(t)>_0 <\Delta \mathcal{H}>_0 \big)$$

$$= \beta f \big(<A(0)A(t)>_0 - <A(t)>_0 <A(0)>_0 \big).$$

$$= \beta f \big(<\delta A(0) \delta A(t)>_0 \big). \qquad (13.76)$$

To provide the detailed derivations of the above expression, one may first consider

$$\bar{A}(t) = \text{Tr}[\exp(-\beta \mathcal{H})A(t)] / \text{Tr}[\exp(-\beta \mathcal{H})]$$

and then

$$\mathcal{H} = \mathcal{H}_0 + \Delta \mathcal{H} = \mathcal{H}_0 - A(0)f;$$

hence,

$$\exp(-\beta \Delta \mathcal{H}) \approx 1 + \beta A(0)f + O(f^2).$$

Then, by Taylor expansion, we have

$$\bar{A}(t) \approx \text{Tr}[\exp(-\beta \mathcal{H}_0)(1 + \beta A(0)f)A(t)] / \text{Tr}[\exp(-\beta \mathcal{H}_0)(1 + \beta A(0)f)]$$

$$= \frac{\text{Tr}[\exp(-\beta \mathcal{H}_0)A(t)]}{\text{Tr}[\exp(-\beta \mathcal{H}_0)]} + \beta f \frac{\text{Tr}[\exp(-\beta \mathcal{H}_0)A(0)A(t)]}{\text{Tr}[\exp(-\beta \mathcal{H}_0)]}$$

$$- \beta f \frac{\text{Tr}[\exp(-\beta \mathcal{H}_0)A(t)]\text{Tr}[\exp(-\beta \mathcal{H}_0)A(0)]}{\text{Tr}[\exp(-\beta \mathcal{H}_0)]^2} \Big) + O(f^2).$$

Reorganizing the above equations, we may write

$$\bar{A}(t) \approx \frac{\text{Tr}[\exp(-\beta \mathcal{H}_0)A(t)]}{\text{Tr}[\exp(-\beta \mathcal{H}_0)]} + \beta f \frac{\text{Tr}[\exp(-\beta \mathcal{H}_0)A(0)A(t)]}{\text{Tr}[\exp(-\beta \mathcal{H}_0)]}$$

$$- \beta f \frac{\text{Tr}[\exp(-\beta \mathcal{H}_0)A(t)]\text{Tr}[\exp(-\beta \mathcal{H}_0)A(0)]}{\text{Tr}[\exp(-\beta \mathcal{H}_0)]^2} \Big)$$

Simplifying the above expression, we obtain

$$\bar{A}(t) = <A(t)> + \beta f (<A(0)A(t)> - <A(0)><A(t)>). \qquad (13.77)$$

Finally, we have

$$\bar{A}(t) - <A>_0 = \delta \bar{A}(t) = \beta f \big(\delta A(0) \delta A(t) \big). \qquad (13.78)$$

In general, we may assume that

$$\mathcal{H} = \mathcal{H}_0 + \Delta \mathcal{H}, \text{ and } \Delta \mathcal{H} = -f(t)B(0), \; B \neq A.$$

One can show that

$$\delta \bar{A}(t) = \frac{\partial \bar{A}}{\partial f} f = -\beta[<(-fB(0))A(t)> - <A><(-fB)>],$$
$$= -\beta[<\Delta \mathcal{H} A(t)> - <A><\Delta \mathcal{H}>],$$

which is called the *susceptibility* that describes the first-order response to the perturbation.

In particular, one may look at the expression,

$$-\beta\big(<\Delta \mathcal{H} A(t)> - <A><\Delta \mathcal{H}>\big)$$
$$= f\beta <(B(0)-<B>)(A(t)-<A>)>$$
$$= f\beta <\delta B(0)\delta A(t)>,$$

where $\delta B(0) := B(0) - <B>$, and $\delta A(t) = A(t) - <A>$.
Hence, if $\Delta \mathcal{H} = -f A(0)$, i.e., $B(0) = A(0)$,

$$\delta \bar{A}(t) = \beta f <\delta A(0)\delta A(t)> + O(f^2).$$

When $t = 0$

$$\delta \bar{A}(0) = \beta f <(\delta A)^2>.$$

In general, we may have

$$\Delta \mathcal{H} = -\sum_i f_i A_i \text{ . for example, } \Delta \mathcal{H} = -\int d\mathbf{r} \Phi_{ext} \rho(\mathbf{r}).$$

To associate $f_i \leftrightarrow \Phi_{ext}$ and $A_i \leftrightarrow \rho$, this case gives,

$$\delta \bar{A}_j(t) = \beta \sum_i f_i <\delta A_i(0)\delta A_j(t)> + O(f^2).$$

Now, we can calculate the ratio,

$$\frac{\delta \bar{A}(t)}{\delta \bar{A}(0)} = \frac{\bar{A}(t) - <A>}{\bar{A}(0) - <A>} = \frac{<\delta A(0)\delta A(t)>}{<(\delta A(0))^2>}, \tag{13.79}$$

which is exactly the same as Onsager's regression hypothesis on the fluctuation decay. Therefore, Onsager's regression hypothesis can be formulated as

$$\frac{\bar{A} - <A>}{\bar{A}(0) - <A>} = \frac{<\delta A(0)\delta A(t)>}{<(\delta A(0))^2>} = \frac{C(t)}{C(0)}, \tag{13.80}$$

where

$$\bar{A} = \frac{T_r\{A(t)\exp(-\beta(H_0 + \Delta H))\}}{T_r\{\exp-\beta(H_0 + \Delta H)\}}, \text{ and } \text{Tr}[\cdot] := \int_\Gamma (\cdot) d\mathbf{q}^N d\mathbf{p}^N \tag{13.81}$$

describes the *"macroscopic relaxation"* of the observable value of $A(t)$, which evolves toward its equilibrium value $<A>$, while the system is initially in a state that is not far away from the equilibrium, and gradually approaches equilibrium with a thermal reservoir.

So far, we mainly discussed how the system will relax to a new equilibrium state, but we still have not discussed how the system responds to the external field.

Assume that a system's response to the external field is a linear relation, i.e., it is obeying the following time convolution relation,

$$\delta \bar{A}(t) = \int_{-\infty}^{\infty} d\tau \chi(t-\tau) f(\tau), \text{ C.f. } \mathbf{J} = \mathbf{L} : \mathbf{F}.$$

We would like to link the response function $\chi(t)$ with the fluctuation $\delta A(t)$. We know the fact that $f(t)$ has been turned off at $t=0$, and thus the system will not respond after $t \geq 0$. Therefore,

$$f(t) = 0, \forall t \geq 0; \text{ and } \chi(t-\tau) = 0, \forall t \leq \tau;$$

and then

$$\delta \bar{A}(t) = \int_{-\infty}^{0} d\tau \chi(t-\tau) f(\tau).$$

Changing of variable $s = t - \tau$, we have

$$\delta \bar{A}(t) = \int_{t}^{\infty} ds \chi(s) f(t-s) = \beta f(t) < \delta A(0) \delta A(t) >.$$

Considering special case $f = const.$, and differentiation gives

$$\chi(t) = \begin{cases} -\dfrac{d}{dt} \beta (< \delta A(0) \delta A(t) >), & t > 0 \\ 0, & t \leq 0. \end{cases} \quad (13.82)$$

The above expression is called **the Kubo expression**, which links a system's response to the perturbation that moves the system away from the equilibrium in terms of spontaneous fluctuation at the equilibrium state.

For the case $\mathcal{H} = \mathcal{H}_0 - f B(0)$, $B \neq A$, the response function becomes

$$\chi_{AB}(t) = -\beta \frac{d}{dt} < \delta B(0) \delta A(t) >. \quad (13.83)$$

Considering the some identities in the Fourier transform, we have

$$\mathcal{F}\left[\frac{d}{dt} A(t)\right] = (i\omega) \hat{A}(\omega); \quad (13.84)$$

$$\mathcal{F}[R_{AA}(t)] = S_{AA}(\omega) = \lim_{T \to \infty} \mathbf{E}\left[|\hat{A}(\omega)|^2\right], \quad (13.85)$$

where $\mathbf{E}[\bullet] = <\bullet>$ is the expectation of a random function or random variable, and

$$R_{AA}(\tau) = < A(t) A(t+\tau) > = \mathbf{E}\left[A(t) A(t+\tau)\right]. \quad (13.86)$$

Taking the Fourier transform of Eq. (13.82) and applying in Eqs. (13.84) – (13.86), we recover the classical version of the FDT

$$\hat{\chi}(\omega) = (-i\omega)S_{AA}(\omega). \tag{13.87}$$

13.4 Mori–Zwanzig Formalism

One of the major achievements of nonequilibrium statistical mechanics or computational statistical physics in twentieth century is the development of the Mori–Zwanzig formulation or formalism. It is a operator projection method that projects the fast variables of a Hamiltonian dynamic system in the slow variables of the system, and subsequently provides a set of coarse-graining dynamic equations with the drag force or memory term and the random force or stochastic terms.

In this section, we provide a simple derivation of the Mori–Zwanzig formulation, because it has become the theoretical foundation for a class of generalized Langevin equations, such as the dissipative particle dynamics.

13.4.1 Operator Projection

Considering two variables or function X and Y in the phase space and assuming that the equilibrium state probability density $f(\mathbf{p}, \mathbf{q})$ is known, we can then define an inner product of a Hilbert space $\mathcal{H}$ as

$$(X, Y) = \int f(\mathbf{p}, \mathbf{q}) X(\mathbf{p}, \mathbf{q}) Y(\mathbf{p}, \mathbf{q}) d\mathbf{p}^N d\mathbf{q}^N. \tag{13.88}$$

In linear algebra, we know that if we want to project a vector $\mathbf{y}$ onto another vector $\mathbf{x}$, we can write the projection as

$$\mathbb{P}_x \mathbf{y} = \frac{\mathbf{x} \cdot \mathbf{y}}{\mathbf{x} \cdot \mathbf{x}} \mathbf{x} = \frac{(\mathbf{x}, \mathbf{y})}{(\mathbf{x}, \mathbf{x})} \mathbf{x}.$$

Since the Hilbert space is a special type of vector space, we can also define the projection of the variable Y project to the variable X as follows,

$$\mathbb{P} Y = \frac{(X, Y)}{(X, X)} X, \tag{13.89}$$

here we drop the subscript of the projection operator.

We can then define the orthogonal complement projection operator as

$$\mathbb{Q} + \mathbb{P} = \mathbb{I} \quad \rightarrow \quad \mathbb{Q} = \mathbb{I} - \mathbb{P}. \tag{13.90}$$

One may readily verify that

$$\mathbb{P}\mathbb{P} = \frac{(X, Y)}{(X, X)} \frac{(X, X)}{(X, X)} X = \frac{(X, Y)}{(X, X)} X \quad \rightarrow \quad \mathbb{P}^2 = \mathbb{P} \tag{13.91}$$

and
$$\mathbb{Q}\mathbb{P}Y = (\mathbb{I} - \mathbb{P})\mathbb{P} = \mathbf{0} = \mathbb{P}\mathbb{Q} \tag{13.92}$$

and
$$\mathbb{Q}\mathbb{Q} = \mathbb{Q}(\mathbb{I} - \mathbb{P}) = \mathbb{Q} \quad \rightarrow \quad \mathbb{Q}\mathbb{Q} = \mathbb{Q}. \tag{13.93}$$

Moreover, one may find that $\forall\, Y, Z \in \mathcal{H}$,

$$(Z, \mathbb{P}Y) = \frac{(X, Y)}{(X, X)}(Z, X) \leftrightarrow (\mathbb{P}Z, Y) = \frac{(X, Z)}{(X, X)}(X, Y) \;\rightarrow\; (Z, \mathbb{P}Y) = (\mathbb{P}Z, Y). \tag{13.94}$$

Furthermore, we have

$$(Z, \mathbb{Q}Y) = (Z, Y) - (Z, \mathbb{P}Y) \leftrightarrow (\mathbb{Q}Z, Y) = (Z, Y) - (\mathbb{P}Z, Y), \tag{13.95}$$

and by using Eq. (13.94) we conclude that

$$\rightarrow (Z, \mathbb{Q}Y) = (\mathbb{Q}Z, Y). \tag{13.96}$$

Equations (13.94) and (13.96) indicate that both operators $\mathbb{P}$ and $\mathbb{Q}$ are self-adjoined operators in $\mathcal{H}$ under the norm defined in Eq. (13.88).

Considering a slow variable in a Hamiltonian system, $\mathbf{X}$, we will have its governing equation in the phase space,

$$\frac{d\mathbf{X}}{dt} = \frac{\partial \mathbf{X}}{\partial t} + i\mathbf{L}\mathbf{X},$$

where $i\mathbf{L}$ is the Liouville operator, i.e.,

$$i\mathbf{L}\mathbf{X} = \{\mathbf{X}, H\},$$

where $\{\cdot,\cdot\}$ is the Poisson's bracket, and hence the Liouville operator is anti-commutative, i.e., $\{\mathbf{X}, H\} = -\{H, \mathbf{X}\}$, which leads to the asymmetric self-adjoint property of the Liouville operator, i.e.,

$$(i\mathbf{L}\mathbf{X}, Y) = -(\mathbf{X}, i\mathbf{L}Y).$$

For the slow variable, we assume that $\mathbf{X} = \mathbf{X}(\mathbf{p}, \mathbf{q}, t')$, where t' is the faster scale time. This is to say that the slow variable does not explicitly depend on the slow scale time. Thus, we have

$$\frac{\partial \mathbf{X}}{\partial t} = 0.$$

Subsequently, the dynamic equation for the slow variable $\mathbf{X}$ becomes

$$\frac{d\mathbf{X}}{dt} = i\mathbf{L}\mathbf{X}. \tag{13.97}$$

Since that

$$\frac{d}{dt}\exp(i\mathbf{L}t) = \frac{d}{dt}\left(\sum_{n=0}^{\infty}\frac{1}{n!}(i\mathbf{L}t)^n\right) = i\mathbf{L}\left(\sum_{n=0}^{\infty}\frac{1}{n!}(i\mathbf{L}t)^n\right) = i\mathbf{L}\exp(i\mathbf{L}t),$$

it is straightforward to show that the solution of Eq. (13.97) is

$$\mathbf{X}(\mathbf{p},\mathbf{q}) = \exp(i\mathbf{L}t)\mathbf{X}(0,0), \qquad (13.98)$$

where $\exp(i\mathbf{L}t)$ is the propagator of the evolution equation, Eq. (13.97), and it is a semigroup of linear operators.

In the later text, without causing confusion, we denote $\mathbf{X}(0) = \mathbf{X}(0,0)$. It may be noted that $\mathbf{X}(0)$ indicates the slow variable configuration at the origin of a slow timescale, which should be understood as an equilibrium configuration. Whereas at this moment, the fast time t' is not necessarily zero, because it does not make much sense to make the origin of the fast time marker, because all atoms are constantly vibrating around their equilibrium positions.

Now we consider the Laplace transform of a semi-group of linear operator,

$$\int_0^\infty \exp(-st)\exp(-\mathbb{A}t)dt = \int_0^\infty \exp(-st)\left(\sum_{n=0}^\infty \frac{(-1)^n}{n!}\mathbb{A}^n t^n\right)dt$$

$$= \sum_{n=0}^\infty \frac{\mathbb{A}^n}{n!}\int_0^\infty \exp(-st)t^n dt = \sum_{n=0}^\infty \frac{\mathbb{A}^n}{n!}\mathcal{L}(t^n), \qquad (13.99)$$

where $\mathcal{L}$ is the notation of Laplace transform. According to Laplace transform, one may find that

$$F(s) = \mathcal{L}(t^n) = \frac{n!}{s^{n+1}},$$

we then have

$$\int_0^\infty \exp(-st)\exp(-\mathbb{A}t)dt = \sum_{n=0}^\infty (-1)^n \frac{\mathbb{A}^n}{s^{n+1}} = \frac{1}{s}\frac{1}{\left(1+\mathbb{A}/s\right)},$$

$$= \frac{1}{(s+\mathbb{A})}, \qquad (13.100)$$

which has the exact the same form as if the operator is a parameter.

This means that for an operator $\mathbb{A}$ is Laplace transform is

$$F_\mathbb{A}(s) = \mathcal{L}(\exp(-\mathbb{A}t)) = \int_0^\infty \exp(-st)\exp(-s\mathbb{A}) = \frac{1}{(s+\mathbb{A})}, \qquad (13.101)$$

and its inverse Laplace transform is

$$\mathcal{L}^{-1}\{F_\mathbb{A}(s)\} = \exp(-\mathbb{A}t). \qquad (13.102)$$

Next, we consider the following operator identity,

$$\mathbb{U}^{-1} = \mathbb{U}^{-1}(\mathbb{U}+\mathbb{V})(\mathbb{U}+\mathbb{V})^{-1} = (\mathbb{I}+\mathbb{U}^{-1}\mathbb{V})(\mathbb{U}+\mathbb{V})^{-1}$$
$$\rightarrow (\mathbb{U}+\mathbb{V})^{-1} = \mathbb{U}^{-1} - \mathbb{U}^{-1}\mathbb{V}(\mathbb{U}+\mathbb{V})^{-1}. \qquad (13.103)$$

Let $\mathbb{U} = s + \mathbb{A}$ and $\mathbb{V} = \mathbb{B}$. Eq. (13.103) yields

$$(s+\mathbb{A}+\mathbb{B})^{-1} = (s+\mathbb{A})^{-1} - (s+\mathbb{A})^{-1}\mathbb{B}(s+\mathbb{A}+\mathbb{B})^{-1}.$$

Taking the inverse Laplace transform of the above identity and using the result of Eq. (13.102), we derive at the following Dyson's operator identity,

$$\exp(-(\mathbb{A}+\mathbb{B})t) = \exp(-\mathbb{A}t) - \int_0^t \exp(-\mathbb{A}\tau)\mathbb{B}\exp(-(\mathbb{A}+\mathbb{B})(t-\tau))d\tau. \quad (13.104)$$

Now, we let $\mathbb{P}$ as the projection operator on the slow variable set $\mathbf{X}$, and we assign

$$\mathbb{A} = i\mathbf{L}\mathbb{Q}, \text{ and } \mathbb{B} = i\mathbf{L}\mathbb{P}.$$

Recalling $\mathbb{P} + \mathbb{Q} = \mathbb{I}$ and $i\mathbf{L}^T = -i\mathbf{L}$, from Eq. (13.104) we have

$$\exp(i\mathbf{L}t) = \exp(\mathbb{Q}i\mathbf{L}t) + \int_0^t \exp(i\mathbf{L}(t-\tau))\mathbb{P}(i\mathbf{L}\exp(\mathbb{Q}i\mathbf{L})(\tau))d\tau. \quad (13.105)$$

13.4.2 Mori-Zwanzig Equation

Now, we come back to Eqs. (13.97) and (13.98),

$$\frac{d\mathbf{X}(t)}{dt} = i\mathbf{L}\mathbf{X}(t) = i\mathbf{L}\exp(i\mathbf{L}t)\mathbf{X}(0),$$

here we used the short-hand notation $\mathbf{X}(t) = \mathbf{X}(\mathbf{p}(t), \mathbf{q}(t))$ and $\mathbf{X}(0) = \mathbf{X}(0,0)$.
By using the Taylor expansion, one can readily show that

$$i\mathbf{L}\exp(i\mathbf{L}t) = i\mathbf{L}\left(\sum_{n=0}^{\infty}\frac{(i\mathbf{L})^n t^n}{n!}\right) = \exp(i\mathbf{L}t)i\mathbf{L}. \quad (13.106)$$

Thus, we have

$$\frac{d\mathbf{X}(t)}{dt} = i\mathbf{L}\mathbf{X}(t) = \exp(i\mathbf{L}t)(\mathbb{Q}+\mathbb{P})(i\mathbf{L})\mathbf{X}(0). \quad (13.107)$$

The second term in the right-hand side of the above equation may be rewritten as

$$\exp(i\mathbf{L}t)\mathbb{P}(i\mathbf{L})\mathbf{X}(0) = \exp(i\mathbf{L}t)\frac{(i\mathbf{L}\mathbf{X}(0), \mathbf{X}(0))}{(\mathbf{X}(0), \mathbf{X}(0))}\mathbf{X}(0) = \frac{(i\mathbf{L}\mathbf{X}(0), \mathbf{X}(0))}{(\mathbf{X}(0), \mathbf{X}(0))}\exp(i\mathbf{L}t)\mathbf{X}(0)$$

$$= \frac{(i\mathbf{L}\mathbf{X}(0), \mathbf{X}(0))}{(\mathbf{X}(0), \mathbf{X}(0))}\mathbf{X}(t), \quad (13.108)$$

and we define its equilibrium part as

$$\Omega := \frac{(i\mathbf{L}\mathbf{X}(0), \mathbf{X}(0))}{(\mathbf{X}(0), \mathbf{X}(0))}. \quad (13.109)$$

Note that Ω is a matrix in general.
Thus, the slow variable dynamic equation becomes

$$\frac{d\mathbf{X}(t)}{dt} = \Omega\mathbf{X}(t) + \exp(i\mathbf{L}t)\mathbb{Q}(i\mathbf{L})\mathbf{X}(0). \quad (13.110)$$

Using Eq. (13.104), we rewrite the dynamic equations of the slow variable vector as

$$\frac{d\mathbf{X}(t)}{dt} = \Omega\mathbf{X}(t) + \int_0^t d\tau \exp(i\mathbf{L}(t-\tau))(\mathbb{P}i\mathbf{L}\exp(\mathbb{Q}i\mathbf{L})(\tau))(\mathbb{Q}i\mathbf{L}\mathbf{X}(0))$$
$$+ \exp(\mathbb{Q}i\mathbf{L}t)(\mathbb{Q}i\mathbf{L})\mathbf{X}(0). \tag{13.111}$$

The physical meaning of the last term in Eq. (13.111) is a random force, which we denote as

$$\mathbf{R}(t) := \exp(\mathbb{Q}i\mathbf{L}t)(\mathbb{Q}i\mathbf{L})\mathbf{X}(0). \tag{13.112}$$

Because $\mathbb{Q} = \mathbb{Q}$, one can derive the following identity

$$\exp(\mathbb{Q}i\mathbf{L}t) = \exp(\mathbb{Q}^2 i\mathbf{L}t) = \sum_{n=0}^{\infty} \frac{\mathbb{Q}^{n+1}(i\mathbf{L}^n t^n)}{n!} = \mathbb{Q}\exp(\mathbb{Q}i\mathbf{L}t). \tag{13.113}$$

Thus, one can show that

$$(\mathbf{R}(t), \mathbf{X}(0)) = (\exp(\mathbb{Q}i\mathbf{L}t)\mathbb{Q}(i\mathbf{L}\mathbf{X}(0), \mathbf{X}(0)) = (\mathbb{Q}\mathbf{R}(t), \mathbf{X}(0)) = 0, \tag{13.114}$$

because $\mathbb{Q}\mathbf{R}(t)$ is orthogonal to $\mathbf{X}(0)$.

The physical meaning of the above equation is that the random force is uncorrelated with the slow variable, and thus, we justify its labeling as a random force, even though it is a colored noise in general.

Now, we consider the middle term in Eq. (13.111). The part after the first exponential term may be written as

$$(\mathbb{P}i\mathbf{L}\exp((\mathbb{Q}i\mathbf{L})t)(\mathbb{Q}i\mathbf{L})\mathbf{X}(0) = (\mathbb{P}i\mathbf{L}\mathbf{R}(t).$$

As shown in Eq. (13.114) that $\mathbf{R}(t) = \mathbb{Q}\mathbf{R}(t)$, we then have

$$\mathbb{P}i\mathbf{L}\mathbf{R}(t) = \mathbb{P}i\mathbf{L}\mathbb{Q}\mathbf{R}(t).$$

Utilizing the self-adjoint property of $\mathbb{Q}$ and asymmetric self-adjoint property of $i\mathbf{L}$, we have

$$\mathbb{P}i\mathbf{L}\mathbf{R}(t) = \mathbb{P}i\mathbf{L}\mathbb{Q}\mathbf{R}(t) = \frac{(i\mathbf{L}\mathbb{Q}\mathbf{R}(t), \mathbf{X}(0))}{(\mathbf{X}(0), \mathbf{X}(0))}\mathbf{X}(0) = -\frac{(\mathbb{Q}\mathbf{R}(t), i\mathbf{L}\mathbf{X}(0))}{(\mathbf{X}(0), \mathbf{X}(0))}\mathbf{X}(0)$$
$$= -\frac{(\mathbf{R}(t), i\mathbf{L}\mathbf{X}(0))}{(\mathbf{X}(0), \mathbb{Q}\mathbf{X}(0))}\mathbf{X}(0) = -\frac{(\mathbf{R}(t), \mathbf{R}(0))}{(\mathbf{X}(0), \mathbf{X}(0))}\mathbf{X}(0). \tag{13.115}$$

We define a memory kernel as

$$\mathbf{K}(t) = \frac{(\mathbf{R}(t), \mathbf{R}(0))}{(\mathbf{X}(0), \mathbf{X}(0))}. \tag{13.116}$$

Substituting the definition of the memory kernel into Eq. (13.111), we have derived the so-called Mori–Zwanzig equation,

$$\frac{d\mathbf{X}(t)}{dt} = \mathbf{\Omega}\mathbf{X}(t) - \int_0^t \exp(i\mathbf{L}(t-\tau))\mathbf{K}(\tau)\mathbf{X}(0)d\tau + \mathbf{R}(t)$$

$$= \mathbf{\Omega}\mathbf{X}(t) - \int_0^t \mathbf{K}(\tau)\mathbf{X}(t-\tau)d\tau + \mathbf{R}(t). \tag{13.117}$$

We can multiply $\mathbf{X}(0)$ with the above equation, and calculate the self-correlation of $\mathbf{C}(t) = <\mathbf{X}(t), \mathbf{X}(0)>$. Considering the fact that $<\mathbf{R}(t), \mathbf{X}(0)> = 0$, one may find that

$$\frac{d\mathbf{C}(t)}{dt} = \mathbf{\Omega}\mathbf{C}(t) - \int_0^t \mathbf{K}(\tau))\mathbf{C}(t-\tau)d\tau + \mathbf{R}(t). \tag{13.118}$$

In recent years, the data-driven approach has been developed to construct the memory kernel for various coarse-grained models, such as Wang et al. (2019, 2020) and Rudzinski (2019).

Part III

Multiscale Modeling and Simulation

14 Virial Theorem and Virial Stress

The concept of stress in classical continuum mechanics is defined as the internal force intensity, i.e., the force per unit of area at a given material point. The first measure of the intensity of the internal force is the traction vector that is defined on a specific cutting plane with out-normal $\mathbf{n}$ passing through the macroscale point A as shown in Fig. 14.1.

As shown in Fig. 14.1, the traction vector at the spatial point A is defined as

$$\mathbf{T}^{(n)} = \lim_{\Delta S \to 0} \frac{\Delta \mathbf{F}}{\Delta S} = \sigma \mathbf{n} + \tau \mathbf{t},$$

where $\mathbf{n}$ is the out-normal of the cutting plane, σ is the normal stress, $\mathbf{t}$ is a tangential vector on the cutting plane, and τ is the so-call shear stress. However, the traction vector, $\mathbf{T}^{(n)}$, cannot completely describe the stress state at the material point A, because the traction vector depends on the cutting plane that passes through the material point A, and there are infinitely many planes with different normals or orientations passing through point A.

It was Augustin-Louis Cauchy who found that one can choose any three independent traction vectors on three mutually orthogonal cutting planes passing through a material point to represent any traction vector on an arbitrary plane passing through that material point. Hence, we can use those three traction vectors to represent the stress state at that material point. This definition of stress is illustrated in Fig. 14.2, and this stress is what we call the Cauchy stress or the Cauchy stress tensor, which is precisely defined as the following mathematical expression:

$$\mathbf{T}^{e_1} = T_1^{e_1}\mathbf{e}_1 + T_2^{e_1}\mathbf{e}_2 + T_3^{e_1}\mathbf{e}_3 = \sigma_{11}\mathbf{e}_1 + \sigma_{12}\mathbf{e}_2 + \sigma_{13}\mathbf{e}_3$$
$$\mathbf{T}^{e_2} = T_1^{e_2}\mathbf{e}_1 + T_2^{e_2}\mathbf{e}_2 + T_3^{e_3}\mathbf{e}_3 = \sigma_{21}\mathbf{e}_1 + \sigma_{22}\mathbf{e}_2 + \sigma_{23}\mathbf{e}_3$$
$$\mathbf{T}^{e_3} = T_1^{e_3}\mathbf{e}_1 + T_2^{e_3}\mathbf{e}_2 + T_3^{e_3}\mathbf{e}_3 = \sigma_{31}\mathbf{e}_1 + \sigma_{32}\mathbf{e}_2 + \sigma_{33}\mathbf{e}_3$$

and

$$\begin{aligned}\sigma &= \mathbf{e}_i \otimes \mathbf{T}^{e_i} = \sigma_{ij}\mathbf{e}_i \otimes \mathbf{e}_j \\ &= \sigma_{11}\mathbf{e}_1 \otimes \mathbf{e}_1 + \sigma_{12}\mathbf{e}_1 \otimes \mathbf{e}_2 + \sigma_{13}\mathbf{e}_1 \otimes \mathbf{e}_3 \\ &\quad + \sigma_{21}\mathbf{e}_2 \otimes \mathbf{e}_1 + \sigma_{22}\mathbf{e}_2 \otimes \mathbf{e}_2 + \sigma_{23}\mathbf{e}_2 \otimes \mathbf{e}_3 \\ &\quad + \sigma_{31}\mathbf{e}_3 \otimes \mathbf{e}_1 + \sigma_{32}\mathbf{e}_3 \otimes \mathbf{e}_2 + \sigma_{33}\mathbf{e}_3 \otimes \mathbf{e}_3. \end{aligned} \quad (14.1)$$

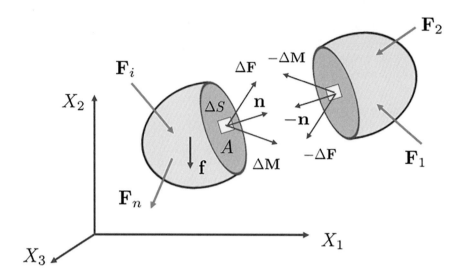

Figure 14.1 Method of section: Traction vector is the intensity of internal force

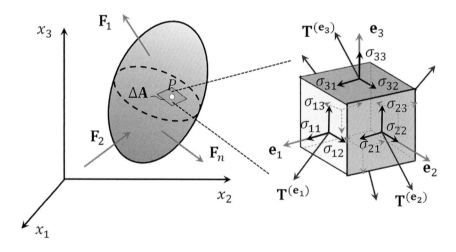

Figure 14.2 Definition of Cauchy stress

The mechanical stress is a macroscale concept or a macroscale mathematical quantity. There is a physical ambiguity on what is the internal force. If we assume that the origin of the internal force is the atomistic bond force, then we should be able to calculate the Cauchy stress from first-principles, provided that we know the precise positions and motions of every atoms in a small volume without even resorting to continuum mechanics. In fact, one of the contemporaries of Cauchy, Jean Claude Saint-Venant, who was a convinced atomist, believed that the stress should be calculated based on averaging the overall molecular and atomistic forces at microscale. This was an impossible task at the time of Cauchy and Saint-Venant. However, the atomistic characterization of stress has not only become possible, but also become essential in many fields of science and technology, in particular, the nanotechnology. Now, in this chapter, we discuss the atomistic counterpart of the "Cauchy stress," which we call the virial stress.

14.1 What Is Virial?

We first consider an atomistic system with N atoms or particles, and we define a scalar moment of inertia of the system as follows,

$$I = \frac{1}{2} \sum_{i=1}^{N} m_i \mathbf{r}_i \cdot \mathbf{r}_i, \tag{14.2}$$

and we can further define a scalar quantity G as the first-order time derivative of I,

$$G = \frac{d}{dt} I = \sum_{i=1}^{N} \mathbf{p}_i \cdot \mathbf{r}_i. \tag{14.3}$$

Subsequently, we have

$$\frac{d^2 I}{dt^2} = \frac{dG}{dt} = \sum_{i=1}^{N} m \dot{\mathbf{r}}_i^2 + \sum_{i=1}^{N} m \ddot{\mathbf{r}}_i \cdot \mathbf{r}_i. \tag{14.4}$$

Considering Newton's second law, we can covert the above equation into the following equation:

$$\frac{d^2 I}{dt^2} = \sum_{i=1}^{N} m \dot{\mathbf{r}}_i^2 + \sum_{i=1}^{N} \mathbf{F}_i \cdot \mathbf{r}_i. \tag{14.5}$$

Note that if the origin of the coordinate of the system is the center of mass, the velocities in Eq. (14.5) should be replaced by

$$\dot{\mathbf{r}}_i \rightarrow \dot{\mathbf{r}}_i - \dot{\mathbf{r}}_{COM}.$$

Since $\mathbf{F}_i$ are interatomic forces, one can show that

$$\sum_{i=1}^{N} \mathbf{F}_i \cdot \mathbf{r}_i = \sum_{i=1}^{N} \mathbf{r}_i \cdot \sum_{j \neq i} \mathbf{F}_{ji}.$$

This is because

$$\sum_i \mathbf{F}_i \cdot \mathbf{r}_i = \sum_i \sum_{j<i} \mathbf{F}_{ji} \cdot \mathbf{r}_i + \sum_i \sum_{j>i} \mathbf{F}_{ji} \cdot \mathbf{r}_i. \tag{14.6}$$

Consider $\mathbf{F}_{ji} = -\mathbf{F}_{ij}$, we have

$$\sum_i \sum_{j>i} \mathbf{F}_{ji} \cdot \mathbf{r}_i = -\sum_i \sum_{j>i} \mathbf{F}_{ij} \cdot \mathbf{r}_i = -\sum_j \sum_{i>j} \mathbf{F}_{ji} \cdot \mathbf{r}_j = -\sum_i \sum_{j<i} \mathbf{F}_{ji} \cdot \mathbf{r}_j. \tag{14.7}$$

Hence,

$$\sum_i \mathbf{F}_i \cdot \mathbf{r}_i = \sum_i \sum_{j<i} \mathbf{F}_{ji} \cdot \mathbf{r}_i - \sum_i \sum_{j<i} \mathbf{F}_{ji} \cdot \mathbf{r}_j = \sum_i \sum_{j<i} \mathbf{F}_{ji} \cdot (\mathbf{r}_i - \mathbf{r}_j)$$
$$= \sum_i \sum_{j<i} \mathbf{F}_{ji} \cdot \mathbf{r}_{ji}. \tag{14.8}$$

For the cases of pair interaction, we have

$$\sum_i \mathbf{F}_i \cdot \mathbf{r}_i = \sum_i \sum_{j<i} \mathbf{F}_{ji} \cdot \mathbf{r}_{ji} = \frac{1}{2} \sum_i \sum_{j \neq i} \mathbf{F}_{ji} \cdot \mathbf{r}_{ji}.$$

Equation (14.7) can be rewritten as

$$\frac{dG}{dt} = 2 < E_{kin} > + \left\langle \sum_i \sum_{j>i} \mathbf{F}_{ji} \cdot \mathbf{r}_{ji} \right\rangle = 3Nk_BT + \left\langle \sum_i \sum_{j>i} \mathbf{F}_{ji} \cdot \mathbf{r}_{ji} \right\rangle. \tag{14.9}$$

Comparing with the expression,

$$p = \frac{2 < E_{kin} >}{3V} + \frac{1}{3V} \left\langle \sum_i \sum_{j>i} \mathbf{F}_{ji} \cdot \mathbf{r}_{ji} \right\rangle, \tag{14.10}$$

we find that the pressure is related with quantity G as

$$p = \frac{1}{3V} \frac{dG}{dt}. \tag{14.11}$$

The time average of pressure becomes

$$<p>_\tau = \frac{1}{\tau} \int_0^\tau \frac{dG}{dt} dt = \frac{1}{\tau} \int_0^\tau dG = \frac{G(\tau) - G(0)}{\tau}. \tag{14.12}$$

Then, the classical virial theorem can be stated as: *Under the equilibrium condition, the average stress of the thermodynamic system is zero*, i.e.,

$$<p>_\tau = \left\langle \frac{dG}{dt} \right\rangle = 0.$$

This leads to

$$\left\langle \frac{dG}{dt} \right\rangle = 2 < E_{kin} > + \left\langle \sum_i \sum_{j>i} \mathbf{F}_{ji} \cdot \mathbf{r}_{ji} \right\rangle = 3Nk_BT + \left\langle \sum_i \sum_{j>i} \mathbf{F}_{ji} \cdot \mathbf{r}_{ji} \right\rangle = 0,$$

which leads to

$$K = < E_{kin} > = -\frac{1}{2} \sum_i \sum_{j>i} \langle \mathbf{F}_{ji} \cdot \mathbf{r}_{ji} \rangle, \tag{14.13}$$

where the term $\sum_i \sum_{j>i} \langle \mathbf{F}_{ji} \cdot \mathbf{r}_{ji} \rangle$ is called **virial**, and this is the *Clausius virial theorem* (see Clausius (1870)).

In summary, in the Clausius virial theorem, the scalar G is defined as,

$$G = \sum_k \mathbf{r}_k \cdot \mathbf{p}_k$$

and

$$\frac{dG}{dt} = \sum_{k=1}^{N} \mathbf{p}_k \cdot \frac{d\mathbf{r}_k}{dt} + \sum_{k=1}^{N} \frac{d\mathbf{p}_k}{dt} \cdot \mathbf{r}_k = \sum_{k=1}^{N} m_k \frac{d\mathbf{r}_k}{dt} \cdot \frac{d\mathbf{r}_k}{dt} + \sum_{k=1}^{N} \mathbf{F}_k \cdot \mathbf{r}_k$$

$$= 2K + \sum_{k=1}^{N} \mathbf{F}_k \cdot \mathbf{r}_k$$

The virial theorem is:

$$<\frac{dG}{dt}>_\tau = 0 \quad \Rightarrow \quad 2<K>_\tau = -\sum_{k=1}^{N} <\mathbf{F}_k \cdot \mathbf{r}_k>_\tau, \quad (14.14)$$

where $<\cdot>_\tau$ indicates the time average.

Rudolf Clausius and Virial Theorem

During his study of the foundations of thermodynamics, Rudolf Clausius (Fig. 14.3) discovered the virial theorem that enables us to calculate pressure, the trace of the stress tensor, of a molecular system, so that he can relate the absolute temperature of an ideal gas to the mean kinetic energy of its molecules.

In molecular mechanics, the virial theorem provides a general equation that relates the time average of the kinetic energy of a particle system under equilibrium, only bound by internal forces generated from the potential energy of the system. Mathematically, the virial theorem is expressed as

$$<K>_t = -\frac{1}{2} \sum_{k=1}^{N} <\mathbf{F}_k \cdot \mathbf{r}_k>. \quad (14.15)$$

Note that the factor 1/2 means that the time average of kinetic energy is one half of the virial, and this is not only restricted to the pair potential here.

The word virial for the right-hand side of the equation derives from vis, the Latin word for "force" or "energy," and was given its technical definition by Rudolf Clausius in his 1870 landmark paper.

14.2 Virial Stress via Tensorial Viral Theorem

In the classical virial theorem, only pressure is involved, which is the (negative) trace of the stress tensor. To derive the atomistic virial stress tensor, we define a tensorial quantity $\mathbf{G}$ as follows,

$$\mathbf{G} = \sum_k \mathbf{r}_k \otimes \mathbf{p}_k, \quad (14.16)$$

where $\otimes$ is the tensor product.

Figure 14.3 Rudolf Julius Emanuel Clausius (1822–1888) (Photo courtesy of Wikipedia.org)

It is straightforward to derive that

$$\frac{d\mathbf{G}}{dt} = \sum_{k=1}^{N} \mathbf{r}_k \otimes \frac{d\mathbf{p}_k}{dt} + \sum_{k=1}^{N} \frac{d\mathbf{r}_k}{dt} \otimes \mathbf{p}_k = \sum_{k=1}^{N} m_k \frac{d\mathbf{r}_k}{dt} \otimes \frac{d\mathbf{r}_k}{dt} + \sum_{k=1}^{N} \mathbf{r}_k \otimes \mathbf{F}_k$$

$$= 2\mathbf{K} + \sum_{k=1}^{N} \mathbf{r}_k \otimes \mathbf{F}_k,$$

where the tensor

$$\mathbf{K} := \sum_k \frac{1}{2m_k} \mathbf{p}_k \otimes \mathbf{p}_k$$

is the tensorial kinetic energy.

Thus, the tensorial virial theorem states that

$$\left\langle \frac{d\mathbf{G}}{dt} \right\rangle = 2 < \mathbf{K} > + \sum_{k=1}^{N} < \mathbf{r}_k \otimes \mathbf{F}_k > = 0. \qquad (14.17)$$

It is noted that for pair interaction we have

$$\sum_{k=1}^{N} \mathbf{r}_k \otimes \mathbf{F}_k = \sum_{k=1}^{N} \sum_{j<k} \mathbf{r}_k \otimes \mathbf{F}_{jk} + \sum_{j=1}^{N} \sum_{j>k} \mathbf{r}_j \otimes \mathbf{F}_{kj}$$

$$= \sum_{k=1}^{N} \sum_{j<k} (\mathbf{r}_k - \mathbf{r}_j) \otimes \mathbf{F}_{jk} = \frac{1}{2} \sum_{k=1} \sum_{j \neq k} \mathbf{r}_{jk} \otimes \mathbf{F}_{jk}, \qquad (14.18)$$

where $\mathbf{F}_{jk} = -\mathbf{F}_{kj}$ is used, and

$$\mathbf{F}_{jk} = -\frac{dV}{dr}\left(\frac{\mathbf{r}_k - \mathbf{r}_j}{r_{jk}}\right) = -V'\left(\frac{\mathbf{r}_{jk}}{r_{jk}}\right).$$

Finally, for pair potentials we have

$$\sum_{k=1}^{N} \mathbf{r}_k \otimes \mathbf{F}_k = \sum_k \sum_{j<k} -\frac{V'(r_{jk})}{r_{jk}}(\mathbf{r}_k - \mathbf{r}_j) \otimes (\mathbf{r}_k - \mathbf{r}_j)$$

$$= -\frac{1}{2}\sum_k \sum_{j\neq k} \frac{V'(r_{jk})}{r_{jk}} \mathbf{r}_{jk} \otimes \mathbf{r}_{jk}. \quad (14.19)$$

In the following, we prove a virial theorem for the stress tensor.

THEOREM 14.1 (Tensorial Virial Stress Theorem) *Assume that an N-atom system in a volume Ω is under thermal mechanical equilibrium condition. The average Cauchy stress of the atomistic system can be expressed as*

$$\bar{\sigma} = \frac{1}{\Omega}\int_\Omega \sigma dv. \quad (14.20)$$

Then we claim that

$$\bar{\sigma} = \frac{1}{\Omega}\sum_{i=1}^{N}\left\{-m_i \mathbf{v}'_i \otimes \mathbf{v}'_i + \frac{1}{2}\sum_{j\neq i} V'(r_{ji})\frac{\mathbf{r}_{ji} \otimes \mathbf{r}_{ji}}{r_{ij}}\right\}, \quad (14.21)$$

*where $\mathbf{v}'_i = \mathbf{v}_i - \bar{\mathbf{v}}$ is the relative velocity or fluctuation velocity of the atom i and $\bar{\mathbf{v}} = \sum_i \mathbf{v}_i/N$. The average Cauchy stress in the cell Ω (Eq. (14.20)) is the called the **virial stress tensor**.*

This theorem provides a direct link between the macroscale Cauchy stress and the atomistic virial stress.

Proof Assume that the atomistic system is under equilibrium state, the atom systems at different time are essentially an atomistic ensemble. The equilibrium condition implies that

$$\frac{d\mathbf{G}}{dt} = 0. \quad (14.22)$$

Note that the above condition is stronger that Eq. (14.17).

Then, the tensorial virial theorem (14.17) states that

$$\frac{d\mathbf{G}}{dt} = \sum_{k=1}^{N}\left\{\frac{1}{m_i}\mathbf{p}_k \otimes \mathbf{p}_k + \frac{1}{2}\sum_{\ell \neq k} \mathbf{r}_{k\ell} \otimes \mathbf{F}_{k\ell}\right\} = 0. \quad (14.23)$$

Consider the multiscale decomposition,

$$\mathbf{p}_k = \bar{\mathbf{p}} + \mathbf{p}'_k, \text{ and } \frac{d\mathbf{p}_k}{dt} = \frac{d\bar{\mathbf{p}}}{dt} + \frac{d\mathbf{p}'_k}{dt}, \quad (14.24)$$

where $\bar{\mathbf{p}}$ is the continuum scale momentum and $\mathbf{p}'_i$ is the atomistic momentum fluctuation for atom i.

Since we may write $m_k = \rho dv$, we may write

$$\sum_k \frac{1}{m_k} \mathbf{p}_k \otimes \mathbf{p}_k = \sum_k \frac{1}{m_k} (\bar{\mathbf{p}} + \mathbf{p}'_k) \otimes (\bar{\mathbf{p}} + \mathbf{p}'_k)$$

$$= \int_V \rho \bar{\mathbf{v}} \otimes \bar{\mathbf{v}} dv + \sum_k \frac{1}{m_k} \mathbf{p}'_k \otimes \mathbf{p}'_k. \qquad (14.25)$$

This is because that

$$\sum_k \mathbf{p}'_k = 0 \;\rightarrow\; \sum_k \frac{1}{m_k} \bar{\mathbf{p}} \otimes \mathbf{p}'_k = \sum_k \frac{1}{m_k} \mathbf{p}'_k \otimes \bar{\mathbf{p}}_k = 0.$$

The first term in Eq. (14.25) is the tensorial kinetic energy at the continuum scale. At the continuum scale, we may let $\rho = \frac{Nm}{\Omega}$ for a control (closed) mass system as shown in Fig. 14.4.

Without body force, we postulate that at the macroscale the continuum body Ω satisfies the (conservation of tensorial angular momentum), i.e.,

$$\frac{D}{Dt} \int_\Omega \rho \mathbf{r} \otimes \bar{\mathbf{v}} dv = \oint_{\partial \Omega} \mathbf{r} \otimes \sigma \cdot \mathbf{n} ds. \qquad (14.26)$$

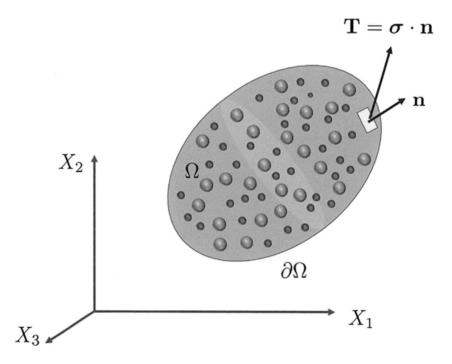

Figure 14.4 A controlled mass atomistic system with prescribed traction

By the divergence theorem, Eq. (14.26) becomes

$$\int_\Omega \rho \bar{\mathbf{v}} \otimes \bar{\mathbf{v}} dv + \int_\Omega \rho \mathbf{r} \otimes \frac{d\bar{\mathbf{v}}}{dt} dv = \int_\Omega \left(\mathbf{I}^{(2)} \cdot \sigma + \mathbf{r} \otimes \text{div}\sigma \right) dv. \quad (14.27)$$

Note that in the above controlled mass system $dm = const$. Since the local equilibrium at macroscale, i.e., the balance of the linear momentum requires that

$$\text{div}\sigma = \rho \frac{d\bar{\mathbf{v}}}{dt}, \quad (14.28)$$

which leads the following expression for the first term of Eq. (14.25) via the mean value theorem,

$$\int_\Omega \rho \bar{\mathbf{v}} \otimes \bar{\mathbf{v}} dv = \bar{\sigma}\Omega, \quad (14.29)$$

where

$$\bar{\sigma} = \frac{1}{\Omega} \int_\Omega \sigma dv.$$

Then the tensorial virial theorem (14.21) can be re-written as follows,

$$\frac{d\mathbf{G}}{dt} = \int_V \rho \bar{\mathbf{v}} \otimes \bar{\mathbf{v}} dv + \sum_{k=1}^N \left\{ \frac{1}{m_i} \mathbf{p}'_k \otimes \mathbf{p}'_k + \frac{1}{2} \sum_{\ell \neq k} \mathbf{r}_{k\ell} \otimes \mathbf{F}_{k\ell} \right\} = 0. \quad (14.30)$$

Considering that

$$\int_\Omega \rho \bar{\mathbf{v}} \otimes \bar{\mathbf{v}} dv = \bar{\sigma}\Omega,$$

we then have

$$0 = \bar{\sigma}\Omega + \sum_k \left\{ m_k \mathbf{v}'_k \otimes \mathbf{v}'_k + \frac{1}{2} \sum_{\ell \neq k} \mathbf{r}_{\ell k} \otimes \mathbf{F}^{int}_{\ell k} \right\}.$$

Finally, we obtain

$$\bar{\sigma} = -\frac{1}{\Omega} \sum_k \left\{ m_k (\mathbf{v}_k - \bar{\mathbf{v}}) \otimes (\mathbf{v}_k - \bar{\mathbf{v}}) + \frac{1}{2} \sum_{\ell \neq k} \mathbf{r}_{\ell k} \otimes \mathbf{F}_{\ell k} \right\}. \quad (14.31)$$

For the pair potential,

$$\mathbf{F}_k = \sum_{k \neq \ell} \mathbf{F}_{\ell k} = -\frac{\partial}{\partial \mathbf{r}_k} V(r_{\ell k}) = -V'(r_{\ell k}) \frac{\mathbf{r}_{\ell k}}{r_{\ell k}}, \quad (14.32)$$

we then arrive at the most popular expression of the virial stress in the literature,

$$\bar{\sigma} = \frac{1}{\Omega} \sum_{i=1}^N \left\{ -m_i \mathbf{v}'_i \otimes \mathbf{v}'_i + \frac{1}{2} \sum_{j \neq i} V'(r_{ji}) \frac{\mathbf{r}_{ji} \otimes \mathbf{r}_{ji}}{r_{ij}} \right\}. \quad (14.33)$$

□

It may be noted that Eq. (14.33) is only valid for pair potentials or pair interactions. In general, the expression for the virial stress is

$$\bar{\sigma} = \frac{-1}{\Omega} \sum_{i=1}^{N} \{m_i \mathbf{v}'_i \otimes \mathbf{v}'_i + \mathbf{r}_i \otimes \mathbf{F}_i\}. \tag{14.34}$$

Since,

$$\mathbf{F}_k = \sum_{\ell \neq k} \mathbf{F}_{\ell k},$$

we can also write the virial as

$$\sum_k \mathbf{r}_k \otimes \mathbf{F}_k = \frac{1}{2} \sum_k \sum_{\ell \neq k} \mathbf{r}_{\ell k} \otimes \mathbf{F}_{\ell k} \tag{14.35}$$

as shown in Eq. (14.31), where the factor $1/2$ is because each interaction is counted twice. The indices ℓk in Eq. (14.35) imply the bond or the interaction between the ℓth atom and the kth atom.

Unlike Eq. (14.33), the virial stress expressed in Eq. (14.34) is not symmetric in general. This is because the tensorial virial that we defined in Eq. (13.16) is not symmetric. In fact, if we define the tensorial virial as

$$\mathbf{G} = \sum_k \mathbf{p}_k \otimes \mathbf{r}_k, \tag{14.36}$$

it can be shown that

$$\bar{\sigma} = \frac{-1}{\Omega} \sum_{i=1}^{N} \{m_i \mathbf{v}'_i \otimes \mathbf{v}'_i + \mathbf{F}_i \otimes \mathbf{r}_i\} = \frac{-1}{\Omega} \sum_{i=1}^{N} \left\{ m_i \mathbf{v}'_i \otimes \mathbf{v}'_i + \frac{1}{2} \sum_{j \ni} \mathbf{F}_{ji} \otimes \mathbf{r}_{ji} \right\}, \tag{14.37}$$

and this expression of the virial stress is also extensively used in the literature. Since, in this book, we illustrate most examples by using pair potentials, there is no difference between Eqs. (14.35) and (14.37) in this particular case. Thus, in the rest of the book, we use both Eqs. (14.35) and (14.37) interchangeably without discrimination.

In the previous expressions of the virial or the virial stress, the summation is on the number of atoms or particles. In some literature, e.g., in LAMMPS developed by National Laboratories (2022), instead of counting for each atom's contribution to the virial or the virial stress, some researchers compute the virial by counting the contribution from each bond or each pair interaction.

For example, we consider a representative atom in a molecular system, and it is involved with N_p numbers of pair interactions; N_b numbers of translation bonds; N_a numbers of angle bonds; N_d numbers of dihedral bonds, N_i number of improper bonds, and N_f number of internal constraints. The contribution to the virial stress for this representative atom will be,

$$\sigma_k = -m_k \mathbf{v}'_k \otimes \mathbf{v}'_k + \frac{1}{2} \sum_{p=1}^{N_p} \left(\mathbf{F}^p_{p1} \otimes \mathbf{r}_{p1} + \mathbf{F}^p_{p2} \otimes \mathbf{r}_{p2} \right) + \frac{1}{2} \sum_{b=1}^{N_b} \left(\mathbf{F}^b_{b1} \otimes \mathbf{r}_{b1} + \mathbf{F}^b_{b2} \otimes \mathbf{r}_{b2} \right)$$

$$+ \frac{1}{3} \sum_{a=1}^{N_a} \left(\mathbf{F}^a_{a1} \otimes \mathbf{r}_{a1} + \mathbf{F}^a_{a2} \otimes \mathbf{r}_{a2} + \mathbf{F}^a_{a3} \otimes \mathbf{r}_{a3} \right)$$

$$+ \frac{1}{4} \sum_{d=1}^{N_d} \left(\mathbf{F}^d_{d1} \otimes \mathbf{r}_{d1} + \mathbf{F}^d_{d2} \otimes \mathbf{r}_{d2} + \mathbf{F}^d_{d3} \otimes \mathbf{r}_{d3} + \mathbf{F}^d_{d4} \otimes \mathbf{r}_{d4} \right)$$

$$+ \frac{1}{4} \sum_{i=1}^{N_i} \left(\mathbf{F}^i_{i1} \otimes \mathbf{r}_{i1} + \mathbf{F}^i_{i2} \otimes \mathbf{r}_{i2} + \mathbf{F}^i_{i3} \otimes \mathbf{r}_{i3} + \mathbf{F}^i_{i4} \otimes \mathbf{r}_{i4} \right) + \sum_{f=1}^{N_f} \mathbf{F}^f_f \otimes \mathbf{r}_f,$$

(14.38)

in which we first write out the total bond and interaction energy of each category, and then give 1/2 of the total pair bond and total translation bond energies to the involved atom k, give 1/3 of the total angle bond energy to the involved atom k, give 1/4 of the total dihedral bond energy and 1/4 of the total improper bound energy to the involved atom k, and last give the total internal constraint energy to the involved atom k. In LAMMPS, there is also an additional term of the K-Space contribution from long-range Coulombic interactions. Readers may find the details in LAMMPS command manual for *compute stress/atom command* (see *https://lammps.sandia.gov/doc/compute_stress_atom.html*).

14.3 Virial Stress via Liouville Theorem: Irving–Kirkwood Formalism

In a landmark paper published in 1950 by Irving and Kirkwood (1950), they derived the expression of the virial stress through statistical physics. The most important of this work is not the expression of the virial stress tensor, but the way how they derived the expression of the virial stress. Today, the statistical mechanics approach that Irving and Kirkwood developed is often referred to as the Irving–Kirkwood formalism that has been used to derive various macroscale thermodynamics property expressions, e.g., Evans and Morriss (2008); Eringen and Maugin (2012); Zhou (2003); Lehoucq and Sears (2011).

The Irving–Kirkwood formalism is also a multiscale approach. This is because that the stress is a macroscale quantity, and we would like to illustrate it from homogenization of microscale atomistic forces and linear momentums. Thus, before we start we first review the linear momentum conservation law at macroscale or continuum scale.

14.3.1 Conservation of Linear Momentum at Continuum Scale

Consider a control volume Ω in the space. The increase of linear momentum in Ω is

$$\int_\Omega \frac{\partial}{\partial t}(\rho \mathbf{v}) dv. \tag{14.39}$$

The linear momentum flux going outside the control volume Ω is

$$\int_{\partial\Omega} (\rho\mathbf{v})\mathbf{v}\cdot \mathbf{n}\,dS = \int_{\Omega} \nabla_x \cdot (\rho\mathbf{v}\otimes\mathbf{v})\,dv. \tag{14.40}$$

Therefore, the net increase of linear momentum is

$$\frac{D\mathbf{L}}{Dt} = \int_{\Omega} \left(\frac{\partial}{\partial t}(\rho\mathbf{v}) + \nabla_x \cdot (\rho\mathbf{v}\otimes\mathbf{v})\right)dv, \tag{14.41}$$

which should be equal to the force exerted on the control volume, if Newton's second law is additive,

$$\frac{D\mathbf{L}}{Dt} = \int_{\Omega} \mathbf{b}\,dv + \int_{\Omega} \mathbf{t}\cdot\mathbf{n}\,dS, \tag{14.42}$$

where $\mathbf{b}$ is the external body force acting in, and $\mathbf{t}$ is the external traction acting on $\partial\Omega$.
We then have the equation of motion

$$\int_{\Omega} \left(\frac{\partial(\rho\mathbf{v})}{\partial t} + \nabla\cdot(\rho\mathbf{v}\otimes\mathbf{v})\right)dv = \int_{\Omega}(\nabla\cdot\boldsymbol{\sigma}+\mathbf{b})\,dv. \tag{14.43}$$

If Ω is arbitrary, the above expression leads to the equation of motion

$$\frac{\partial(\rho\mathbf{v})}{\partial t} + \nabla\cdot(\rho\mathbf{v}\otimes\mathbf{v}) = \nabla\cdot\boldsymbol{\sigma} + \mathbf{b}. \tag{14.44}$$

In the derivation of the virial stress, we often assume that $\mathbf{b}=0$.
The Irving–Kirkwood approach, one will first write the following expression:

$$\frac{\partial(\rho\mathbf{v})}{\partial t} + \nabla\cdot(\rho\mathbf{v}\times\mathbf{v}) = \nabla\cdot\boldsymbol{\sigma} \tag{14.45}$$

in terms of molecular coordinates and then identify $\boldsymbol{\sigma}$.

How to write Eq. (14.45) in terms of molecular coordinates? We use the Liouville Theorem.

14.3.2 Conservation of Linear Momentum via Liouville Theorem

In the following, we briefly review the Liouville theorem. Consider the following probability distribution density function, $f(\mathbf{r}^N,\mathbf{p}^N)$, where

$$\mathbf{r}^N = (\mathbf{r}_1,\mathbf{r}_2,\cdot,\mathbf{r}_N), \text{ and } \mathbf{p}^N = (\mathbf{p}_1,\mathbf{p}_2,\ldots,\mathbf{p}_N).$$

The net increase of the probability in the phase space region Ω will be

$$\int_{\Omega} \frac{\partial f}{\partial t}\,dv + \sum_{i=1}^{N} \int_{\partial\Omega} \left(f(\dot{\mathbf{r}}_i\cdot\mathbf{n}+\dot{\mathbf{p}}_i\cdot\mathbf{n})\right)dS = 0, \tag{14.46}$$

where $dv = d\mathbf{r}_1 d\mathbf{r}_2\cdots d\mathbf{r}_N d\mathbf{p}_1\cdots d\mathbf{p}_N = d\mathbf{r}^N d\mathbf{p}^N$.

By divergence theorem, we have

$$\sum_{k=1}^{N} \int_{\partial \Omega} \left(f(\dot{\mathbf{r}}_k \cdot \mathbf{n} + \dot{\mathbf{p}}_k \cdot \mathbf{n}) \right) dS$$

$$= \sum_{k=1}^{N} \int_{\Omega} \left(\frac{\partial f}{\partial \mathbf{r}_k} \cdot \dot{\mathbf{r}}_k + f \frac{\partial \dot{\mathbf{r}}_k}{\partial \mathbf{r}_k} + \frac{\partial f}{\partial \mathbf{p}_k} \cdot \dot{\mathbf{p}}_k + f \frac{\partial \dot{\mathbf{p}}_k}{\partial \mathbf{p}_k} \right) dv. \qquad (14.47)$$

If the atomistic system is a Hamiltonian system, we can write

$$\dot{\mathbf{r}}_k = \frac{\partial H}{\partial \mathbf{p}_k} \Rightarrow \frac{\partial \dot{\mathbf{r}}_k}{\partial \mathbf{r}_k} = \frac{\partial^2 H}{\partial \mathbf{r}_k \partial \mathbf{p}_k} \qquad (14.48)$$

and hence

$$\dot{\mathbf{p}}_k = -\frac{\partial H}{\partial \mathbf{r}_k} \Rightarrow \frac{\partial \dot{\mathbf{p}}_k}{\partial \mathbf{p}_k} = -\frac{\partial^2 H}{\partial \mathbf{r}_k \mathbf{p}_k}. \qquad (14.49)$$

Thus, the second term and the fourth term of in the right-hand side of Eq. (14.47) cancel each other, which leads to the following equation:

$$\int_{\Omega} \left\{ \frac{\partial f}{\partial t} + \sum_{k=1}^{N} \left(\frac{\partial f}{\partial \mathbf{r}_k} \dot{\mathbf{r}}_k + \frac{\partial f}{\partial \mathbf{p}_k} \dot{\mathbf{p}}_k \right) \right\} dv = 0. \qquad (14.50)$$

The macroscale "strong form" of the Liouville equation will be

$$\frac{\partial f}{\partial t} + \sum_{k=1}^{N} \left(\frac{\partial f}{\partial \mathbf{r}_k} \dot{\mathbf{r}}_k + \frac{\partial f}{\partial \mathbf{p}_k} \dot{\mathbf{p}}_k \right) = \frac{\partial f}{\partial t} + \sum_{k=1}^{N} \left(\frac{\partial f}{\partial \mathbf{r}_k} \frac{\partial H}{\partial \mathbf{p}_k} - \frac{\partial f}{\partial \mathbf{p}_k} \frac{\partial H}{\partial \mathbf{r}_k} \right) = 0, \qquad (14.51)$$

which we often write in a compact form,

$$\frac{\partial f}{\partial t} + [f, H] = 0, \quad \Leftarrow \text{Poisson's bracket}. \qquad (14.52)$$

Note that Eq. (14.51) is satisfied at any spatial point of the phase space.

Now, we define the ensemble average for any arbitrary random variable $\alpha(\mathbf{r}^N, \mathbf{p}^N)$,

$$<\alpha; f> := \int_{\Omega} \alpha(\mathbf{r}^N, \mathbf{p}^N) f(\mathbf{r}^N, \mathbf{p}^N, t) dv. \qquad (14.53)$$

Note that $\alpha(\mathbf{r}^N, \mathbf{p}^N)$ is a field function in the phase space, which can be a scaler field, a vector field, or a tensorial field.

Considering

$$\frac{\partial f}{\partial t} = -\sum_{k=1}^{N} \left(\frac{\partial f}{\partial \mathbf{r}_k} \dot{\mathbf{r}}_k + \frac{\partial f}{\partial \mathbf{p}_k} \dot{\mathbf{p}}_k \right),$$

we can write

$$\frac{\partial}{\partial t} <\alpha; f> = <\alpha; \frac{\partial f}{\partial t}> = \sum_{k=1}^{N} \left[<\alpha; -\frac{\mathbf{p}_k}{m_k} \cdot \nabla_{r_k} f> + <\alpha; -\mathbf{F}_k \cdot \nabla_{p_k} f> \right],$$

$$(14.54)$$

where the gradient operators are defined as

$$\nabla_{r_k} := \frac{\partial}{\partial \mathbf{r}_k}, \text{ and } \nabla_{p_k} := \frac{\partial}{\partial \mathbf{p}_k}, \quad k = 1, 2, \ldots, N.$$

Since the probability density $f \to 0$, $\forall (\mathbf{r}^N, \mathbf{p}^N) \in \partial \Omega$, integration by parts yields,

$$< \alpha; -\frac{\mathbf{p}_i}{m_i} \cdot \nabla_{r_k} f > = < \frac{\mathbf{p}_i}{m_i} \cdot \nabla_{r_k} \alpha; f > \text{ and} \quad (14.55)$$

$$< \alpha; -\mathbf{F}_k \cdot \nabla_{p_k} f > = < \mathbf{F}_k \cdot \nabla_{p_k} \alpha; f >. \quad (14.56)$$

Substituting Eqs. (14.55) and (14.56) into Eq. (14.54), we have

$$\frac{\partial}{\partial t} < \alpha; f > = \sum_{k=1}^{N} \left\langle \frac{\mathbf{p}_k}{m_k} \cdot \nabla_{r_k} \alpha + \mathbf{F}_k \cdot \nabla_{p_k} \alpha; f \right\rangle. \quad (14.57)$$

To convert a discrete mass function to a continuous mass distribution function, Irving and Kirkwood used a nonuniform "Dirac sampling" to define a continuous mass function,

$$M(\mathbf{x}) = \sum_{k=1}^{N} m_k \delta(\mathbf{r}_k - \mathbf{x}), \quad (14.58)$$

where m_k is the mass for the k-th atom. Now the mass function $M(\mathbf{x})$ may be viewed as a continuous mass function in the sense of the generalize function theory. By doing so, we can take the derivative of the mass density function.

We can further define the mass density function by calculating its ensemble average of a molecular system,

$$\rho(\mathbf{x}; t) := < M; f > = \sum_{k=1}^{N} m_k < \delta(\mathbf{r}_k - \mathbf{x}); f > = < M(\mathbf{x}); f >. \quad (14.59)$$

This can be verified by calculating the total mass of the molecular system, M, i.e.,

$$M := \int_V \rho(\mathbf{x}; t) dV = \int_V \sum_{k=1}^{N} m_k < \delta(\mathbf{r}_k - \mathbf{x}); f > dV = \int_V < M(\mathbf{x}); f > dV,$$

where V is the spatial domain or volume of the molecular system and M is the total mass of the molecular system. Note that V is not a volume in the ensemble phase space. Since

$$\int_V M(\mathbf{x}) dV = \sum_{k=1}^{N} m_k = M$$

and $< 1; f > = 1$, thus, we have

$$\int_V \rho(\mathbf{x}; t) dV = M. \quad (14.60)$$

Based on this approach, we can also use the "Dirac comb" to define the continuous linear momentum function as

$$\mathbf{L}(\mathbf{x},t) := \sum_{k=1}^{N} \mathbf{p}_k \delta(\mathbf{r}_k - \mathbf{x}), \quad (14.61)$$

where $\mathbf{p}_k = m_k \mathbf{v}_k$ is the linear momentum of the kth atom.

In doing so, we can further define the density function of the linear momentum as

$$\rho(\mathbf{x};t)\mathbf{v}(\mathbf{x};t) := \sum_{k=1}^{N} <\mathbf{p}_k \delta(\mathbf{r}_k - \mathbf{x}); f> = <\mathbf{L}; f>. \quad (14.62)$$

Note that the macroscale velocity $\mathbf{v}(\mathbf{x},t) = \bar{\mathbf{v}}$ is the average velocity defined in the previous section.

14.3.3 Balance of Linear Momentum: From Microscale to Macroscale

Now, we are in a position to express macroscale linear momentum in terms of molecular coordinates.

First, we consider $\alpha = \mathbf{L}$, and hence the macroscale linear momentum density is the ensemble average of the continuous linear momentum function,

$$\rho(\mathbf{x};t)\mathbf{v}(\mathbf{x};t) = <\alpha; f> = <\mathbf{L}; f>. \quad (14.63)$$

Recalling Eq. (14.57), i.e.,

$$\frac{\partial}{\partial t} <\alpha; f> = \sum_{k=1}^{N} \left\langle \frac{\mathbf{p}_k}{m_k} \cdot \nabla_{r_k} \alpha + \mathbf{F}_k \cdot \nabla_{p_k} \alpha; f \right\rangle,$$

and applying Eq. (14.57) to the vector fields $\alpha = \mathbf{L}$, we have

$$\frac{\partial}{\partial t} <\mathbf{L}; f> = \sum_{k=1}^{N} \left\langle \frac{\mathbf{p}_k}{m_k} \cdot \nabla_{r_k} \otimes \mathbf{L} + \mathbf{F}_k \cdot \nabla_{p_k} \otimes \mathbf{L}; f \right\rangle$$

$$= \sum_{k=1}^{N} \left\langle \left(\frac{\mathbf{p}_k}{m_k} \cdot \nabla_{r_k}\right) \mathbf{L} + \mathbf{F}_k \cdot \nabla_{p_k}\right) \mathbf{L}; f \right\rangle, \quad (14.64)$$

where $\mathbf{L} = \sum_{i=1}^{N} \mathbf{p}_i \delta(\mathbf{r}_i - \mathbf{x})$.

One may find that the first term in the ensemble average expression (14.64) can be rewritten as,

$$\sum_{k=1}^{N}\left\{\left(\frac{\mathbf{p}_k}{m_k}\cdot\nabla_{\mathbf{r}_k}\right)L+\mathbf{F}_k\cdot\nabla_{\mathbf{p}_k})L\right\}$$

$$=\sum_{k=1}^{N}\left\{\left(\frac{\mathbf{p}_k}{m_k}\cdot\nabla_{\mathbf{r}_k}\right)\mathbf{p}_k\delta(\mathbf{r}_k-\mathbf{x})+(\mathbf{F}_k\cdot\nabla_{\mathbf{p}_k})\mathbf{p}_k\delta(\mathbf{r}_k-\mathbf{x})\right\}$$

$$=-\sum_{k=1}^{N}\left\{\nabla_{\mathbf{x}}\cdot\left(\frac{\mathbf{p}_k\otimes\mathbf{p}_k}{m_k}\delta(\mathbf{r}_k-\mathbf{x})\right)-\mathbf{F}_k\cdot\delta(\mathbf{r}_k-\mathbf{x})\right\}, \quad (14.65)$$

where $\nabla_{\mathbf{p}_k}\mathbf{p}_k=\mathbf{I}^{(2)}$ is the second-order identity tensor. Note that in Eq. (14.65) we have used the chain rule,

$$\nabla_{\mathbf{r}_k}\delta(\mathbf{r}_k-\mathbf{x})=\frac{\partial}{\partial\mathbf{r}_k}\delta(\mathbf{r}_k-\mathbf{x})=\frac{\partial}{\partial(\mathbf{x}-\mathbf{r}_k)}\delta(\mathbf{r}_k-\mathbf{x})\frac{\partial(\mathbf{x}-\mathbf{r}_k)}{\partial\mathbf{r}_k}$$

$$=-\frac{\partial}{\partial\mathbf{x}}\delta(\mathbf{r}_k-\mathbf{x})=-\nabla_{\mathbf{x}}\delta(\mathbf{r}_k-\mathbf{x}).$$

Substituting Eq. (14.65) back to Eq. (14.64), we then have,

$$\frac{\partial}{\partial t}[\rho(\mathbf{x};t)\mathbf{v}(\mathbf{x};t)]=-\nabla_{\mathbf{x}}\cdot\sum_{k=1}^{N}\left\langle\frac{\mathbf{p}_k\otimes\mathbf{p}_k}{m_k}\delta(\mathbf{r}_k-\mathbf{x});f\right\rangle$$

$$+\sum_{k=1}^{N}\sum_{j\neq k}<\mathbf{F}_{jk}\delta(\mathbf{r}_k-\mathbf{x});f>. \quad (14.66)$$

Note that for the pair interaction

$$\mathbf{F}_k=\sum_{j\neq k}\mathbf{F}_{jk}.$$

Consider the fact

$$-\nabla_{\mathbf{x}}\sum_{k=1}^{N}m_k\left\langle\left(\frac{\mathbf{p}'_k}{m_k}+\mathbf{v}\right)\otimes\left(\frac{\mathbf{p}'_k}{m_k}+\mathbf{v}\right)\delta(\mathbf{r}_k-\mathbf{x});f\right\rangle$$

$$=-\nabla_{\mathbf{x}}\sum_{k=1}^{N}m_k\left\langle\left(\frac{\mathbf{p}'_k\otimes\mathbf{p}'_k}{m_k}\right)\delta(\mathbf{r}_k-\mathbf{x});f\right\rangle-\nabla_{\mathbf{x}}(\rho\mathbf{v}\otimes\mathbf{v}). \quad (14.67)$$

We then have

$$\frac{\partial}{\partial t}[\rho(\mathbf{x};t)\mathbf{v}(\mathbf{x};t)]+\nabla_{\mathbf{x}}(\rho\mathbf{v}\otimes\mathbf{v})=-\nabla_{\mathbf{x}}\cdot\sum_{k=1}^{N}\left\langle\left(\frac{\mathbf{p}_k}{m_k}-\mathbf{v}\right)\otimes\left(\frac{\mathbf{p}_k}{m_k}-\mathbf{v}\right)\delta(\mathbf{r}_k-\mathbf{x});f\right\rangle$$

$$+\sum_{k=1}^{N}\sum_{j\neq k}<\mathbf{F}_{jk}\delta(\mathbf{r}_k-\mathbf{x});f>, \quad (14.68)$$

where $\mathbf{v}=\bar{\mathbf{v}}$. Now we have express the macroscale "local" form linear momentum conservation in terms of molecular coordinates.

For the last term in Eq. (14.68), if we only consider the pair interaction, using the identity,

$$\sum_j \sum_{j \neq k} \{\} = \sum_k \sum_{j \neq k} \{\},$$

we then have

$$\sum_{k=1}^{N} \sum_{j \neq k} < \mathbf{F}_{jk} \delta(\mathbf{r}_k - \mathbf{x}); f >$$

$$= \frac{1}{2} \sum_{k=1}^{N} \sum_{j \neq k} < \mathbf{F}_{jk} \delta(\mathbf{r}_k - \mathbf{x}) + \mathbf{F}_{kj} \delta(\mathbf{r}_j - \mathbf{x}); f >$$

$$= \frac{1}{2} \sum_{k=1}^{N} \sum_{j \neq k} < \mathbf{F}_{jk} (\delta(\mathbf{r}_k - \mathbf{x}) - \delta(\mathbf{r}_j - \mathbf{x})); f >,$$

where Newton's third law $\mathbf{F}_{jk} = -\mathbf{F}_{kj}$ is used.

For a given function $f(\mathbf{r})$, we can expand it into a Taylor series at point $\mathbf{r}_j$,

$$f(\mathbf{r}_k) = f(\mathbf{r}_j) + \nabla_r f(\mathbf{r}_j)(\mathbf{r}_k - \mathbf{r}_j) + \frac{1}{2} \nabla_r^2 f(\mathbf{r}_j)(\mathbf{r}_k - \mathbf{r}_j)^2 + \cdots$$

Letting

$$f(\mathbf{r}_k) = \delta(\mathbf{r}_k - \mathbf{x}) \text{ and } f(\mathbf{r}_j) = \delta(\mathbf{r}_j - \mathbf{x})$$

and recalling that $\nabla_r = -\nabla_x$, we then find that

$$\delta(\mathbf{r}_k - \mathbf{x}) - \delta(\mathbf{r}_j - \mathbf{x}) = -\mathbf{r}_{jk} \cdot \nabla_x \delta(\mathbf{r}_j - \mathbf{x}) + \frac{1}{2!} (\mathbf{r}_{jk} \cdot \nabla_x)^2 \delta(\mathbf{r}_j - \mathbf{x}) + \cdots$$

$$= \nabla_x \cdot \left[\mathbf{r}_{jk} \left\{ -1 + \frac{1}{2!} \mathbf{r}_{jk} \cdot \nabla_x + \cdots + \frac{1}{n!} (-\mathbf{r}_{jk} \cdot \nabla_x)^{n-1} + \cdots \right\} \delta(\mathbf{r}_j - \mathbf{x}) \right].$$

Hence, the last term in Eq. (14.68) becomes

$$\frac{1}{2} \sum_{k=1}^{N} \sum_{j \neq k} < \mathbf{F}_{jk} (\delta(\mathbf{r}_k - \mathbf{x}) - \delta(\mathbf{r}_j - \mathbf{x})); f >$$

$$= \nabla_x \cdot \left[\frac{1}{2} \sum_{k=1}^{N} \sum_{j \neq k} < \mathbf{F}_{jk} \otimes \mathbf{r}_{jk} \left\{ -1 + \frac{1}{2} \mathbf{r}_{jk} \cdot \nabla_x + \cdots \right\} \delta(\mathbf{r}_j - \mathbf{x}); f > \right]. \tag{14.69}$$

Finally, we can rewrite Eq. (14.68) as

$$\frac{\partial}{\partial t} [\rho \mathbf{v}] + \nabla_x \cdot [\rho \mathbf{v} \otimes \mathbf{v}]$$

$$= \nabla_x \cdot \left[-\sum_{k=1}^{N} \left\langle m_k \left(\frac{\mathbf{p}_k}{m_k} - \mathbf{v} \right) \otimes \left(\frac{\mathbf{p}_k}{m_k} - \mathbf{v} \right) \delta(\mathbf{r}_k - \mathbf{x}); f \right\rangle \right.$$

$$\left. + \frac{1}{2} \sum_{k=1}^{N} \sum_{j \neq k} \left\langle \mathbf{F}_{jk} \otimes \mathbf{r}_{jk} \left\{ -1 + \frac{1}{2} \mathbf{r}_{jk} \cdot \nabla_x + \cdots \right\} \delta(\mathbf{r}_j - \mathbf{x}); f \right\rangle \right]. \tag{14.70}$$

If we assume that the probability density distribution is uniform, i.e., $f = \frac{1}{\Omega}$, the ensemble average will become trivial. We can then have an explicit continuum scale linear momentum equation in terms of molecular coordinates as follows:

$$\frac{D(\rho \mathbf{v})}{Dt} = \frac{\partial}{\partial t}[\rho \mathbf{v}] + \nabla_x \cdot [\rho \mathbf{v} \otimes \mathbf{v}]$$

$$= \nabla_x \cdot \left\{ -\frac{1}{\Omega} \sum_{k=1}^{N} \left(m_k \left(\frac{\mathbf{p}_k}{m_k} - \mathbf{v} \right) \otimes \left(\frac{\mathbf{p}_k}{m_k} - \mathbf{v} \right) - \frac{1}{2} \sum_{j \neq k} \mathbf{F}_{jk} \otimes \mathbf{r}_{jk} \right) \right\}. \tag{14.71}$$

By comparing Eq. (14.71) with the local form of the balance equation of linear momentum in continuum mechanics, i.e.,

$$\frac{\partial(\rho \mathbf{v})}{\partial t} + \nabla \cdot (\rho \mathbf{v} \times \mathbf{v}) = \nabla \cdot \boldsymbol{\sigma},$$

we can then identify the virial stress tensor as

$$\boldsymbol{\sigma} = \frac{1}{\Omega} \sum_{k=1}^{N} \left\{ -m_k \left(\frac{\mathbf{p}_k}{m_k} - \bar{\mathbf{v}} \right) \otimes \left(\frac{\mathbf{p}_k}{m_k} - \bar{\mathbf{v}} \right) + \frac{1}{2} \sum_{j \neq k} V'(r_{jk}) \frac{\mathbf{r}_{jk} \otimes \mathbf{r}_{jk}}{r_{jk}} \right\}, \tag{14.72}$$

where σ is $\bar{\sigma}$ in the previous section. This is because Eq. (14.71) is obtained through ensemble average over the whole phase space.

It may be noted that the Irving–Kirkwood's derivation of the virial stress is based on the pair interaction, so that the virial stress expression in Eq. (14.72) is symmetric, and this is consistent with that of the macroscale Cauchy stress. In the following box and Fig. 14.5, we present a short biographical note for John H. Irving.

Joseph Liouville
(1809–1882)
(a)

A book owned
by John H. Irving
(b)

John G. Kirkwood
(1907–1959)
(c)

Figure 14.5 (a) Joseph Liouville was a French mathematician (Photo courtesy of Wikipedia.org); (b) a nuclear physics lecture book originally from the Manhattan Project at Los Alamos owned by John H. Irving; and (c) John G. Krikwood was a professor at Caltech (Photo courtesy of American Institute of Physics)

> **John "Jack" Howard Irving (1920–2008)**
>
> IRVING, Jack Howard Died peacefully on November 11th with his family at his side. He was 87. Jack was born in Cleveland, Ohio to Lottie and William Irving on December 31, 1920. When he was eleven, the family moved to Los Angeles. Jack attended Beverly Hills High where he won an academic scholarship to study physics at Caltech. Upon graduation from Caltech in 1942, Jack joined the staff of the MIT Radiation Lab, where he was involved in the design of the analog computer for precision radar. From 1946–48, Jack was a graduate student in physics at Princeton University. Jack interrupted his studies to accept a fellowship at Caltech, where he developed innovative methods in statistical mechanics. Following his fellowship at Caltech, Jack became Head of Systems Planning and Analysis at Hughes Aircraft. His group at Hughes developed the first airborne digital computer for fighter aircraft. In 1954, Jack joined the Ramo-Wooldridge Corporation (later TRW). At Ramo-Wooldridge, he directed early theoretical studies of synchronous orbit satellites (which became crucial to global communications systems) and of rocket navigation and control (precursor to the Apollo program). In 1960, Jack joined the newly formed not-for-profit Aerospace Corporation as head of the Systems Research and Planning Division. In 1963, he took a sabbatical from Aerospace to complete his Ph.D. in physics at Princeton. Following his return to Aerospace, Jack pioneered the concept of personal rapid transit — an automated light-rail taxi system for nonstop transportation from origin to destination. In retirement, Jack started a partnership, Applied Research & Technology, to develop energy-efficient pumps and compressors. Jack's interests were wide-ranging; he loved music, art, and travel. ······
> Excerpt From Los Angeles Times on Nov. 16, 2008

14.4 Hardy Stress

In actual computer simulations, the computed virial stress based on the Irving–Kirkwood formalism may fluctuate significantly. To have a stable virial stress formulation, in 1982 Robert J. Hardy proposed the so-called Hardy stress. Today, most of the atomistic virial stress computed in the computer simulation are actually the Hardy stress, or equivalent stress measures.

The essence of the Hardy stress is to replace the Dirac delta function in the Irving–Kirkwood formulation by a smooth function in a finite support. To illustrate the procedure of derivation of the Hardy stress, in the following, for simplicity, we illustrate Hardy's treatment of the micro-to-macro transition for the conservation of linear momentum, without considering the kinetic part of the virial stress. In the literature, this procedure is also called as the Hardy–Murdoch procedure (see: Murdoch (1983)).

Replacing the Dirac's delta function by a smooth function $\varphi(\mathbf{x})$ in Eq. (14.61), we let the linear momentum of the atomistic system expressed as,

$$\mathbf{L}(\mathbf{x},t) := \sum_{i=1}^{N} m_i \mathbf{v}_i \varphi(\mathbf{x} - \mathbf{x}_i) = \sum_{i=1}^{N} \mathbf{p}_i \varphi(\mathbf{x} - \mathbf{x}_i), \qquad (14.73)$$

where the smoothing function must satisfy the following conditions:

- $\varphi(\mathbf{x} - \mathbf{x}_i)$ has its maximum at $\mathbf{x} = \mathbf{x}_i$;
- $\varphi(\mathbf{x} - \mathbf{x}_i) \to \delta(\mathbf{x} - \mathbf{x}_i)$ as $\mathbf{x} - \mathbf{x}_i \to 0$;
- $\int_{\mathbb{R}^3} \varphi(\mathbf{x} - \mathbf{x}_i) dV_\mathbf{x} = 1$.

Taking the material time derivative on Eq. (14.73), we have

$$\frac{\partial}{\partial t} \mathbf{L}(\mathbf{x},t) = \frac{\partial}{\partial t} \sum_{i=1}^{N} m_i \mathbf{v}_i \varphi(\mathbf{x} - \mathbf{x}_i) = \sum_{i=1}^{N} \mathbf{f}_i \varphi(\mathbf{x} - \mathbf{x}_i). \qquad (14.74)$$

In the last equality, the Newton's second law is used.

If we only consider the pair force, we have

$$\mathbf{f}_i = \sum_{j \neq i} \mathbf{f}_{ij}.$$

Considering the antisymmetric property of the interatomic force, i.e., $\mathbf{f}_{ij} = -\mathbf{f}_{ji}$ we can rewrite Eq. (14.74) as

$$\begin{aligned}
\frac{\partial}{\partial t} \mathbf{L}(\mathbf{x},t) &= \sum_{i=1}^{N} \sum_{j \neq i} \mathbf{f}_i \varphi(\mathbf{x} - \mathbf{x}_i) = \frac{1}{2} \sum_{i=1}^{N} \sum_{j \neq i} \mathbf{f}_{ij} (\varphi(\mathbf{x} - \mathbf{x}_i) + \varphi(\mathbf{x} - \mathbf{x}_i)) \\
&= \frac{1}{2} \sum_{i=1}^{N} \sum_{j \neq i} \mathbf{f}_{ij} \varphi(\mathbf{x} - \mathbf{x}_i) + \frac{1}{2} \sum_{j=1}^{N} \sum_{i \neq j} \mathbf{f}_{ji} \varphi(\mathbf{x} - \mathbf{x}_j) \\
&= \frac{1}{2} \sum_{i=1}^{N} \sum_{j \neq i} \mathbf{f}_{ij} \varphi(\mathbf{x} - \mathbf{x}_i) - \frac{1}{2} \sum_{i=1}^{N} \sum_{j \neq i} \mathbf{f}_{ij} \varphi(\mathbf{x} - \mathbf{x}_j) \\
&= \frac{1}{2} \sum_{i=1}^{N} \sum_{j \neq i} \mathbf{f}_{ij} \big(\varphi(\mathbf{x} - \mathbf{x}_i) - \varphi(\mathbf{x} - \mathbf{x}_j) \big).
\end{aligned} \qquad (14.75)$$

Define a so-called bond function,

$$B_{ij}(\mathbf{x}) := \int_0^1 \varphi(\mathbf{x} - (\mathbf{x}_i + \alpha \mathbf{x}_{ij})) d\alpha, \qquad (14.76)$$

where $\mathbf{x}_{ij} := \mathbf{x}_j - \mathbf{x}_i$ and $0 \leq \alpha \leq 1$.

Hardy (1982) showed that

$$\frac{\partial}{\partial \alpha} \varphi(\mathbf{x} - (\mathbf{x}_i + \alpha \mathbf{x}_{ij})) = -\mathbf{x}_{ij} \cdot \nabla_\mathbf{x} \varphi(\mathbf{x} - (\mathbf{x}_i + \alpha \mathbf{x}_{ij})). \qquad (14.77)$$

Integrating the above expression from $\alpha = 0$ to 1, one obtains

$$\varphi(\mathbf{x} - \mathbf{x}_i) - \varphi(\mathbf{x} - \mathbf{x}_j) = -\mathbf{x}_{ij} \cdot \nabla_{\mathbf{x}} B_{ij}(\mathbf{x}). \tag{14.78}$$

Subsequently, we have

$$\frac{\partial}{\partial t}\mathbf{L} = \frac{1}{2}\sum_{i=1}^{N}\sum_{j \neq i}\mathbf{f}_{ij}\left(\varphi(\mathbf{x} - \mathbf{x}_i) - \varphi(\mathbf{x} - \mathbf{x}_j)\right) = -\frac{1}{2}\sum_{i=1}^{N}\sum_{j \neq i}\mathbf{f}_{ij}\mathbf{x}_{ij} \cdot \nabla_{\mathbf{x}} B_{ij}(\mathbf{x})$$

$$= \nabla \cdot \sigma^{Hardy}. \tag{14.79}$$

From Eq. (14.79), we can identify that

$$\sigma^{Hardy} = -\frac{1}{2}\sum_{i=1}^{N}\sum_{j \neq i}\mathbf{f}_{ij} \otimes \mathbf{r}_{ij} B_{ij}(\mathbf{x}). \tag{14.80}$$

Note that we always choose $\mathbf{x}$ as the center of the unit cell.

Now, we consider a few examples.

Example 14.2 Let $\varphi(\mathbf{x})$ as the radial step function,

$$\varphi(r) = \begin{cases} \frac{1}{\Omega_x}, & r \leq \delta \\ 0, & \text{otherwise,} \end{cases} \tag{14.81}$$

where Ω_x is the volume of the unit cell as $\mathbf{x}$ at its center.

If we choose $w(\mathbf{x})$ as the spherical radial step function as shown in Eq. (14.81), one can see that

$$\varphi(\alpha(\mathbf{x}_j - \mathbf{x}_i) + \mathbf{x}_i - \mathbf{x}) = \varphi(\alpha(\mathbf{x}_j - \mathbf{x}) + (1 - \alpha)(\mathbf{x}_I - \mathbf{x})).$$

Let the unit cell $\mathcal{H}_x$ as a spherical ball with radius δ. If $\mathbf{x}_i, \mathbf{x}_j \in \mathcal{H}_x$, it is always true that

$$\mathbf{x}_{x\alpha} := \alpha(\mathbf{x}_j - \mathbf{x}) + (1 - \alpha)(\mathbf{x}_i - \mathbf{x}) \in \mathcal{H}_x.$$

This is because that if $|\mathbf{x}_j - \mathbf{x}| \geq |\mathbf{x}_i - \mathbf{x}|$

$$|\alpha(\mathbf{x}_j - \mathbf{x}) + (1 - \alpha)(\mathbf{x}_i - \mathbf{x})| \leq |\alpha(\mathbf{x}_j - \mathbf{x}) + (1 - \alpha)(\mathbf{x}_j - \mathbf{x})| = |\mathbf{x}_j - \mathbf{x}| \leq \delta$$

and vice vera if $|\mathbf{x}_j - \mathbf{x}| \leq |\mathbf{x}_i - \mathbf{x}|$

$$|\alpha(\mathbf{x}_j - \mathbf{x}) + (1 - \alpha)(\mathbf{x}_i - \mathbf{x})| \leq |\alpha(\mathbf{x}_I - \mathbf{x}) + (1 - \alpha)(\mathbf{x}_i - \mathbf{x})| = |\mathbf{x}_i - \mathbf{x}| \leq \delta.$$

This can be made clear by inspection of Fig. 14.6.

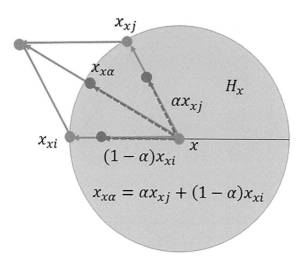

Figure 14.6 Graphical illustration of the bond integration variable $\mathbf{x}_{x\alpha} \in \mathcal{H}_x$, where $\mathbf{x}_{x\alpha} = \alpha \mathbf{x}_{xj} + (1-\alpha)\mathbf{x}_{xi}$.

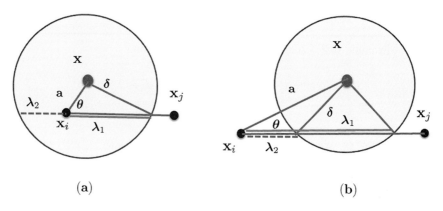

Figure 14.7 Graphical illustration of the bond integration variable for (a) $\mathbf{x}_i \in \mathcal{H}_x$ but $\mathbf{x}_j \notin \mathcal{H}_x$ and (b) $\mathbf{x}_i, \mathbf{x}_j \notin \mathcal{H}_x$ where $\mathbf{x}_{x\alpha} = \alpha \mathbf{x}_{xj} + (1-\alpha)\mathbf{x}_{xi}$

Since $\mathbf{x}_{x\alpha} \in \mathcal{H}_x$, by the definition of the radial step function, i.e., Eq. (14.81), $\phi(|\mathbf{x}_{x\alpha}|) = 1/\Omega_x$, and hence

$$B_{ij}(\mathbf{x}) = \frac{1}{\Omega_x}, \quad \text{if } \mathbf{x}_i, \mathbf{x}_j \in \mathcal{H}_x, \tag{14.82}$$

which recovers the static Irving–Kirkwood virial stress expression.

In the above expression, i.e., Eq. (14.82), we exclude the cases when (a) $\mathbf{x}_i \in \mathcal{H}_x$ but $\mathbf{x}_j \notin \mathcal{H}_x$ and (b) $\mathbf{x}_i, \mathbf{x}_j \notin \mathcal{H}_x$, as shown in Fig. 14.7.

If we include these two cases, the Hardy stress will become

$$\sigma^{Hardy} = -\frac{1}{2\Omega_x} \sum_{i\in\Omega_x} \sum_{j\in\Omega_x} \mathbf{f}_{ij} \otimes \mathbf{r}_{ij}(\mathbf{x}) - \sum_{i\in\Omega_x} \sum_{j\neq\Omega_x} \mathbf{f}_{ij} \otimes \mathbf{r}_{ij} B_{ij}^{(a)}(\mathbf{x})$$

$$-\frac{1}{2} \sum_{i\notin\Omega_x} \sum_{j\notin\Omega_x} \mathbf{f}_{ij} \otimes \mathbf{r}_{ij} B_{ij}^{(b)}(\mathbf{x}). \tag{14.83}$$

Note that in both cases (a) and (b) the sphere with the radius that equals to the cutoff distance of the atomistic potential at atom i or atom j, i.e., S_i or S_j, must satisfy the conditions: $S_k \cap \Omega_x \neq \emptyset$, $k = i, j$, and $S_j \cap S_j \neq \emptyset$.

To find $B_{ij}^{(a)}$ and $B_{ij}^{(b)}$, we notice that

$$B_{ij}^{(a)} = \frac{1}{\Omega_x} \frac{\ell_{ij}^{(a)}}{\|\mathbf{x}_j - \mathbf{x}_j\|} \quad \text{or} \quad B_{ij}^{(b)} = \frac{1}{\Omega_x} \frac{\ell_{ij}^{(b)}}{\|\mathbf{x}_j - \mathbf{x}_j\|}, \tag{14.84}$$

where ℓ_{ij} is the segment of ij bond inside Ω_x.

To find the length of ℓ for both cases (a) and (b), we first let $\mathbf{a} = \mathbf{x}_i - \mathbf{x}$ and then calculate the cosine between the vectors $\mathbf{a}$ and $\mathbf{x}_j - \mathbf{x}_i$,

$$\cos\theta = \frac{\mathbf{a} \cdot (\mathbf{x}_j - \mathbf{x}_i)}{\|\mathbf{a}\| \|\mathbf{x}_j - \mathbf{x}_i\|}. \tag{14.85}$$

Then based on law of cosine, we can find λ as

$$\lambda^2 - 2a\lambda \cos\theta + (a^2 - \delta^2) = 0 \rightarrow \lambda_{1,2} = a\cos\theta \pm \sqrt{\delta^2 - a^2 \sin^2\theta}. \tag{14.86}$$

Consequently, we can find that

$$\ell_{ij}^{(a)} = \lambda_1 = a\cos\theta + \sqrt{\delta^2 - a^2 \sin^2\theta}, \tag{14.87}$$

as shown in Fig. 14.7(a).

Subsequently for case (b), we have

$$\ell_{ij}^{(b)} = \lambda_1 - \lambda_2 = 2\sqrt{\delta^2 - a^2 \sin^2\theta}. \tag{14.88}$$

For detailed discussions, the readers are referred to Elder et al. (2019).

Example 14.3 In the second example, we consider the following Gaussian type smoothing function

$$\varphi(x) = \frac{1}{\delta^3 (2\pi)^{3/2}} \exp\left(-\frac{1}{2}(x/\delta)^2\right).$$

We then have

$$\varphi(\alpha \mathbf{x}_{ij} + \mathbf{x}_i - \mathbf{x}) = \frac{1}{\delta^3 (2\pi)^{3/2}} \exp\left(-\frac{1}{2\delta^2}\left(\alpha^2 x_{ij}^2 + 2\alpha \cos\theta x_{ij} x_{xI} + X_{xI}^2\right)\right),$$

where $\mathbf{x}_{ij} = \mathbf{x}_j - \mathbf{x}_i$, $x_{ij} = |\mathbf{x}_{ij}|$; $\mathbf{x}_{xi} = \mathbf{x}_i - \mathbf{x}$ and $x_{xI} = |\mathbf{x}_{xi}|$, and $\cos\theta = \mathbf{x}_{ij} \cdot \mathbf{x}_{xi}/(x_{ij} x_{xi})$.

Using the formula

$$\int \frac{1}{\sqrt{2\pi}} \exp(-x^2) dx = \Phi(x) + C,$$

where

$$\Phi(x) = \frac{1}{2}\left(1 + \mathrm{erf}\left(\frac{x}{\sqrt{2}}\right)\right),$$

we then have

$$B_{ij}(\mathbf{x}) = \int_0^1 \varphi(\alpha \mathbf{x}_{ij} + \mathbf{x}_i - \mathbf{x}) d\alpha$$

$$= \frac{1}{\delta^3 (2\pi)^{3/2}} \exp\left(\frac{1}{2\delta^2}\left(-\sin^2 \theta^2 x_{xi}^2\right)\right) \frac{1}{b} \Phi(a + b\alpha) \Big|_0^1$$

$$= \frac{1}{\delta^3 (2\pi)^{3/2} b} \exp\left(\frac{1}{2\delta^2}\left(-\sin^2 \theta^2 x_{xi}^2\right)\right) (\Phi(a+b) - \Phi(a)), \quad (14.89)$$

where

$$a = \frac{\cos \theta x_{xi}}{\delta}, \text{ and } b = \frac{x_{ij}}{\delta}.$$

14.5 Homework Problems

Problem 14.1 Derive the virial stress formula based on conservation of linear momentum in phase space by using the Irving–Kirkwood procedure, i.e.,
Derive the expression of the virial stress,

$$\sigma_V = \frac{1}{\Omega} \sum_{k=1}^N \left\{ -m_k (\mathbf{v}_k - \bar{\mathbf{v}}) \otimes (\mathbf{v}_k - \bar{\mathbf{v}}) + \frac{1}{2} \sum_{j \neq k} U'(r_{jk}) \frac{\mathbf{r}_{jk} \otimes \mathbf{r}_{jk}}{r_{jk}} \right\}$$

based on the following equation:

$$\frac{\partial}{\partial t} <\mathbf{L}; f> = <\mathbf{L}; \frac{\partial f}{\partial t}> = \sum_{k=1}^N \left[<\mathbf{L}; -\frac{\mathbf{p}_k}{m_k} \nabla_{r_k} f> + <\mathbf{L}; \nabla_{r_k} U \cdot \nabla_{p_k} f> \right],$$

where

$$\mathbf{L} := \sum_{k=1}^N \mathbf{p}_k \delta(\mathbf{r}_k - \mathbf{x}) \text{ and } \rho(\mathbf{x}; t)\mathbf{v}(\mathbf{x}; t) := \sum_{k=1}^N <\mathbf{p}_k \delta(\mathbf{r}_k - \mathbf{x}); f> = <\mathbf{L}; f>.$$

Hints: Read Irving, J.H. and Kirkwood, J.G. (1950) The statistical mechanical theory of transport processes. IV. The equations of hydrodynamics. *Journal of Chemical Physics*, **18**, 817–829.

15 Cauchy–Born Rule and Multiscale Methods

In this chapter, we shall discuss the statistical theory of the Cauchy continuum. Even with today's computer and computational technologies, to conduct an all-atom simulation of a macroscale object is still out of reach in terms of computational resources. On the other hand, one may ask whether it is possible to find macroscale thermodynamic quantities without carrying out all-atom large-scale molecular dynamics (MD) simulations with reasonable accuracy.

For example, can we approximate the virial stress from the expression,

$$\sigma_V = \frac{-1}{\Omega} \sum_{k=1}^{N} \left\{ m_k \left(\frac{\mathbf{p}_k}{m_k} - \bar{\mathbf{v}} \right) \otimes \left(\frac{\mathbf{p}_k}{m_k} - \bar{\mathbf{v}} \right) + \frac{1}{2} \sum_{j \neq k} \mathbf{F}_{jk} \otimes \mathbf{r}_{jk} \right\},$$

$$\text{where } \bar{\mathbf{v}} = \frac{1}{\sum_k m_k} \sum_i m_i \mathbf{v}_i, \tag{15.1}$$

without carrying out a full-atom MD simulation?

The essential question is how can we make a scale transition from microscale to macroscale without knowing every details of each atom's trajectory and velocity at any time. In general, this is an impossible task, because Eq. (15.1) requires the information of each atom's position and velocity. However, under some circumstances, we may have ways to make some approximations in calculating the virial stress by using limited atomistic information. In recent years, classes of these methods have been developed, which are labeled as the multiscale methods, which can calculate macroscale stress with limited but essential atomistic information at fine scale-based kinematics as well as thermodynamics principles. These multiscale methods have been extensively used in material modelings and simulations.

We start the chapter by discussing how to construct an approximation for the second term of the virial stress for crystalline materials, and then come back to discuss how to approximate both terms of the virial stress under some specific conditions for given materials.

15.1 Cauchy–Born Rule

The Cauchy–Born rule, or Cauchy–Born approximation, states that when a crystalline solid is under uniform deformation the positions of the atoms within the crystal lattice follow the overall strain of the medium. In other words, the relative displacement of two atoms scale with the strain the solid. It was Augustin-Louis Cauchy who proposed the idea that this is the only case that can scale atomistic motion from microscale to macroscale, and it was Born (see Born and Huang (1954)) who precisely formulated the approximation in terms of mathematical formulation.

This approximation generally holds for face-centered and body-centered cubic crystal systems. For complex lattices such as diamond, or even amorphous solids, the rule has to be modified to allow for internal degrees of freedom among different sublattice units.

To start, we first introduce the original version of the Cauchy–Born rule. The original Cauchy–Born rule model has two approximations: (1) neglect the virial stress contribution from the kinetic energy and (2) scale the microscale deformation with the overall macroscale strain. Because of the first approximation, only the second term of Eq. (15.1) is considered in the Cauchy–Born rule stress calculation,

$$\sigma = -\frac{1}{2\Omega} \sum_i \sum_{j \neq i} \mathbf{F}_{ij} \otimes \mathbf{r}_{ij}, \qquad (15.2)$$

where $r_{ij} = \mathbf{r}_j - \mathbf{r}_i$ and for the pair interaction

$$\mathbf{F}_{ij} = -\frac{\partial V(r_{ij})}{\partial r_{ij}} \frac{\mathbf{r}_{ij}}{r_{ij}}, \quad \text{with } r_{ij} = |\mathbf{r}_{ij}|$$

at the given atom site. This approximation may be used when temperature is low. However, even under this condition, Eq. (15.2) still requires knowing the spatial positions for every atoms in the system. This still requires a full atom simulation.

The essence of the Cauchy–Born rule is the so-called strain scaling. It assumes that the lattice is under uniform deformation, meaning both the macroscale strain and microscale strain are the same. Under the condition of finite deformation, this condition requires that the deformation of the solid is

$$\mathbf{F} = \frac{\partial \mathbf{x}}{\partial \mathbf{X}} = Const.,$$

where $\mathbf{x} = \varphi(\mathbf{X})$ is the position vector of the material point $\mathbf{X} \in \mathcal{B}_0$.

Since the deformation gradient $\mathbf{F}$ is constant, for a fixed atom in a unit cell the distance vector $\mathbf{r}_{ij}$ between the ith atom and the jth atom in the deformed configuration can be expressed in terms of its undeformed state $\mathbf{R}$ as

$$\mathbf{r}_{ij} = \mathbf{F} \cdot \mathbf{R}_{ij}, \qquad (15.3)$$

where $\mathbf{r}_{ij} = \mathbf{r}_j - \mathbf{r}_i$ and $\mathbf{R}_{ij} = \mathbf{R}_j - \mathbf{R}_i$, which is usually a lattice constant. Equation (15.3) is valid because by definition $d\mathbf{x} = \mathbf{F} \cdot d\mathbf{X}$, and

$$\mathbf{F} = const. \quad \rightarrow \quad \Delta \mathbf{x} = \mathbf{F} \cdot \Delta \mathbf{X} \quad \rightarrow \quad \mathbf{r}_{ij} = \mathbf{F} \cdot \mathbf{R}_{ij}.$$

This means that the deformed lattice vector $\mathbf{r}$ can be obtained by scaling the undeformed lattice vector $\mathbf{R}$ without MD or quantum dynamics calculations. One can find the relative position or the deformation of a pair of atoms by simply scaling

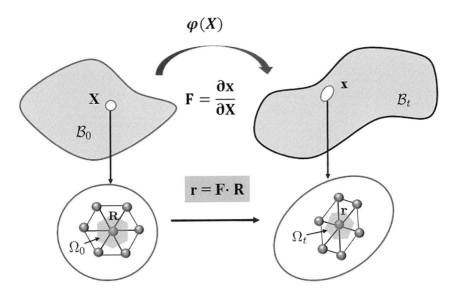

Figure 15.1 Illustration of strain scaling principle of the Cauchy–Born rule

their undeformed relative position vector. The approximation process is illustrated in Fig. 15.1.

Now we assume that the elastic energy density in a unit cell Ω_t is the average of the total bond potential in the unit cell, i.e.,

$$W_0 = \frac{1}{2\Omega_t} \sum_{i=1}^{N_b} \sum_{j \neq i} V(r_{ij}), \tag{15.4}$$

where $r_{ij} = |\mathbf{r}_j - \mathbf{r}_i|$ and N_b is the total number of bonds in the unit cell.

For a Bravais lattice, we can write the expression of elastic potential energy density as

$$W_0 = W_0(r_{i1}, r_{i2}, \ldots, r_{iN_b}) = W_0(|\mathbf{F} \cdot \mathbf{R}_{i1}|, \ldots |\mathbf{F} \cdot \mathbf{R}_{iN_b}|) = W_0(\mathbf{F}),$$

where ith atom is the center atom inside the Wigner–Seitz cell.

If we only consider the Bravais lattices, there is only one atom inside each Wigner–Seitz unit cell. Thus, the elastic density of each Wigner–Seitz cell will be,

$$W = \frac{1}{2\Omega_0} \sum_{k=1}^{N_b} V(r_k),$$

where N_b is the number of bonds in the unit cell.

By definition, we can find the first Piola–Kirchhoff stress as follows:

$$\mathbf{P} = \frac{\partial W}{\partial \mathbf{F}} = \frac{\partial}{\partial \mathbf{F}} \left(\frac{1}{2\Omega_0} \sum_k V(r_k) \right) = \frac{1}{2\Omega_0} \sum_{k=1}^{N_b} \frac{\partial V}{\partial r_k} \frac{\partial r_k}{\partial \mathbf{F}}.$$

One can find the first Piola–Kirchhoff stress as follows:

$$\mathbf{P} = \frac{\partial W}{\partial \mathbf{F}} = \frac{1}{2\Omega_0} \sum_k \frac{\partial V}{\partial r_k} \frac{\mathbf{r}_k \otimes \mathbf{R}_k}{r_k}. \tag{15.5}$$

Considering the fact that

$$r_k^2 = \mathbf{r}_k \cdot \mathbf{r}_k = (\mathbf{F} \cdot \mathbf{R}_k) \cdot (\mathbf{F} \cdot \mathbf{R}_k), \tag{15.6}$$

we can take derivative of both sides of Eq. (15.6) with respect to $\mathbf{F}$, it yields,

$$2r_k \frac{\partial r_k}{\partial \mathbf{F}} = 2(\mathbf{F}\mathbf{R}_k) \otimes \mathbf{R}_k = 2\mathbf{r}_k \otimes \mathbf{R}_k, \quad \rightarrow \quad \frac{\partial r_k}{\partial \mathbf{F}} = \frac{\mathbf{r}_k \otimes \mathbf{R}_k}{r_k},$$

and hence,

$$\mathbf{P} = \frac{1}{2\Omega_0} \sum_k \frac{\partial V}{\partial r_k} \frac{\partial r_k}{\partial \mathbf{F}} = \frac{1}{2\Omega_0} \sum_k \frac{\partial V}{\partial r_k} \frac{\mathbf{r}_k \otimes \mathbf{R}_k}{r_k}. \tag{15.7}$$

Similarly, we consider the right Cauchy–Green tensor $\mathbf{C} = \mathbf{F}^T \mathbf{F}$,

$$\mathbf{S} = 2\frac{\partial W}{\partial \mathbf{C}} = \frac{2}{2\Omega_0} \sum_k \frac{\partial V(r_k)}{\partial \mathbf{C}} = \frac{1}{\Omega_0} \sum_k \frac{\partial V}{\partial r_k} \frac{\partial r_k}{\partial \mathbf{C}}.$$

Since we know that

$$r_k^2 = \mathbf{r}_k \cdot \mathbf{r}_k = \mathbf{R}_k \mathbf{F}^T \mathbf{F} \mathbf{R}_k = \mathbf{R}_k \cdot \mathbf{C} \cdot \mathbf{R}_k, \quad \rightarrow \quad 2r_k \frac{\partial r_k}{\partial \mathbf{C}} = \mathbf{R}_k \otimes \mathbf{R}_k,$$

hence

$$\frac{\partial r_k}{\partial \mathbf{C}} = \frac{\mathbf{R}_k \otimes \mathbf{R}_k}{2r_k}, \quad \Rightarrow \quad \mathbf{S} = \frac{1}{2\Omega_0} \sum_k V'(r_k) \frac{\mathbf{R}_k \otimes \mathbf{R}_k}{r_k}, \quad \text{and}$$

$$\sigma = \frac{1}{J} \mathbf{F} \cdot \mathbf{S} \mathbf{F}^T = \frac{1}{2J\Omega_0} \sum_k V'(r_k) \frac{(\mathbf{F} \cdot \mathbf{R}_k \otimes \mathbf{R}_k \cdot \mathbf{F}^T)}{r_k} = \frac{1}{2\Omega} \sum_k V'(r_k) \frac{\mathbf{r}_k \otimes \mathbf{r}_k}{r_k},$$

where $\Omega = J\Omega_0$ and $J = det(\mathbf{F})$.

In fact, if one considers the left Cauchy–Green tensor,

$$\mathbf{b} = \mathbf{F}\mathbf{F}^T,$$

one can find that

$$\sigma = \frac{2}{J} \frac{\partial W}{\partial \mathbf{b}} \mathbf{b} = \beta \sum_{k \in S_c} \frac{\partial V}{\partial r_k} \frac{\mathbf{r}_k \otimes \mathbf{r}_k}{r_k} \tag{15.8}$$

and for pair potentials,

$$\mathbb{C} = \frac{4}{J} \mathbf{b} \frac{\partial^2 W}{\partial \mathbf{b}^2} \mathbf{b} = \beta \sum_{k \in S_c} \left(\frac{\partial^2 V}{\partial r_k^2} - \frac{1}{r_k} \frac{\partial V}{\partial r_k} \right) \frac{\mathbf{r}_k \otimes \mathbf{r}_k \otimes \mathbf{r}_k \otimes \mathbf{r}_k}{r_k^2}. \tag{15.9}$$

In summary, we have

$$\mathbf{P} = \frac{\partial W}{\partial \mathbf{F}} = \frac{1}{2\Omega_0} \sum_k \frac{\partial V}{\partial r_k} \frac{\mathbf{r}_k \otimes \mathbf{R}_k}{r_k}; \tag{15.10}$$

$$\sigma = \frac{1}{2\Omega} \sum_k \frac{\partial V}{\partial r_k} \frac{\mathbf{r}_k \otimes \mathbf{r}_k}{r_k}; \tag{15.11}$$

$$\mathbf{S} = \frac{1}{2\Omega_0} \sum_k \frac{\partial V}{\partial r_k} \frac{\mathbf{R}_k \otimes \mathbf{R}_k}{r_k}; \tag{15.12}$$

$$\mathbb{C} = \frac{J}{4} \sum_{k \in S_c} \left(\frac{\partial^2 V}{\partial r_k^2} - \frac{1}{r_k} \frac{\partial V}{\partial r_k} \right) \frac{\mathbf{r}_k \otimes \mathbf{r}_k \otimes \mathbf{r}_k \otimes \mathbf{r}_k}{r_k^2}. \tag{15.13}$$

In the following, we are examining a few examples of the Cauchy–Born rule in details.

Example 15.1 In this example, we calculate the second Piola–Kirchhoff stress tensor for a 2D hexagonal lattice unit cell based on the Cauchy–Born rule, i.e.,

$$\mathbf{S} = \frac{1}{2\Omega_0} \sum_k \frac{\partial V}{\partial r_k} \frac{\mathbf{R}_k \otimes \mathbf{R}_k}{r_k}, \tag{15.14}$$

where $\mathbf{r}_k = \mathbf{F}\mathbf{R}_k$. This is a Bravais lattice, and these is only one atom at the center of the Wigner–Seitz cell. The center atom is connecting to six (first) neighboring atoms. That is, the unit cell has six bonds. Since the six bonds are shared by the neighboring unit cells, each bond only contributes half of the energy to the unit cell. That is why there is an one half factor in front of Eq. (15.14).

Assume that the center atom is at the origin of the coordinates. The position vectors of the six bonds in a 2D hexagonal lattice can be expressed as follows (see Fig. 15.2):

$$\mathbf{R}_1 = (a, 0),$$
$$\mathbf{R}_2 = (a \times \cos(\pi/3), a \times \sin(\pi/3)),$$
$$\mathbf{R}_3 = (a \times \cos(2\pi/3), a \times \sin(2\pi/3)),$$
$$\mathbf{R}_4 = (-a, 0),$$
$$\mathbf{R}_5 = (a \times \cos(-2\pi/3), a \times \sin(-2\pi/3)),$$
$$\mathbf{R}_6 = (a \times \cos(-\pi/3), a \times \sin(-\pi/3)).$$

where a is the equilibrium bond length. Thus, we have

$$\mathbf{R}_1 \otimes \mathbf{R}_1 = \begin{bmatrix} a \\ 0 \end{bmatrix} [a \ 0] = \begin{bmatrix} a^2 & 0 \\ 0 & 0 \end{bmatrix};$$

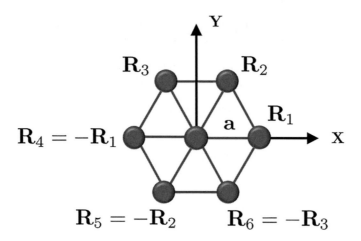

Figure 15.2 Two-dimensional hexagonal lattice Wigner–Seitz unit cell.

$$\mathbf{R}_2 \otimes \mathbf{R}_2 = \begin{bmatrix} a\cos(\pi/3) \\ a\sin(\pi/3) \end{bmatrix} [a\cos(\pi/3) \ a\sin(\pi/3)]$$

$$= a^2 \begin{bmatrix} \cos^2(\pi/3) & \sin(\pi/3)\cos(\pi/3) \\ \sin(\pi/3)\cos(\pi/3) & \sin^2(\pi/3) \end{bmatrix};$$

$$\mathbf{R}_3 \otimes \mathbf{R}_3 = \begin{bmatrix} a\cos(2\pi/3) \\ a\sin(2\pi/3) \end{bmatrix} [a\cos(2\pi/3) \ a\sin(2\pi/3)]$$

$$= a^2 \begin{bmatrix} \cos^2(2\pi/3) & \sin(2\pi/3)\cos(2\pi/3) \\ \sin(2\pi/3)\cos(2\pi/3) & \sin^2(2\pi/3) \end{bmatrix};$$

$$\mathbf{R}_4 \otimes \mathbf{R}_4 = \begin{bmatrix} -a \\ 0 \end{bmatrix}[-a \ 0] = \begin{bmatrix} a^2 & 0 \\ 0 & 0 \end{bmatrix}$$

$$= \mathbf{R}_1 \otimes \mathbf{R}_1;$$

$$\mathbf{R}_5 \otimes \mathbf{R}_5 = \begin{bmatrix} a\cos(-2\pi/3) \\ a\sin(-2\pi/3) \end{bmatrix} [a\cos(-2\pi/3) \ a\sin(-2\pi/3)]$$

$$= a^2 \begin{bmatrix} \cos^2(\pi/3) & \sin(\pi/3)\cos(\pi/3) \\ \sin(\pi/3)\cos(\pi/3) & \sin^2(\pi/3) \end{bmatrix}$$

$$= \mathbf{R}_2 \otimes \mathbf{R}_2;$$

$$\mathbf{R}_6 \otimes \mathbf{R}_6 = \begin{bmatrix} a\cos(-\pi/3) \\ a\sin(-\pi/3) \end{bmatrix} [a\cos(-\pi/3) \ a\sin(-\pi/3)]$$

$$= a^2 \begin{bmatrix} \cos^2(2\pi/3) & \sin(2\pi/3)\cos(2\pi/3) \\ \sin(2\pi/3)\cos(2\pi/3) & \sin^2(2\pi/3) \end{bmatrix}$$

$$= \mathbf{R}_3 \otimes \mathbf{R}_3.$$

Substituting the above expressions into Eq. (15.14) we can obtain an explicit expression of PK-II stress in the unit cell,

$$[\mathbf{S}] = \frac{a^2}{\Omega_0} \left(\frac{V'(r_1)}{r_1} \begin{bmatrix} 1 & 0 \\ 0 & 0 \end{bmatrix} + \frac{V'(r_2)}{r_2} \begin{bmatrix} \cos^2(\pi/3) & \sin(\pi/3)\cos(\pi/3) \\ \sin(\pi/3)\cos(\pi/3) & \sin^2(\pi/3) \end{bmatrix} \right.$$
$$\left. + \frac{V'(r_3)}{r_3} \begin{bmatrix} \cos^2(2\pi/3) & \sin(2\pi/3)\cos(2\pi/3) \\ \sin(2\pi/3)\cos(2\pi/3) & \sin^2(2\pi/3) \end{bmatrix} \right). \quad (15.15)$$

In the above expression, we used the identities,

$$|\mathbf{r}_1| = |\mathbf{FR}_1| = |\mathbf{F}(-\mathbf{R}_1)| = |\mathbf{r}_4|; \quad |\mathbf{r}_2| = |\mathbf{FR}_2| = |\mathbf{F}(-\mathbf{R}_2)| = |\mathbf{r}_5|, \text{ and}$$
$$|\mathbf{r}_3| = |\mathbf{FR}_3| = |\mathbf{F}(-\mathbf{R}_3)| = |\mathbf{r}_6|.$$

Example 15.2 In this example, we calculate the PK-II stress for a 3D FCC lattice based on the Cauchy–Born rule,

$$\mathbf{S} = \frac{1}{2\Omega_0} \sum_{k=1}^{12} \frac{\partial \phi}{\partial r_k} \frac{\mathbf{R}_k \otimes \mathbf{R}_k}{r_k}, \quad (15.16)$$

where $\mathbf{r}_k = \mathbf{FR}_k$.

Let the center atom at the origin of the coordinates. As shown in Fig. 15.3, the position vectors of the twelve bonds in a 3D FCC unit cell can be expressed as follows:

$$\mathbf{R}_1 = a(\cos(\pi/4), 0, \sin(\pi/4)),$$
$$\mathbf{R}_2 = a(-\cos(\pi/4), 0, \sin(\pi/4)),$$
$$\mathbf{R}_3 = a(-\cos(\pi/4), 0, -\sin(\pi/4)),$$
$$\mathbf{R}_4 = a(\cos(\pi/4), 0, -\sin(\pi/4)),$$
$$\mathbf{R}_5 = a(\cos(\pi/4), \sin(\pi/4), 0),$$
$$\mathbf{R}_6 = a(-\cos(\pi/4), \sin(\pi/4), 0),$$

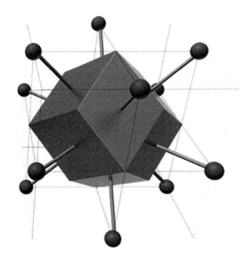

Figure 15.3 FCC lattice unit cell

$$\mathbf{R}_7 = a(-\cos(\pi/4), -\sin(\pi/4), 0),$$
$$\mathbf{R}_8 = a(0, \cos(\pi/4), -\sin(\pi/4)),$$
$$\mathbf{R}_9 = a(0, \cos(\pi/4), \sin(\pi/4)),$$
$$\mathbf{R}_{10} = a(0, -\cos(\pi/4), \sin(\pi/4)),$$
$$\mathbf{R}_{11} = a(0, -\cos(\pi/4), -\sin(\pi/4)),$$
$$\mathbf{R}_{12} = a(0, \cos(\pi/4), -\sin(\pi/4)).$$

We can then calculate the bond dyads as follows:

$$\mathbf{R}_1 \otimes \mathbf{R}_1 = \begin{bmatrix} a\cos(\pi/4) \\ 0 \\ a\sin(\pi/4) \end{bmatrix} \begin{bmatrix} a\cos(\pi/4) & 0 & a\sin(\pi/4) \end{bmatrix}$$

$$= a^2 \begin{bmatrix} \cos^2(\pi/4) & 0 & \sin(\pi/4)\cos(\pi/4) \\ 0 & 0 & 0 \\ \sin(\pi/4)\cos(\pi/4) & 0 & \sin^2(\pi/4) \end{bmatrix};$$

$$\mathbf{R}_2 \otimes \mathbf{R}_2 = \begin{bmatrix} -a\cos(\pi/4) \\ 0 \\ a\sin(\pi/4) \end{bmatrix} \begin{bmatrix} -a\cos(\pi/4) & 0 & a\sin(\pi/4) \end{bmatrix}$$

$$= a^2 \begin{bmatrix} \cos^2(\pi/4) & 0 & -\sin(\pi/4)\cos(\pi/4) \\ 0 & 0 & 0 \\ -\sin(\pi/4)\cos(\pi/3) & 0 & \sin^2(\pi/4) \end{bmatrix};$$

$$\mathbf{R}_3 \otimes \mathbf{R}_3 = \begin{bmatrix} -a\cos(\pi/4) \\ 0 \\ -a\sin(\pi/4) \end{bmatrix} \begin{bmatrix} -a\cos(\pi/4) & 0 & -a\sin(\pi/4) \end{bmatrix}$$

$$= a^2 \begin{bmatrix} \cos^2(\pi/4) & 0 & \sin(\pi/4)\cos(\pi/4) \\ 0 & 0 & 0 \\ \sin(\pi/4)\cos(\pi/4) & 0 & \sin^2(\pi/4) \end{bmatrix}$$

$$= \mathbf{R}_1 \otimes \mathbf{R}_1;$$

$$\mathbf{R}_4 \otimes \mathbf{R}_4 = \begin{bmatrix} a\cos(\pi/4) \\ 0 \\ -a\sin(\pi/4) \end{bmatrix} \begin{bmatrix} a\cos(\pi/4) & 0 & -a\sin(\pi/4) \end{bmatrix}$$

$$= a^2 \begin{bmatrix} \cos^2(\pi/4) & 0 & -\sin(\pi/4)\cos(\pi/4) \\ 0 & 0 & 0 \\ -\sin(\pi/4)\cos(\pi/4) & 0 & \sin^2(\pi/4) \end{bmatrix}$$

$$= \mathbf{R}_2 \otimes \mathbf{R}_2;$$

$$\mathbf{R}_5 \otimes \mathbf{R}_5 = \begin{bmatrix} a\cos(2\pi/4) \\ a\sin(2\pi/4) \\ 0 \end{bmatrix} \begin{bmatrix} a\cos(\pi/4) & a\sin(\pi/4) & 0 \end{bmatrix}$$

$$= a^2 \begin{bmatrix} \cos^2(\pi/4) & \sin(\pi/4)\cos(\pi/4) & 0 \\ \sin(\pi/4)\cos(\pi/4) & \sin^2(\pi/4) & 0 \\ 0 & 0 & 0 \end{bmatrix};$$

$$\mathbf{R}_6 \otimes \mathbf{R}_6 = \begin{bmatrix} -a\cos(\pi/4) \\ a\sin(\pi/4) \\ 0 \end{bmatrix} \begin{bmatrix} -a\cos(\pi/4) & a\sin(\pi/4) & 0 \end{bmatrix}$$

$$= a^2 \begin{bmatrix} \cos^2(\pi/4) & -\sin(\pi/4)\cos(\pi/4) & 0 \\ -\sin(\pi/4)\cos(\pi/4) & \sin^2(\pi/4) & 0 \\ 0 & 0 & 0 \end{bmatrix};$$

$$\mathbf{R}_7 \otimes \mathbf{R}_7 = \begin{bmatrix} -a\cos(2\pi/4) \\ -a\sin(2\pi/4) \\ 0 \end{bmatrix} \begin{bmatrix} -a\cos(\pi/4) & -a\sin(\pi/4) & 0 \end{bmatrix}$$

$$= a^2 \begin{bmatrix} \cos^2(\pi/4) & \sin(\pi/4)\cos(\pi/4) & 0 \\ \sin(\pi/4)\cos(\pi/4) & \sin^2(\pi/4) & 0 \\ 0 & 0 & 0 \end{bmatrix}$$

$$= \mathbf{R}_5 \otimes \mathbf{R}_5;$$

$$\mathbf{R}_8 \otimes \mathbf{R}_8 = \begin{bmatrix} a\cos(\pi/4) \\ -a\sin(\pi/4) \\ 0 \end{bmatrix} \begin{bmatrix} a\cos(\pi/4) & -a\sin(\pi/4) & 0 \end{bmatrix}$$

$$= a^2 \begin{bmatrix} \cos^2(\pi/4) & -\sin(\pi/4)\cos(\pi/4) & 0 \\ -\sin(\pi/4)\cos(\pi/4) & \sin^2(\pi/4) & 0 \\ 0 & 0 & 0 \end{bmatrix}$$

$$= \mathbf{R}_6 \otimes \mathbf{R}_6;$$

$$\mathbf{R}_9 \otimes \mathbf{R}_9 = \begin{bmatrix} 0 \\ a\cos(\pi/4) \\ a\sin(\pi/4) \end{bmatrix} \begin{bmatrix} 0 & a\cos(\pi/4) & a\sin(\pi/4) \end{bmatrix}$$

$$= a^2 \begin{bmatrix} 0 & 0 & 0 \\ 0 & \cos^2(\pi/4) & \sin(\pi/4)\cos(\pi/4) \\ 0 & \sin(\pi/4)\cos(\pi/4) & \sin^2(\pi/4) \end{bmatrix};$$

$$\mathbf{R}_{10} \otimes \mathbf{R}_{10} = \begin{bmatrix} 0 \\ -a\cos(\pi/4) \\ a\sin(\pi/4) \end{bmatrix} \begin{bmatrix} 0 & -a\cos(\pi/4) & a\sin(\pi/4) \end{bmatrix}$$

$$= a^2 \begin{bmatrix} 0 & 0 & 0 \\ 0 & \cos^2(\pi/4) & -\sin(\pi/4)\cos(\pi/4) \\ 0 & -\sin(\pi/4)\cos(\pi/4) & \sin^2(\pi/4) \end{bmatrix};$$

$$\mathbf{R}_{11} \otimes \mathbf{R}_{11} = \begin{bmatrix} 0 \\ -a\cos(\pi/4) \\ -a\sin(\pi/4) \end{bmatrix} \begin{bmatrix} 0 & -a\cos(\pi/4) & -a\sin(\pi/4) \end{bmatrix}$$

$$= a^2 \begin{bmatrix} 0 & 0 & 0 \\ 0 & \cos^2(\pi/4) & \sin(\pi/4)\cos(\pi/4) \\ 0 & \sin(\pi/4)\cos(\pi/4) & \sin^2(\pi/4) \end{bmatrix}$$

$$= \mathbf{R}_9 \otimes \mathbf{R}_9;$$

$$\mathbf{R}_{12} \otimes \mathbf{R}_{12} = \begin{bmatrix} 0 \\ a\cos(\pi/4) \\ -a\sin(\pi/4) \end{bmatrix} \begin{bmatrix} 0 & a\cos(\pi/4) & -a\sin(\pi/4) \end{bmatrix}$$

$$= a^2 \begin{bmatrix} 0 & 0 & 0 \\ 0 & \cos^2(\pi/4) & -\sin(\pi/4)\cos(\pi/4) \\ 0 & -\sin(\pi/4)\cos(\pi/4) & \sin^2(\pi/4) \end{bmatrix}$$

$$= \mathbf{R}_{10} \otimes \mathbf{R}_{10}.$$

Substituting the above expressions into Eq. (15.16), we have an explicit expression of the second Piola–Kirchhoff stress in an FCC unit cell,

$$[\mathbf{S}] = \frac{a^2}{\Omega_0} \left(\frac{V'(r_1)}{r_1} \begin{bmatrix} \cos^2(\pi/4) & 0 & \sin(\pi/4)\cos(\pi/4) \\ 0 & 0 & 0 \\ \sin(\pi/4)\cos(\pi/4) & 0 & \sin^2(\pi/4) \end{bmatrix} \right.$$

$$+ \frac{V'(r_2)}{r_2} \begin{bmatrix} \cos^2(\pi/4) & 0 & -\sin(\pi/4)\cos(\pi/4) \\ 0 & 0 & 0 \\ -\sin(\pi/4)\cos(\pi/4) & 0 & \sin^2(\pi/4) \end{bmatrix}$$

$$+ \frac{V'(r_5)}{r_5} \begin{bmatrix} \cos^2(\pi/4) & \sin(\pi/4)\cos(\pi/4) & 0 \\ \sin(\pi/4)\cos(\pi/4) & \sin^2(\pi/4) & 0 \\ 0 & 0 & 0 \end{bmatrix}$$

$$+ \frac{V'(r_6)}{r_6} \begin{bmatrix} \cos^2(\pi/4) & -\sin(\pi/4)\cos(\pi/4) & 0 \\ -\sin(\pi/4)\cos(\pi/4) & \sin^2(\pi/4) & 0 \\ 0 & 0 & 0 \end{bmatrix}$$

$$+ \frac{V'(r_9)}{r_9} \begin{bmatrix} 0 & 0 & 0 \\ 0 & \cos^2(\pi/4) & \sin(\pi/4)\cos(\pi/4) \\ 0 & \sin(\pi/4)\cos(\pi/4) & \sin^2(\pi/4) \end{bmatrix}$$

$$\left. + \frac{V'(r_{10})}{r_{10}} \begin{bmatrix} 0 & 0 & 0 \\ 0 & \cos^2(\pi/4) & -\sin(\pi/4)\cos(\pi/4) \\ 0 & -\sin(\pi/4)\cos(\pi/4) & \sin^2(\pi/4) \end{bmatrix} \right). \quad (15.17)$$

REMARK 15.3 In the above example, we find that the Cauchy–Born rule-based constitutive relation derived from the pair potential has some drawbacks, such as without shear stress components. The deficiency is also reflected in the elastic tensor. Recalling that

$$C_{IJKL} = \frac{\partial^2 W}{\partial E_{IJ} \partial E_{KL}} = \frac{\partial S_{IJ}}{\partial E_{KL}}$$

$$= \frac{1}{4\Omega_0} \sum_k \left(\phi''(r_k) - \frac{1}{r_k} \phi'(r_k) \right) \frac{R_I R_J R_K R_L}{r_k}. \quad (15.18)$$

One may immediately find that Eq. (15.18) leads to a so-called **Cauchy relation**, for example,

$$C_{1122} = C_{1212}.$$

Table 15.1 Elastic constants for real materials (MP_a)

	C_{1122}	C_{1212}
AL	61.3	28.5
Ag	93.4	46.1
Au	157.0	42.0
Cu	121.4	75.4

Let
$$C_{IJLM} = \frac{\partial^2 W}{\partial E_{IJ} \partial E_{LM}}.$$

Among different types of symmetries in the elastic tensor, the Cauchy relation is referred to the last equality relation,

1. $(IJ) \Leftrightarrow (LM)$; the major symmetry;
2. $(IJ) \Leftrightarrow (JI)$; the minor symmetry;
3. $(LM) \Leftrightarrow (ML)$; the minor symmetry;
4. $(JL) \Leftrightarrow (IM)$; the Cauchy relation.

In general, the Cauchy relation has the following six equations:

$$C_{1122} = C_{1212},\ C_{1133} = C_{1313},\ C_{2233} = C_{2323};$$
$$C_{1213} = C_{1123},\ C_{2213} = C_{2123},\ C_{3312} = C_{3132},$$

which is not true for real materials as shown in Table 15.1

To remedy this situation, we must use more realistic atomistic potential to construct the atomistic informed model, such as adopting the atomistic potential of the embedded-atom method (EAM) for metals.

The EAM potential is for metallic materials. In the EAM theory, we assume that the total energy of the atomistic system consists two parts: the electron density contribution and contribution from nucleus,

$$E_{tot} = \sum_{i=1}^{N} E_i,\quad E_i = F(\bar{\rho}_i) + \frac{1}{2} \sum_{j(j \neq i)} \phi(r_{ij}),$$

where F is *the embedding function*, which is the energy required to embed an atom into the background electron cloud at the atom site i. Here, $\bar{\rho}_i$ is the electron density at site i, which is a linear superposition of spherically averaged atomic electron density of different atoms,

$$\bar{\rho}_i = \sum_{j(j \neq i)} \rho_j(r_{ij}).$$

Example 15.4 In this example, we employ the Cauchy–Born rule to calculate stress and elastic tensor based on the embedded atom potential.

Consider the EAM potential in a unit cell centered at the atom site i,

$$E_i = \left(F_i(\bar{\rho}_i) + \frac{1}{2}\sum_{j\neq i}\phi(r_{ij})\right), \quad W = \frac{1}{\Omega_0}E_i, \tag{15.19}$$

where $\Omega_0 = \frac{\sqrt{2}}{2}|\mathbf{R}_0|^3$ for FCC lattices.

(1) As same as the previous example, the second Piola–Kirchhoff stress can be found as

$$S_{IJ} = \frac{1}{\Omega_0}\sum_{i\neq j}\frac{\partial E_i}{\partial r_{ij}}\frac{R_I R_J}{r_{ij}} = \frac{1}{\Omega_0}\sum_{j\neq i}\left(F'(\bar{\rho}_i)\frac{\partial \bar{\rho}_i}{\partial r_{ij}} + \frac{1}{2}\sum_{j\neq i}\phi'(r_{ij})\right)\frac{R_I R_J}{r_{ij}}. \tag{15.20}$$

We can also find the elastic tensor,

$$C_{IJKL} = \frac{\partial^2 W}{\partial E_{IJ}\partial E_{KL}} = \frac{\partial S_{IJ}}{\partial E_{KL}}$$

$$= \frac{1}{2\Omega_0}\sum_{j\neq i}\left(\phi''(r_{ij}) - \frac{1}{r_{ij}}\phi'(r_{ij})\right)\frac{R_I R_J R_K R_L}{r_{ij}}$$

$$+ \frac{1}{\Omega_0}\left(F''(\bar{\rho}_i)\left(\sum_{j\neq i}\frac{\partial \bar{\rho}_i}{\partial r_{ij}}\frac{R_I R_J}{r_{ij}}\right)\left(\sum_{i\neq j}\frac{\partial \bar{\rho}_i}{\partial r_{ij}}\frac{R_K R_L}{r_{ij}}\right)\right)$$

$$+ \frac{1}{\Omega_0}\left(F'(\bar{\rho}_i)\sum_{i\neq j}\left(\frac{\partial \bar{\rho}_i}{\partial r_{ij}} - \frac{1}{r_{ij}}\frac{\partial \bar{\rho}_i}{\partial r_{ij}}\right)\frac{R_I R_J R_K R_L}{r_{ij}}\right). \tag{15.21}$$

Since in EAM formulation, in general $\bar{\rho}_i \neq \bar{\rho}_j$, from Eqs. (15.20) and (15.21), one can find that the stress tensor should have non-zero shear stress components, and that the Cauchy relation will not be held anymore.

To illustrate the computation details, we choose the Lennard-Jones EAM potential (see: Baskes (1999)) as an example, we then have

$$E_i = F(\bar{\rho}_i) + \frac{1}{2}\sum_{i\neq j}\phi(r_{ij});$$

$$F(\bar{\rho}_i) = \frac{AZ_0}{2}\bar{\rho}_i(\ln(\bar{\rho}_i) - 1);$$

$$\bar{\rho}_i = \frac{1}{Z_0}\sum_{i\neq j}\rho(r_{ij}), \quad r_{ij} = |\mathbf{r}_i - \mathbf{r}_j|;$$

$$\phi(r) = \phi_{LJ}(r) - \frac{2}{Z_0}F(\rho(r));$$

$$\rho(r_{ij}) = \exp(-\beta(r_{ij} - 1));$$

$$\phi_{LJ}(r) = \frac{1}{r^{12}} - \frac{2}{r^6}.$$

Considering an FCC lattice and the nearest neighbor interaction, we can have

$$\frac{\partial \bar{\rho}_i}{\partial r_i} = \frac{1}{Z_0} \sum_{j \neq i} -\beta \exp(-\beta(r_{ij} - 1)) = \frac{1}{Z_0} \sum_{j \neq i} \frac{\partial \rho_i}{\partial r_{ij}};$$

$$\frac{\partial \rho_i}{\partial r_i} = -\beta \exp(-\beta(r_i - 1));$$

$$\frac{\partial \phi_{LJ}}{\partial r_i} = \frac{6}{r_i^7} - \frac{12}{r_i^{13}};$$

$$\frac{\partial F(\bar{\rho}_i)}{\partial r_i} = \frac{AZ_0}{2} \ln(\bar{\rho}_i) \frac{\partial \bar{\rho}_i}{\partial r_i};$$

$$\frac{\partial F(\rho_i)}{\partial r_i} = \frac{AZ_0}{2} \ln(\rho_i) \frac{\partial \rho_i}{\partial r_i};$$

$$\frac{\partial E_i}{\partial r_{ij}} = \frac{A}{2} \ln(\bar{\rho}_i) \sum_{j \neq i} \frac{\partial \bar{\rho}_i}{\partial r_{ij}} + \frac{1}{2} \sum_{j \neq i} \left(\frac{6}{r_{ij}} - \frac{12}{r_{ij}^{13}} \right) - \frac{A}{2} \ln(\rho_i) \sum_{j \neq i} \frac{\partial \rho_i}{\partial r_{ij}}.$$

In this example, one may choose $Z_0 = 12$, $\beta = 6$, and $A = 0.5$.

To illustrate how to implement the computation, we display the nearest neighbor shell for an FCC lattice in Fig. 15.4.

In Fig. 15.5, we show the elastic constants of copper calculated by using the so-called Mishin embedded-atom potential (Mishin et al. (2001)). The interested readers may find the calculation details in Lyu and Li (2017).

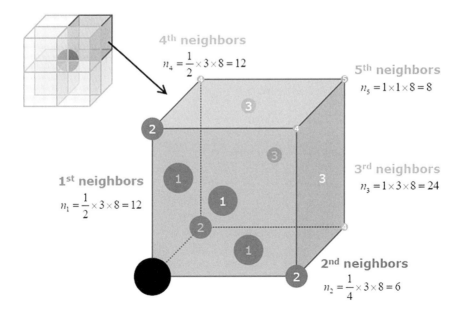

Figure 15.4 The nearest neighbor number for an FCC lattice

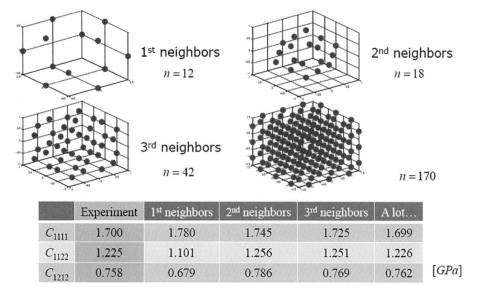

Figure 15.5 The elastic constants of cooper calculated by using the Mishin embedded-atom potential

15.2 Higher-Order Cauchy–Born Rule

As we have seen from the last section, the Cauchy–Born rule is a very coarse approximation in stress calculation. There are many ways to improve the approximation, and one of them is the so-called *higher-order Cauchy–Born rule*.

If we consider the Taylor expansion of deformed atom position into higher order, i.e.,

$$\mathbf{r}_j = \mathbf{r}_i + \frac{\partial \mathbf{r}}{\partial \mathbf{R}}(\mathbf{R}_j - \mathbf{R}_i) + \frac{1}{2!}\frac{\partial^2 \mathbf{r}}{\partial \mathbf{R}^2} : [(\mathbf{R}_j - \mathbf{R}_i) \otimes (\mathbf{R}_j - \mathbf{R}_i)]$$
$$+ \frac{1}{3!}\frac{\partial^3 \mathbf{r}}{\partial \mathbf{R}^3} \vdots [(\mathbf{R}_j - \mathbf{R}_i) \otimes (\mathbf{R}_j - \mathbf{R}_i) \otimes (\mathbf{R}_j - \mathbf{R}_i)]$$
$$+ \cdots \qquad (15.22)$$

Rearranging the order of terms in Eq. (15.22) and considering that $\mathbf{r}_{ij} = \mathbf{r}_j - \mathbf{r}_i$, we have the following higher-order approximation of relative position vector in the deformed configuration,

$$\mathbf{r}_{ij} = \frac{\partial \mathbf{r}}{\partial \mathbf{R}} \cdot \mathbf{R}_{ij} + \frac{1}{2!}\frac{\partial^2 \mathbf{r}}{\partial \mathbf{R}^2} : [\mathbf{R}_{ij} \otimes \mathbf{R}_{ij}] + \frac{1}{3!}\frac{\partial^3 \mathbf{r}}{\partial \mathbf{R}^3} \vdots [\mathbf{R}_{ij} \otimes \mathbf{R}_{ij} \otimes \mathbf{R}_{ij}] + \cdots$$

Note that in the above equations the indices i, j are labels of atoms but not the index of the bonds.

Thus, we can employ the higher-order Cauchy–Born rule to develop a higher strain gradient-based constitutive relation. For instance, there are $i = 1, 2, \ldots, N_b$ pair bonds in a unit cell, and a fourth-order Cauchy–Born rule may be written in terms of each bond as,

$$\mathbf{r}_i = \mathbf{F} \cdot \mathbf{R}_i + \frac{1}{2!}\mathbf{G} : (\mathbf{R}_i \otimes \mathbf{R}_i) + \frac{1}{3!}\mathbf{H} \vdots (\mathbf{R}_i \otimes \mathbf{R}_i \otimes \mathbf{R}_i) + \frac{1}{4!}\mathbf{K} \vdots (\mathbf{R}_i \otimes \mathbf{R}_i \otimes \mathbf{R}_i \otimes \mathbf{R}_i) \quad (15.23)$$

where

$$\mathbf{F} = \frac{\partial \mathbf{x}}{\partial \mathbf{X}}; \quad (15.24)$$

$$\mathbf{G} := \frac{\partial^2 \mathbf{x}}{\partial \mathbf{X}^2} = \frac{\partial \mathbf{F}}{\partial \mathbf{X}}; \quad (15.25)$$

$$\mathbf{H} := \frac{\partial^3 \mathbf{x}}{\partial \mathbf{X} \otimes \partial \mathbf{X} \otimes \partial \mathbf{X}} = \frac{\partial^2 \mathbf{F}}{\partial \mathbf{X} \otimes \partial \mathbf{X}} = \frac{\partial \mathbf{G}}{\partial \mathbf{X}}; \quad (15.26)$$

$$\mathbf{K} := \frac{\partial^4 \mathbf{x}}{\partial \mathbf{X} \otimes \partial \mathbf{X} \otimes \partial \mathbf{X} \otimes \partial \mathbf{X}} = \frac{\partial^3 \mathbf{F}}{\partial \mathbf{X} \otimes \partial \mathbf{X} \otimes \partial \mathbf{X}} = \frac{\partial^2 \mathbf{G}}{\partial \mathbf{X} \otimes \partial \mathbf{X}} = \frac{\partial \mathbf{H}}{\partial \mathbf{X}}. \quad (15.27)$$

This implies that the elastic potential energy density may be written as

$$W = W(\mathbf{F}, \mathbf{G}, \mathbf{H}, \mathbf{K}).$$

Subsequently, we can find the atomistic-informed relations between higher stresses and higher-order strain tensors,

$$\mathbf{P} = \frac{\partial W}{\partial \mathbf{F}} = \frac{1}{2\Omega_0} \sum_{i=1}^{n_{bond}} \phi'(r_i) \frac{\mathbf{r}_i \otimes \mathbf{R}_i}{r_i}; \quad (15.28)$$

$$\mathbf{Q} = \frac{\partial W}{\partial \mathbf{G}} = \frac{1}{4\Omega_0} \sum_{i=1}^{n_{bond}} \phi'(r_i) \frac{\mathbf{r}_i \otimes \mathbf{R}_i \otimes \mathbf{R}_i}{r_i}; \quad (15.29)$$

$$\mathbf{R} = \frac{\partial W}{\partial \mathbf{H}} = \frac{1}{12\Omega_0} \sum_{i=1}^{n_{bond}} \phi'(r_i) \frac{\mathbf{r}_i \otimes \mathbf{R}_i \otimes \mathbf{R}_i \otimes \mathbf{R}_i}{r_i}; \quad (15.30)$$

$$\mathbf{T} = \frac{\partial W}{\partial \mathbf{K}} = \frac{1}{48\Omega_0} \sum_{i=1}^{n_{bond}} \phi'(r_i) \frac{\mathbf{r}_i \otimes \mathbf{R}_i \otimes \mathbf{R} \otimes \mathbf{R}_i \otimes \mathbf{R}_i}{r_i}. \quad (15.31)$$

In the following example, we show the detailed derivations for the second-order Cauchy–Born model for a pair interaction system.

Example 15.5 For the second-order Cauchy–Born rule, we have

$$\mathbf{r}_i = \mathbf{F} \cdot \mathbf{R}_i + \frac{1}{2}\mathbf{G} : [\mathbf{R}_i \otimes \mathbf{R}_i].$$

Assume that there are N_b number of pair bonds associated with the center atom in a Wigner–Seitz cell. Thus, we can write the elastic potential energy density as

$$W = W(r_1, \ldots, r_i, \ldots r_{N_b})$$
$$= W\left(|\mathbf{F} \cdot \mathbf{R}_1 + \frac{1}{2}\mathbf{G} : [\mathbf{R}_1 \otimes \mathbf{R}_1]|, \ldots, |\mathbf{F} \cdot \mathbf{R}_{N_b} + \frac{1}{2}\mathbf{G} : [\mathbf{R}_{N_b} \otimes \mathbf{R}_{N_b}]|\right)$$
$$= W(\mathbf{F}, \mathbf{G}), \quad (15.32)$$

where
$$W = \frac{1}{2\Omega_0} \sum_{i=1}^{N_b} \phi(r_i). \tag{15.33}$$

Thus,
$$\frac{\partial W}{\partial \mathbf{G}} = \sum_k \frac{\partial \phi}{\partial r_k} \frac{\partial r_k}{\partial \mathbf{G}}.$$

Recalling that
$$r_k^2 = \mathbf{r}_k \cdot \mathbf{r}_k \rightarrow 2r_k \frac{\partial r_k}{\partial \mathbf{G}} = 2\mathbf{r}_k \cdot \frac{\partial \mathbf{r}_k}{\partial \mathbf{G}} = 2\mathbf{r}_k \cdot \left(\frac{1}{2}\mathbf{I}^{(2)} \otimes \mathbf{R}_k \otimes \mathbf{R}_k\right)$$
$$= \mathbf{r}_k \otimes \mathbf{R}_k \otimes \mathbf{R}_k.$$

Finally, we have,
$$\frac{\partial r_k}{\partial \mathbf{G}} = \frac{\mathbf{r}_k \otimes \mathbf{R}_k \otimes \mathbf{R}_k}{2r_k} \rightarrow \mathbf{Q} = \frac{\partial W}{\partial \mathbf{G}} = \frac{1}{4\Omega_0} \sum_{i=1}^{n_{bond}} \phi'(r_i) \frac{\mathbf{r}_i \otimes \mathbf{R}_i \otimes \mathbf{R}_i}{r_i}. \tag{15.34}$$

Furthermore, we can calculate the second derivative of W with respect to $\mathbf{G}$ as
$$\mathbb{M}_{GG} := \frac{\partial^2 W}{\partial \mathbf{G} \otimes \partial \mathbf{G}} = \frac{\partial \mathbf{Q}}{\partial \mathbf{G}} = \frac{1}{4\Omega_0} \sum_{i=1}^{N_b} \left(\phi''(r_i) - \frac{1}{r_i}\phi'(r_i)\right) \frac{\partial r_i}{\partial \mathbf{G}} \otimes \mathbf{r}_i \otimes \mathbf{R}_i \otimes \mathbf{R}_i$$
$$+ \sum_{i=1}^{N_b} \phi'(r_i) \frac{\partial \mathbf{r}_i}{\partial \mathbf{G}} \otimes \frac{\mathbf{R}_i \otimes \mathbf{R}_i}{r_i}. \tag{15.35}$$

Considering that
$$\frac{\partial r_i}{\partial \mathbf{G}} = \frac{\mathbf{r}_i \otimes \mathbf{R}_i \otimes \mathbf{R}_i}{2r_i} \quad \text{and} \quad \frac{\partial \mathbf{r}_i}{\partial \mathbf{G}} = \frac{1}{2}\mathbf{I}^{(2)} \otimes \mathbf{R}_i \otimes \mathbf{R}_i,$$

we have
$$\mathbb{M}_{GG} = \frac{1}{8\Omega_0} \sum_{i=1}^{N_b} \left(\left(\phi''(r_i) - \frac{1}{r_i}\phi'(r_i)\right) \frac{\mathbf{r}_i \otimes \mathbf{R}_i \otimes \mathbf{R}_i \otimes \mathbf{r}_i \otimes \mathbf{R}_i \otimes \mathbf{R}_i}{r_i}\right.$$
$$\left. + \phi'(r_i) \frac{\mathbf{I}^{(2)} \otimes \mathbf{R}_i \otimes \mathbf{R}_i \otimes \mathbf{R}_i \otimes \mathbf{R}_i}{r_i}\right).$$

Similarly, we can calculate the mixed higher-order elastic stiffness tensor such as
$$\mathbb{M}_{FG}$$
$$= \frac{\partial^2 W}{\partial \mathbf{F} \otimes \partial \mathbf{G}} = \frac{\partial \mathbf{Q}}{\partial \mathbf{F}}$$
$$= \frac{1}{4\Omega_0} \sum_{i=1}^{N_b} \left(\phi''(r_i) - \frac{1}{r_i}\phi'(r_i)\right) \frac{\partial r_i}{\partial \mathbf{F}} \otimes \mathbf{r}_i \otimes \mathbf{R}_i \otimes \mathbf{R}_i + \sum_{i=1}^{N_b} \phi'(r_i) \frac{\partial \mathbf{r}_i}{\partial \mathbf{F}} \otimes \frac{\mathbf{R}_i \otimes \mathbf{R}_i}{r_i}.$$
$$\tag{15.36}$$

Considering that

$$\frac{\partial r_i}{\partial \mathbf{F}} = \frac{\mathbf{r}_i \otimes \mathbf{R}_i}{r_i} \quad \text{and} \quad \frac{\partial \mathbf{r}_i}{\partial \mathbf{F}} = \mathbf{I}^{(2)} \otimes \mathbf{R}_i,$$

we have

$$\mathbb{M}_{FG} = \frac{1}{4\Omega_0} \sum_{i=1}^{N_b} \left(\left(\phi''(r_i) - \frac{1}{r_i} \phi'(r_i) \right) \frac{\mathbf{r}_i \otimes \mathbf{R}_i \otimes \mathbf{r}_i \otimes \mathbf{R}_i \otimes \mathbf{R}_i}{r_i} \right.$$
$$\left. + \phi'(r_i) \frac{\mathbf{I}^{(2)} \otimes \mathbf{R}_i \otimes \mathbf{R}_i \otimes \mathbf{R}_i}{r_i} \right). \tag{15.37}$$

We leave the calculation of mixed higher-order elastic tensor $\mathbb{M}_{GF}$ to the readers as a homework.

Now, we look at a simple 2D example of higher-order Cauchy–Born model.

Example 15.6 Consider an inhomogeneous 2D motion as

$$[\mathbf{x}] = \chi(\mathbf{X}) = \begin{bmatrix} X_1 + AX_2 + BX_1X_2 \\ X_2 + AX_1 + BX_1X_2 \end{bmatrix},$$

in a unit cell of the 2D hexagonal lattice.

Thus, the deformation gradient is

$$\mathbf{F} = \frac{\partial \chi}{\partial \mathbf{X}} = \begin{bmatrix} 1 + BX_2 & A + BX_1 \\ A + BX_2 & 1 + BX_1 \end{bmatrix}.$$

The strain gradient $\mathbf{G}$ has the components

$$\mathbf{G}_{,1} = \frac{\partial \mathbf{F}}{\partial X_1} = \begin{bmatrix} 0 & B \\ 0 & B \end{bmatrix} \quad \text{and} \quad \mathbf{G}_{,2} = \frac{\partial \mathbf{F}}{\partial X_2} = \begin{bmatrix} B & 0 \\ B & 0 \end{bmatrix}.$$

Therefore, the second-order Cauchy–Born rule,

$$\mathbf{r}_i = \mathbf{F} \cdot \mathbf{R}_i + \frac{1}{2} \mathbf{G} : [\mathbf{R}_i \otimes \mathbf{R}_i],$$

can be explicitly written as

$$\begin{bmatrix} r_{i1} \\ r_{i2} \end{bmatrix} = \begin{bmatrix} 1 + BX_2 & A + BX_1 \\ A + BX_2 & 1 + BX_1 \end{bmatrix} \begin{bmatrix} R_{i1} \\ R_{i2} \end{bmatrix} + \frac{B}{2} \begin{bmatrix} R_{i1} + R_{i2} \\ R_{i1} + R_{i2} \end{bmatrix}.$$

Suppose that the center atom is located at the origin of the coordinate, i.e., $\mathbf{X}_0 = (0,0)^T$. Then we have

$$\begin{bmatrix} r_{i1} \\ r_{i2} \end{bmatrix} = \begin{bmatrix} 1 & A \\ A & 1 \end{bmatrix} \begin{bmatrix} R_{i1} \\ R_{i2} \end{bmatrix} + \frac{B}{2} \begin{bmatrix} R_{i1} + R_{i2} \\ R_{i1} + R_{i2} \end{bmatrix}, \quad i = 1, 2 \ldots 6.$$

15.3 Cauchy–Born Rule for Non-Bravais Lattices

In the previous section, the Cauchy–Born rule is only applied to the Bravais lattice, in which there is one single atom inside its Wigner–Seitz cell. There are many crystal solid materials that have more than one atom inside the Wigner–Seitz cells such as semiconductor materials, silicon (Si), germanium (Ge), and gallium arsenide (GaAs); diamond structure materials, such as diamond and boron nitride (BN); and among many others.

For these types of materials, we have more than one atom inside the Wigner–Seitz cell, the usual Cauchy–Born rule does not apply. In crystallography theory, we distinguish the concept of lattice point with the atom nucleus. At each lattice point, there could be several atoms. We call the list of atoms in a single lattice point as the *motif* (see Fig. 15.6).

To illustrate how to apply Cauchy–Born rule to non-Bravais lattices, we first consider a special case – the lattice of diamond structure. The diamond structure lattice may be viewed as two FCC lattices interpenetrating each other, which is shown as the light color atoms and dark color atoms in Figs. 15.6 and 15.7. One may also view the diamond structure as a single FCC lattice, but, however, there are two atoms (light color atom + dark color atom) at each lattice point. In this case, the choice of the unit

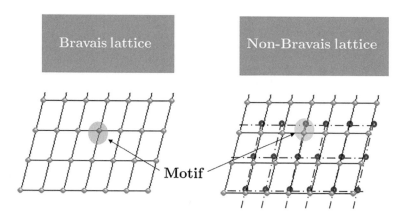

Figure 15.6 Bravais and non-Bravais lattices and lattice motif

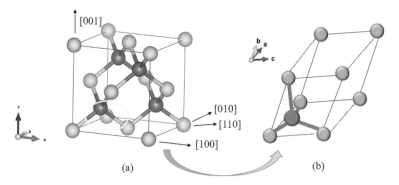

Figure 15.7 (a) Diamond lattice structure and (b) the unit cell of the diamond lattice

cell is a controversial issue. As one may find in Fig. 15.7, each dark color atom has four nearest light color atoms, while each light color atom has four nearest dark color atoms. This is obvious for dark color atoms shown in Fig. 15.7. Since the positions of the light color atoms and the positions of the dark color atoms are symmetric, if we change the light color atoms to the dark color atoms and the dark color atoms to the light color atoms, we can also see that each light color atom has four nearest dark color neighbors. Thus, one of the most common way to choose the unit cell of the diamond lattice is illustrated in Fig. 15.7(b), in which one dark color atom is surrounded by four light color atoms. Thus, if we label the center dark color atom as atom number 1, then there will be four adjacent light color atoms: atom 2, atom 3, atom 4, and atom 5 surrounding around it. If we follow this choice, we must recognize that the light color atoms and the dark color atoms are from different FCC lattices, and when the bond vectors between the light color atoms and the dark color atom, i.e., $\mathbf{R}_{1j} = \mathbf{R}_j - \mathbf{R}_1$, $j = 2, 3, 4, 5$, deform, there is an extra degree of freedom between the dark color atom lattice and the light color atom lattice, which we may denote as $\mathbf{v}$. Thus,

$$\mathbf{r}_{1j} = \mathbf{r}_j - \mathbf{r}_1 + \mathbf{v}, \quad j = 2, 3, 4, 5,$$

which implies the following Cauchy–Born rule,

$$\mathbf{r}_{1j} = \mathbf{F}\mathbf{R}_{1j} + \mathbf{v}, \quad j = 2, 3, 4, 5 \tag{15.38}$$

or in terms of the number of bonds, i.e.,

$$\mathbf{r}_i = \mathbf{F}\mathbf{R}_i + \mathbf{v}, \quad i = 1, 2, 3, 4, \tag{15.39}$$

where $\mathbf{v}$ is an unknown internal vector degree freedom.

Based on the atomistic-informed strain energy density, one can find the PK-II stress as

$$\mathbf{S} = \frac{\partial W}{\partial \mathbf{E}} = \frac{1}{2\Omega_0} \frac{\partial}{\partial \mathbf{E}} \left(\sum_{j=2}^{5} V_{1j} \right) = \frac{1}{2\Omega_0} \sum_{j=2}^{5} \left(\frac{\partial V_{1j}}{\partial \mathbf{E}} + \frac{\partial V_{1j}}{\partial \mathbf{v}} \frac{\partial \mathbf{v}}{\partial \mathbf{E}} \right)$$

or in terms of bonds

$$\mathbf{S} = \frac{\partial W}{\partial \mathbf{E}} = \frac{1}{2\Omega_0} \frac{\partial}{\partial \mathbf{E}} \left(\sum_{i=1}^{4} V(r_i) \right) = \frac{1}{2\Omega_0} \sum_{i=1}^{4} \left(\frac{\partial V(r_i)}{\partial \mathbf{E}} + \frac{\partial V(r_i)}{\partial \mathbf{v}} \frac{\partial \mathbf{v}}{\partial \mathbf{E}} \right). \tag{15.40}$$

The inner displacement $\mathbf{v}$ can be evaluated by minimizing the strain energy density W, i.e.,

$$\left. \frac{\partial W}{\partial \mathbf{v}} \right|_{\mathbf{E}} = \sum_{i=1}^{4} \left. \frac{\partial V(r_i)}{\partial \mathbf{v}} \right|_{\mathbf{E}} = 0, \tag{15.41}$$

or formally

$$\mathbf{v}_{opt} = \arg \min W(\mathbf{v}) \Big|_{\mathbf{E}}. \tag{15.42}$$

In actual computations, we may use the Newton–Raphson method to solve $\mathbf{v}$ such that it satisfies Eq. (15.42). That is,

$$\mathbf{v}^{n+1} = \mathbf{v}^n + \Delta \mathbf{v}, \quad \text{until} \quad \|\mathbf{v}^{n+1} - \mathbf{v}^n\| \leq \epsilon, \tag{15.43}$$

where n is the iteration index, ϵ is the tolerance, and

$$\Delta \mathbf{v} = -\left(\frac{\partial^2 W}{\partial \mathbf{v}^2}\right)^{-1}\left(\frac{\partial W}{\partial \mathbf{v}}\right), \tag{15.44}$$

in which,

$$W = \frac{1}{2\Omega_0}\sum_{i=1}^{4} V(r_i);$$

the underlying lattice structure is a diamond lattice structure.

This is because the necessary condition for achieving minimum is

$$\frac{\partial W}{\partial \mathbf{v}}\bigg|_{\mathbf{E}=0}(\mathbf{v}^{n+1}) = 0 \rightarrow \frac{\partial W}{\partial \mathbf{v}}(\mathbf{v}^{n+1}) \approx \frac{\partial W}{\partial \mathbf{v}}(\mathbf{v}^n) + \frac{\partial^2 W}{\partial \mathbf{v}^2}(\mathbf{v}^n) \cdot \Delta \mathbf{v} = 0,$$

which results in Eq. (15.44).

To calculate PK-II stress (15.40), we take a derivative of Eq. (15.41) with respect to $\mathbf{E}$,

$$\sum_{i=1}^{4}\frac{\partial^2 V(r_i)}{\partial \mathbf{E}\partial \mathbf{v}}\bigg|_{\mathbf{E}} = \sum_{i=1}^{4}\frac{\partial^2 V(r_i)}{\partial \mathbf{E}\partial \mathbf{v}} + \sum_{i=1}^{4}\frac{\partial^2 V(r_i)}{\partial \mathbf{v}\partial \mathbf{v}}\frac{\partial \mathbf{v}}{\partial \mathbf{E}} = 0.$$

Subsequently, we can find that

$$\frac{\partial \mathbf{v}}{\partial \mathbf{E}} = -\left(\sum_{i=1}^{4}\frac{\partial^2 V(r_i)}{\partial \mathbf{v}\partial \mathbf{v}}\right)^{-1}\left(\sum_{i=1}^{4}\frac{\partial^2 V(r_i)}{\partial \mathbf{E}\partial \mathbf{v}}\right),$$

which is a constant second-order tensor inside the unit cell, i.e., it is the same for every bonds. Finally, we can find the PK-II stress as follows:

$$\mathbf{S} = \frac{\partial W}{\partial \mathbf{E}} = \frac{1}{2\Omega_0}\sum_{i=1}^{4}\left(\frac{\partial V(r_i)}{\partial \mathbf{E}} + \frac{\partial V(r_i)}{\partial \mathbf{v}}\frac{\partial \mathbf{v}}{\partial \mathbf{E}}\right) = \frac{1}{2\Omega_0}\sum_{i=1}^{4}\left(\frac{\partial V(r_i)}{\partial \mathbf{E}}\right). \tag{15.45}$$

In the last line of the above equation the following conditions are used

$$\frac{\partial \mathbf{v}}{\partial \mathbf{E}} = const. \text{ and } \sum_{i=1}^{4}\frac{\partial V(r_i)}{\partial \mathbf{v}}\bigg|_{\mathbf{E}} = 0.$$

Taking the second derivative of the elastic potential energy with respect to the Green–Lagrangian strain $\mathbf{E}$, we can also find the elastic tensor as follows:

$$\mathbb{C} = \frac{\partial^2 W}{\partial \mathbf{E}\partial \mathbf{E}} = \frac{\partial \mathbf{S}}{\partial \mathbf{E}} = \frac{1}{2\Omega_0}\sum_{i=1}^{4}\left(\frac{\partial^2 V(r_i)}{\partial \mathbf{E}\partial \mathbf{E}} + \frac{\partial^2 V(r_i)}{\partial \mathbf{v}\partial \mathbf{E}}\frac{\partial \mathbf{v}}{\partial \mathbf{E}}\right)$$

$$= \frac{1}{2\Omega_0}\left\{\left(\sum_{i=1}^{4}\frac{\partial^2 V(r_i)}{\partial \mathbf{E}\partial \mathbf{E}}\right) - \left(\sum_{i=1}^{4}\frac{\partial^2 V(r_i)}{\partial \mathbf{v}\partial \mathbf{E}}\right)\left(\sum_{i=1}^{4}\frac{\partial^2 V(r_i)}{\partial \mathbf{v}\partial \mathbf{v}}\right)^{-1}\left(\sum_{i=1}^{4}\frac{\partial^2 V(r_i)}{\partial \mathbf{E}\partial \mathbf{v}}\right)\right\}.$$

$$\tag{15.46}$$

15.4 Cauchy–Born Rule for Amorphous Solids

The Cauchy–Born rule and the related multiscale algorithms discussed above are only applicable to crystalline solids that have definite lattice structure or microstructure. The method was not applicable to amorphous solids, when the Cauchy-Born rule was first proposed.

In order to develop a multiscale modeling method for amorphous solids, Urata and Li (2017b, 2018) extended the Cauchy–Born rule to amorphous materials, e.g., the Lennard-Jones binary glass and amorphous polymeric materials (see Fig. 15.8).

Since amorphous materials have no unit cells with definite microstructure, we have to first select a representative unit cell by conducting a series MD simulations to find out a proper size of the simulation cell beyond which the size effect becomes negligible (see Urata and Li (2017b)). Once the size of the unit cells were selected, one may have to conduct series MD simulations for various molecular configurations to set up a database of different molecular configurations, because the molecular configuration in amorphous materials has some randomness and uncertainties. This and other issues related to multiscale modeling of amorphous materials will be discussed in next chapter. For the moment, we assume that after some preparations and simulations we can acquire a representative sampling unit cell (RS-cell) for a given amorphous material from a database (see Urata and Li (2017b), (2018); Murashima et al. (2019)).

After we set up a unit cell for a given amorphous material, we can choose a center atom and compute the bonds between the center atom and the rest of the atoms in the unit cell. Unlike crystalline solids, the volume and shape of the unit cell of an amorphous solid is not determined by its microstructure, but the three edge vectors of the unit cell, i.e., $\mathbf{a}, \mathbf{b}$, and $\mathbf{c}$. Following the approach adopted in PR-MD, we can then define the RS-cell shape tensor as follows:

$$\mathbf{h}(0) = [\mathbf{a}, \mathbf{b}, \mathbf{c}], \tag{15.47}$$

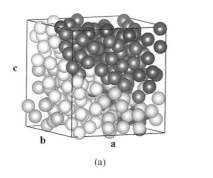
(a)

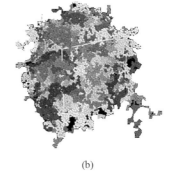
(b)

Figure 15.8 (a) Lennard-Jones binary glass and (b) amorphous polymeric materials

where **a**, **b**, and **c** are three edge vectors of the unit cell at initial time, as suggested by Parrinello and Rahman (1981). Following Parrinello and Rahman (1981), we may use the cell shape tensor **h** to rescale the initial positions of all atoms in the unit cell, i.e.,

$$\mathbf{R}_i = \mathbf{h}(0)\mathbf{S}_i, \quad i = 1, 2, \ldots N_c, \tag{15.48}$$

where $\mathbf{S}_i$ is the scaled atom positions that have the property

$$|\mathbf{S}_i| \leq 1.$$

Readers may consult reference Li and Urata (2016) for detailed discussions.

Thus, the Cauchy–Born rule for amorphous solids may be modified as

$$\mathbf{r}_{1i}^g = \mathbf{F} \cdot \mathbf{R}_{1i} = \mathbf{F} \cdot \mathbf{h}(0) \cdot \mathbf{S}_{1i}, \quad i = 2, \ldots N_c, \tag{15.49}$$

where $\mathbf{r}_{1i}^g = \mathbf{r}_i^g - \mathbf{r}_1^g$, $\mathbf{R}_{1i} = \mathbf{R}_i - \mathbf{R}_1$, and $\mathbf{S}_{1i} = \mathbf{S}_i - \mathbf{S}_1$ are various relative position vectors, i.e., bond vectors in the current configuration, in the reference configuration, and in the scaled or parametric configuration, as shown in Fig. 15.9. The superscript g indicates that this is the guessed position vectors. In the following, we may count the bond index $i = 1, 2, \ldots N_c - 1$ instead of counting on each individual atom.

Since in amorphous solids, there is no definite microstructure, and the atom to atom connection cannot be defined by an arbitrarily chosen center atom. Therefore, there will be $N_c - 1$ internal degrees of freedom of motions that cannot be captured by a simple affine mapping, which we label as $\mathbf{v}_k, k = 1, 2, \ldots N_c - 1$. In other words, each bond in the amorphous unit cell has one vector internal degree of freedom. Thus, the total elastic potential energy may be expressed as

$$W = \frac{1}{2\Omega_0} \sum_{k=1}^{N_c-1} V(r_k) = W(\mathbf{F}, \{\mathbf{v}_k\}). \tag{15.50}$$

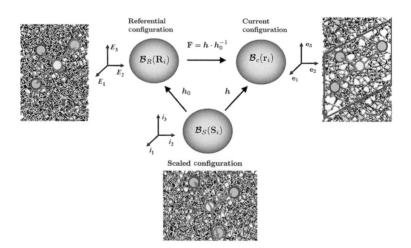

Figure 15.9 Schematic illustration of Cauchy–Born rule in amorphous materials

The first Piola–Kirchhoff stress can be obtained as

$$\mathbf{P} = \frac{1}{2\Omega_0} \sum_{k=1}^{N_c-1} \left(\frac{\partial V}{\partial \mathbf{r}_k} \frac{\partial \mathbf{r}_k}{\partial \mathbf{F}} + \frac{\partial V}{\partial \mathbf{v}_k} \frac{\partial \mathbf{v}_k}{\partial \mathbf{F}} \right). \quad (15.51)$$

The original Cauchy–Born rule is

$$\mathbf{r}_k^g = \mathbf{F} \cdot \mathbf{R}_k,$$

and kinematic condition requires that

$$\mathbf{r}_k^{opt} = \mathbf{r}_k^g + \mathbf{v}_k.$$

Thus, it leads to

$$\mathbf{P} = \frac{1}{2\Omega_0} \sum_{k=1}^{N_c=1} \left(\frac{\partial V}{\partial \mathbf{r}_k} \frac{\partial \mathbf{r}_k}{\partial \mathbf{F}} + \frac{\partial V}{\partial \mathbf{r}_k} \frac{\partial \mathbf{v}_k}{\partial \mathbf{F}} \right) = \frac{1}{2\Omega_0} \sum_{k=1}^{N_c=1} V'\left(r_k^{opt}\right) \frac{\mathbf{r}_k^{opt} \otimes \mathbf{R}_k^{opt}}{r_k^{opt}}. \quad (15.52)$$

This is because that:
(**1**) based on the kinematic condition, we have

$$\mathbf{r}_k^{opt} = \mathbf{r}_k^g + \mathbf{v}_k, \quad \rightarrow \quad \frac{\partial V}{\partial \mathbf{v}_k} = \frac{\partial V}{\partial \mathbf{r}_k}, \quad (15.53)$$

and
(**2**) to solve the internal degrees of freedom of motions, we postulate that the following Cauchy–Born rule for internal degree of freedom, i.e.,

$$\mathbf{v}_k = \mathbf{F} \cdot \mathbf{V}_k \quad \rightarrow \quad \frac{\partial \mathbf{v}_k}{\partial \mathbf{F}} = \mathbf{I}^{(2)} \otimes \mathbf{V}_k. \quad (15.54)$$

With Eqs. (15.53) and (15.54), we have

$$\mathbf{P} = \frac{1}{2\Omega_0} \sum_{k=1}^{N_c=1} \left(V'(r_k^{opt}) \frac{\mathbf{r}_k \otimes \mathbf{R}_k}{r_k} + V'(r_k^{opt}) \frac{\mathbf{r}_k \otimes \mathbf{V}_k}{r_k} \right)$$

$$= \frac{1}{2\Omega_0} \sum_{k=1}^{N_c=1} \left(V'\left(r_k^{opt}\right) \frac{\mathbf{r}_k^{opt} \otimes (\mathbf{R}_k^g + \mathbf{V}_k)}{r_k} \right) = \frac{1}{2\Omega_0} \sum_{k=1}^{N_c=1} \left(V'\left(r_k^{opt}\right) \frac{\mathbf{r}_k^{opt} \otimes \mathbf{R}_k^{opt}}{r_k} \right), \quad (15.55)$$

where

$$\mathbf{R}_k^{opt} = \mathbf{R}_k + \mathbf{V}_k = \mathbf{h} \mathbf{S}_k^{opt} \quad (15.56)$$

and

$$\mathbf{S}_k^{opt} = (\mathbf{F} \cdot \mathbf{h})^{-1} \cdot \mathbf{r}_k^{opt}. \quad (15.57)$$

From Eq. (15.57), we can obtain

$$\mathbf{h} \mathbf{S}_k^{opt} = \mathbf{h}(\mathbf{h}^{-1} \cdot \mathbf{F}^{-1})(\mathbf{r}_k + \mathbf{v}_k) = \mathbf{R}_k + \mathbf{V}_k,$$

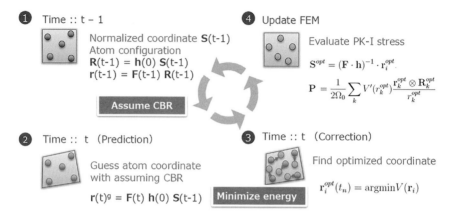

Figure 15.10 Algorithmic flowchart of multiscale simulation by using the Cauchy–Born rule for amorphous solids

and hence we recover Eq. (15.56). It may be noted that $\{\mathbf{R}_k\}$ are not the original atom positions in the referential configuration, but they are the optimum atom positions in the referential configuration corresponding to the deformed equilibrium atom positions in the current configuration.

To find the equilibrium bond vectors $\mathbf{r}_k^{opt}, k = 1, 2, \ldots, N_c - 1$, we apply the optimum condition,

$$\mathbf{r}_i^{opt} = \underset{\{\mathbf{v}_i\}}{\operatorname{argmin}} \frac{1}{2\Omega_0} \sum_k V(r_k), \qquad (15.58)$$

from which, we obtain

$$\sum_k \frac{\partial V(r_k^{opt})}{\partial \mathbf{v}_k} = 0 \rightarrow \sum_k \frac{\partial V(r_k^n)}{\partial \mathbf{v}_k^n} + \sum_k \frac{\partial^2 V(r_k^n)}{\partial \mathbf{v}_k^n \partial \mathbf{v}_k^n} \cdot \Delta \mathbf{v}_k^n \approx 0. \qquad (15.59)$$

By using Eq. (15.59), we can solve $\Delta \mathbf{v}_k^n$, so that we can move to the next atom configuration by setting $\mathbf{r}_k^{n+1} = \mathbf{r}_k^n + \Delta \mathbf{v}_k^n$, where

$$\Delta \mathbf{v}_k^n = -\left(\sum_k \frac{\partial^2 V(r_k^n)}{\partial \mathbf{v}_k^n \partial \mathbf{v}_k^n} \right)^{-1} \left(\sum_k \frac{\partial V(r_k^n)}{\partial \mathbf{v}_k^n} \right), \qquad (15.60)$$

where $\mathbf{r}_k^n = \mathbf{r}_k^0 + \mathbf{v}_k^n$, and when $n = 0$, we choose $\mathbf{r}_k^0 = \mathbf{r}_k^g$. Once we found $\{\mathbf{r}_k^{opt}\}$, we can find

$$\mathbf{R}_k^{opt} = \mathbf{h} \cdot \mathbf{S}_k^{opt}, \quad \text{where} \quad \mathbf{S}_k^{opt} = (\mathbf{F} \cdot \mathbf{h})^{-1} \cdot \mathbf{r}_k^{opt}.$$

It may be noted that we may be able to optimize cell shape tensor $\mathbf{h}$ by minimizing the potential energy $V(r)$ in terms of $\mathbf{h}$ as well. This is useful when we start the simulation and we do not if $\mathbf{h}(0)$ is the right choice of the shape tensor. To find the optimum shape tensor for the unit cell, we can set an optimization procedure such

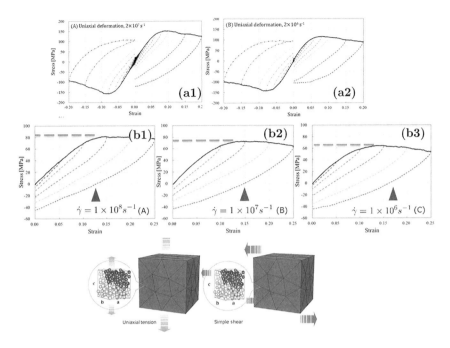

Figure 15.11 Multiscale modeling of inelastic stress–strain relations for Lennard-Jones binary glass via the coarse-grained Parrinello–Rahman (CG-PR) method: (a1)–(a2) Stress–strain relation in uniaxial tension and compression and (b1)–(b3) stress–strain relation in pure shear (Data source is from Urata and Li [2018], Acta Materialia, **155**, 153–165.).

that,

$$\mathbf{h}^{opt} = \underset{\{\mathbf{h}\}}{\operatorname{argmin}} \frac{1}{2\Omega_0} \sum_k V(r_k). \tag{15.61}$$

If we consider $\mathbf{r}_k^m = \mathbf{F}\mathbf{h}^m \mathbf{S}_k^m$, we can have

$$\sum_k \frac{\partial V(r_k^{opt})}{\partial \mathbf{h}^{opt}} = 0 \rightarrow \sum_k \frac{\partial V(r_k^m)}{\partial \mathbf{h}^m} + \sum_k \frac{\partial^2 V(r_k^m)}{\partial \mathbf{h}^m \partial \mathbf{h}^m} \cdot \Delta \mathbf{h}^m \approx 0.$$

Thus, we can find $\mathbf{h}^{opt}$ by iterative solution $\mathbf{h}^{m+1} = \mathbf{h}^m + \Delta \mathbf{h}^m$, where

$$\Delta \mathbf{h}^m = -\left(\sum_k \frac{\partial^2 V(r_k^m)}{\partial \mathbf{h}^m \partial \mathbf{h}^m} \right)^{-1} \left(\sum_k \frac{\partial V(r_k^m)}{\partial \mathbf{h}^m} \right). \tag{15.62}$$

A complete computational algorithmic flowchart is shown in Fig. 15.10. In Urata and Li (2018), the amorphous Cauchy–Born rule has been used in a multiscale finite element analysis to study plasticity of amorphous solids (see Fig. 15.11). The authors coined the method as the coarse-grained Parrinello–Rahman (CG-PR) method. It has been demonstrated in Urata and Li (2017b, 2018); Murashima et al. (2019) that the Cauchy–Born rule for amorphous solids can predict inelastic behaviors of the amorphous solids at macroscale by using the atomistic-informed elastic potential energy at microscale. This is because that the complex microstructure of the amorphous solids have many complicated and nonconvex local potential energy, and the optimization procedure in the CG-PR method or in the Cauchy–Born rule of the amorphous solids will predict a path-dependent macroscale stress–strain relation for general amorphous materials.

15.5 Homework Problems

Problem 15.1 Use the Cauchy–Born rule to calculate the Cauchy stress and elastic tensor based on the following formula, i.e.,

$$\sigma = 2J^{-1}\frac{\partial W}{\partial \mathbf{b}}\mathbf{b} = \frac{1}{2\Omega}\sum_{k \in S_c}\frac{\partial \phi}{\partial r_k}\frac{\mathbf{r}_k \otimes \mathbf{r}_k}{r_k}$$

and

$$\mathbb{C} = \frac{4}{J}\mathbf{b}\frac{\partial^2 W}{\partial \mathbf{b}^2}\mathbf{b} = \frac{1}{2\Omega}\sum_{k \in S_c}\left(\frac{\partial^2 \phi}{\partial r_k^2} - \frac{1}{r_k}\frac{\partial \phi}{\partial r_k}\right)\frac{\mathbf{r}_k \otimes \mathbf{r}_k \otimes \mathbf{r}_k \otimes \mathbf{r}_k}{r_k^2},$$

where $\mathbf{b} = \mathbf{F}\cdot\mathbf{F}^T$, $J = det\mathbf{F}$, and

$$W = \frac{1}{2\Omega}\sum_k \phi(r_k).$$

Problem 15.2 Use the second-order Cauchy–Born rule to calculate the second-order stress,

$$\mathbf{Q} = \frac{\partial W}{\partial \mathbf{G}}, \quad \text{where } \mathbf{G} = \nabla \otimes \mathbf{F},$$

where

$$W = \frac{1}{2\Omega_0}\sum_{i \neq j}\Phi(r_{ij}), \quad r_{ij} = |\mathbf{r}_j - \mathbf{r}_i|,$$

where Φ is an atomistic pair potential and Ω_0 is the volume of the unit cell.

Hint: Read the article: Sunyk, R. and Steinnmann, P. (2003). On higher gradients in continuum-atomistic modeling. *International Journal of Solids and Structures*, **40**, 6877–6886.

16 Statistical Theory of Cauchy Continuum

In this chapter, we discuss the statistical theory of the Cauchy continuum, in which the Cauchy–Born rule is extended to finite temperature case. In other words, we provide an approximation of the virial stress with finite temperature.

16.1 Quasi-Harmonic Approximation

In the original Cauchy–Born rule, we do not take into account the contribution from kinetic energy. More precisely speaking, we do not consider the contribution from temperature. Then how do we take into account the effect of temperature in stress calculation? From a slightly different angle, we may ask *How can we approximate the Virial stress by considering the contribution from both kinetic and potential energies?*

We now present a multiscale quasi-harmonic approximation theory to extend the static Cauchy–Born rule theory to include finite temperature cases. For convenience of the presentation, we present the theory in the context of pair potential even though the theory is applicable to general atomic potentials. Moreover, we slightly changed the notation. Consider the force acting on the atom i as,

$$\mathbf{F}_i = -\frac{\partial V}{\partial \mathbf{r}_i}, \tag{16.1}$$

where V is the overall atomistic potential energy.

For the Bravais lattices, we have

$$V(\mathbf{r}_1, \mathbf{r}_2, \ldots, \mathbf{r}_N) = \frac{1}{2}\sum_{i=1}^{N}\sum_{j\neq i}\phi(r_{ij}) = \sum_{i=1}^{N}\sum_{i<j}\phi(r_{ij}), \tag{16.2}$$

where $\phi(r)$ is the atomistic potential.

In Eq. (16.2), the atom positions in the current configuration may be expressed as

$$\mathbf{r}_i = \mathbf{R}_i + \mathbf{u}_i, \tag{16.3}$$

where $\mathbf{R}_i$ is the undeformed position vector of the ith atom and $\mathbf{u}_i$ is its displacement. Consider the additive multiscale decomposition,

$$\mathbf{r}_i = \bar{\mathbf{r}} + \mathbf{r}'_i, \text{ where } \bar{\mathbf{r}}_i = \bar{\mathbf{R}} + \bar{\mathbf{u}}_i, \text{ and } \mathbf{r}'_i = \mathbf{R}'_i + \mathbf{u}'_i, \tag{16.4}$$

where

$$\bar{\mathbf{r}} = \frac{1}{\sum_{j=1}^{N} m_j} \sum_{j=1}^{N} m_j \mathbf{r}_j \text{ and } \bar{\mathbf{R}} = \frac{1}{\sum_{j=1}^{N} m_j} \sum_{j=1}^{N} m_j \mathbf{R}_j \qquad (16.5)$$

are the centers of the deformed atomistic system and its undeformed system.

With above definitions, we can derive

$$\sum_i \mathbf{r}'_i = 0, \text{ and } \sum_i \mathbf{R}'_i = 0. \qquad (16.6)$$

In principle, the Cauchy–Born rule may be only applied at coarse scale,

$$\bar{\mathbf{r}} = \mathbf{F} \cdot \bar{\mathbf{R}}. \qquad (16.7)$$

We assume that the total potential energy may be written as,

$$V(\mathbf{r}^N) = V(\mathbf{r}'^N, \bar{\mathbf{r}}^N) = V(\mathbf{r}'^N, \mathbf{F}), \qquad (16.8)$$

where $\mathbf{r}' = \{\mathbf{r}'_i\}$, and

$$\mathbf{F} = \frac{\partial \mathbf{r}}{\partial \mathbf{R}}\bigg|_{\mathbf{R}', \mathbf{r}'=0} = \frac{\partial \bar{\mathbf{r}}}{\partial \bar{\mathbf{R}}},$$

is coarse scale deformation gradient.

In the rest of the presentation, we shall adopt the following variables with so-called mass-reduced units,

$$\mathbf{q}_i = \sqrt{m_i} \mathbf{r}'_i, \text{ and } \mathbf{p}_i = \sqrt{m_i} \mathbf{v}_i = \sqrt{m_i} \dot{\mathbf{r}} + \sqrt{m_i} \dot{\mathbf{r}}' = (\sqrt{m_i} \bar{\mathbf{v}} + \dot{\mathbf{q}}_i), \qquad (16.9)$$

and the Hamiltonian of the atomistic system may be written as

$$\mathcal{H} = \frac{1}{2} \sum_{i=1}^{N} \mathbf{p}_i \cdot \mathbf{p}_i + V(\mathbf{q}, \mathbf{F}). \qquad (16.10)$$

Considering the small lattice vibrations around the ground state, i.e.,

$$|\mathbf{q}_i| \ll 1,$$

we may expand the potential energy around its deformed initial equilibrium configuration,

$$V(\mathbf{q}^N; \mathbf{F}) = V(0; \mathbf{F}) + \sum_{i=1}^{N} \frac{\partial V}{\partial \mathbf{q}_i}\bigg|_{\mathbf{q}^N=0} \cdot \mathbf{q}_i + \frac{1}{2} \frac{\partial^2 V}{\partial \mathbf{q}_i \partial \mathbf{q}_j}\bigg|_{\mathbf{q}^N=0} : (\mathbf{q}_i \otimes \mathbf{q}_j) + \cdots$$

Considering equilibrium conditions,

$$\frac{\partial V}{\partial \mathbf{q}_i}\bigg|_{\mathbf{q}^N=0} = 0,$$

the second term of the above expression will vanish.

We call the following expansion as the **Harmonic approximation**:

$$V(\mathbf{q}^N; \mathbf{F}) \approx V_h(\mathbf{q}^N; \mathbf{F}) = V_0(\mathbf{F}) + \frac{1}{2} \sum_i \sum_j V_{,ij} \mathbf{q}_i \cdot \mathbf{q}_j, \text{ where } V_{,ij} := V_{,q_i q_j}. \quad (16.11)$$

Considering a special case that $m_i = 1$ and

$$\bar{\mathbf{v}} = 0, \text{ and } \mathbf{p}_i = \dot{\mathbf{q}}_i, \quad (16.12)$$

the Hamiltonian may be written as,

$$\mathcal{H} = \frac{1}{2} \sum_{i=1}^N \dot{\mathbf{q}}_i \cdot \dot{\mathbf{q}}_i + V_0(\mathbf{F}) + \frac{1}{2} \sum_i \sum_j V_{,ij} \mathbf{q}_i \cdot \mathbf{q}_j \quad (16.13)$$

under the harmonic approximation.

Example 16.1 We consider an one-dimensional (1D) harmonic chain, whose governing equation is as follows:

$$\ddot{q}_i + V_{,ij} q_j = 0, \quad (16.14)$$

where $i, j = 1, 2, \ldots n$, and

$$V_{,ij} = \frac{\partial^2 V}{\partial q_i \partial q_j} > 0. \quad (16.15)$$

We seek for a particular solution of the form,

$$q_i = R_i \cos \omega t. \quad (16.16)$$

By substituting Eq. (16.16) into Eq. (16.15), we have the following eigenvalue problem,

$$[V_{,ij}][R_j] = \omega^2 [R_i],$$

in a N-dimensional linear space.

Since $[V_{,ij}]$ is real and symmetric, we have n real eigenvalues, $\omega_r^2, r = 1, 2, \ldots, n$, which correspond to n eigenvectors of the system, $[R_i^r]$, i.e.,

$$[V_{,ij}][R_{jr}] = \omega_r^2 [R_{ir}], \quad r = 1, 2, \ldots, n. \quad (16.17)$$

Since $[R_{ij}]$ forms an orthonormal transformation matrix, i.e.,

$$[R_{ij}]^{-1} = [R_{ij}]^T \quad \rightarrow \quad [R_{ri}][R_{is}]^T = [\delta_{rs}],$$

we have desired the result,

$$[R_{si}]^T [V_{ij}][R_{jr}] = \omega^2 [\delta_{sr}].$$

In fact, here we are using the property

$$[R_{si}]^T [R_{ir}] = [\delta_{sr}].$$

In structural dynamics, $[R_{ij}]$ often serves as transformation matrix, which transforms a tensor from the original Cartesian coordinates to the so-called principal coordinates, or principal directions.

Let $[P_i] = [R_{ij}][q_j] \to [q_i] = [R_{ij}^T[P_j]] = [R_{ji}][P_j]$. We have

$$p_i^2 \to [q_i]^T[q_i] = [P_j]^T[R_{ji}^T][R_{ji}][P_j] = [P_j]^T[\delta_{ij}][P_j] \to \sum_i P_i^2$$

and let $[Q_i] = [R_{ij}][q_j] \to [q_i] = [R_{ij}^T][Q_j]$, which yields,

$$\sum_i \sum_j V_{,ij} q_i q_j \to [Q_s]^T[R_{si}][V_{,ij}][R_{jr}][Q_r] = [Q_s]^T \omega^2 [\delta_{rs}][Q_r] \to \sum_r \omega_r^2 Q_r^2.$$

Thus, we can express the Hamiltonian in terms of the normal mode coordinates, i.e.,

$$H_h = \frac{1}{2}\sum_{i=1}^n p_i^2 + V_0(\mathbf{F}) + \frac{1}{2}\sum_{i=1}^n \sum_{j=1}^n V_{,ij} q_i q_j = V_0(\mathbf{F}) + \frac{1}{2}\sum_{r=1}^n (P_r^2 + \omega_r^2 Q_r^2). \quad (16.18)$$

These normal modes are harmonic oscillators, i.e., they obey the following equations:

$$\ddot{Q}_r + \omega_r^2 Q_r = 0, \quad r = 1, \ldots, n \quad (16.19)$$

which are also called **phonon**.

REMARK 16.2 1. In physics, a phonon is a collective excitation in a periodic, elastic arrangement of atoms or molecules in condensed matter. It represents an excited state in the quantum mechanical quantization of vibration modes of elastic structures of interacting particles.

2. In general, an arbitrary Hamiltonian can be written as the sum of a harmonic Hamiltonian and an anharmonic part in terms of the normal coordinates,

$$H(\mathbf{Q}^N, \mathbf{P}^N; \mathbf{F}) = H_h(\mathbf{Q}^N, \mathbf{P}^N; \mathbf{F}) + A(\mathbf{Q}^N; \mathbf{F}).$$

In the above case, we set the coarse-scale velocity field $\bar{\mathbf{v}} = 0$. That is, we did not consider the coarse-scale or continuum-scale dynamics. If you wish to consider the coarse-scale dynamics, i.e., $\bar{\mathbf{v}} \neq 0$, you should consult the work of coarse-grained molecular dynamics (CG-MD) developed by Rudd and Broughton (1998, 2005).

Here, we are focusing on how to find an atomistic-informed Cauchy stress tensor that includes the thermal fluctuations.

16.1.1 Thermal-Mechanical Canonical Ensemble

Consider the partition function of the canonical ensemble as,

$$Z(\mathbf{F}, T) = \int_\Gamma \exp(-\beta H(\mathbf{q}, \mathbf{p}; \mathbf{F})) \, d\mathbf{q}^N d\mathbf{p}^N = Z_p(T) Z_q(\mathbf{F}, T),$$

where $\beta = 1/(k_B T)$, and

$$Z_p(T) = \int_{\Gamma_p} \exp\left(-\frac{\beta}{2}\sum_{i=1}^{N} \mathbf{p}_i^2\right) d\mathbf{p}_i^N,$$

$$Z_q(\mathbf{F}; T) = \int_{\Gamma_q} \exp\left(-\beta V(\mathbf{q}^N; \mathbf{F})\right) d\mathbf{q}^N.$$

We are especially interested in the partition function for the harmonic systems,

$$Z_h(\mathbf{F}; T) = \int_{\Gamma} \exp\left(-\beta H_h(\mathbf{Q}^N, \mathbf{P}^N; \mathbf{F})\right) d\mathbf{Q}^N d\mathbf{P}^N.$$

To calculate the partition function for a harmonic system, we choose to integrate it with normal mode coordinates. Recalling that

$$H_h = V_0(\mathbf{F}) + \frac{1}{2}\sum_{r=1}^{N}(P_r^2 + \omega_r^2 Q_r^2),$$

we have

$$Z_h(\mathbf{F}; T) = \int_{\Gamma} \exp\left(-\beta H_h(\mathbf{Q}^N, \mathbf{P}^N; \mathbf{F})\right) d\mathbf{Q}^N d\mathbf{P}^N,$$

$$= \exp(-\beta V_0(\mathbf{F})) \prod_{r=1}^{N} \int \exp\left(-\frac{\beta}{2}P_r^2\right) dP_r \int \exp\left(-\frac{\beta}{2}\omega_r^2 Q_r^2\right) dQ_r,$$

$$= \left(\frac{2\pi}{\beta}\right)^N \exp(-\beta V_0(\mathbf{F})) \prod_{r=1}^{N} \frac{1}{\omega_r}, \quad \leftarrow \text{ for 1D systems.}$$

where $\beta = (k_B T)^{-1}$. Here, the free energy formulation is expressed in terms of phonon normal modes.

Note that the following identity is used in the above integration,

$$\int_{-\infty}^{\infty} \exp\left(-\frac{\beta x^2}{2}\right) dx = \sqrt{\frac{2\pi}{\beta}}.$$

In Chapter 6, we have shown that the Helmholtz energy can be represented by the partition function,

$$F = -k_B T \ln Z \quad \leftarrow \text{ Partition function.}$$

Hence, the free energy of the harmonic system can be written as

$$F_h(\mathbf{F}, T) = -k_B T \ln Z_h(\mathbf{F}; T)$$

$$= V_0(\mathbf{F}) - N k_B T \ln(2\pi k_B T) + k_B T \sum_{r=1}^{N} \ln \omega_r.$$

Considering that

$$F_h(\mathbf{F}, T) = V_0(\mathbf{F}) - Nk_B T \ln(2\pi k_B T) + k_B T \sum_{r=1}^{N} \ln \omega_r(\mathbf{F}),$$

we can calculate stress in a continuum by using the Cauchy–Born rule,

$$\mathbf{P} = \frac{1}{\Omega} \frac{\partial F_h}{\partial \mathbf{F}}\bigg|_T = \frac{1}{\Omega} \left[\frac{\partial V_0}{\partial \mathbf{F}} + k_B T \sum_{r=1}^{N} \frac{\partial \ln \omega_r}{\partial \mathbf{F}} \right], \quad (16.20)$$

where the second term is due to thermal fluctuation. One can also derive elastic constants, thermal coefficient, and thermal expansion constant, for instance,

$$C_{LMNK}^T = \frac{1}{\Omega_0} \frac{\partial^2 F_h}{\partial E_{LM} \partial E_{NK}} = \frac{1}{\Omega_0} \left[\frac{\partial^2 V_0}{\partial E_{LM} \partial E_{NK}}\bigg|_{q=0} + k_B T \sum_{r=1}^{N} \frac{\partial^2 \ln \omega_r}{\partial E_{LM} \partial E_{NK}}\bigg|_{q=0} \right]. \quad (16.21)$$

16.1.2 How to Find $\omega_r(\mathbf{F})$?

For a general 3D lattice, one simple way to compute ω_{ik} is to calculate the following eigenvalue problem,

$$\left| m_r \omega_r^2 \mathbf{I}^{3\times 3} - \frac{\partial^2 V_0}{\partial \mathbf{r}_i \partial \mathbf{r}_j}\bigg|_{\mathbf{r}_i = \mathbf{R}_i} \right| = 0, \quad i, j = 1, \ldots, n_c,$$

where n_c is the number of atoms inside the unit cell.

If the unit cell is a 1D lattice chain, one can show that

$$\omega_r = 2 \left(\frac{\phi''}{m} \right)^{1/2} \sin\left(\frac{r\pi}{2(N+1)} \right). \quad r = 1, 2, \ldots N+1.$$

Example 16.3 (a) The 1D chain

In this example, we assume that there are total $N+1$ atoms, and N bonds; and the original lattice constant is b_0,

$$\bar{\mathbf{R}}_j = jb_0, \quad j = 0, \ldots, N+1,$$

as shown in Fig. 16.1.

The deformation gradient $\mathbf{F} = F_{11}$, and hence the bond length,

$$\bar{r}_k = F_{11} \bar{R}_k, \quad j = 1, 2, \cdots N+1.$$

The current position of the atom j is,

$$r_j(t) = \bar{r}_j + q_j(t), \quad j = 0, 1, 2, \ldots, N+1,$$

with the boundary conditions,

$$r_0 = R_0 = 0, \text{ and } r_{N+1} = R_{N+1} = 0.$$

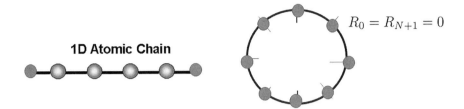

Figure 16.1 Illustration of one-dimensional lattice chain with periodic boundary condition

In 1D case, the deformation gradient is basically a scalar, i.e., $\mathbf{F} \sim F_{11}$, and thus,

$$\mathbf{r}_j = \mathbf{F}\mathbf{R}_j \quad \rightarrow \quad \bar{r}_j = F_{11}\bar{R}_j \quad \rightarrow \quad jb = F_{11}(jb_0).$$

The total potential energy then becomes,

$$V(q_0, q_1, \ldots, q_{N+1}; F_{11}) = \sum_{k=0}^{N} \phi(|r_{k+1} - r_k|).$$

(b) Normal Modes

Consider the harmonic vibrations around the coarse-scale equilibrium position, $\bar{r}_j = jb$ with the nearest neighbor interaction,

$$V_h = V_0(b) + \frac{1}{2}\sum_i \sum_j V_{,ij} q_i q_j,$$

where $V_0(b) = (N+1)\phi(b)$ and

$$\frac{\partial V}{\partial r_j} = -\phi'(r_{j+1} - r_j) + \phi'(r_j - r_{j-1}),$$

$$\frac{\partial^2 V}{\partial r_j^2} = \phi''(r_{j+1} - r_j) + \phi''(r_j - r_{j-1}),$$

$$\frac{\partial^2 V}{\partial r_{j+1} \partial r_j} = -\phi''(r_{j+1} - r_j),$$

$$\frac{\partial^2 V}{\partial r_{j+k} \partial r_j} = 0, \quad k > 1,$$

in which

$$\phi'(x) = \frac{d\phi}{dx}, \text{ and } \phi''(x) = \frac{d^2\phi}{dx^2}.$$

The resulting stiffness matrix is a block diagonal matrix, i.e.,

$$V_{ii}\big|_{q=0} = 2\phi''(b), \quad i = 1, 2, \ldots, N;$$

$$V_{i,i+1}\big|_{q=0} = -\phi''(b), \quad i = 1, \ldots, N-1;$$

$$V_{i-1,i}\big|_{q=0} = -\phi''(b), \quad i = 2, \ldots, N.$$

If $q_i = r'_i$ (not $q_i = \sqrt{m_i} r'_i$), we have

$$m_i \ddot{q}_i = V_{ij} q_j.$$

Substituting the normal mode solution,

$$q_i = R_i \cos \omega t,$$

into the equations of motion, we have

$$V_{ij} R_j = m \omega^2 R_i, \text{ with } R_0 = R_{N+1} = 0.$$

They may be rewritten as

$$-\phi''(b)(R_2 - R_1) = m\omega^2 R_1;$$
$$-\phi''(b)(R_{j+1} - 2R_j + R_{j-1}) = m\omega^2 R_j, \ j = 2, \ldots, N-1;$$
$$-\phi''(b)(-2R_N + R_{N-1}) = m\omega^2 R_N,$$

which can be written in a single equation,

$$-\phi''(b)(R_{j+1} - 2R_j + R_{j-1}) = m\omega^2 R_j, \ j = 1, \ldots, N, \quad (16.22)$$

with the boundary conditions, $R_0 = R_{N+1} = 0$.
If the lattice structure is periodic,

$$q_i = A \exp(ikj) \cos \omega t; \ i = \sqrt{-1},$$

By Bloch's theorem, we choose the following form of the Bloch solution:

$$R_j = A \exp(ikj), \text{ where } i = \sqrt{-1}, \text{ and } j = 1, \ldots, N$$

and substitute it into

$$-\phi''(b)(R_{j+1} - 2R_j + R_{j-1}) = m\omega^2 R_j, \ j = 2, \ldots, N-1.$$

Subsequently, we can then have

$$-\phi''(b)(\exp(ik) - 2 + \exp(-ik)) = m\omega^2.$$

Finally, we derived the following dispersive relation,

$$\omega^2 = \frac{2\phi''(b)}{m}(1 - \cos k) = \frac{4\phi''(b)}{m} \sin^2\left(\frac{k}{2}\right). \quad (16.23)$$

However, we still do not know the wave number k yet. Substituting the Bloch solution,

$$R_j = Re(A) \cos kj - Im(A) \sin kj,$$

into the boundary conditions,

$$R_0 = 0, \ \rightarrow \ Re(A) = 0; \text{ and } R_{N+1} = 0, \ \rightarrow \ \sin k(N+1) = 0.$$

We then obtain

$$k = k_r = \pm \frac{r\pi}{N+1}, \ r = 1, 2, \ldots, N.$$

Therefore, the eigenvectors or the normal modes are

$$R_{rj} = C_r \sin\left(\frac{r\pi j}{N+1}\right), \quad r = 1, \ldots, N.$$

Therefore, the eigenvectors or the normal modes are

$$R_{rj} = C_r \sin\left(\frac{r\pi j}{N+1}\right), \quad r = 1, \ldots, N,$$

with the corresponding normal mode frequencies,

$$\omega_r = 2\left(\frac{\phi''}{m}\right)^{1/2} \sin\left(\frac{r\pi}{2(N+1)}\right).$$

The orthogonality of the eigenvectors can be shown as

$$\sum_{j=1}^{N} R_{rj} R_{sj} = \sum_{j=1}^{N} C_r C_s \sin\left(\frac{r\pi j}{N+1}\right) \sin\left(\frac{s\pi j}{N+1}\right) = C_r C_s \left(\frac{N+1}{2}\right) \delta_{rs}.$$

Using the identities

$$\sum_{j=1}^{N} \sin\left(\frac{r\pi j}{N+1}\right) \sin\left(\frac{s\pi j}{N+1}\right) = \left(\frac{N+1}{2}\right) \delta_{rs},$$

one can find that the normalization factor C_r should be,

$$C_r = \left(\frac{2}{N+1}\right)^{1/2}. \tag{16.24}$$

16.1.3 Free Energy and Stress Calculations for Harmonic Systems

Now consider the harmonic free energy (in fact quasi-harmonic) for the system,

$$F_h(b, T) = V(b) - Nk_B T \ln(2\pi k_B T) + k_B T \sum_{r=1}^{N} \ln \omega_r(b).$$

We can find the stress,

$$P_{11} = \frac{1}{\Omega_0} \frac{\partial F_h}{\partial F_{11}} = \frac{1}{(N+1)b_0^3} \left[\frac{\partial V_0}{\partial F_{11}} + k_B T \sum_r \frac{\partial \ln \omega_r}{\partial F_{11}}\right],$$

where $\Omega_0 = (N+1)b_0^3$.

Using the relation,

$$\omega_r = 2\left(\frac{\phi''(F_{11}b_0)}{m}\right)^{1/2} \sin\left(\frac{r\pi}{2(N+1)}\right),$$

we can find that

$$P_{11}(b, T) = \frac{1}{b_0^2}\left(\phi'(b) + \frac{Nk_B T \phi'''(b)}{2(N+1)\phi''(b)}\right).$$

Similarly, one may find the elastic constant,

$$C_{1111}^T = \frac{1}{\Omega_0} \frac{\partial^2 F_h}{\partial E_{11}^2}\bigg|_{q=0} = \frac{1}{(N+1)b_0^3} \left[\left(\frac{b_0^2}{b^2} \frac{\partial^2 F}{\partial F_{11}^2} \right) - \left(\frac{b_0^3}{b} \frac{\partial F}{\partial F_{11}} \right) \bigg|_{q=0} \right]$$

$$= \frac{1}{b_0} \left[\left(\phi''(b_0) - \frac{1}{b_0} \phi'(b_0) \right) + \frac{Nk_B T}{2(N+1)} \left(\frac{\phi^{IV}(b_0)}{\phi''(b_0)} \right) - \left(\frac{\phi'''(b_0)}{\phi''(b_0)} \right)^2 \right.$$

$$\left. - \frac{1}{b_0} \frac{\phi'''(b_0)}{\phi''(b_0)} \right],$$

and the thermal coefficient,

$$\kappa_{11} = \frac{1}{\Omega_0} \frac{\partial^2 F_h}{\partial E_{11} \partial T}\bigg|_{q=0} = \frac{1}{b_0} \frac{Nk_B}{2(N+1)} \frac{\phi'''(b_0)}{\phi''(b_0)}.$$

Note that

$$E_{11} = \frac{1}{2}(F_{11}^2 - 1) = \frac{1}{2}\left[\left(\frac{b}{b_0} \right)^2 - 1 \right].$$

REMARK 16.4 The above example solution shows the following features:

(i) If

$$\phi''' \neq 0,$$

indicates that this is not a purely harmonic approximation; and we call it as the quasi-harmonic approximation.

(ii) The initial stress and initial temperature are related by,

$$P_{11}^0 = \frac{1}{b_0^2} \left[\phi'(b_0) + \frac{Nk_B T_0}{2(N+1)} \frac{\phi'''(b_0)}{\phi''(b_0)} \right].$$

(iii) One can calculate the acoustic wave speed

$$c_0 = \left(\frac{C_{1111}}{\rho} \right)^{1/2}, \quad \text{and} \quad \rho = m/b^3.$$

16.1.4 Quantum Effects

Since the lattice structure is periodic, the following lattice motion is compatible with the lattice structure,

$$q_j = A \exp(ikj) \cos \omega t, \quad i = \sqrt{-1},$$

which may represent a set of particle vibrations, or stationary waves.
For a classical system, the Hamiltonian of 1D single harmonic oscillator is

$$H_q = \frac{1}{2M} p^2 + \frac{K}{2} x^2.$$

However, the phonon in solids is not a classical harmonic oscillator. Phonons are bosons, and they are quantum particles. Thus, the quantum effect must be taken into account.

In quantum mechanics, the 1D quantum harmonic oscillator obeys the following steady-state Schrödinger equation,

$$\frac{\hbar^2}{2m}\frac{d^2\psi}{dx^2} + \frac{1}{2}m\omega^2 x^2 \psi = E\psi, \tag{16.25}$$

where $\hbar = \frac{h}{2\pi}$ is the reduced Planck constant.

The above equation can also be written as,

$$H_q \psi = E_n \psi, \quad E_n = \left(n + \frac{1}{2}\right)\hbar\omega. \tag{16.26}$$

The energy levels for a quantum harmonic oscillator are,

$$E_n = \left(n + \frac{1}{2}\right)\hbar\omega.$$

Recall that in canonical ensemble, the occupation number is,

$$a_j(E_j) = A \exp(-\beta E_j), \quad \beta = \frac{1}{k_B T}.$$

Thus, the partition function for **one** quantum harmonic oscillator is

$$Z_{qu} = \sum_n \exp(-\beta E_n) = \exp(-\hbar\omega/(2k_B T)) \sum_{n=0}^{\infty} \exp(-n\hbar\omega/(k_B T))$$

$$= \frac{\exp(-\hbar\omega/(2k_B T))}{1 - \exp(-\hbar\omega/(k_B T))} = \left(2\sinh\frac{\hbar\omega}{2k_B T}\right)^{-1}.$$

Note that

$$\sum_{n=0}^{\infty} \exp(-nx) = \frac{1}{1 - \exp(-x)}.$$

We can now write the total energy of the multiscale system in terms of phonon modes,

$$H_q = V_0(\mathbf{F}) + \sum_r E_r, \quad \to \quad Z_{qm} = \prod_r \exp\left(-\frac{(V_0(\mathbf{F}) + E_r)}{k_B T}\right). \tag{16.27}$$

Recall that for **one** quantum harmonic oscillator is

$$Z_r = \exp(-E_r/k_B T) = \frac{\exp(-\hbar\omega_r/(2k_B T))}{1 - \exp(-\hbar\omega_r/(k_B T))}.$$

The partition function for N 3D quantum harmonic oscillators is then,

$$Z_{qm}(\mathbf{F}, T) = \exp\left(-\frac{V_0(\mathbf{F})}{k_B T}\right) \prod_{r=1}^{3N} Z_r(\omega_r, T),$$

where
$$Z_r(\mathbf{F}, T) = \frac{\exp(-x_r/2)}{1 - \exp(-x_r)} = \left(2 \sinh \frac{x_r}{2}\right)^{-1},$$

with
$$x_r = \frac{\hbar \omega_r}{k_B T}.$$

Finally, the free energy is (see: Li and Sheng (2010)),
$$F = -k_B T \log Z_{qm} = V_0(\mathbf{F}) + k_B T \sum_{i=1}^{N} \sum_{k=1}^{3} \log\left[2 \sinh\left(\frac{\hbar \omega_{ik}(\mathbf{F})}{2\pi k_B T}\right)\right]. \quad (16.28)$$

Moreover, we can then find the entropy of the system as,
$$S = -\frac{\partial F_h}{\partial T} = \frac{h}{2\pi T} \sum_{i=1}^{n_c} \sum_{k=1}^{3} \omega_{ik}(\mathbf{F}) \coth\left(\frac{\hbar \omega_{ik}(\mathbf{F})}{2\pi k_B T_c}\right)$$
$$- k_B \sum_{i=1}^{n_c} \sum_{k=1}^{3} \log\left[2 \sinh\left(\frac{\hbar \omega_{ik}(\mathbf{F})}{2\pi k_B T_c}\right)\right]. \quad (16.29)$$

Similarly, we can find the internal energy E,
$$E_h = F_h + TS = V_0(\mathbf{F}) + \frac{h}{4\pi} \sum_{i=1}^{n_c} \sum_{k=1}^{3} \omega_{ik}(\mathbf{F}) \coth\left(\frac{\hbar \omega_{ik}(\mathbf{F})}{2\pi k_B T}\right),$$

and the first Piola–Kirchhoff stress $\mathbf{P}$,
$$\mathbf{P}(\mathbf{F}, T) = \frac{1}{\Omega} \frac{\partial F_h}{\partial \mathbf{F}} = \frac{1}{\Omega} \left\{ \frac{1}{2} \sum_{i=1}^{n_c} \sum_{\alpha=1}^{n_b} \varphi'(\bar{r}_{i\alpha}) \frac{\bar{r}_{i\alpha} \otimes \mathbf{R}_{i\alpha}}{\bar{r}_{i\alpha}} \right.$$
$$\left. + \frac{h}{4\pi} \sum_{i=1}^{n_c} \sum_{k=1}^{3} \left[\coth\left(\frac{\hbar \omega_{ik}(\mathbf{F})}{2\pi k_B T_c}\right) \sum_{\alpha=1}^{n_b} \omega'_{ik}(\bar{r}_{i\alpha}) \frac{\bar{r}_{i\alpha} \otimes \mathbf{R}_{i\alpha}}{\bar{r}_{i\alpha}}\right] \right\}.$$

16.2 Homework Problems

Problem 16.1 Consider the following multiscale (quasi-harmonic) free energy,
$$F_{qh} = -k_B T \log Z_{qm} = V_0(\mathbf{F}) + k_B T \sum_{i=1}^{N} \sum_{k=1}^{3} \log\left[2 \sinh\left(\frac{\hbar \omega_{ik}(\mathbf{F})}{2\pi k_B T}\right)\right].$$

Find the first Piola–Kirchhoff stress, i.e.,
$$\mathbf{P} = \frac{1}{\Omega_0} \frac{\partial F_{qh}}{\partial \mathbf{F}} \; ?$$

Problem 16.2 Show that

$$H_h = \frac{1}{2}\sum_{i=1}^{n} p_i^2 + V_0(\mathbf{F}) + \frac{1}{2}\sum_{i=1}^{n}\sum_{j=1}^{n} V_{ij} q_i q_j = V_0(\mathbf{F}) + \frac{1}{2}\sum_{r=1}^{n}(P_r^2 + \omega_r^2 Q_r^2),$$

where

$$Q_r = R_{ri} q_i, \quad P_r = R_{ri} p_i, \quad \text{and} \quad [V_{ij}][R_{jr}] = \omega_r^2 [R_{ir}], \; r = 1, 2, \ldots, n.$$

Hint:
Choose $[R_{ij}]$ as an orthonormal transformation matrix, i.e.,

$$[R_{ij}]^{-1} = [R_{ij}]^T \quad \rightarrow \quad [R_{ri}][R_{is}]^T = [\delta_{ij}];$$

we have,

$$[R_{si}]^T [V_{ij}][R_{jr}] = \omega_r^2 [\delta_{sr}].$$

In fact, here we are using the property

$$[R_{ri}]^T [R_{is}] = [\delta_{rs}],$$

which is often called completeness.

In structural mechanics, $[R_{ij}]$ often serves as transformation matrix, which transforms the original Cartesian coordinates to the so-called principal coordinates, or principal directions.

17 Multiscale Method (I): Multiscale Micromorphic Molecular Dynamics

As an extension of PR-MD, Li and his co-workers have developed an atomistic-to-continuum molecular dynamics (MD), which is coined as *multiscale micromorphic molecular dynamics* (MMMD) (see: Li and Tong (2015); Li and Urata (2016); Tong and Li (2016)). In this chapter, we discuss how to use this method to construct multiscale methods.

17.1 Multiscale Partition of First-Principles MD Lagrangian

We first partition the Lagrangian of first-principles MD based on domain decomposition and multiscale kinematics of the atom motion.

17.1.1 Scale Decomposition

The first part of multiscale partition is the domain decomposition, i.e., we divide the finite size simulation domain into a finite number of supercells or local ensembles, and we choose the αth cell as the representative supercell, to illustrate formulation, and in the whole simulation domain, we have $\alpha = 1, 2, \ldots, N$ number of cells.

Assume that in the αth cell, it has N_α number of atoms. The position of the center of mass (COM) of this cell is defined as,

$$\mathbf{r}_\alpha(t) = \frac{1}{\sum_i m_i} \sum_i m_i \mathbf{r}_i(t), \tag{17.1}$$

where $\mathbf{r}_i, i = 1, 2, \ldots N_\alpha$ are the spatial position of each atom in the current cell configuration at current time t. Inversely, we can express the spatial position of ith atom in the αth MD cell as,

$$\mathbf{r}_i = \mathbf{r}_\alpha + \mathbf{r}_{\alpha i}, \quad \alpha = 1, 2, \ldots, N; \ i = 1, 2, \ldots, N_\alpha, \tag{17.2}$$

where $\mathbf{r}_{\alpha i}$ is the relative position for the ith atom in αth cell and

$$\mathbf{r}_\alpha = \frac{1}{\sum_i^{N_\alpha} m_i} \sum_i^{N_\alpha} m_i \mathbf{r}_i$$

is the position of the center of mass. By this definition, we have

$$\sum_i m_i \mathbf{r}_{\alpha i} = \mathbf{0}. \qquad (17.3)$$

Now, we introduce a so-called micromorphic multiplicative decomposition to describe the relative position of each atom,

$$\mathbf{r}_{\alpha i} = \boldsymbol{\phi}_\alpha \cdot \mathbf{S}_i, \text{ and } \boldsymbol{\phi}_\alpha := \mathbf{F}_\alpha \cdot \boldsymbol{\chi}_\alpha, \qquad (17.4)$$

where $\boldsymbol{\phi}_\alpha$ is the total deformation gradient of the αth supercell; $\mathbf{S}_i$ are the scaled atom position vectors for every atoms inside the αth cell (we use $\mathbf{S}_i$ instead of $\mathbf{S}_{\alpha i}$ for simplicity); the second-order tensor $\boldsymbol{\chi}_\alpha$ is the micro deformation tensor of the αth cell, and its physical meaning may be interpreted as the shape tensor of the αth cell,

$$\boldsymbol{\chi}_\alpha(t)\mathbf{S}_i(t) = \xi_i(t)\mathbf{a}(t) + \eta_i(t)\mathbf{b}(t) + \zeta_i(t)\mathbf{c}(t), \quad i = 1, 2, \ldots, N_\alpha,$$

where $\mathbf{a}, \mathbf{b}$, and $\mathbf{c}$ are the MD cell edge vectors and ξ_i, η_i, and ζ_i are the components of the local position vector $\mathbf{S}_i$ projecting onto the MD cell edge. They can be expressed into the following matrix form:

$$\boldsymbol{\chi}_\alpha(t) = \begin{pmatrix} a_x(t) & b_x(t) & c_x(t) \\ a_y(t) & b_y(t) & c_y(t) \\ a_z(t) & b_z(t) & c_z(t) \end{pmatrix}, \quad \mathbf{S}_i(t) = \begin{pmatrix} \xi_i(t) \\ \eta_i(t) \\ \zeta_i(t) \end{pmatrix}.$$

If we place the center of the supercell at the origin of the local coordinate, the components of the local position vector of $\mathbf{S}_i$ oscillate in the range,

$$-0.5 \leq \xi_i(t), \eta_i(t), \zeta_i(t) \leq 0.5,$$

which characterizes the internal degrees of freedom of atom motions. When the total number of atoms increases, the motion of $\mathbf{S}_i$ may become random. Therefore, we refer the local position vectors $\mathbf{S}_i$ as the statistical variables. Borrowing the terminology in macroscale continuum mechanics, we call the assemble of $\{\mathbf{S}_i\}$ as the statistical parametric configuration, i.e., $\mathbf{S}_i \in \mathcal{B}_S$.

Thus, Eq. (17.3) leads to the condition,

$$\sum_i m_i \mathbf{S}_{\alpha i} = \mathbf{0}. \qquad (17.5)$$

Here, we used $\mathbf{S}_{\alpha i}$ to distinguish the numbering index in different supercells. However, if without confusion, we denote αi as i or $i \in \alpha$ in the rest of book without clarification.

The key part of the multiplicative micromorphic decomposition is the introduction of a coarse-scale deformation gradient $\mathbf{F}_\alpha = \mathbf{F}_\alpha(\{\mathbf{r}_\beta\})$ that are determined by the overall motion of all centers of mass of every cells, which are completely determined by the spatial distribution of centers of mass (implied by the dependence of $\{\mathbf{r}_\beta\}$) of all cells.

Figure 17.1 shows a schematic spatial cell division and the corresponding distribution of centers of mass.

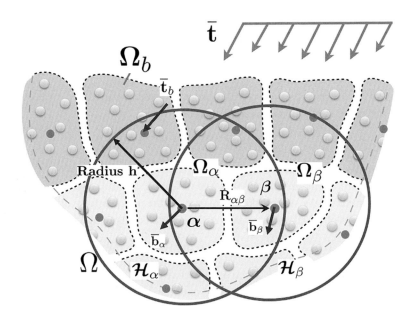

Figure 17.1 Schematic illustration of supercell domain decomposition; the dark points are centers of mass of supercells that may be subjected by external body force; the dark supercells are boundary cells that may be subjected external traction. The center of mass for each supercell has a support $\mathcal{H}_\alpha$ that contains some other centers of mass of adjacent supercells

$\mathbf{F}_\alpha$ together with χ_α constitute the total deformation gradient ϕ_α, whereas $\mathbf{F}_\alpha$ are determined by the distribution of $\{\mathbf{r}_\alpha\}$. For example, we can determine $\mathbf{F}_\alpha$ by using an approach adopted in the reproducing kernel particle method, e.g., Li and Liu (1999) or in the state-based peridynamics (PD) (Silling et al. (2007)).
Denote

$$\mathbf{r}_{\alpha\beta}(t) = \mathbf{r}_\beta(t) - \mathbf{r}_\alpha(t), \quad \mathbf{R}_\alpha := \mathbf{r}_\alpha(0), \text{ and } \mathbf{R}_{\alpha\beta} = \mathbf{R}_\beta - \mathbf{R}_\alpha, \quad (17.6)$$

where the Greek subscripts are indices of the centers of mass of MD cells. As shown in Fig. 17.1, each supercell center has a compact support, say $\mathcal{H}_\alpha$ for the cell α. We can then construct the so-called moment matrix or shape tensor for the cell center α by calculating following discrete sum over the other cell centers inside the support $\mathcal{H}_\alpha$ (see: Li and Liu (1999)),

$$\mathbf{M}_\alpha := \sum_{\beta=1}^{N_h} \omega(|\mathbf{R}_{\alpha\beta}|)\mathbf{R}_{\alpha\beta} \otimes \mathbf{R}_{\alpha\beta}\Omega_{\beta 0}, \quad (17.7)$$

where $\otimes$ is the tensor product operator; N_h is the number of supercell centers inside $\mathcal{H}_\alpha$, $\Omega_{\beta 0}$ is the volume of the βth cell in the referential configuration $\mathcal{B}_R$, and $\omega(|\mathbf{R}_{\alpha\beta}|)$ is a localized positive window function.

To construct the macroscale deformation gradient tensor, we first define a two-point second-order tensor by the following discrete sum,

17 Multiscale Method (I): Multiscale Micromorphic Molecular Dynamics

$$\mathbf{N}_\alpha := \left(\sum_{\beta=1}^{N_h} \omega(|\mathbf{R}_{\alpha\beta}|) \mathbf{r}_{\alpha\beta} \otimes \mathbf{R}_{\alpha\beta} \Omega_{\beta 0} \right). \quad (17.8)$$

Considering the Cauchy–Born rule (Tadmor et al. (1996)), we have

$$\mathbf{r}_{\alpha\beta} = \mathbf{F}_\alpha \cdot \mathbf{R}_{\alpha\beta} \quad \to \quad \mathbf{N}_\alpha = \mathbf{F}_\alpha \cdot \mathbf{M}_\alpha,$$

and we can then derive the coarse-scale deformation gradient as

$$\mathbf{F}_\alpha = \left(\sum_{\beta=1}^{N_h} \omega(|\mathbf{R}_{\alpha\beta}|) \mathbf{r}_{\alpha\beta} \otimes \mathbf{R}_{\alpha\beta} \Omega_{\beta 0} \right) \cdot \mathbf{M}_\alpha^{-1}, \quad (17.9)$$

where $\omega(|\mathbf{R}_{\alpha\beta}|)$ is a localized positive window function, and a common choice is the Gaussian function,

$$\omega_h(\mathbf{x}) = \frac{1}{(\pi h^2)^{d/2}} \exp\left(-\frac{\mathbf{x} \cdot \mathbf{x}}{h^2}\right), \quad (17.10)$$

where d is the number of space dimension and h is the radius of the support. Obviously, $\mathbf{F}_\alpha$ depends on the relative positions of all other atoms in the support of the αatom, i.e., Ω_α. Note that in the configurations $\mathcal{B}_R$ and $\mathcal{B}_I$ the center of the supercell occupies the same position, i.e., $\mathbf{R}_\alpha \equiv \mathcal{R}_\alpha$, but not for each atom, i.e., $\mathbf{R}_{\alpha i} = \chi_\alpha(0) \mathbf{S}_i \neq \mathcal{R}_{\alpha i} = \chi_\alpha(t) \mathbf{S}_i$.

At the beginning of the coarse-scale deformation, the initial position of the supercell center $\mathbf{r}_\alpha(0) = \mathbf{R}_\alpha$, therefore,

$$\mathbf{F}_\alpha = \left(\sum_{\beta=1}^{N_h} \omega(|\mathbf{R}_{\alpha\beta}|) \mathbf{r}_{\alpha\beta} \otimes \mathbf{R}_{\alpha\beta} \Omega_{\beta 0} \right) \cdot \mathbf{M}_\alpha^{-1}$$

$$= \left(\sum_{\beta=1}^{N_h} \omega(|\mathbf{R}_{\alpha\beta}|) \mathbf{R}_{\alpha\beta} \otimes \mathbf{R}_{\alpha\beta} \Omega_{\beta 0} \right) \cdot \mathbf{M}_\alpha^{-1} = \mathbf{I},$$

where $\mathbf{I}$ is the unit of second-order tensor.

Initially, the supercell shape tensor, χ_0, which is spanned by the three edges of the MD cell, $\mathbf{a}(0), \mathbf{b}(0)$, and $\mathbf{c}(0)$, may be expressed as,

$$\chi_0 := \chi(0) = [\mathbf{a}(0), \mathbf{b}(0), \mathbf{c}(0)].$$

For all the atoms in a representative MD cell, say the αth cell, its referential position are defined as

$$\mathbf{R}_i := \mathbf{R}_\alpha + \mathbf{R}_{\alpha i}, \quad \text{with } \mathbf{R}_{\alpha i} = \chi_0 \mathbf{S}_i. \quad (17.11)$$

In the rest of this book, we refer this molecular configuration as the referential configuration $\mathcal{B}_R$. One can find that this definition is consistent with kinematic assumption,

$$\mathbf{r}_i = \mathbf{r}_\alpha + \mathbf{r}_{\alpha i} = \mathbf{r}_\alpha + \mathbf{F}_\alpha \cdot \chi_\alpha \cdot \mathbf{S}_i. \quad (17.12)$$

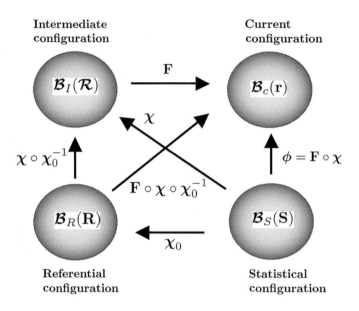

Figure 17.2 Deformation map that connects four different kinematic configuration spaces: (1) statistical configuration; (2) referential configuration; (3) intermediate configuration, and (4) current configuration.

Based on Eq. (17.12), we may define an intermediate referential configuration,

$$\mathcal{R}_i = \mathbf{R}_\alpha + \mathcal{R}_{\alpha i} = \mathbf{R}_\alpha + \chi_\alpha(t) \cdot \mathbf{S}_i, \tag{17.13}$$

so that the coarse-scale deformation gradient is a continuum-scale deformation gradient.

In this multiscale method, we have introduced four different configurations for a representative cell α, as shown in Fig. 17.2: (1) the spatial configuration, or the current configuration $\mathcal{B}_c(\mathbf{r})$ configuration, in which $\mathbf{r}_\alpha \in \Omega_\alpha$; (2) the intermediate equilibrium configuration, i.e., $\mathcal{B}_I(\mathcal{R})$ configuration; (3) the referential equilibrium configuration, i.e., $\mathcal{B}_I(\mathbf{R})$ configuration, in which $\mathbf{R}_i \in \Omega_{\alpha 0}$, and (4) the statistical configuration $\mathcal{B}_S$, in which the scaled atom coordinate $\mathbf{S}_i$ are primary variables. Figure 17.2 shows a deformation map that connects these four configuration spaces.

Since for monoatom systems $\sum_i \mathbf{S}_i = 0$ and $\mathbf{S}_i$ are oscillating in the range so that $\|\mathbf{S}_i\| < 1$, as atoms oscillate around their equilibrium positions, we may interpret $\mathbf{S}_i$ as a statistical variable. For a given macroscale kinematic variable or point, $\mathbf{r}_\alpha(t)$ (the center of cell), there are many sets of $\{\mathbf{S}_i\}$ corresponding to it, and this is the case because as long as $\sum_i m_i \mathbf{S}_i = 0$ satisfies. Therefore, a given macroscale material point $\mathbf{r}_\alpha(t)$ does not correspond to a single set of $\{\mathbf{S}_i\}$ in a unique one-to-one manner, i.e., such correspondence is not unique. A macroscale state $\mathbf{r}_\alpha$ corresponds to many microstates $\{\mathbf{S}_i\}$, and a microstate is determined by the distribution of statistical variable $\mathbf{S}_i$.

It may be also noted that on the configuration spaces $\mathcal{B}_I, \mathcal{B}_R$, the center of mass coordinate are the same, i.e., $\mathbf{R}_\alpha$, whereas the configuration space $\mathcal{B}_S$ is defined cell by cell, and the coordinate of the center of mass of each cell is zero.

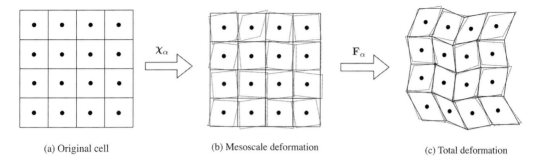

(a) Original cell (b) Mesoscale deformation (c) Total deformation

Figure 17.3 Cell deformation in different configurations: (a) The original undeformed system that is divided into several unit cells. The atomic positions are given by $\mathbf{R}_i$. The solid circles represent the centers of mass; (b) the cells (dashed parallelograms) further undergoe microscale deformation χ_α around their own center of mass separately without connection (with the cell center fixed). This is because that continuity may not be valid at fine scale. (c) The configuration undergoes macroscale deformation $\mathbf{F}_\alpha$, i.e., now the cell centers start to move. Notice that the continuity is a macroscale concept. Based on this concept, we can use the positions of all cell centers to construct, e.g., interpolate, a coarse-scale displacement field.

A pictorial illustration of three physical configurations is displayed in Fig. 17.3, where physical interpretations of different kinematic mappings at different scales are graphically illustrated. In passing, we note that the continuum compatibility condition is a macroscale condition, and at both microscale and mesoscale this condition is not necessarily satisfied. For crystalline solids, this may be linked to the defect states or quasi-crystal states. Even though the referential configuration is not essential in the fine-scale calculations, but its information is needed in coarse-scale computations.

To clearly specify displacement decomposition at each scale, we consider the following atomic position decomposition,

$$\mathbf{r}_i = \mathbf{r}_\alpha + \mathbf{r}_{\alpha i} \Rightarrow \mathbf{r}_i = \mathbf{r}_\alpha + \mathbf{F}_\alpha \cdot \mathcal{R}_{\alpha i} = \mathbf{r}_\alpha + \mathbf{F}_\alpha \cdot \chi_\alpha \cdot \mathbf{S}_i, \quad (17.14)$$

where

$$\mathcal{R}_{\alpha i} = \chi_\alpha \cdot \mathbf{S}_i.$$

These relations are schematically illustrated in Fig. 17.4. Subsequently, we can decompose the atomistic displacement into three different scales,

$$\bar{\mathbf{u}} = \mathbf{r}_\alpha - \mathbf{R}_\alpha, \quad \tilde{\mathbf{u}}_i = (\mathbf{F}_\alpha \cdot \chi_\alpha - \chi_\alpha)\mathbf{S}_i, \text{ and } \mathbf{u}'_i = (\chi_\alpha - \chi_{\alpha 0})\mathbf{S}_i, \quad (17.15)$$

where $\bar{\mathbf{u}}, \tilde{\mathbf{u}}_i, \mathbf{u}'_i$ denote macroscale, mesoscale, and microscale displacements. Apparently, three of them together constitute the total displacement $\mathbf{u}_i$, which can be expressed as,

$$\mathbf{u}_i = \bar{\mathbf{u}} + \tilde{\mathbf{u}}_i + \mathbf{u}'_i = \mathbf{r}_i - \mathbf{R}_i. \quad (17.16)$$

In the scale decomposition, we select three independent kinematic variables, $\{\mathbf{S}_i, \chi_\alpha, \text{ and } \mathbf{r}_\alpha\}$, to represent three different scales. The novelty of the proposed multiscale decomposition is the **multiplicative multiscale decomposition**,

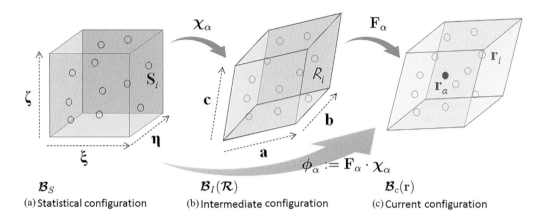

Figure 17.4 Schematic illustrations of relations among (a) statistical configuration, (b) iIntermediate configuration, and (c) current configuration.

$$\boldsymbol{\phi}_\alpha = \mathbf{F}_\alpha \cdot \boldsymbol{\chi}_\alpha. \tag{17.17}$$

This decomposition is in sharp contrast with the additive multiscale decomposition proposed by Wagner and Liu (2003). The total deformation gradient $\boldsymbol{\phi}_\alpha$ consists of a macroscale deformation gradient $\mathbf{F}_\alpha$ and a micromorphic (meso)deformation gradient $\boldsymbol{\chi}_\alpha$. The macroscale deformation gradient at a local position is completely determined by the local distribution of the supercell centers, which can be quantified, e.g., by Eq. (17.9). On the other hand, the microscale deformation gradient, which is often called as the micro-deformation in the literature of micromorphic theory, is an independent mesoscale field variable that quantifies the independent supercell rotation or microscale deformation. The macro-deformation $\mathbf{F}_\alpha$ may be considered to be **long-range order**, whereas the micro-deformation $\boldsymbol{\chi}_\alpha$ may be considered to be **short-range order**.

17.1.2 Statistical Conditions

Before deriving the dynamic equations of MMMD, we first examine the kinetic energy of the αth cell:

$$K_\alpha = \frac{1}{2}\sum_i m_i \dot{\mathbf{r}}_i \cdot \dot{\mathbf{r}}_i = \frac{1}{2}\sum_i m_i (\dot{\mathbf{r}}_\alpha + \dot{\boldsymbol{\phi}}_\alpha \cdot \mathbf{S}_i + \boldsymbol{\phi}_\alpha \cdot \dot{\mathbf{S}}_i) \cdot (\dot{\mathbf{r}}_\alpha + \dot{\boldsymbol{\phi}}_\alpha \cdot \mathbf{S}_i + \boldsymbol{\phi}_\alpha \cdot \dot{\mathbf{S}}_i)$$

$$= \underbrace{\frac{M_\alpha}{2}\dot{\mathbf{r}}_\alpha \cdot \dot{\mathbf{r}}_\alpha}_{K_1} + \underbrace{\frac{1}{2}\dot{\boldsymbol{\phi}}_\alpha^T \dot{\boldsymbol{\phi}}_\alpha \sum_i m_i \mathbf{S}_i \otimes \mathbf{S}_i}_{K_2} + \underbrace{\frac{1}{2}\sum_i m_i \dot{\mathbf{S}}_i \cdot \mathbf{C} \cdot \dot{\mathbf{S}}_i}_{K_3}$$

$$+ \underbrace{\frac{1}{2}\dot{\boldsymbol{\phi}}_\alpha^T \boldsymbol{\phi}_\alpha \sum_i m_i \mathbf{S}_i \otimes \dot{\mathbf{S}}_i + \frac{1}{2}\boldsymbol{\phi}_\alpha^T \dot{\boldsymbol{\phi}}_\alpha \sum_i m_i \dot{\mathbf{S}}_i \otimes \mathbf{S}_i}_{K_4}, \tag{17.18}$$

where $M_\alpha = \sum_i m_i$ is the mass of the cell and $\mathbf{C} = \boldsymbol{\phi}_\alpha^T \boldsymbol{\phi}_\alpha$ is the multiscale right Cauchy–Green tensor.

We postulate the following statistical condition,

$$\mathbf{J}_\alpha^S = \sum_i m_i \mathbf{S}_i \otimes \mathbf{S}_i = \boldsymbol{\chi}_{\alpha 0}^{-1} \cdot \mathbf{J}_\alpha \cdot \boldsymbol{\chi}_{\alpha 0}^{-T} = constant\ tensor, \quad (17.19)$$

where

$$\mathbf{J}_\alpha = \sum_i m_i \mathbf{R}_{\alpha i} \otimes \mathbf{R}_{\alpha i} \quad (17.20)$$

is a second-order tensor that is similar to the moment of inertia tensor of the cell in referential space. In mechanics, the moment of inertia tensor of a rigid body is defined as,

$$\mathbf{I}_{Euler} := \int_V \rho \left(\|\mathbf{R}\|^2 \mathbf{I}^{(2)} - \mathbf{R} \otimes \mathbf{R} \right) dV,$$

where ρ is mass density and $\mathbf{I}^{(2)}$ is the second-order unit tensor. It is obvious that $\mathbf{I}_{Euler}$ is dependent on both the shape and size of the rigid body. Similarly, the tensor $\mathbf{J}_\alpha$ is also dependent on both the shape and size of the micromorphic MD cell. On the other hand, the second-order tensor $\mathbf{J}_\alpha^S$ is not dependent on the MD cell size and shape. In fact, $\mathbf{J}_\alpha^S$ is the push-back of $\mathbf{J}_\alpha$ (see Marsden and T. Hughes (1983)), and $\boldsymbol{\chi}_{\alpha 0} = \boldsymbol{\chi}(0)$ is a constant second-order tensor, i.e., the geometric shape tensor of the original cell.

We can replace the discrete summation in Eq. (17.19) with a continuous integration, i.e.,

$$\mathbf{J}_\alpha^S = \int_\omega \rho \mathbf{S} \otimes \mathbf{S} d\omega_S, \quad (17.21)$$

where ω is the volume of a unit cell. Let $\mathbf{E}_I$ be the principal axes of $\mathbf{J}_\alpha^S$. We can then have the spectral decomposition of $\mathbf{J}_\alpha^S$,

$$\mathbf{J}_\alpha^S = \sum_{I=1}^3 \left(\int_\omega \rho S_I^2 d\omega \right) \mathbf{E}_I \otimes \mathbf{E}_I. \quad (17.22)$$

If we assume that $\mathbf{J}_\alpha^S$ is a spherical tensor, we have

$$\int_\omega \rho S_1^2 d\omega = \int_\omega \rho S_2^2 d\omega = \int_\omega \rho S_3^2 d\omega =: W_\alpha. \quad (17.23)$$

It should be noted that the supercell is not a necessarily unit cube, and it is related to the physical shape of the lattice representative cell. Therefore, the related Euler's moment of inertia tensor, $\mathbf{I}_{Euler}$, is in general anisotropic. On the other hand, the moment tensor $\mathbf{J}_\alpha^S$ in Eq. (17.21) is defined in the statistical configuration. The assumption that $\mathbf{J}_\alpha^S$ is a spherical tensor is based on the general ergodicity postulate that the space and time that a MD simulation is involved are statistically isotropic and homogeneous, which is one of the fundamental hypothesis in statistical physics and molecular physics. Note that we are discussing a quantity in a statistical configuration

$\mathcal{B}_S$, not the lattice configuration $\mathcal{B}_R, \mathcal{B}_I$ nor the current configuration $\mathcal{B}_c$. The anisotropy of a material in the lattice space is not in conflict with the ergodic assumption stated here.

Next, we consider the fourth term of the kinetic energy K_4. Since we may label $\mathbf{S}_i$ statistical variables, we may find that $\mathbf{J}_\alpha^S$ is in fact a tensorial autocorrelation function of statistical positions,

$$\mathbf{J}^S(\tau) = <\mathbf{S}_i(t) \otimes \mathbf{S}_i(t+\tau)> := \sum_i m_i \mathbf{S}_i(t) \otimes \mathbf{S}_i(t+\tau). \tag{17.24}$$

One can prove that the autocorrelation tensor is an even function of τ, if $\mathbf{J}^S$ is spherical. This implies the derivation at the origin $\tau = 0$ is zero. Hence,

$$\frac{d}{d\tau}\mathbf{J}_\alpha^S|_{\tau=0} = \sum_i m_i \mathbf{S}_i(t) \otimes \dot{\mathbf{S}}_i(t) = \mathbf{0}, \tag{17.25}$$

and similarly,

$$\sum_i m_i \dot{\mathbf{S}}_i(t) \otimes \mathbf{S}_i(t) = \mathbf{0}. \tag{17.26}$$

Equations (17.25) and (17.26) imply that $K_4 = 0$. In fact, the second statistical condition states that $\mathbf{J}_\alpha^S$ is a spherical tensor, which implies that

$$\mathbf{J}_\alpha^S(\tau) = <\mathbf{S}_i(t) \otimes \mathbf{S}_i(t+\tau)> = \sum_i m_i \mathbf{S}_i(t) \otimes \mathbf{S}_i(t+\tau)$$

$$= \left(\sum_i m_i S_i(t) S_i(t+\tau)\right) \mathbf{E}_I \otimes \mathbf{E}_I,$$

where $\mathbf{E}_I, I = 1, 2, 3$ are the basis of a Cartesian coordinate. It then follows,

$$\mathbf{J}_\alpha^S(-\tau) = <\mathbf{S}_i(t) \otimes \mathbf{S}_i(t-\tau)> = \sum_i m_i \mathbf{S}_i(t) \otimes \mathbf{S}_i(t-\tau)$$

$$= \left(\sum_i m_i S_i(t) S_i(t-\tau)\right) \mathbf{E}_I \otimes \mathbf{E}_I = \left(\sum_i m_i S_i(t+\tau) S_i(t)\right) \mathbf{E}_I \otimes \mathbf{E}_I$$

$$= <\mathbf{S}_i(t) \otimes \mathbf{S}_i(t+\tau)> = \mathbf{J}_\alpha^S(\tau).$$

Utilizing the two statistical conditions, we can write the Lagrangian of the MMMD system as,

$$\mathcal{L}_m = \frac{1}{2}\sum_\beta M_\beta \dot{\mathbf{r}}_\beta \cdot \dot{\mathbf{r}}_\beta + \frac{1}{2}\sum_\beta \mathbf{J}_\beta^S : (\dot{\boldsymbol{\phi}}_\beta^T \dot{\boldsymbol{\phi}}_\beta)$$

$$+ \frac{1}{2}\sum_\beta \sum_i m_i \dot{\mathbf{S}}_i \cdot \mathbf{C}_\beta \cdot \dot{\mathbf{S}}_i - \frac{1}{2}\sum_\beta \sum_\gamma \sum_{i\in\beta, j\in\gamma} V(r_{ij}) - \sum_\beta \sum_{i\in\beta} \mathbf{b}_i \cdot \mathbf{r}_i,$$

$$\tag{17.27}$$

where β, γ are cell indices and the abbreviation $i \in \beta$ means that the ith atom belongs to the βth cell. $\mathbf{C}_\beta := \boldsymbol{\phi}_\beta^T \boldsymbol{\phi}_\beta$ is the microscale right Cauchy–Green tensor for the total deformation.

In the rest of the book, we choose three independent field variables, $\mathbf{r}_\alpha, \boldsymbol{\phi}_\alpha$, and $\mathbf{S}_i$ representing kinematic variables for three scales, instead of the set of original variables $\mathbf{r}_\alpha, \boldsymbol{\chi}_\alpha, \mathbf{S}_i$, which will be equivalent to each other. Thus, $\mathcal{L}_m = \mathcal{L}_m(\mathbf{r}_\alpha, \boldsymbol{\phi}_\alpha, \mathbf{S}_i)$. The equations of motion are stated as

$$\frac{d}{dt}\frac{\partial \mathcal{L}_m}{\partial \dot{\mathbf{r}}_\alpha} - \frac{\partial \mathcal{L}_m}{\partial \mathbf{r}_\alpha} = 0; \quad \frac{d}{dt}\frac{\partial \mathcal{L}_m}{\partial \dot{\boldsymbol{\phi}}_\alpha} - \frac{\partial \mathcal{L}_\alpha}{\partial \boldsymbol{\phi}_\alpha} = 0, \text{ and } \frac{d}{dt}\frac{\partial \mathcal{L}_m}{\partial \dot{\mathbf{S}}_i} - \frac{\partial \mathcal{L}_\alpha}{\partial \mathbf{S}_i} = 0. \quad (17.28)$$

17.2 Multiscale Micromorphic MD

In this section, we derive and discuss the dynamic equations for MMMD molecular dynamics.

17.2.1 Coarse-Scale Dynamic Equation

We start by deriving some relations that are needed in subsequent derivations. Considering

$$\mathbf{F}_\beta = \mathbf{F}_\beta(\{\mathbf{r}_\alpha\}) \text{ and } \dot{\mathbf{F}}_\beta = \dot{\mathbf{F}}_\beta(\{\mathbf{r}_\alpha\}, \{\dot{\mathbf{r}}_\alpha\}), \quad (17.29)$$

we then have

$$\dot{\mathbf{F}}_\beta = \sum_\alpha \frac{\partial \mathbf{F}_\beta}{\partial \mathbf{r}_\alpha} \dot{\mathbf{r}}_\alpha, \quad (17.30)$$

which leads to the relation,

$$\frac{\partial \dot{\mathbf{F}}_\beta}{\partial \dot{\mathbf{r}}_\alpha} = \frac{\partial \mathbf{F}_\beta}{\partial \mathbf{r}_\alpha}. \quad (17.31)$$

Moreover,

$$\ddot{\mathbf{F}}_\beta = \sum_\alpha \left(\frac{d}{dt}\left(\frac{\partial \mathbf{F}_\beta}{\partial \mathbf{r}_\alpha}\right) \dot{\mathbf{r}}_\alpha + \frac{\partial \mathbf{F}_\beta}{\partial \mathbf{r}_\alpha} \ddot{\mathbf{r}}_\alpha \right), \quad (17.32)$$

on the other hand,

$$\ddot{\mathbf{F}}_\beta = \sum_\alpha \left(\frac{\partial \dot{\mathbf{F}}_\beta}{\partial \mathbf{r}_\alpha} \dot{\mathbf{r}}_\alpha + \frac{\partial \dot{\mathbf{F}}_\beta}{\partial \dot{\mathbf{r}}_\alpha} \ddot{\mathbf{r}}_\alpha \right). \quad (17.33)$$

Comparing Eqs. (17.32) and (17.33) and utilizing Eq. (17.31), we obtain

$$\frac{d}{dt}\left(\frac{\partial \mathbf{F}_\beta}{\partial \mathbf{r}_\alpha}\right) = \frac{\partial \dot{\mathbf{F}}_\beta}{\partial \mathbf{r}_\alpha}. \quad (17.34)$$

Furthermore, considering,

$$\dot{\phi}_\beta = \dot{F}_\beta \chi_\beta + F_\beta \cdot \dot{\chi}_\beta, \tag{17.35}$$

we can obtain,

$$\frac{\partial \dot{\phi}_\beta}{\partial \dot{r}_\alpha} = \frac{\partial \dot{F}_\beta}{\partial \dot{r}_\alpha} \chi_\beta + \frac{\partial F_\beta}{\partial \dot{r}_\alpha} \dot{\chi}_\beta = \frac{\partial F_\beta}{\partial r_\alpha} \chi_\beta = \frac{\partial \phi_\beta}{\partial r_\alpha}. \tag{17.36}$$

According to Eq. (17.29), F_β is independent from the velocity field, and we have

$$\frac{\partial F_\beta}{\partial \dot{r}_\alpha} = 0.$$

By virtue of Eqs. (17.34) $\sim$ (17.36), we find that

$$\frac{\partial \dot{\phi}_\beta}{\partial r_\alpha} = \frac{\partial \dot{F}_\beta}{\partial r_\alpha} \chi_\beta + \frac{\partial F_\beta}{\partial r_\alpha} \dot{\chi}_\beta = \frac{\partial F_\beta}{\partial r_\alpha} \dot{\chi}_\beta + \frac{d}{dt}\left(\frac{\partial F_\beta}{\partial r_\alpha}\right) \chi_\beta$$

$$= \frac{d}{dt}\left(\frac{\partial \phi_\beta}{\partial r_\alpha}\right) = \frac{d}{dt}\left(\frac{\partial \dot{\phi}_\beta}{\partial \dot{r}_\alpha}\right). \tag{17.37}$$

This relation is needed in the subsequent derivation. Considering the coarse-scale Lagrangian equation and utilizing the above relation, we can derive

$$\frac{d}{dt}\left(\frac{\partial \mathcal{L}_m}{\partial \dot{r}_\alpha}\right) = \frac{d}{dt}\left(M_\alpha \dot{r}_\alpha + \sum_\beta \frac{\partial \mathcal{L}_m}{\partial \dot{\phi}_\beta} \cdot \frac{\partial \dot{\phi}_\beta}{\partial \dot{r}_\alpha}\right)$$

$$= M_\alpha \ddot{r}_\alpha + \sum_\beta \frac{d}{dt}\left(\frac{\partial \mathcal{L}_m}{\partial \dot{\phi}_\beta}\right) \cdot \frac{\partial \dot{\phi}_\beta}{\partial \dot{r}_\alpha} + \sum_\beta \frac{\partial \mathcal{L}_m}{\partial \dot{\phi}_\beta} \cdot \frac{d}{dt}\left(\frac{\partial \dot{\phi}_\beta}{\partial \dot{r}_\alpha}\right)$$

$$= M_\alpha \ddot{r}_\alpha + \sum_\beta \frac{\partial \mathcal{L}_m}{\partial \phi_\beta} \cdot \frac{\partial \phi_\beta}{\partial r_\alpha} + \sum_\beta \frac{\partial \mathcal{L}_m}{\partial \dot{\phi}_\beta} \cdot \frac{d}{dt}\left(\frac{\partial \phi_\beta}{\partial r_\alpha}\right)$$

$$= M_\alpha \ddot{r}_\alpha + \sum_\beta \frac{\partial \mathcal{L}_m}{\partial \phi_\beta} \cdot \frac{\partial \phi_\beta}{\partial r_\alpha} + \sum_\beta \frac{\partial \mathcal{L}_m}{\partial \dot{\phi}_\beta} \cdot \frac{\partial \dot{\phi}_\beta}{\partial r_\alpha}, \tag{17.38}$$

and furthermore,

$$\frac{\partial \mathcal{L}_m}{\partial r_\alpha} = \sum_{\beta \neq \alpha} \sum_{i \in \alpha, j \in \beta} V'(r_{ij}) \frac{r_{ij}}{|r_{ij}|} - \sum_{i \in \alpha} b_i + \sum_\beta \frac{\partial \mathcal{L}_m}{\partial \phi_\beta} \cdot \frac{\partial \phi_\beta}{\partial r_\alpha} + \sum_\beta \frac{\partial \mathcal{L}_m}{\partial \dot{\phi}_\beta} \cdot \frac{\partial \dot{\phi}_\beta}{\partial r_\alpha}. \tag{17.39}$$

Substituting Eqs. (17.38) and (17.39) into the coarse-scale Lagrangian equation, we derive and recover the classical molecular dynamics equations,

$$\frac{d}{dt}\left(\frac{\partial \mathcal{L}_m}{\partial \dot{r}_\alpha}\right) - \frac{\partial \mathcal{L}_m}{\partial r_\alpha} = 0 \rightarrow M_\alpha \ddot{r}_\alpha = \sum_{\beta \neq \alpha} \sum_{i \in \alpha, j \in \beta} V'(r_{ij}) \frac{r_{ij}}{|r_{ij}|} + \mathcal{B}_\alpha, \tag{17.40}$$

where $\mathcal{B}_\alpha = \sum_i \mathbf{b}_i$. Note that $-V'(r_{ij})\frac{\mathbf{r}_{ij}}{|\mathbf{r}_{ij}|} = \mathbf{f}_{ij}$ is the pair force between atoms. If one would like to include the macroscale traction boundary condition as indicated in Fig. 17.1, one needs to add another term on the center of mass of the boundary cells. In this case, we can redefine the external force density as,

$$\mathcal{B}_\alpha = S_\alpha \bar{\mathbf{t}}_\alpha + \Omega_\alpha \bar{\mathbf{b}}_\alpha, \quad (17.41)$$

where S_α is the area of macroscale traction boundary of αth cell, $\bar{\mathbf{t}}_\alpha$ is the traction vector on S_α, and $\bar{\mathbf{b}}_\alpha = \sum_i \mathbf{b}_i / \Omega_\alpha$.

Then, we can rewrite the above equation as,

$$M_\alpha \ddot{\mathbf{r}}_\alpha = -\sum_{\beta \neq \alpha} \sum_{i \in \alpha, j \in \beta} \mathbf{f}_{ij} + \mathcal{B}_\alpha. \quad (17.42)$$

The first term in the right-hand side (RHS) of Eq. (17.42) represents the cell-to-cell interactions. We can also rewrite the first term of RHS of Eq. (17.42) as

$$\sum_{\beta \neq \alpha} \sum_{i \in \alpha, j \in \beta} \mathbf{f}_{ij} = \sum_{\beta \neq \alpha} \mathbf{f}_{\alpha\beta}, \text{ where } \mathbf{f}_{\alpha\beta} := \sum_{i \in \alpha} \sum_{j \in \beta} \mathbf{f}_{ij}. \quad (17.43)$$

Then, Eq. (17.42) can be written in terms of coarse grain variables completely,

$$M_\alpha \ddot{\mathbf{r}}_\alpha = -\sum_{\beta \neq \alpha} \mathbf{f}_{\alpha\beta} + \mathcal{B}_\alpha, \quad (17.44)$$

where the Greek letters α, β denote the centers of mass of MD cells.

In Eq. (17.43), the cell-to-cell interaction term consists of all the atomic bond forces that pass through the surface of the cell, which is actually the divergence of the Cauchy stress ($\nabla \cdot \sigma_\alpha$) generated by the atomic interaction from outside the cell multiplied by the volume Ω_α. This can be seen as follows:

$$-\sum_{\beta \neq \alpha} \sum_{i \in \alpha, j \in \beta} \mathbf{f}_{ij} \approx \int_{\partial \Omega_\alpha} \sigma \cdot \mathbf{n} dS = \int_{\Omega_\alpha} \nabla \cdot \sigma dV \approx \nabla \cdot \sigma_\alpha \Omega_\alpha.$$

Thus, Eq. (17.42) may be interpreted as,

$$M_\alpha \ddot{\mathbf{r}}_\alpha = (\nabla \cdot \sigma_\alpha) \Omega_\alpha + \mathcal{B}_\alpha, \quad (17.45)$$

which is a coarse scale balance of linear momentum. Equation (17.45) not only establishes a link from MD to continuum mechanics, but also provides an alternative computational formulation, which is explained as follows.

First, we can rewrite the divergence of the Cauchy stress as

$$\Omega_\alpha \nabla_r \cdot \sigma_\alpha = \tilde{\Omega}_\alpha J \nabla_\mathcal{R} \cdot \sigma_\alpha \left(\frac{\partial \mathcal{R}}{\partial \mathbf{r}}\right) = \tilde{\Omega}_\alpha \left(\nabla_\mathcal{R} \cdot \left(J \sigma_\alpha \mathbf{F}_\alpha^{-T}\right) - \sigma_\alpha \cdot \nabla_\mathcal{R} \cdot (J \mathbf{F}_\alpha^{-T})\right)$$

$$= \tilde{\Omega}_\alpha \nabla_\mathcal{R} \cdot \left(J \sigma_\alpha \mathbf{F}_\alpha^{-T}\right) = \tilde{\Omega}_\alpha \nabla_\mathcal{R} \cdot \mathbf{P}_\alpha,$$

where $\tilde{\Omega}_\alpha$ is the volume of the MD cell in the intermediate configuration, i.e., $\mathcal{R}$-configuration and $J := det(\mathbf{F}_\alpha) = \Omega_\alpha/\tilde{\Omega}_\alpha$. In the above derivation, the Piola identity $\nabla_\mathcal{R} \cdot (J\mathbf{F}^{-T}) = 0$ is used, and moreover $\mathcal{P}_\alpha = J\sigma_\alpha \mathbf{F}_\alpha^{-T}$ is the first Piola–Kirchhoff stress with respect to the current configuration (r-configuration) and the intermediate referential configuration ($\mathcal{R}$-configuration), which can be shown as

$$\mathcal{P}_\alpha = \mathcal{P}_\alpha^{Virial} = \frac{1}{\tilde{\Omega}_\alpha} \sum_{i \in \alpha} \left(-\phi_\alpha m_i \dot{\mathbf{S}}_i \otimes \dot{\mathbf{S}}_i + \frac{1}{2} \sum_{j \in \alpha, j \neq i} \mathbf{f}_{ij} \otimes \mathbf{S}_{ij} \right) \cdot \chi_\alpha^T, \quad (17.46)$$

when the equilibrium state is reached. The detailed derivation of Eq. (17.46) will be discussed in the subsequent sections, and our purpose here is to derive the macroscale dynamic equation.

Now, we can rewrite Eq. (17.45) as

$$M_\alpha \ddot{\mathbf{r}}_\alpha = \tilde{\Omega}_\alpha \nabla_\mathcal{R} \cdot \mathcal{P}_\alpha + \mathcal{B}_\alpha. \quad (17.47)$$

Recall that the coarse-scale deformation gradient may be approximated as Eq. (17.9),

$$\mathbf{F}_\alpha = \nabla_\mathcal{R} \otimes \mathbf{r}_\alpha \approx \left(\sum_{\beta=1}^{N_h} \omega(|\mathbf{R}_{\alpha\beta}|) \mathbf{r}_{\alpha\beta} \otimes \mathbf{R}_{\alpha\beta} \tilde{\Omega}_\beta \right) \cdot \mathbf{M}_\alpha^{-1}, \quad (17.48)$$

which indicates that the structure of the discrete differential operator may be expressed as follows:

$$\nabla_\mathcal{R} \otimes (\mathbf{f}_\alpha) \approx \left(\sum_{\beta=1}^{N_h} \omega(|\mathbf{R}_{\alpha\beta}|)(\mathbf{f}_\beta - \mathbf{f}_\alpha) \otimes \mathbf{R}_{\alpha\beta} \tilde{\Omega}_\beta \right) \cdot \mathbf{M}_\alpha^{-1}.$$

Following the same procedure, we can define the discrete gradient operator as,

$$\nabla_\mathcal{R} \cdot (\mathbf{f}_\alpha) \approx \left(\sum_{\beta=1}^{N_h} \omega(|\mathbf{R}_{\alpha\beta}|)(\mathbf{f}_\beta - \mathbf{f}_\alpha) \cdot \mathbf{R}_{\alpha\beta} \tilde{\Omega}_\beta \right) \cdot \mathbf{M}_\alpha^{-1}. \quad (17.49)$$

Note that the center of mass positions, $\mathbf{R}_\alpha$ or $\mathbf{R}_\beta$, are the same in the intermediate configuration ($\mathcal{R}$) as well as in the referential configuration ($\mathbf{R}$).

Finally, we can approximate Eq. (17.47) as

$$M_\alpha \ddot{\mathbf{r}}_\alpha \approx \left(\sum_{\beta=1}^{N_h} \omega(|\mathbf{R}_{\alpha\beta}|)(\mathcal{P}_\beta - \mathcal{P}_\alpha) \cdot \mathbf{R}_{\alpha\beta} \tilde{\Omega}_\beta \right) \cdot \mathbf{M}_\alpha^{-1} + \mathcal{B}_\alpha. \quad (17.50)$$

This is an alternative computation formulation in comparison to Eq. (17.42), because one does not need to calculate interactions among different cells but only calculate the virial stress in each cell. This formulation provides a natural passage bridging MD and continuum mechanics coupling.

17.2.2 Mesoscale Dynamic Equations

We now exam, the mesoscale Lagrangian equation. Considering the derivative terms with respect to the chosen mesoscale variable, i.e., $\dot{\boldsymbol{\phi}}_\alpha$ and $\boldsymbol{\phi}_\alpha$, we can first write,

$$\frac{\partial \mathcal{L}_m}{\partial \dot{\boldsymbol{\phi}}_\alpha} = \dot{\boldsymbol{\phi}}_\alpha \cdot \mathbf{J}_\alpha^S \rightarrow \frac{d}{dt}\left(\frac{\partial \mathcal{L}_m}{\partial \dot{\boldsymbol{\phi}}_\alpha}\right) = \frac{d}{dt}(\dot{\boldsymbol{\phi}}_\alpha \cdot \mathbf{J}_\alpha^S) = \ddot{\boldsymbol{\phi}}_\alpha \cdot \mathbf{J}_\alpha^S, \quad (17.51)$$

and then we have,

$$\frac{\partial \mathcal{L}_m}{\partial \boldsymbol{\phi}_\alpha} = \frac{1}{2}\sum_i m_i \dot{\mathbf{S}}_i \frac{\partial \mathbf{C}_\alpha}{\partial \boldsymbol{\phi}_\alpha} \dot{\mathbf{S}}_i - \frac{1}{2}\sum_{i,j\in\alpha} V'(r_{ij})\frac{\mathbf{r}_{ij}}{|\mathbf{r}_{ij}|} \cdot \frac{\partial \mathbf{r}_{ij}}{\partial \boldsymbol{\phi}_\alpha}$$
$$- \sum_\beta \sum_{\substack{i\in\alpha \\ j\in\beta\neq\alpha}} V'(r_{ij})\frac{\mathbf{r}_{ij}}{|\mathbf{r}_{ij}|} \cdot \frac{\partial \mathbf{r}_{ij}}{\partial \boldsymbol{\phi}_\alpha} - \sum_{i\in\alpha} \mathbf{b}_i \cdot \frac{\partial \mathbf{r}_i}{\partial \boldsymbol{\phi}_\alpha}$$
$$= \dot{\boldsymbol{\phi}}_\alpha \sum_i m_i \dot{\mathbf{S}}_i \otimes \dot{\mathbf{S}}_i - \frac{1}{2}\sum_{i,j\in\alpha} \mathbf{f}_{ij} \otimes \mathbf{S}_{ij} + \sum_\beta \sum_{\substack{i\in\alpha \\ j\in\beta\neq\alpha}} \mathbf{f}_{ij} \otimes \mathbf{S}_i + \sum_{i\in\alpha} \mathbf{b}_i \otimes \mathbf{S}_i. \quad (17.52)$$

Finally, the mesoscale Lagrangian equations have the form,

$$\ddot{\boldsymbol{\phi}}_\alpha \cdot \mathbf{J}_\alpha^S = \dot{\boldsymbol{\phi}}_\alpha \sum_i m_i \dot{\mathbf{S}}_i \otimes \dot{\mathbf{S}}_i - \frac{1}{2}\sum_{i,j\in\alpha} \mathbf{f}_{ij} \otimes \mathbf{S}_{ij} + \sum_\beta \sum_{\substack{i\in\alpha \\ j\in\beta\neq\alpha}} \mathbf{f}_{ij} \otimes \mathbf{S}_{ij} + \sum_{i\in\alpha} \mathbf{b}_i \otimes \mathbf{S}_i. \quad (17.53)$$

To understand physical meanings of the above equation, we can define the mesoscale first Piola–Kirchhoff stress tensors as,

$$\mathbf{P}_\alpha^{Virial} := \sum_{i\in\alpha}\left(-\dot{\boldsymbol{\phi}}_\alpha m_i \dot{\mathbf{S}}_i \otimes \dot{\mathbf{S}}_i + \frac{1}{2}\sum_{j\in\alpha, j\neq i} \mathbf{f}_{ij} \otimes \mathbf{S}_{ij}\right) \quad (17.54)$$

$$\mathbf{P}_\alpha^{ext} = \sum_{\beta\neq\alpha}\sum_{i\in\alpha, j\in\beta} \mathbf{f}_{ij} \otimes \mathbf{S}_{ij} = \boldsymbol{\phi}_\alpha \cdot \left(\sum_{\beta\neq\alpha}\sum_{i\in\alpha, j\in\beta} V'(r_{ij})\frac{\mathbf{S}_{ij}\otimes\mathbf{S}_{ij}}{r_{ij}}\right), \quad (17.55)$$

where $\mathbf{P}_\alpha^{Virial}$ and $\mathbf{P}_\alpha^{ext}$ are the first Piola–Kirchhoff stress of the αth cell, which are the two-point tensors defined on the current configuration, i.e., **r**-configuration and the statistical configuration, i.e., the **S**-configuration.

Figuratively speaking, these tensors have two legs, one is in **r**-configuration and the other is in **S**-configuration. If we push the leg in **S**-configuration forward to the intermediate configuration $\mathcal{R}$-configuration, i.e., $\mathcal{R}_i = \chi \cdot \mathbf{S}_i$, the corresponding first Piola–Kirchhoff stress will have the following expression:

$$\mathcal{P}_\alpha^{Virial} = \frac{1}{\tilde{\Omega}_\alpha}\sum_{i\in\alpha}\left(-\dot{\boldsymbol{\phi}}_\alpha m_i \dot{\mathbf{S}}_i \otimes \dot{\mathbf{S}}_i + \frac{1}{2}\sum_{j\in\alpha, j\neq i} \mathbf{f}_{ij} \otimes \mathbf{S}_{ij}\right) \cdot \chi_\alpha^T, \quad (17.56)$$

which is used in Eq. (17.46).

With above definitions of stresses, we can recast the mesoscale dynamics equations as

$$\ddot{\boldsymbol{\phi}}_\alpha \cdot \mathbf{J}_\alpha^S = -\left(\mathbf{P}_\alpha^{Virial} - \mathbf{P}_\alpha^{ext}\right) + \mathbf{M}_\alpha, \tag{17.57}$$

where $\mathbf{P}_\alpha^{Virial}$ is given by Eq. (17.54), $\mathbf{P}_\alpha^{ext}$ is given by Eq. (17.55), and $\mathbf{M}_\alpha = \sum_{i\in\alpha} \mathbf{b}_i \otimes \mathbf{S}_i$ is the mesoscale external couple. Note that Eqs. (17.54) and (17.55) are insightful, because it resolves one of the outstanding debates on the definition of the virial stress. Equation (17.54) is basically the mathematical definition of the virial stress, see, e.g., Irving and Kirkwood (1950) and Tsai (1979). However, Zhou (2003) argued that the kinetic energy part should be dropped out in the stress calculation, even though many disagreed, e.g., Murdoch (2007), Subramaniyan and Sun (2008). We now see from Eqs. (17.54) and (17.55) that if the stress is internally generated, the definition of the virial stress is the original definition of the virial stress, but if the stress is an external stress, then the kinetic energy part should drop out from its expression. This is because that the current formulation of the MMMD is a adiabatic formulation, which does not consider the heat exchange among the cells. If in Eq. (17.57), $\ddot{\boldsymbol{\phi}}_\alpha = 0$, we have $\mathbf{P}_\alpha^{ext} = \mathbf{P}_\alpha^{Virial} \rightarrow \mathcal{P}_\alpha^{ext} = \mathcal{P}_\alpha^{Virial}$, which is the proof of Eq. (17.46).

17.2.3 Microscale Dynamic Equations

After evaluating the fine-scale Lagrange equation for $i \in \alpha$, we first have

$$\frac{d}{dt}\frac{\partial \mathcal{L}_m}{\partial \dot{\mathbf{S}}_i} = m_i \left(\mathbf{C}_\alpha \ddot{\mathbf{S}}_i + \dot{\mathbf{C}}_\alpha \cdot \dot{\mathbf{S}}_i\right). \tag{17.58}$$

The derivative of fine-scale Lagrange equation with respect to $\mathbf{S}_i$ has two cases,

$$(a)\ \alpha = \beta: \quad \frac{\partial \mathcal{L}_m}{\partial \mathbf{S}_i} = -\frac{1}{2}\sum_{j\neq i}\left(\frac{V'(r_{ij})}{r_{ij}}\mathbf{C}_\alpha \cdot \mathbf{S}_{ij}\right)$$

$$(b)\ \alpha \neq \beta: \quad \frac{\partial \mathcal{L}_m}{\partial \mathbf{S}_i} = -\frac{1}{2}\sum_{\alpha\neq\beta}\sum_{j\neq i}\left(\frac{V'(r_{ij})}{r_{ij}}\boldsymbol{\phi}_\alpha^T \cdot \mathbf{r}_{ij}\right),$$

where $\mathbf{r}_{ij} = \mathbf{r}_{\alpha\beta} + \boldsymbol{\phi}_\beta \cdot \mathbf{S}_j - \boldsymbol{\phi}_\alpha \cdot \mathbf{S}_i$.

Combining the above two equations and Eq. (17.58), we finally obtain

$$m_i \ddot{\mathbf{S}}_i + \frac{1}{2}\boldsymbol{\phi}_\alpha^{-1}\sum_\beta\sum_{i\neq j}\left(\frac{V'(r_{ij})}{r_{ij}}(\mathbf{r}_{\alpha\beta} + \boldsymbol{\phi}_\beta \cdot \mathbf{S}_j - \boldsymbol{\phi}_\alpha \cdot \mathbf{S}_i)\right)$$
$$+ m_i \mathbf{C}_\alpha^{-1} \dot{\mathbf{C}}_\alpha \cdot \dot{\mathbf{S}}_i + \boldsymbol{\phi}_\alpha^{-1} \cdot \mathbf{b}_i = 0, \tag{17.59}$$

where $i \in \alpha$. One may note that the second term in Eq. (17.59) contains both interaction of atoms within the αth cell and between two different cells, i.e., the case $i \in \alpha$, $j \in \beta$, and $\beta \neq \alpha$.

In summary, the three scale governing equations of MMMD are as follows:

$$M_\alpha \ddot{\bar{\mathbf{r}}}_\alpha = -\sum_{\beta \neq \alpha} \sum_{i \in \alpha, j \in \beta} \mathbf{f}_{ij} + S_\alpha \bar{\mathbf{t}}_\alpha + \Omega_\alpha \bar{\mathbf{b}}_\alpha, \qquad (17.60)$$

$$\ddot{\boldsymbol{\phi}}_\alpha \cdot \mathbf{J}_\alpha^S = -(\mathbf{P}_\alpha^{Virial} - \mathbf{P}_\alpha^{ext}) + \mathbf{M}_\alpha, \qquad (17.61)$$

$$m_i \ddot{\mathbf{S}}_i = -m_i \mathbf{C}_\alpha^{-1} \cdot \dot{\mathbf{C}}_\alpha \cdot \dot{\mathbf{S}}_i + \boldsymbol{\phi}_\alpha^{-1} \left(\sum_\beta \sum_{j \in \beta \neq i \in \alpha} \mathbf{f}_{ji} + \mathbf{b}_i \right), \qquad (17.62)$$

where the micromorphic deformation tensor is $\boldsymbol{\phi}_\alpha = \mathbf{F}_\alpha \cdot \boldsymbol{\chi}_\alpha$. One can see from Eq. (17.60) that the macroscale traction boundary condition is embedded in the coarse-scale dynamic equation.

The MMMD derived here was inspired by the Anderson–Parrinello–Rahman MD (APR-MD) (Andersen (1980); Parrinello and Rahman (1980), (1981)), in which the supercell or unit cell of atoms is subjected to periodic boundary condition. Whereas in MMMD framework, different from APR-MD, the supercell overall is free to move without periodic restriction. Moreover, MMMD can take into account both displacement and traction boundary conditions at macroscale.

17.3 Numerical Examples

In this section, we present two examples that simulate the structure phase transition of single crystal Nickel (Ni) and Iron (Fe).

17.3.1 Structure Phase Transition of Single Crystal Ni

In the first example, we apply MMMD to simulate structure phase transition of Ni, which was also used as the validation example for PR MD (Parrinello and Rahman (1981)). Different from Parrinello and Rahman (1981), the validation test is done in a finite size specimen without imposing periodic boundary condition.

By imposing an uniaxial compression load, the original FCC lattice of single crystal Ni will go through structure change (see Milstein and Farber (1980)), as shown in Fig. 17.5. The compression load is applied to the centers of mass of the boundary MD cells. The interaction between atoms is modeled by the Morse potential,

$$\phi(r) = D(e^{-2\alpha(r-r_0)} - 2e^{-\alpha(r-r_0)}). \qquad (17.63)$$

The interaction force is given by

$$F(r) = -\frac{\partial \phi(r)}{\partial r} = 2D\alpha(-e^{-2\alpha(r-r_0)} + e^{-\alpha(r-r_0)}), \qquad (17.64)$$

with the parameter constants chosen as $D = 3.5059 \times 10^{-20}$ J, $\alpha = 8.766/a_0$, and $r_0 = 0.71727$Å, where a_0 denotes the constants of the FCC lattice of Ni, i.e., $a_0 = 3.52$Å (Parrinello and Rahman 1981), with which the single crystal Ni particle

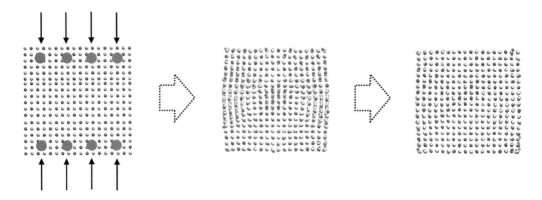

Figure 17.5 Uniaxial compression of a finite size (nanoscale) Nickel particle: The external load is applied by using a macroscale boundary condition, i.e., prescribing the displacement of the centers of mass of the boundary cells (dark dots)

is in a global elastic energy minimum state. Therefore, the initial starting point is a stable equilibrium state, which is internal stress free. The atomic weight of Ni atom is 58.69 u.

For the microscale calculation, the reversible NVT integrator (Martyna et al. (1996)) was implemented for time integration, and temperature was controlled by using the Nosé–Hoover thermostat (Nosé (1984); Hoover (1985)). In the case of the mesoscale computation, the Gear six-order predictor–corrector (Frenkel and Smit (1996)) was used. Contrarily, coarse-scale integration was accomplished by using velocity Verlet algorithm with velocity scaling method as thermostat (Frenkel and Smit 1996). Time intervals for microscale, mesoscale, and coarse scale are set to 0.1, 1.0, and 10.0 [fs], respectively.

In the simulation, we set the temperature higher than room temperature to create an environment favorable to structure phase transition. The temperature is controlled by rescaling the atom velocity in each MD cell at every micro time step, so that the kinetic temperature is controlled at 350°K. Before the calculation, random perturbation of positions and velocities were assigned to each atom to ensure an equilibrium state at the desired temperature. In simulations, we have constructed three different cell partitions: (1) each micromorphic cell contains $3 \times 3 \times 3 = 27$ unit cells; (2) each micromorphic cell contains $4 \times 4 \times 4 = 64$ unit cells, and (3) each micromorphic cell contains $5 \times 5 \times 5 = 125$ unit cells, and in each coordinate direction there are three or five micromorphic cell (supercell). We have two types of simulation systems: $3 \times 3 \times 3 = 27$ micromorphic cells and $5 \times 5 \times 5 = 125$ micromorphic cells. For both systems, there is one internal micromorphic cell at the center of the simulation system, with which we investigate its phase transformation process. Since each micromorphic cell or supercell contains either 27, 64, or 125 FCC unit cells, the simulation model may have different numbers of FCC unit cells ranging from 729 to 15,625 FCC unit cells. We know that each unit cell of a FCC lattice has four atoms, hence the total number of atoms in the whole simulation system ranges from 2916 atoms to 62,500 atoms.

In the simulation, we first prescribed displacement boundary condition on the centers of mass of the boundary cells by compressing the molecular system on [100] direction.

Subsequently, we control the coarse-scale stretch λ_1 at the loading boundary of the simulation system at the specified value. By doing so, we can gradually control the stretch state, λ_1, of all MD cells, at boundary as well as in the interior, to the value $\lambda_1 \approx 0.7$ by compressing the displacements of the COMs of the boundary cells. During the load steps, the other components of $\mathbf{F}_\alpha$ may change according to complicated atomistic interaction, and as the prescribed boundary condition, we only control the principal stretch in [100] direction as the load parameter. Figure 17.6 displays the snapshots of deformation process of Ni block, and it also reveals the history of lattice structure change of the center MD cell. Note that the zoom-in picture is the center MD cell.

When the stretch is relatively small, i.e., $0.93 < \lambda_1 < 1$, the Ni block is going through some elastic deformation. The particles deform uniformly. This is even true for much larger stretch in infinite lattices. Since the MD simulation domain has finite size, boundary effects will affect the overall pattern of deformation. Therefore, when $\lambda \approx 0.925$, there is lattice glide or slip initiating at the lateral boundary of the system. It may be noted that the original PR MD cannot predict lattice slip when it simulates phase transform in an infinite lattice.

After slip or dislocation motion, the lattice structure start to change at boundary. As principal stretch decreases from $\lambda \approx 0.82$ to $\lambda \approx 0.7$, the structure will undergo dramatic structure change, and settles at a new stable state other than the initial configuration as shown in Figs. 17.6 and 17.7.

We plot the deformation sequence of MMMD simulation system in Figs. 17.6 and 17.7 to show the nonequilibrium transient process. The deformation and lattice structure change process in the center cell are shown in the small zoom-in box in Figs. 17.6 and 17.7.

It can be found that the center supercell is going through several stages of elastic deformation ($\lambda_{[100]} = 1 \sim 0.7$), and then some closely packed planes of {010} are gradually formed. When $\lambda_{[100]} = 0.9 \sim 0.8$, the relative positions of atoms are held between different planes. Finally, dislocations start to form between those closely packed planes, and we observed that some interplanar slips, in order to achieve a new stable position ($\lambda_{[100]} = 0.8 \sim 0.7$), and the stable lattice structure pattern is an HCP structure.

Zooming into the deformed lattice microstructure and comparing it with the initial lattice structure, we can see clearly that the lattice structure changes from FCC to HCP. The phase transition occurs when principal stretch reaches to the critical value: $\lambda_{[100]} \approx 0.7$, which is an optimal stretch to form a HCP lattice. Under the compression, the bond angle between two Ni atoms changes from $90°$ to $60°$, while the bond length hold fixed. The original {010} plane of FCC lattice turned into the closely packed plane {001} of HCP. Thus, the stretch can be calculated by $\lambda_{[100]} = a_0/\sqrt{2}a_0 \approx 0.7$ which is consistent with the result in our simulation.

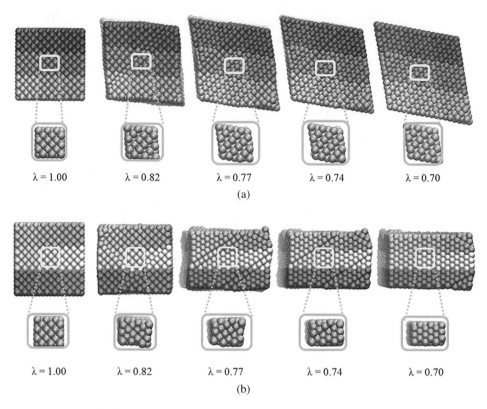

Figure 17.6 Deformation and structure transition history: (a) (100) plane (top view) and (b) (010) plane (side view) (Data source is from Li and Urata (2016)).

Figure 17.8 displays the supercell configurations and structures, which are represented by the total deformation gradient ϕ_α and the positions of the center of mass of the supercell $\mathbf{r}_\alpha$. It is possible to see visually that every cells change their shape from cubic to distorted diamond-shaped hexahedron with decreasing the stretch ratio λ. MMMD method enables us to recognize local deformation by the individual deformation tensor, that distinguished MMMD from conventional MD.

Different from the PR MD or APR MD simulation, which is an equilibrium ensemble MD simulation with the prescribed constant stress, MMMD is a type of nonequilibrium MD that can be used to calculate the nonuniform stress distribution at different locations. To illustrate the MMMD's capacity to capture nonequilibrium state of the MD system, we plot stress distribution in Fig. 17.9. Figure 17.9(a) shows the axial stress σ_{11} distribution in renormalized unit for each local cell, where the stress distribution is in (010) plane. However, this graph only provides a general trend of stress distribution. To obtain continuum stress distribution, a larger system with more cells is needed. To quantitatively assess the accuracy of MMMD simulation, we have studied the stress–strain relation of the center supercell cell, and compared it with both references (Milstein and Farber (1980); Parrinello and Rahman (1981)). The

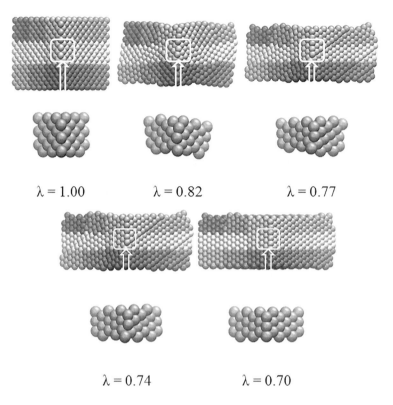

Figure 17.7 Deformation and structure transition history at the corner of (010) plane and (001) plane (along [100] axis) (Data source is from Li and Urata (2016)).

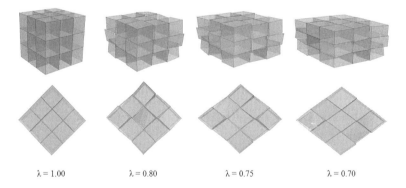

Figure 17.8 Visualization of cell structures during transition of single crystal Nickel (Data source is from Li and Urata (2016)).

quantitative comparisons with Milstein and Farber (1980) and Parrinello and Rahman (1981) are displayed in Fig. 17.10. Our simulation results show that the calculated critical stress value of the phase transition is almost the same as that of the infinite lattice, and it seems that MMMD provides more accurate prediction on the bifurcation of the primary and secondary load deflection paths than that of PR-MD.

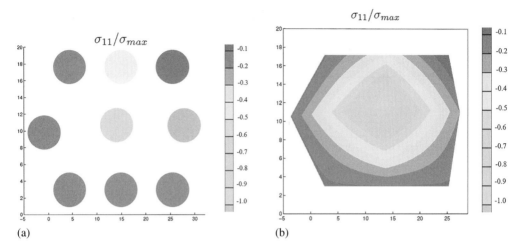

Figure 17.9 Nonuniform stress distribution (σ_{11}): (a) σ_{11} in each super cells at horizontal surface and (b) the smoothed global stress field (Data source is from Li and Urata (2016)).

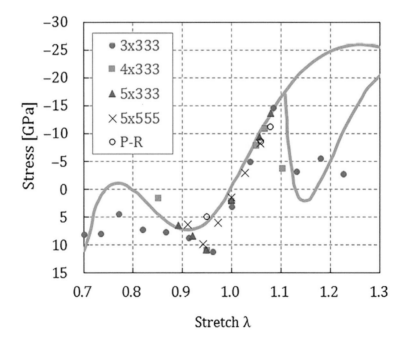

Figure 17.10 Load deflection path: σ_{11} vs. the principal stretch λ_1 for a nanoscale Nickel block under structure phase transition during compression–tension test (Data source is from Li and Urata (2016)).

Figure 17.10 also demonstrates the size effects of total system and supercell. Since all models show almost the same stress in the range from 0.9 to 1.1 of λ, it is concluded that the minimum supercell composed of $3 \times 3 \times 3$ unit cells is large enough to capture the phase transition.

17.3.2 Structure Phase Transition of Single Crystal Fe

In the second example, we apply MMMD to simulate structure phase transition of a single crystal BCC Fe to FCC structure under uniaxial loading. The structure phase transition of single crystal iron from BCC structure to FCC structure under the uniaxial tension in [0 0 1] direction is first reported by Clatterbuck et al. (2003) based on an observation from ab-initio calculations. This was later confirmed by Zeng (2011) by using a coarse-grained MD field theory.

In the MMMD simulation, we used the same geometry and the size that are reported in Zeng (2011). The test specimen is a single crystal iron nanorod, with two different sizes: (1) $3 \times 3 \times 11$ supercells and (2) $5 \times 5 \times 21$ supercells. Each supercell contains $3 \times 3 \times 3$ BCC unit cells with two atoms per unit cell, so that the total number of atoms are 5346 and 28,350, respectively. Both macroscale traction and macroscale displacement boundary conditions are used in the simulation, i.e., we prescribe either traction or displacement at the centers of mass of $2 \times 9 = 18$ and $2 \times 25 = 50$ supercells at the two ends of the Fe nanorod for $3 \times 3 \times 11$ supercells and (2) $5 \times 5 \times 21$ supercells, respectively.

In this simulation, we adopt the Finnis–Sinclair model (FSM) (1984) for material BCC iron. The potential energy of the FSM and the EAM has the following general form:

$$U = \frac{1}{2} \sum_{i=1}^{N} \sum_{j=1}^{N} V(r_{ij}) + \sum_{i=1}^{N} F(\rho_i), \qquad (17.65)$$

where $F(\rho_i)$ is a functional describing the energy of embedding an atom into the background electron cloud, and it is defined as

$$\rho_i = \sum_{j \neq i}^{N} \rho(r_{ij}), \ \mathbf{r}_{ij} = \mathbf{r}_i - \mathbf{r}_j, \ r_{ij} = |\mathbf{r}_{ij}|. \qquad (17.66)$$

The Finnis–Sinclair potential is defined as

$$V(r_{ij}) = (r_{ij} - c)^2 (c_0 + c_1 r_{ij} + c_2 r_{ij}^2), \ \rho(r_{ij})$$
$$= (r_{ij} - d)^2 + \beta \frac{(r_{ij} - d)^3}{d}, \ F(\rho_i) = -A \sqrt{\rho_i}, \qquad (17.67)$$

with parameters $c_0, c_1, c_2, c, A, d,$ and β taken from Finnis and Sinclair (1986). Note that both c and d are cutoff distances.

The results of MMMD simulation confirmed a similar result as shown in Fig. 17.11. In the case of traction force boundary condition, it can be observed that before the necking of nanorod both twinning and dislocation occurred as the tensile strain increased, and then FCC structure domain occurs at the necking region before the specimen breaks into two parts. On the other hand, clear phase transition was not observed in the simulation by using constant displacement boundary condition. To confirm size dependency, we also computed larger system using $5 \times 5 \times 21$ supercells with traction force boundary condition. Figure 17.12 clearly shows phase transition

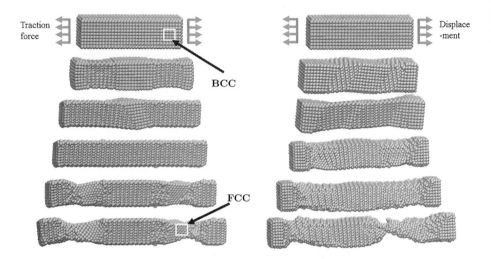

Figure 17.11 Snapshots of single crystal nanorod undergoes structure phase transition from BCC to FCC during uniaxial tension. $3 \times 3 \times 11$ supercells. (left) Traction force boundary condition; (right) displacement boundary condition (Data source is from Li and Urata (2016)).

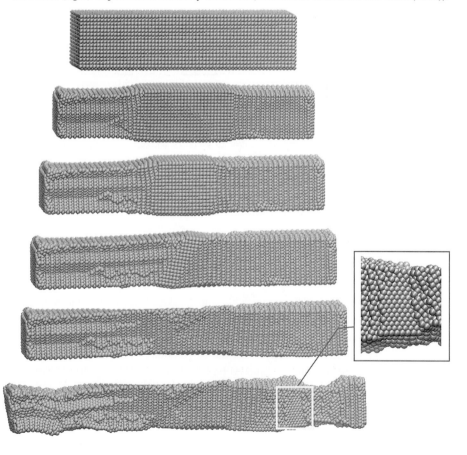

Figure 17.12 Snapshots of single crystal nanorod undergoes structure phase transition from BCC to FCC during uniaxial tension. $5 \times 5 \times 21$ supercells (Data source is from Li and Urata (2016)).

from BCC to FCC around the neck just before break off of Fe crystal. According to these results, we can judge an appropriate boundary condition is necessary to reproduce physical phenomena. It is relatively easy for MMMD to apply both displacement and traction force boundary conditions on the cell center position and it should be one of the advantages of this method.

17.4 Multiscale Coupling between MMMD and PD

In this section, we introduce the current multiscale coupling method between MMMD and PD. The following contents presented here are mainly based on the work of Tong and Li (2016, 2020).

This concurrent atomistic continuum two-scale model consists of three parts: an atomistic region, a continuum region, and a translation zone, which is used to pass the information between different scales, as shown in Fig. 17.13. The challenge of such type of concurrent multiscale model is the message passing, since the physical variables are interpreted differently at different scales, e.g., the material displacement

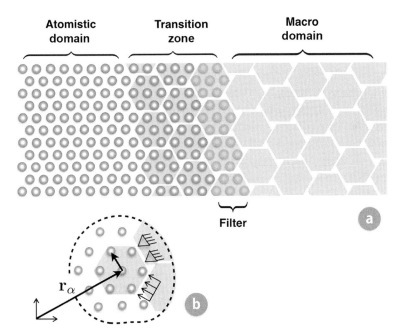

Figure 17.13 (a) The multiscale model consists of three part: atomistic region, macro region, and transition zone. The essential part is the transition zone, which serves as a messenger to translate information from both regions. A filter is constructed near the interface to solve the issue of high-frequency wave reflection. (b) The adaptive multiscale element in the transition zone. This element is an assemble of atoms, which has macroscale properties such as shape and average displacement while the atomistic resolution is retained. The element is capable of carrying and translating information from different scales.

is measured by atomistic positions and trajectories in microscale but continuum displacement field in macroscale. The simplest treatment is to glue the atoms on the continuum body in the transition zone, to let them move together, which is called "hand-shaking" (see Abraham et al. (1998); Tang et al. (2006)). However, the difficulty of message transition is where nonuniform atomistic motion such as crystal defects or phase transformation exists. when a continuum element in the transition zone is subject to compressive or shear stress, the underlying crystal lattice might go through phase transition or lattice glide. The "hand-shaking" approach is not able to pass such information from one scale to another. In this regard, (Tong and Li (2016), (2020)) developed a concurrent multiscale coupling approach that utilizes the advantage of MMMD to a transition zone that can smoothly pass the information from one scale to another scale. In the following sections, we shall introduce this multiscale method.

For the purpose of better presentation, we first recall and summarize the MMMD derived in the previous sections.

17.4.1 Micromorphic Multiscale MD

As shown in Fig. 17.13(a), between the fine-scale MD region and the coarse-scale PD region there is a transition zone. The transition zone has atomistic resolution same as atomistic region, however, we divide atoms in this zone into finite number of supercells. Each supercell may be viewed as a material point at macroscale, and it has a shape as an assemble of atoms. Furthermore, each atom inside the supercell is free to move as the internal degree of freedom. Having the same geometric property of the underlined lattice structure, the supercell is able to describe both mesoscale and macroscale mechanical motions such as deformation and cell-level displacement. Therefore, it is possible to apply associated macroscale force field such as stress on the supercell. On the other hand, with the atomistic resolution, all information of atomistic scale is retained. As a consequence, the supercell has multiscale structure and property. A detailed extensive discussion of MMMD and how to choose the supercell size can be found in Li and Tong (2015) and Tong and Li (2015).

To describe the global motion, the center of mass of the supercell is introduced in addition to the deformation gradient and the internal degrees of freedom. Therefore, as shown in Fig. 17.13(b), the atomistic position $\mathbf{r}_i(t)$ in the current configuration is decomposed as:

$$\mathbf{r}_i(t) = \mathbf{r}_\alpha(t) + \boldsymbol{\phi}_\alpha(t) \cdot \mathbf{S}_i(t), \qquad (17.68)$$

where $\mathbf{r}_\alpha$ is the center of mass of αth cell is calculated as,

$$\mathbf{r}_\alpha = \frac{\sum_{i \in S_\alpha} m_i \mathbf{r}_i}{\sum_{i \in S_\alpha} m_i}, \qquad (17.69)$$

with m_i as the mass of ith atom inside the αth cell and S_α is the index set of all atoms inside the αth cell.

From the perspective of a single cell, the motion of the center of mass may represent its rigid body translation of the supercell. However, the aggregated motion of all centers of mass of every supercells describes the coarse-scale deformation at the continuum level. In Eq. (17.68), $\boldsymbol{\phi}_\alpha$ is the total deformation gradient of the αth cell and is uniform throughout the cell. $\mathbf{S}_i$ is the internal degree of freedom, which represents the atomistic distribution inside the cell. $\boldsymbol{\phi}_\alpha \cdot \mathbf{S}_i$ is the relative position compared to the center of mass. This operation is different from the uniform deformation described by the Cauchy–Born rule, and it is a multiscale micromorphic deformation (Li and Tong (2015)). In fact, $\boldsymbol{\phi}_\alpha$ can be further decomposed to

$$\boldsymbol{\phi}_\alpha(t) = \mathbf{F}_\alpha(t) \cdot \boldsymbol{\chi}_\alpha(t), \tag{17.70}$$

where $\mathbf{F}_\alpha$ is related to macroscale continuum deformation and depending on the distribution or the aggregated motion of centers of mass of supercells (Tong and Li (2015)), whereas $\boldsymbol{\chi}_\alpha$ is an independent mesoscale deformation tensor for each supercell (or for the representative αth supercell), which includes local stretch and local rotation. By introducing the center of mass and deformation gradient for supercells, each supercell obtains the properties of a material point in macroscale continuum mechanics. The internal degrees of freedom then enables the interaction between particles from atomistic domain.

The equation of motion for three quantities is then derived based on the Lagrangian of each supercell, where the potential energy is written as

$$V_\alpha = \sum_{i \in \alpha, j \notin \alpha} \varphi(r_{ij}) + \frac{1}{2} \sum_{i, j \in \alpha} \varphi(r_{ij}) - S_\alpha^0 \bar{\mathbf{t}}_\alpha^0 \cdot \mathbf{r}_\alpha, \tag{17.71}$$

where V_α is the total potential energy in the αth supercell and first term of the RHS is the potential energy contribution from the atoms outside the α-th cell. Here, the index i represents the atoms in the αth supercell, while j is the index of atoms from the outside. The second term is the internal potential energy. The last term is a continuum-scale potential energy that is expressed as surface traction, where S_α^0 is the surface area exposed to the external traction $\bar{\mathbf{t}}_\alpha^0$.

The kinetic energy can be calculated based on the multiscale kinematic decomposition (see Eq. (17.68)) as discussed in Li and Tong (2015) and Tong and Li (2015),

$$K_\alpha = \frac{1}{2} \sum_{i \in S_\alpha} m_i \dot{\mathbf{r}}_i \cdot \dot{\mathbf{r}}_i = K_\alpha^{rigid} + K_\alpha^{cell} + K_\alpha^{atom}$$

$$= \frac{1}{2} M_\alpha \dot{\mathbf{r}}_\alpha \cdot \dot{\mathbf{r}}_\alpha + \frac{1}{2} \dot{\boldsymbol{\phi}}_\alpha^T \dot{\boldsymbol{\phi}}_\alpha : \mathbf{J}_\alpha + \frac{1}{2} \mathbf{C}_\alpha : \sum_{i \in S_\alpha} m_i \dot{\mathbf{S}}_i \otimes \dot{\mathbf{S}}_i, \tag{17.72}$$

where M_α is the mass of the whole cell, $\mathbf{C}_\alpha = \boldsymbol{\phi}_\alpha^T \boldsymbol{\phi}_\alpha$ is the right Cauchy–Green tensor, and $\mathbf{J}_\alpha = \sum_{i \in S_\alpha} m_i \mathbf{S}_i \otimes \mathbf{S}_i$ is the moment inertia tensor, which is approximated as a constant spherical tensor that is independent from time. The above kinetic energy is slightly different from the original MD kinetic energy from first principles.

However, the additional terms are dropped out because of imposed statistical constraints. Readers may find the detailed discussion in Li and Tong (2015) and Tong and Li (2015).

The Lagrangian for the αth supercell can then be written as,

$$\mathcal{L}_\alpha = K_\alpha - V_\alpha = K_\alpha^{rigid} + K_\alpha^{cell} + K_\alpha^{atom} - V_\alpha^{int} - V_\alpha^{ext}$$

$$= \frac{1}{2} M_\alpha \dot{\mathbf{r}}_\alpha \cdot \dot{\mathbf{r}}_\alpha + \frac{1}{2} \dot{\boldsymbol{\phi}}_\alpha^T \dot{\boldsymbol{\phi}}_\alpha : \mathbf{J}_\alpha + \frac{1}{2} \mathbf{C}_\alpha : \sum_{i \in S_\alpha} m_i \dot{\mathbf{S}}_i \otimes \dot{\mathbf{S}}_i$$

$$- \frac{1}{2} \sum_{i,j \in S_\alpha} \varphi(r_{ij}) - \sum_{i \in S_\alpha, j \notin S_\alpha} \varphi(r_{ij}) + S_\alpha^0 \bar{\mathbf{t}}_\alpha^0 \cdot \mathbf{r}_\alpha. \qquad (17.73)$$

The above Lagrangian has three independent variables: $\mathbf{r}_\alpha$, $\boldsymbol{\phi}_\alpha$ (or χ_α), and $\mathbf{S}_i$. Through the standard derivation procedure (see Li and Urata (2016)), the equations of motion for these variables are obtained as,

$$M_\alpha \ddot{\mathbf{r}}_\alpha = \sum_{i \in S_\alpha, j \notin S_\alpha} \mathbf{f}_{ij} + S_\alpha^0 \bar{\mathbf{t}}_\alpha^0, \qquad (17.74)$$

$$\ddot{\boldsymbol{\phi}}_\alpha \cdot \mathbf{J}_\alpha = \left(\mathcal{P}_\alpha^{ext} - \mathcal{P}_\alpha^{int} \right) \Omega_\alpha^0, \text{ and} \qquad (17.75)$$

$$m_i \mathbf{C}_\alpha \cdot \ddot{\mathbf{S}}_i = \sum_j \mathbf{f}_{ij} \cdot \boldsymbol{\phi}_\alpha - m_i \dot{\mathbf{C}}_\alpha \cdot \dot{\mathbf{S}}_i, \qquad (17.76)$$

where $\mathbf{f}_{ij}$ is the interaction force on ith atom from jth atom, Ω_α^0 is the volume of the supercell in the referential configuration, and

$$\mathcal{P}_\alpha^{int} = \frac{1}{\Omega_\alpha^0} \left(\frac{1}{2} \sum_{i,j \in S_\alpha} \mathbf{f}_{ij} \otimes \mathbf{S}_{ij} - \boldsymbol{\phi}_\alpha \cdot \sum_{i \in S_\alpha} m_i \dot{\mathbf{S}}_i \otimes \dot{\mathbf{S}}_i \right), \qquad (17.77)$$

$$\mathcal{P}_\alpha^{ext} = \frac{1}{\Omega_\alpha^0} \sum_{i \in \alpha, j \notin S_\alpha} \mathbf{f}_{ij} \otimes \mathbf{S}_i. \qquad (17.78)$$

They are defined as the internal and external first Piola–Kirchhoff (PK-I) stresses.

17.4.2 Peridynamics

In this concurrent coupling method, we employ PD (see Silling (2000); Silling and Askari (2005); Silling et al. (2007); Silling and Lehoucq (2010)) as the physical model to model the continuum region, because PD is essentially a mesoscale model. Different from classic continuum mechanics, the material interaction in PD is nonlocal. To be consistent with MMMD, we denote the representative PD material point as the αth material point to be consistent with the center of mass in the αth supercell in MMMD, and all other particles in the horizon of the αth particle are denoted by the index $\beta = 1, 2, \ldots N_\alpha$. We use uppercase letter $\mathbf{R}_\alpha$ to represent the position vector of the PD particle α in referential configuration. $\mathcal{H}^\alpha$ denotes the horizon of the αth PD particle with radius δ. For the state-based PD, the stress state of a macroscale material point is described by strain tensor, or the deformation state gradient $\mathbf{F}_\alpha$, which in turn determines the force state $\mathbf{T}_\alpha$ at point $\mathbf{r}_\alpha$.

Thus, the first step in the state-based PD is how to approximate the deformation gradient. To construct a discrete nonlocal deformation gradient, we first associate a compact support with each center of mass of a supercell, say the αth cell. Then, we can construct a shape tensor for each horizon that is based on the distribution of all centers of mass inside the compact support of the center of mass of the αth cell, i.e.,

$$\mathbf{K}_\alpha := \int_{\mathcal{H}_{\mathbf{R}_\alpha}} \omega(|\mathbf{R}_\alpha|)\mathbf{R}_{\alpha\beta} \otimes \mathbf{R}_{\alpha\beta} dV_\beta \approx \sum_{\beta \in S_N} \omega(|\mathbf{R}_\alpha|)\mathbf{R}_{\alpha\beta} \otimes \mathbf{R}_{\alpha\beta} \Delta V_\beta, \quad (17.79)$$

where $\mathbf{R}_{\alpha\beta} = \mathbf{R}_\beta - \mathbf{R}_\alpha$ and S_N is the index set for all the centers of supercells. The shape tensor is basically a moment tensor or loosely speaking a moment of inertia tensor. Note that there is no difference between $\mathbf{R}_\alpha$ and $\mathcal{R}_\alpha$.

One can then define a two-point nonlocal second-order tensor $\mathbf{N}$ as

$$\mathbf{N}_\alpha = \int_{\mathcal{H}_{\mathbf{R}_\alpha}} \omega(|\mathbf{R}_\alpha|)\mathbf{r}_{\alpha\beta} \otimes \mathbf{R}_{\alpha\beta} dV_\beta \approx \sum_{\beta=1}^{N} \omega(|\mathbf{R}_\alpha|)\mathbf{r}_{\alpha\beta} \otimes \mathbf{R}_{\alpha\beta} \Delta V_\beta, \quad (17.80)$$

where $\mathbf{r}_{\alpha\beta} = \mathbf{r}_\beta - \mathbf{r}_\alpha$. At the coarse scale, we assume that the following Cauchy–Born rule is held in each compact support of the center of mass of the supercell,

$$\mathbf{r}_{\alpha\beta} = \mathbf{F}_\alpha \mathbf{R}_{\alpha\beta}. \quad (17.81)$$

By substituting Eq. (17.81) into Eq. (17.80), we obtain the expression for the discrete nonlocal deformation gradient,

$$\mathbf{F}_\alpha = \mathbf{N}_\alpha \mathbf{K}_\alpha^{-1} = \int_{\mathcal{H}_{\mathbf{R}_\alpha}} \left(\omega(|\mathbf{R}_\alpha|)\mathbf{r}_{\alpha\beta} \otimes \mathbf{R}_{\alpha\beta} dV_\beta \right) \mathbf{K}_\alpha^{-1}$$

$$\approx \left(\sum_{\beta=1}^{N} \omega(|\mathbf{R}_\alpha|)\mathbf{r}_{\alpha\beta} \otimes \mathbf{R}_{\alpha\beta} \Delta V_\beta \right) \mathbf{K}_\alpha^{-1}. \quad (17.82)$$

Considering the state-based PD, we denote the force state at material point α as $\mathbf{T}_\alpha < \mathbf{R}_\beta - \mathbf{R}_\alpha >$. Assume that there exists a macroscale free energy density at the material point α, i.e., $\Psi(\mathbf{X}^\alpha)$. Then the virtual work or the variation of the free energy density may be written as,

$$\delta\Psi(\mathbf{R}^\alpha) = \mathbf{P}_\alpha : \delta\mathbf{F}_\alpha = \mathbf{P}_\alpha : \int_{\mathcal{H}_{\mathbf{R}_\alpha}} \omega(|\mathbf{R}_{\alpha\beta}|)\delta\mathbf{r}_{\alpha\beta} \otimes \mathbf{R}_{\alpha\beta} dV_\beta \mathbf{K}_\alpha^{-1}$$

$$= \int_{\mathcal{H}_{\mathbf{R}_\alpha}} \mathbf{T}_\alpha < \mathbf{R}_\beta - \mathbf{R}_\alpha > \cdot \delta\mathbf{r}_{\alpha\beta} dV_\beta, \quad (17.83)$$

which leads to the force state expression (see Silling and Lehoucq (2010)) that is determined by the local stress state at the given material point α, i.e.,

$$\mathbf{T}_\alpha < \mathbf{R}_\beta - \mathbf{R}_\alpha > = \omega(|\mathbf{R}_\alpha|)\mathbf{P}_\alpha \cdot \mathbf{R}_{\alpha\beta} \mathbf{K}_\alpha^{-1}, \quad (17.84)$$

where $\mathbf{P}_\alpha$ is the coarse-scale first Piola–Kirchhoff stress at the material point α.

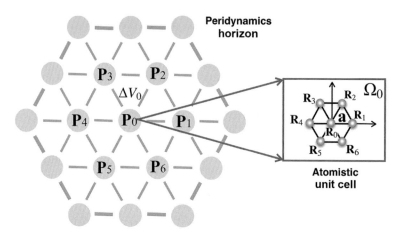

Figure 17.14 Cauchy–Born rule–based multiscale peridynamics

To find the local stress, we employ the atomic potential of the underlined solid and the Cauchy–Born rule. Since this is the coarse-scale calculation, we assume that each PD point is associated with an atomistic unit cell (see Fig. 17.14). For illustration purpose, we assume that the elastic energy density at each PD point may be expressed as a pair potential in a Bravais lattice,

$$W = \frac{1}{2\Omega_0} \sum_{k=1}^{N_b} \varphi(r_k), \tag{17.85}$$

where Ω_0 is the volume of the unit cell, and inside there are total N_b number of bonds connecting to the center atom, $\varphi(r_k)$ is the atomistic potential of kth bond, and r_k is the atomistic bond length in the deformed configuration, $k = 1, \ldots, N_b$, with N_b the number of bonds in a unit cell. Based on the Cauchy–Born rule, we can calculate the first Poila–Kirchhoff stress by taking derivative of elastic energy density with respect to deformation gradient,

$$\mathbf{P}_\alpha = \left.\frac{\partial W}{\partial \mathbf{F}}\right|_{\mathbf{F}=\mathbf{F}_\alpha} = \frac{1}{2\Omega_0} \sum_{k=1}^{N_b} \varphi'(r_k) \frac{\mathbf{r}_k \otimes \mathbf{R}_k}{r_k}, \tag{17.86}$$

where $\mathbf{r}_k$ and $\mathbf{R}_k$ are bond vectors in the current and the referential configurations, respectively. The Cauchy–Born rule implies that $\mathbf{r}_k = \mathbf{F}_\alpha \cdot \mathbf{R}_k$. Therefore, if the deformation gradient is given, we can calculate the first Piola–Kirchhoff stress at the macroscale material point α.

After determining the stress state at every PD particles, we can establish a unified equation of motions for peridynamic particles based on the so-called state-based PD formulation. Considering the balance of linear momentum at the material point $\mathbf{r}_\alpha$, we have the following nonlocal balance equation:

$$\rho_\alpha \ddot{\mathbf{r}}_\alpha = \mathbf{L}(\mathbf{R}_\alpha, t) + \rho_\alpha \mathbf{b}(\mathbf{R}_\alpha), \tag{17.87}$$

where

$$L(\mathbf{R}_\alpha, t) = \int_{\mathcal{H}_\alpha} (\mathbf{T}_\alpha < \mathbf{R}_\beta - \mathbf{R}_\alpha > - \mathbf{T}_\beta < \mathbf{R}_\alpha - \mathbf{R}_\beta >) dV_\beta \quad (17.88)$$

is the nonlocal stress divergence vector acting on the αth material point by neighboring macroscale points β. Its counterpart in classical continuum mechanics is $\nabla_\mathbf{R} \cdot \mathbf{P}_\alpha$, which is a local divergence term. Again, here $\mathbf{P}_\alpha$ is the first Piola–Kirchhoff stress tensor in continuum mechanics. To solve Eq. (17.88), one mainly need to evaluate the nonlocal force density vector or the nonlocal divergence vector $\mathbf{L}(\mathbf{R}_\alpha, t)$. In computation implementation, since the domain is discretized into many material points with volume, the integral can be replaced or approximated by the following summation:

$$L(\mathbf{R}_\alpha, t) = \sum_{\beta=1}^{N} (\mathbf{T}_\alpha < \mathbf{R}_\beta - \mathbf{R}_\alpha > - \mathbf{T}_\beta < \mathbf{R}_\alpha - \mathbf{R}_\beta >) \Delta V_\beta. \quad (17.89)$$

To determine the damage of the bonds, the criterion of equivalent strain is introduced (Warren et al. (2009)). We can measure the stretch $\lambda_{\alpha\beta}$ along the direction of $\mathbf{n}_{\alpha\beta} = \mathbf{r}_{\alpha\beta}/r_{\alpha\beta}$, and if $\lambda_{\alpha\beta} \geq \lambda_c$,

$$\lambda_c = \sqrt{\frac{10 G_0}{\pi c \delta^5}},$$

the material bond between the partial α and the particle β is broken. Note that G_0 is the energy release of the material, and c is the micro-modulus of PD model.

Consider a numerical model of 2D closely packed Al plane (FCC (111) plane) with the lattice constant $a_0 = 2.878$ Å. The 2D lattice is not a realistic one, but it is stable when the out-of-plane displacement is prohibited. The atomic weight of Al is 26.98 u. The Morse interatomic potential $\varphi(r) = D(\exp(-2\alpha(r - r_0)) - 2\exp(-\alpha(r - r_0))$ is used for simplicity, where $D = 0.0965$ eV, $\alpha = 2.71$ Å, and $r_0 = 2.878$ Å. The geometry of the crack model is shown in Fig. 17.15. The three regions are separately modeled including a predefined crack line, which spans half the length of the specimen. The atomistic region has dimensions of 86×30 nm, which consists of 36,165 atoms. The continuum region has 44,901 peridynamic particles and the dimensions are 259×120 nm not including the atomistic region. The radius of horizon δ is 2.6 nm. The transition zone in between the regions has 801 supercells in total. Each supercell consists of nine atoms and the shape is shown in the top-right graph in Fig. 17.15.

Figure 17.16 presents a series of snapshots during the process of crack propagation in the atomistic region under the strain rate $\dot{\epsilon} = 0.00036$ ps^{-1} before the crack reaches to the mesoscale PD region. The potential energy level is shown as contour profile in the figure. As the application of remote loads, the stress wave propagates to the region around the pre-notch tip and a stress concentration region is formed, which is consistent with the distribution of the potential energy. Figure 17.16(a) is at the time when the Griffith critical fracture stress σ_c is reached. At the onset of fracture, a new crack is observed at this moment. The atoms that are initially at the original lattice points move in a nonuniform fashion. The atomic bonds start to break one

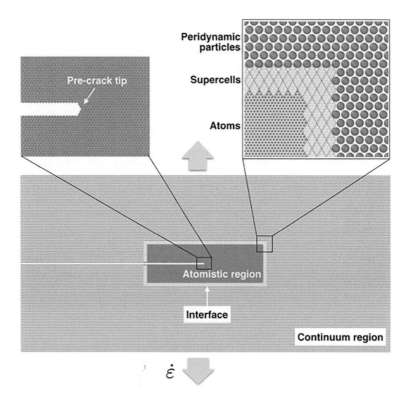

Figure 17.15 Schematic illustration of the computational model. The atomistic region is surrounded by the continuum region, and the interface (transition zone) is in between the regions. A strain rate as boundary condition is applied on top and bottom surfaces of the model. A predefined crack is on the left-half plane. Top-left diagram is the close-up of crack tip area. Top-right diagram is the close-up of the region around the interface

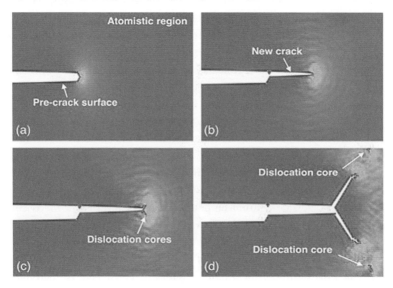

Figure 17.16 Dynamic crack propagation in the atomistic region under the strain rate $\dot{\varepsilon} = 0.00036\,\text{ps}^{-1}$. The color contour is the magnitude of the potential energies (eV): (a) Pre-notched crack; (b) New crack surface; (c) Initial dislocation cores, and (d) Dislocation cores. (The data source is from Tong and Li (2020)).

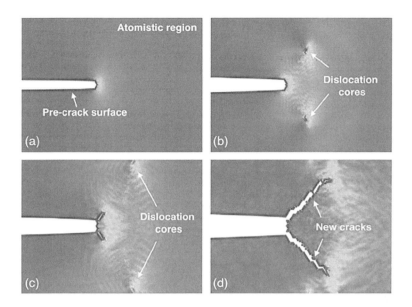

Figure 17.17 Dynamic crack propagation in the atomistic region under the strain rate $\dot{e} = 0.00048$ ps^{-1}. The color contour is the magnitude of the potential energies (eV): (a) Pre-notched crack; (b) Dislocation core initiation without fracture; (c) Crack initiation, and (d) Crack growth. (The data source is from Tong and Li (2020)).

by one from the pre-crack tip, which leads to the formation of new crack surface. Figure 17.16(b) shows the dynamic fracture in a perfectly brittle fashion, which is smooth without any dislocation. The atomic bonds are broken by separation instead of sliding. In this stage, the crack tip quickly accelerates to a terminal constant velocity of $v = 1.8$ km/s, which is about 0.42 time the Rayleigh wave speed. This suggests that the growth velocity for mode I crack is limited by the Rayleigh wave speed.

From Fig. 17.16(c) and (d), one may observe the crack branching occurs along the $\pm 60°$ directions as the energy builds up in the vicinity of the crack tip. In Fig. 17.16(d), one may find the potential energy hotspots, which may indicate the dislocation cores emitting from the crack tips.

Figure 17.17 shows the snapshots of the dynamic crack growth driven by a higher strain rate $\dot{e} = 0.00048$ ps^{-1}. In this case, the loading rate is larger than the previous case. Consequently, the accumulated energy during the pre-opening stage is enough to trigger the crack-branching instability. The branching occurs as the onset of the new crack. Figure 17.17(a) and (b) illustrates the preprocess of dynamic crack growth. By monitoring the potential energy profile in Fig. 17.17(b), we clearly observe a group of dislocations nucleated from the pre-existing crack tip and propagate immediately along the $\pm 60°$ slip planes. In Fig. 17.17(c), as the dislocations travel out to the continuum region, a new group of dislocation cores are generated as the shock wave emitting out from the crack tip. In Fig. 17.17(d), the branched cracks continuously propagate moving out of the atomistic region. We can see the interference fringe of shock waves from the crack tips. Note that the lines of the crack branches do not strictly move in $\pm 60°$ directions. Instead, the zigzag crack surface pattern lower the

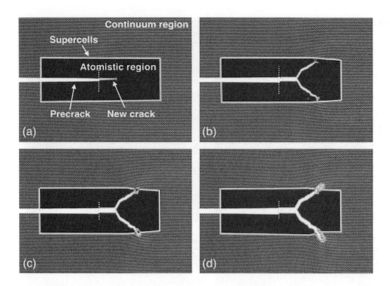

Figure 17.18 The dynamic process of a crack propagating through the atomistic/continuum interphase zone under the applied strain rate $\dot{\epsilon} = 0.00036\,\mathrm{ps}^{-1}$. The color contour indicates the damage level in the materials: (a) Crack initiation in atomistic region; (b) Crack bifurcation in atomistic region; (c) Crack reaching to the boundary between the atomistic region and continuum region, and (c) Cracks reach to continuum region.

angles overall. This may be due to the energy release and energy dissipation under high strain rate condition.

To demonstrate the capability of the multiscale model, we examine the crack propagation across the atomistic continuum interface. The transition is based on the supercell that is able to translate the mechanical information concurrently, i.e., atomic motion and continuum deformation. Figure 17.18 shows the process of crack propagation under the strain rate $\dot{\epsilon} = 0.00036\,\mathrm{ps}^{-1}$, which is the same as in Fig. 17.16. Figure 17.18(a) is a global view of Fig. 17.16(c). The contour profile represents the damage level of the local material, which is calculated as the change of neighboring atoms/peridynamic points. The range of damage level is in between [0] and [1], where "0" means that every bonds at that point are perfect, and the damage level "1" means completely broken bonds at that material point.

Figure 17.18(b) shows the complete crack paths in the atomistic region. The paths can be separated into three stages based on the characters. The first stage of the fracture is a horizontal crack line representing the brittle crack; the second stage of the fracture is a rapid branching perfectly along the $\pm 60°$ directions over a few lattice spaces; last stage is a growth period with the angles decreasing due to the energy release. Fig. 17.14(c) is the snapshot at the critical time when the cracks are propagating through the atomistic continuum interface. The defects of microscale materials are captured by the supercells and passed to the continuum region. In Fig. 17.18(d), the cracks are observed to pass the interface and keep growing in the continuum region. However, the trajectory of the crack growth seems to be altered. This is a combined effect of the slip planes and the enlarged lattice spacing. As discussed previously, crack paths switch between $0°$ and $\pm 60°$, which decays the overall angle and forms

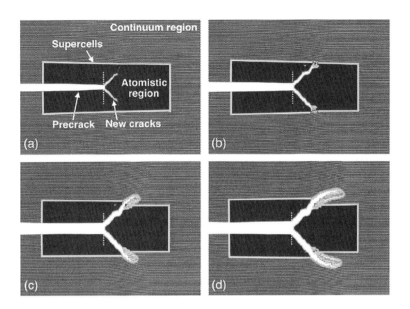

Figure 17.19 The dynamic process of a crack propagating through the atomistic/continuum interphase zone under the applied strain rate $\dot{\epsilon} = 0.00048 \text{ ps}^{-1}$: (a) Crack bifurcation in the atomistic region; (b) Cracks reach to the interface between atomistic region and continuum region; (c) Cracks reach to the continuum region, and (d) Cracks start propagating in continuum region. (Data source is from Tong and Li (2020))

a zigzag pattern. As the cracks enter the continuum region, the lattice spacing is coarsened, and the ±60° angle paths become dominate crack branches than those in the atomistic region. Figure 17.19 illustrates the global view of the process in Fig. 17.17. The applied strain rate is $\dot{\epsilon} = 0.00048 \text{ ps}^{-1}$. In Fig. 17.19(a), the branches grow inside the atomistic region as the stress wave propagates from the continuum region. Figure 17.19(b) is the snapshot when the cracks are crossing the interface, and as shown in Fig. 17.19(c) and (d), as the crack branches propagate to the continuum region, the angles of crack propagation are getting even smaller.

17.5 Homework Problems

Problem 17.1 Consider the following multiscale Lagrangian:

$$\mathcal{L}_m(\mathbf{r}_\beta, \boldsymbol{\phi}_\beta, \mathbf{S}_i) = \frac{1}{2} \sum_\beta M_\beta \dot{\mathbf{r}}_\beta \cdot \dot{\mathbf{r}}_\beta + \frac{1}{2} \sum_\beta \mathbf{J}_\beta \left(\dot{\boldsymbol{\phi}}_\beta^T \cdot \dot{\boldsymbol{\phi}} \right)_\beta$$

$$+ \frac{1}{2} \sum_\beta \sum_i m_i (\dot{\mathbf{S}}_i \cdot \mathbf{C}_\beta \cdot \dot{\mathbf{S}}_i) - \frac{1}{2} \sum_\beta \sum_\gamma \sum_{i \in \beta, j \in \gamma} U(r_{ij}), \quad (17.90)$$

where β and γ are indices of cell number, the abbreviation $i \in \beta$ means that the ith atom in the βth cell, $\mathbf{C}_\beta := \boldsymbol{\phi}_\beta^T \boldsymbol{\phi}_\beta$ is the multiscale right Cauchy–Green tensor for the total deformation of the β-th cell, and $\mathbf{J}_\beta$ is a fixed spherical tensor, i.e., $\mathbf{J}_\beta = W \mathbf{E}_I \otimes \mathbf{E}_I$.

Derive the corresponding Euler–Lagrangian equations for the mesoscale variable $\phi_\alpha = \mathbf{F}_\alpha \cdot \boldsymbol{\chi}_\alpha$ or $\boldsymbol{\chi}_\alpha$,

$$\frac{d}{dt}\left(\frac{\partial \mathcal{L}_m}{\partial \dot{\phi}_\alpha}\right) - \frac{\partial L_m}{\partial \phi_\alpha} = 0,$$

where

$$\mathbf{F}_\alpha = \mathbf{F}_\alpha(\mathbf{r}_\beta) = \left(\sum_{\beta=1}^{N_b} \omega(\mathbf{R}_{\alpha\beta}|)\mathbf{r}_{\alpha\beta} \otimes \mathbf{R}_{\alpha\beta}\Omega_{\beta 0}\right) \cdot \mathbf{M}^{-1}$$

and

$$\mathbf{M}_\alpha = \mathbf{F}(\mathbf{r}_\alpha) = \left(\sum_{\beta=1}^{N_b} \omega(\mathbf{R}_{\alpha\beta}|)\mathbf{R}_{\alpha\beta} \otimes \mathbf{R}_{\alpha\beta}\Omega_{\beta 0}\right).$$

Hint: Read:

(1) Li, S. and Tong, Q. 2015. A concurrent multiscale micromorphic molecular dynamics. *Journal of Applied Physics*, **117**(**154303**), DOI:10.1063/1.4916702.

(2) Tong, Q. and, Li, S. 2015. From molecular systems to continuum solids: A multiscale structure and dynamics. *Journal of Chemical Physics*, **143**(**064101**), DOI: 10.1063/1.4927656.

(3) Li, S. and Urata, S. 2016. An atomistic-to-continuum molecular dynamics: Theory, algorithm, and applications. *Computer Methods in Applied Mechanics and Engineering*, **306**, 452–478.

18 Multiscale Methods (II): Multiscale Finite Element Methods

In this chapter, we introduce several useful multiscale finite element methods (FEMs) that can be used to conduct modeling and simulations of macroscale material and structure failure analysis. For example, we can apply the multiscale FEM to simulate indentation tests of different materials such as metals, ceramics, or polymers. The multiscale simulation strategy adopted here is that in the contact region we use multiscale molecular dynamics (MD) discussed in the previous chapter to simulate the nanoscale contact, while we use a multiscale continuum FEM to simulate the rest of the substrate. At the interface of the MD region and multiscale finite element region, various coupling approaches have been developed to match the two different methods. By doing so, we can save much computational cost while retaining necessary accuracy in modeling and simulation, and we may be able to simulate a multiscale physical process from the macroscale perspective (see Fig. 18.1).

Moreover, we can also apply the multiscale FEM to simulate material and structure failures. Thus, in this chapter we focus on discussing multiscale FEMs and their applications.

18.1 Multiscale Finite Element Formulation

To begin with, we first introduce the multiscale finite element theory and formulation. Based on continuum mechanics theory, The Hamiltonian principle for a continuum body may be expressed as a time integration of the variation of total energy of a mechanical system,

$$\int_{t_0}^{t_1} \left(\delta \mathcal{K} - (\delta \mathcal{W}_{int} + \delta \mathcal{W}_{ext})\right) dt = 0, \tag{18.1}$$

where

$$\int_{t_1}^{t_2} \delta \mathcal{K} dt = \int_{t_1}^{t_2} \int_{\Omega} \rho \dot{\mathbf{u}} \cdot \delta \dot{\mathbf{u}} dV dt = -\int_{t_1}^{t_2} \int_{\Omega} \rho \ddot{\mathbf{u}} \cdot \delta \mathbf{u} dV dt, \tag{18.2}$$

$$\delta \mathcal{W}_{int} = \int_{\Omega} \frac{\partial W}{\partial \mathbf{F}} : \delta \mathbf{F} dV = \int_{\Omega} \mathbf{P} : \delta \mathbf{F} dV, \tag{18.3}$$

$$\delta \mathcal{W}_{ext} = -\int_{\Omega} \mathbf{b} \cdot \delta \mathbf{u} dV - \int_{\delta \Omega_t} \bar{\mathbf{T}} \cdot \delta \mathbf{u} dS. \tag{18.4}$$

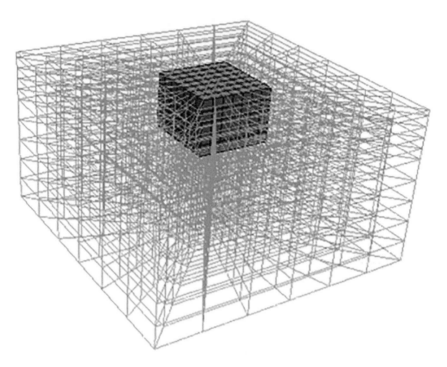

Figure 18.1 Multiscale finite element mesh and computational setting for nano-indentation

In Eq. (18.2), integration by parts is used in the last equality. In Eq. (18.3), W is the strain energy density and $\mathbf{P} = \frac{\partial W}{\partial \mathbf{F}}$ is the first Piola–Kirchhoff stress tensor. In Eq. (18.4), $\mathbf{b}$ and $\bar{\mathbf{T}}$ are body force and prescribed traction vectors.

Considering the finite element interpolation, we can express the displacement field as,

$$\mathbf{u}^h(\mathbf{X}) = \underset{e=1}{\overset{nelem}{A}} \sum_{I=1}^{n_e} N_I(\mathbf{X})\mathbf{u}_I = [\mathbf{N}][\mathbf{d}], \quad (18.5)$$

where $\underset{e=1}{\overset{nelem}{A}}$ is element assemble operator (see Hughes (2012)), $N_I(\mathbf{X})$ are the finite element interpolation functions in an element, $\mathbf{u}_I$ are the nodal displacements, *nelem* is the total number of element, and n_e is the number of FEM shape functions in the element e. In Eq. (18.5), $[\mathbf{N}]$ is the FEM shape function matrix, and $[\mathbf{d}]$ is the FEM nodal displacement vector. The superscript "h" in $\mathbf{u}$ indicates that this is a displacement field interpolated on a finite element mesh.

The discrete FEM equation of motion can be expressed as,

$$[\mathbf{M}][\ddot{\mathbf{d}}] = [\mathbf{f}^{int}(\mathbf{d})] - [\mathbf{f}^{ext}], \quad (18.6)$$

where $\mathbf{d}$ is the nodal displacement vector; $[\mathbf{M}]$ is the mass matrix; $[\mathbf{f}^{int}]$ and $[\mathbf{f}^{ext}]$ are force vectors from each element, respectively. These quantities are precisely defined as follows:

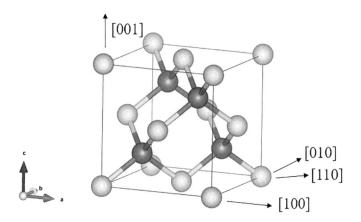

Figure 18.2 Unit cell of silicon crystal diamond cubic crystal. The dark- and light-colored spheres indicate atoms belonging to two different FCC lattices

$$\mathbf{M} = \underset{e=1}{\overset{nelem}{A}} \int_{\Omega_e} \rho_0 [\mathbf{N}_e]^T [\mathbf{N}_e] dV, \tag{18.7}$$

$$[\mathbf{f}^{int}] = \underset{e=1}{\overset{nelem}{A}} \int_{\Omega_e} [\mathbf{B}_e]^T \mathbf{P}^e(\mathbf{d}) dV, \tag{18.8}$$

$$[\mathbf{f}^{ext}] = \underset{e=1}{\overset{nelem}{A}} \left\{ \int_{\Omega_e} [\mathbf{N}_e]^T [\mathbf{B}_e] dV + \int_{\Gamma_t} [\mathbf{N}_e]^T \bar{\mathbf{T}}^e dS \right\}, \tag{18.9}$$

where Ω_e is the domain for the e-th element; $\partial \Gamma_t$ is the traction boundary; $[\mathbf{B}^e]$ are the element strain–displacement matrix and the gradient of the strain–displacement matrix, respectively.

Applying the multiscale FEM to a concrete application example, we demonstrate how to apply it to compute a nontrivial real engineering problem: modeling and simulation of the nano-indentation process of a silicon crystal (see: Urata and Li (2017a)).

In this case, the crystal structure of silicon is diamond cubic, $Fd3m$, which is regarded as a set of two FCC lattices, α and β (see: Fig. 18.2). Hence, the inner displacements between two lattices can be defined as a vector $\mathbf{v}$ from the first lattice α to the second lattice β (see: Park and Klein (2008); Khoei and DorMohammadi (2012); Khoei et al. (2014)). The diamond cubic structure is not a Bravis lattice, and its smallest unit cell only has five atoms. As one can see that each dark color atom in Fig. 18.2 has four light color neighboring atoms, or vice versa, each light color atom has four dark color neighboring atoms.

Common interatomic potentials used to model silicon crystal is the Tersoff potential (see: Dodson (1987)). The Tersoff potential is a three-body potential functional that has an angular contribution of the interatomic force. When evaluating the stress-based Tersoff potential, we only take the nearest four atoms into account to calculate interactions, because silicon crystal has the diamond structure. The strain energy density and its derivative can be calculated by the pseudo-unit cell composed of five atoms.

The nearest four atoms, which composed of a tetrahedral structure with the central atom, all belong to a different FCC lattice unit cell β from the FCC lattice α to which the center silicon resides (see: Fig. 18.2). If we term the central atom as index 1, and others are from 2 to 5, distances of atoms are defined as,

$$r_{1j} = |\mathbf{r}_{1j}| = |\mathbf{r}_1 - \mathbf{r}_j - \mathbf{v}| \; j = 2, 3, \ldots, 5, \quad (18.10)$$

$$r_{jk} = |\mathbf{r}_{jk}| = |\mathbf{r}_j - \mathbf{r}_k| \quad j, k = 2, 3, \ldots, 5, \; \text{and} \; j \neq k. \quad (18.11)$$

The vector $\mathbf{v}$ can be evaluated to minimize strain energy density as shown in Urata and Li (2017). It had been found that the optimization of vector $\mathbf{v}$ is crucial for finding the stablest or global equilibrium configuration of atoms at each deformation state (Khoei et al. (2014)).

Because the overall nonuniform deformation field can be interpreted by a set of bulk elements with piece-wise uniform deformations, we can assume the first-order Cauchy–Born (CB) rule to represent deformation of the bulk element.

$$\mathbf{r}_i = \mathbf{F}\mathbf{R}_i, \; i = 1, 2, \ldots N_b, \quad (18.12)$$

where i is the bond index (not the atom indices), $\mathbf{R}_i$ is original position of atom i in the unit cell, $\mathbf{r}_i$ is the position after the deformation, and $\mathbf{F}$ is the deformation gradient in each element, which is defined as

$$\mathbf{F} = \frac{\partial \mathbf{x}}{\partial \mathbf{X}}. \quad (18.13)$$

18.1.1 Tersoff Potential

Now, we examine the Tersoff potential closely. For an atomistic system, we can write the strain energy density W as follows,

$$W = \frac{U_1}{\Omega_0} = \frac{1}{2\Omega_0} \sum_{j=2}^{5} V_{1j}, \quad (18.14)$$

where U_1 is the potential energy for the central atom 1 in Fig. 18.2, Ω_0 is the volume occupied by an atom in the initial configuration, and V_{1j} is the Tersoff potential energy between the central atom 1 and a surrounding atom j as

$$V_{1j} = f_C(r_{1j})[f_R(r_{1j}) + b_{1j} f_A(r_{1j})]. \quad (18.15)$$

In the Tersoff potential, the functions f_R, f_A are the cutoff functions that are defined as,

$$f_R(r_{1j}) = A \exp(-\lambda_{1j} r_{1j}), \quad (18.16)$$

$$f_A(r_{1j}) = -B \exp(-\mu_{1j} r_{1j}), \quad (18.17)$$

and the function f_C is defined as follows:

$$f_C(r_{1j}) = \begin{cases} 1 & r_{1j} \leq R_{1j} \\ \frac{1}{2} + \frac{1}{2}\cos\left(\frac{\pi(r_{1j} - R_{1j})}{S_{1j} - R_{1j}}\right) & R_{1j} < r_{1j} < S_{1j}. \\ 0 & r_{1j} \geq S_{1j} \end{cases} \quad (18.18)$$

Note here that second differentiation of the original cutoff function expressed in Eq. (18.18) will be discontinuous, it is thus unfavorable to estimate the inner vector $\mathbf{v}$ at the transition points $r = R_{1j}$ and $r = S_{1j}$. To solve this inexpedience, Izumi and Sakai (2004) introduced the dumping cutoff function as,

$$f_C(r_{1j}) = \frac{1}{2} - \frac{1}{2}\tanh\left[\frac{\pi}{2}\left(\frac{r_{1j} - R_{1j}}{S_{1j} - R_{1j}}\right)\right]. \quad (18.19)$$

In this problem, we employ the modified cutoff function of Eq. (18.19) instead of Eq. (18.18) without changing parameters R_{1j} and S_{1j} of the original Tersoff potential.

By taking account three body interaction, the parameter b_{1j} in Eq. (18.15) will explicitly depend on the location of the third atom k as follows:

$$b_{1j} = (1 + \beta^n \zeta_{1j}^n)^{\frac{-1}{2n}}, \quad (18.20)$$

$$\zeta_{1j} = \sum_{k \neq i,j}^{5} f_c(r_{1k}) g(\theta_{1jk}), \quad (18.21)$$

$$g(\theta_{1jk}) = 1 + \frac{c^2}{d^2} - \frac{c^2}{d^2 + (h - \cos\theta_{1jk})^2}, \quad (18.22)$$

$$\cos\theta_{1jk} = \frac{\mathbf{r}_{1j} \cdot \mathbf{r}_{1k}}{r_{1j} r_{1k}} = \frac{(r_{1j}^2 + r_{1k}^2 - r_{jk}^2)}{2 r_{1j} r_{1k}}. \quad (18.23)$$

The parameter set of Tersoff potential used in this study is adopted from Tersoff's 1988 paper, which can reproduce elastic properties of silicon more accurately.

According to the strain energy density, the first Piola–Kirchhoff stress tensor is derived as,

$$\mathbf{P} = \frac{\partial W}{\partial \mathbf{F}} = \frac{1}{2\Omega_0} \frac{\partial}{\partial \mathbf{F}}\left(\sum_{j=2}^{5} V_{1j}\right) = \frac{1}{2\Omega_0} \sum_{j=2}^{5}\left(\frac{\partial V_{1j}}{\partial \mathbf{F}} + \frac{\partial V_{1j}}{\partial \mathbf{v}} \frac{\partial \mathbf{v}}{\partial \mathbf{F}}\right). \quad (18.24)$$

The inner displacement $\mathbf{v}$ can be evaluated by minimizing of the strain energy density W, i.e.,

$$\sum_{j=2}^{5}\left(\frac{\partial V_{1j}}{\partial \mathbf{v}}\right)_\mathbf{F} = 0. \quad (18.25)$$

This condition is achieved by using Newton's method to find the minimum of the strain energy density, and as a result, inner displacement vector $\mathbf{v}$ can be obtained. The procedure is outlined as follows. Since

$$\sum_j \left(\frac{\partial V_{1j}}{\partial \mathbf{v}}\right)_F = 0 \rightarrow \sum_j \frac{\partial V_{1j}}{\partial \mathbf{v}}\bigg|_{F_n} + \sum_j \frac{\partial^2 V_{1j}}{\partial \mathbf{v}^2}\bigg|_{F_n} \cdot \Delta \mathbf{v} \approx 0,$$

we can find the increment of internal displacement vector as

$$\Delta \mathbf{v} = -\left[\sum_j \frac{\partial^2 V_{1j}}{\partial \mathbf{v}^2}\right]^{-1}\bigg|_{F_n} \cdot \left[\sum_j \frac{\partial V_{1j}}{\partial \mathbf{v}}\right]\bigg|_{F_n} \quad (18.26)$$

and

$$\mathbf{v}_{n+1} = \mathbf{v}_n + \Delta \mathbf{v}. \quad (18.27)$$

This process will continue until the convergence criterion is met. Finally, from Eq. (18.24), we can obtain

$$\mathbf{P} = \frac{1}{2\Omega_0}\sum_{j=2}^{5}\left(\frac{\partial V_{1j}}{\partial \mathbf{F}}\right)$$

$$= \frac{1}{2\Omega_0}\sum_{j=2}^{5}\left[\frac{\partial V_{1j}}{\partial \mathbf{r}_{1j}}\frac{\partial \mathbf{r}_{1j}}{\partial \mathbf{F}} + \sum_{k=2, k\neq j}^{5}\left(\frac{\partial V_{1j}}{\partial r_{1k}}\frac{\partial r_{1k}}{\partial \mathbf{F}} + \frac{\partial V_{1j}}{\partial \cos\theta_{1jk}}\frac{\partial \cos\theta_{1jk}}{\partial \mathbf{F}}\right)\right]. \quad (18.28)$$

where

$$\frac{\partial r_{ij}}{\partial \mathbf{F}} = \frac{\partial r_{ij}}{\partial \mathbf{r}_{ij}}\frac{\partial \mathbf{r}_{ij}}{\partial \mathbf{F}} = \frac{\mathbf{r}_{ij}\otimes \mathbf{R}_{ij}}{r_{ij}} \quad (18.29)$$

$$\frac{\partial \cos\theta_{1jk}}{\partial \mathbf{F}} = \left(\frac{1}{r_{1k}} - \frac{\cos\theta_{1jk}}{r_{1j}}\right)\frac{\partial r_{1j}}{\partial \mathbf{F}} + \left(\frac{1}{r_{1j}} - \frac{\cos\theta_{1jk}}{r_{1k}}\right)\frac{\partial r_{1k}}{\partial \mathbf{F}}$$

$$- \left(\frac{r_{jk}}{r_{1j}r_{1k}}\right)\frac{\partial r_{jk}}{\partial \mathbf{F}}. \quad (18.30)$$

According to Eq. (18.15), the derivative of V_{1j} by the position vector r is as follows:

$$\frac{\partial V_{1j}}{\partial r_{1j}} = \frac{\partial f_C}{\partial r_{1j}}(f_R(r_{1j}) + b_{1j}f_A(r_{1j})) + f_C(r_{1j})\left(\frac{\partial f_R(r_{1j})}{\partial r_{1j}} + b_{1j}\frac{\partial f_A(r_{1j})}{\partial r_{1j}}\right), \quad (18.31)$$

where

$$\frac{\partial f_C(r_{1j})}{\partial r_{1j}} = -\frac{\pi}{4(S_{1j} - R_{1j})}\cosh\left[\frac{\pi(r_{1j} - R_{1j})}{2(S_{1j} - R_{1j})}\right], \quad (18.32)$$

$$\frac{\partial f_R(r_{1j})}{\partial r_{1j}} = -A\lambda_{1j}\exp(-\lambda_{1j}r_{1j}), \quad (18.33)$$

$$\frac{\partial f_A(r_{1j})}{\partial r_{1j}} = B\mu_{1j}\exp(-\mu_{1j}r_{1j}). \quad (18.34)$$

Similarly, we can take derivatives of V_{1j} by r_{1k} and θ_{1jk} to estimate Eq. (18.28), and they are,

$$\frac{\partial V_{1j}}{\partial r_{1k}} = f_C(r_{1j})f_A(r_{1j})\left[-\frac{1}{2}(1+\beta^n\zeta_{1j}^n)^{\frac{-1}{2n}-1}\right]\beta^n\zeta_{1j}^{n-1} \cdot \frac{\partial f_C(r_{ik})}{\partial r_{1k}}g(\theta_{1jk}), \tag{18.35}$$

$$\frac{\partial V_{1j}}{\partial \cos\theta_{1jk}} = f_C(r_{1j})f_A(r_{1j})\left[-\frac{1}{2}(1+\beta^n\zeta_{1j}^n)^{\frac{-1}{2n}-1}\right]\beta^n\zeta_{1j}^{n-1},$$

$$\cdot f_C(r_{1k})\frac{2c^2(\cos\theta_{1jk}-h)}{[d^2+(h-\cos\theta_{1jk})^2]^2}. \tag{18.36}$$

18.2 MMMD/FEM Coupling Method

In this section, we present a multiscale coupling method between multiscale micromorphic MD (MMMD) and continuum FEM.

One of the main advantages of MMMD is that it provides a perfect means to design multiscale coupling strategy, because it automatically splits conventional MD into three scales: microscale, mesoscale, and macroscale. Therefore, if one would like to couple MMMD with continuum-scale dynamics, one can directly connect the macroscale MMMD with the continuum dynamics.

To demonstrate such multiscale connection, we are now presenting the multiscale coupling algorithm that couples nonlinear continuum mechanics-based FEM with MMMD.

In Fig. 18.3, we show a spatial coupling partition of finite element domain and MD domain. In the MD domain, we employ the multiscale MMMD, while in the finite element domain we employ a CB based nonlinear FEM to model and simulate the material. For self-completeness, we briefly discuss the CB rule-based finite element in the following subsection.

18.2.1 Multiscale Coupling: From MMMD to FEM

The connection between MMMD and FEM is a two-way coupling. We first discuss how to transfer information from MMMD to FEM. A schematic illustration of the multiscale coupling is shown in Fig. 18.3 to demonstrate the coupling strategy at the boundary of FEM region and MMMD region.

First, in order to pass the mechanical information from MMMD to FEM, the FEM nodes at the multiscale boundary are chosen as the centers of mass of boundary MMMD cells shown in Fig. 18.3, and we call these nodes as the intermediate node. The global arrangement of the intermediate nodes is also shown in Fig. 18.4.

Second, at each center of mass of MMMD cells, we can calculate the first Piola–Kirchhoff (PK-I) virial stress $\mathcal{P}_\alpha^{Virial}$ based on Eq. (17.46). We can then use the PK-I virial stress to calculate traction force of the boundary between FEM region and MMMD region. As an example, for a four-node quadrilateral element surface,

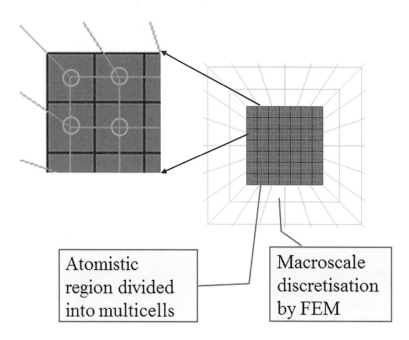

Figure 18.3 The multiscale coupling strategy: From MMMD to FEM

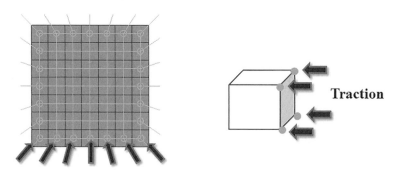

Figure 18.4 The multiscale coupling strategy: From FEM to MMMD

it contains four centers of mass of MMMD cells, and we assume that the normal of the quadrilateral element is **N**. The traction vector on the surface of this quadrilateral element will be,

$$\mathbf{T}_m = \sum_{\alpha=1}^{4} \mathbf{N} \cdot \mathcal{P}_\alpha^{Virial}, \qquad (18.37)$$

where the subscript m denotes multiscale boundary. Thus, the traction boundary Γ_t in Eq. (18.9) has two parts, i.e., $\Gamma_t = \Gamma_T \bigcup \Gamma_m$, where Γ_T is the macroscale traction boundary and Γ_m is the multiscale traction boundary.

As shown in Fig. 18.3, each center of mass of MMMD boundary cell will be in contact with two elements (for 2D quadrilateral FEM mesh). In 3D cases, it is in contact with four or even elements. For each MMMD cell, there is applied external load from it environment, i.e.,

$$M_\alpha \ddot{\mathbf{r}}_\alpha = -\sum_{\beta \neq \alpha} \mathbf{r}_{\alpha\beta} + \mathcal{B}_\alpha, \qquad (18.38)$$

in which the external force,

$$\mathcal{B}_\alpha = S_\alpha \bar{\mathbf{t}}_\alpha + \Omega_\alpha \bar{\mathbf{b}}_\alpha. \qquad (18.39)$$

Suppose that a center of mass of boundary MMMD cell is in contact with two elements and the MMMD cell has two exterior boundaries with normals $\mathbf{N}_1$ and $\mathbf{N}_2$ and surface facts S_1 and S_2.

We choose the following expression of $\mathcal{B}_\alpha$,

$$\mathcal{B}_\alpha = \sum_{I=1}^{2} S_I \mathbf{N}_I \cdot \mathbf{P}_I, \qquad (18.40)$$

where the PK-I stress is the weighted average of all PK-I stress evaluated at each quadrature point of that element, e.g.,

$$\mathbf{P}_I = \sum_{i=1}^{4} \frac{\partial W}{\partial \mathbf{F}}\bigg|_{\mathbf{x}_i} w_i, \qquad (18.41)$$

where w_i are the Gaussian quadrature weights. The explicit expression for $\frac{\partial W}{\partial \mathbf{F}}$ for the Tersoff potential is given in Eq. (18.24) and related equations. A schematic illustration of the multiscale coupling (from FEM to MMMD) is provided in Fig. 18.4.

18.2.2 Numerical Example

In this numerical example, we employ the coupled FEM–MMMD method to simulate nano-indentation of single-crystal silicon. The multiscale system contains the MD region and FEM region. In MD region, the domain is partitioned into $7 \times 7 \times 5 = 245$ cells, and in each cell there are 64 silicon atoms so that there are a total of 15,680 atoms in the MD region. We used the Tersoff potential to model single-crystal silicon. The FEM region is wrapped around the MD region, it contains a total of 1860 brick elements as shown in Fig. 18.5.

Two types of indenters are used in simulations: a 20 nm diameter spherical indenter and Vicker's indenter. In actual simulations, we used analytical expressions to express the indenters' contact surfaces, and we prescribed the motion of the indenters. Since the indentation contact region is only restricted to the MD region, we therefore only apply the indentation contact condition to MD region. We treat the indentation contact boundary as a coarse-scale MMMD boundary condition. Thus, we apply this prescribed indentation boundary condition only on the centers of MMMD surface cells. On the other hand, the induced stress and force on the indenter can be calculated by projecting and summing the virial stress of each MMMD cell that is in contact with the indenter.

Table 18.1 Equilibrium bond length and coordination number

Crystal	Bond [Å]	Coordination No.
Surface/Amorphous	2.35	< 4
Si-I	2.35	4
bct-5	2.31	4
	2.44	1
Si-III, XII	2.39	4
	3.2–3.4	1
Si-II	2.42	4
	2.57	2

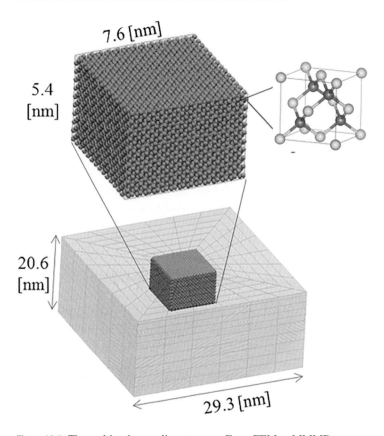

Figure 18.5 The multiscale coupling strategy: From FEM to MMMD

Nano-indentations of various forms of silicon are simulated, and the simulation conditions, i.e., the equilibrium bond distance and coordination number are listed in Table 18.1.

The simulation results for Si–I are displayed in Figs. 18.6 and 18.7. From Fig. 18.6, one may find the simulated relation between indentation depth and the force exerted

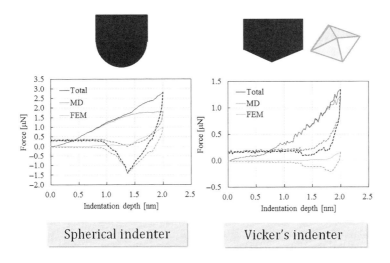

Figure 18.6 Nano-indentation simulation result

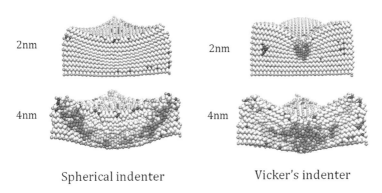

Figure 18.7 Indentation contact region morphology

by the indenter. Since this is a multiscale simulation, the both MMMD and FEM parts contribute to silicon substrate deformation or indentation depth. In Fig. 18.6, we plot the total indentation depth and load curves as well as the contribution from MD (MMMD) and FEM responses. In Fig. 18.7, we show the molecular contact surface morphologies with two different indenters.

18.3 Multiscale Cohesive Interphase Zone Model

In this section, we discuss the multiscale cohesive (interphase) zone model (MCZM), which is an atomistic-informed interphase zone model, and it was developed by Zeng and Li (see: Zeng and Li (2010); Li et al. (2012)).

The motivation in developing the MCZM is to have an atomistic-informed cohesive zone (CZ) model to model fracture and material/structure failures so that we do not

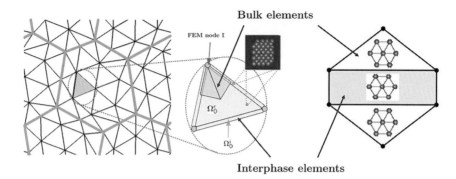

Figure 18.8 Schematic illustration of interphase zone element

need to use the ad-hoc material interface constitutive relation that is extensively used in conventional CZ model, e.g., Xu and Needleman (1994) and Elices et al. (2002). The multiscale CZ model is a finite element analysis-based multiscale method. The basic kinematic assumptions of the MCZM are as follows:

1. Bulk elements undergo uniform deformation, and all the nonuniform deformation is confined inside the interphase elements, i.e., the dark element region in Fig. 18.8.
2. Fracture is no longer deemed solely as cleavage surface or interface separation but material failure due to a general debonding damage over a finite volume.
3. The local material strength is not defined as the threshold of the surface separation but the critical state of the overall constitutive relation of the material.

For crystalline materials, the crystal defects are often associated with persistent slip bands, grain boundaries, twin boundaries, stacking faults, and the like. Based on this observation, we adopt the following kinematic assumptions: *The deformation inside every bulk element is uniform or homogeneous, whereas all defect caused non-uniform deformations are confined inside the interphase element, which is a narrow finite width strip that is either along the slip planes, grain boundaries, or twin boundaries.*

18.3.1 Multiscale Constitutive Modeling of Interphase Zone

There are two types of elements in the multiscale CZ method: the bulk element and the interphase element. Since the deformation inside the bulk element is assumed to be homogeneous, its constitutive relation can be modeled by directly applying the CB rule. Whereas inside the interphase zone element, the deformation is nonuniform, and we cannot use the first-order CB rule to model it.

In MCZM method, we use an interphase depletion potential that is derived from the atomistic potential inside the bulk element. To obtain the depletion potential for the weak interphase zone, we assume that the interphase element is much softer than the adjacent bulk elements, and the intermolecular interaction inside the interphase zone may be treated as a van der Waals interaction that can be linearly superposed.

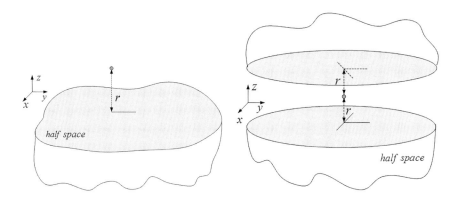

Figure 18.9 Modeling of atomistic potential inside the interphase element

In fact, we may view the bulk medium as rigid with almost no deformation, so the two bulk elements adjacent to the compliant interphase zone may be viewed as two rigid body half spaces (see: Fig. 18.9). If the atomistic potential for a given bulk medium is available, we can obtain the depletion potential in the interphase element by integrating the bulk atomistic potential over two bulk medium half spaces. In general, the interphase depletion potential can be obtained by the following analytical integration (see: Israelachvili (2011), pp. 156–158),

$$\phi_{depl}(r) = \int_{\text{Half Space}} \beta \phi_{bulk}(r - r') dV', \qquad (18.42)$$

where β is parameter related to normalized atom density.

Some interphase depletion potentials may even have close form expressions. For example, if the Lennard-Jones (LJ) potential is chosen as the atomistic potential of bulk medium as shown in Eq. (18.43),

$$\phi_{bulk} = 4\epsilon \left(\left(\frac{\sigma}{r}\right)^{12} - \left(\frac{\sigma}{r}\right)^{6} \right), \qquad (18.43)$$

the corresponding interphase depletion potential will be found as follows (see: Sauer and Li 2007):

$$\phi_{depl} = \frac{\pi \epsilon}{\sqrt{2}} \left(\frac{1}{45} \left(\frac{r_0}{r}\right)^9 - \frac{1}{3} \left(\frac{r_0}{r}\right)^3 \right), \qquad (18.44)$$

where ϵ is the depth of the potential well, σ is the (finite) distance at which the bulk atomistic potential is zero, and $r_0 = \sigma 2^{1/6}$ is the equilibrium bond distance in the bulk material.

In the actual implementation of the multiscale CZ model, we choose the following atomistic potential in the interphase element,

$$\phi_{inpt}(r) = \begin{cases} \phi_{bulk}(r), & r < r^*, \\ \alpha \phi_{bulk}(r) + (1 - \alpha) \phi_{depl}(r), & r \geq r^*, \end{cases} \qquad (18.45)$$

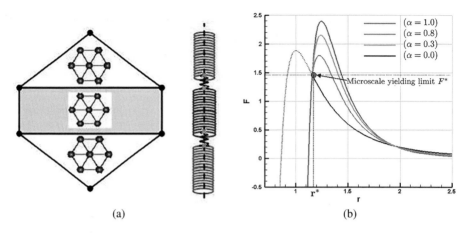

Figure 18.10 (a) Two bulk elements sandwich an interphase element and (b) atomistic force–displacement relation inside interphase elements ($\phi_{intp} = \alpha \phi_{bulk} + (1 - \alpha)\phi_{depl}$)

where r^* is a bond distance at which, $\phi'_{bulk}(r^*) = \phi'_{depl}(r^*)$ (see: Fig. 18.9) and $0 \leq \alpha \leq 1$ is a material parameter. By doing so, we can construct a perfect weak zone by controlling failure point.

Without loss of generality, we illustrate the proposed constitutive modeling inside the interphase element by using the following 2D example. In this case, the multiscale finite element model may be simplified as two triangle bulk elements sandwiching one quadrilateral interphase element (see: Fig. 18.10(a)).

To demonstrate how MCZM works, one may simplify the combination of the two bulk triangle elements/one interphase quadrilateral element as a 1D model of three spring in series connection: the two outside springs representing two bulk elements are at the top and the bottom, and the center spring representing the interphase element is in the middle of the series spring connection (see: Fig. 18.10(a)).

In Fig 18.10(b), we then plot the force–displacement relations inside the interphase element with different values of α. At the beginning, when bond distance $r < r^*$, both the bulk element and the interphase element have the same atomistic potential (the dark line). When $r \rightarrow r^*$, the force inside all three springs will reach to F^*. If we continue to stretch the bond length, the interphase element will start to unload ($\alpha = 0$), which follows the light dark path, whereas the bulk elements will also be in an unloading state, but they will stay in the dark path ($\alpha = 1.0$) below the point F^*.

Therefore, as the material continues to stretch, the bond inside the interphase element will stretch significantly until it breaks while the bulk elements remain in a uniform deformation state, in which the force–displacement relation is almost a linear elastic. From Fig. 18.10, one may observe that no matter what value α is the atomistic potential, the unloading force inside the interphase element will always start at the same point (r^*, F^*), which we label it as the *microscale yielding limit*. Since this point is solely determined as the interception between the depletion potential and the original atomistic potential, it may be regarded as a material parameter. When $\alpha = 0$, after this point, the interphase element starts moving on a softening path; when $\alpha \neq 0$, the interphase element may experience some microscale hardening, which resembles the macroscale material property. Second, the value microscale yields may be controlled by the normalized atom density parameter β, which may be related to the porosity or vacancy density in the interphase element. Third, if $\alpha = 1.0$, then $\phi_{intp} = \phi_{bulk}$, which is the case that there is basically no material degradation in the interphase element. The proposed atomistic potential-based multiscale CZ method will still work in case, and most of the results that have been presented in this chapter are valid in this case. Interested readers may consult He and Li (2012), in which all the simulations are entirely done by taking $\alpha = 1.0$.

18.3.2 Element (Mesh) Fault Energies

In metallic materials, the stacking fault energy is referred to the energy stored between two interrupted layers of a stacking crystal plane sequence. The stacking fault energy may be viewed as a material property to characterize defect evolution such as dislocation motions.

In MCZM, one may view the interphase element as an artificial fault, which may representing a coarse-grained model for actual physical fault when the atomistic potential inside the interphase element switching from the bulk potential to the depletion potential.

As an analogous to stacking fault energy, we study the artificial fault energy between two bulk elements. Let the local coordinate X_1 parallel to the element mesh boundary be considered. Hence, we may denote $\bar{u}$ as the relative effective horizontal (tangential) opening displacement and $\bar{v}$ as the relative effective vertical (normal) opening displacement of the interphase zone. Therefore, one may find that the traction along the element boundary is an explicit function of the effective deformation gradient inside the interphase zone.

We consider the cohesive traction forces along the boundary of the interphase zone, which is the same boundary of the adjacent bulk elements with the opposite out normals,

$$\mathbf{T}^{cohe} = \mathbf{P}^c(\bar{\mathbf{F}}^c) \cdot \mathcal{N}, \qquad (18.46)$$

where $\mathcal{N}$ is the out-normal of adjacent bulk FE elements. The effective deformation gradient inside the interphase element for the cases of pure Mode I and pure Mode II are given separately as follows:

$$\bar{\mathbf{F}}_n = \begin{bmatrix} 1, & 0 & 0 \\ 0, & 1+\bar{v}/R_0 & 0 \\ 0, & 0, & 1 \end{bmatrix} \text{ and } \bar{\mathbf{F}}_t = \begin{bmatrix} 1, & \bar{u}/R_0, & 0 \\ 0, & 1, & 0 \\ 0, & 0, & 1 \end{bmatrix}, \qquad (18.47)$$

where R_0 is the width of the interphase element. Substituting the relations in Eq. (18.47) into Eq. (18.46), one can find the relationship between the element traction and the relative separation (opening displacements) of the interphase zone.

In general, the longitudinal direction of the soft interphase zone may be viewed as a fault, or more precisely an element (mesh) fault, because we assume that total

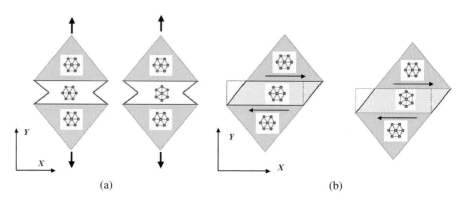

Figure 18.11 Interphase zones under normal and tangential traction: (a) Interface zone under uniaxial tension, and (b) Interface zone under simple shear.

deformation field at macroscale is piece-wise constant. In this context, this kinematic assumption is not just a convenience for finite element discretization, but a physical modeling to mimic the defect distribution, which is very similar to the kinematics of crystal plasticity, e.g., Taylor (1938). The ability to evaluate the element fault energy will help us design multiscale simulations that have predictive power. In this section, we shall focus on analysis and calculation element fault energies.

Consider that the crystalline solid is a single crystal that has hexagonal symmetry. Hence, the lattice orientation in each bulk element will be the same. Let the finite element mesh boundary coincides with the possible slip line directions, so in the slip line strip the original lattice may be rotated, distorted, or remain the same. In the case discussed in this chapter we assume that the symmetry inside the interphase zone remains hexagonal, but it may be rigidly rotated to a fixed angle. We consider two cases: (I) one pair of the hexagonal axes are parallel to the element boundary and (II) one pair of the hexagonal axes are perpendicular to the element boundaries.

This multiscale method allows us to calculate the mesh stacking fault energy in terms of both interface normal opening (mode I) and the tangential opening (model II).

Since the deformation gradient is only the function of interface opening displacement, the bulk atomistic potential may directly be linked to the coarse-grain traction/displacement potential, and the following is their explicit expressions,

$$W_n(\bar{v}) = \beta_c \sum_{i=1}^{6} \phi_{deple}(r_i(\bar{v})), \text{ and } W_t(\bar{u}) = \beta_c \sum_{i=1}^{6} \phi_{depl}(r_i(\bar{u})), \qquad (18.48)$$

in which $\mathbf{r}_i = \bar{\mathbf{F}}_n(\bar{v}) \cdot \mathbf{R}_i$ or $\mathbf{r}_i = \bar{\mathbf{F}}_t(\bar{u}) \cdot \mathbf{R}_i$ and β_c is the atom density in the interphase zone.

Substituting Eq. (18.47) into Eq. (18.48) and considering the relative interphase zone opening displacements as the effective displacements inside the interphase zone, we can find the surface energy and stacking fault energy in terms of the effective displacements. However, the calculation of element fault energy depends on the lattice structure inside the interphase zone. Here, we consider two cases of different lattice orientations of hexagonal lattices:

For Case I,
$$R_i = a\left\{\cos\left(\frac{(i-1)\pi}{3}\right), \sin\left(\frac{(i-1)\pi}{3}\right)\right\}, \quad i = 1, 2, \ldots, 6 \tag{18.49}$$

and for Case II
$$R_i = a\left\{\cos\left(\frac{\pi}{6} + \frac{(i-1)\pi}{3}\right), \sin\left(\frac{\pi}{6} + \frac{(i-1)\pi}{3}\right)\right\}, \quad i = 1, 2, \ldots, 6. \tag{18.50}$$

Two interphase zone loading tests are conducted, which are illustrated in Fig. 18.11, and the corresponding element or mesh staking fault energies calculated are depicted in Fig. 18.12. One may find that in Case I, the unstable stacking fault energy ($\gamma_{us} = 3.25$) is smaller than the corresponding surface energy ($2\gamma_s = 6.85$), the ratio between the two is

$$\alpha_I = \frac{\gamma_{us}}{2\gamma_s} = 0.47 < 1, \tag{18.51}$$

whereas in the second case, the unstable stacking fault energy ($\gamma_{us} = 843.27$), which is much larger (in Fig 18.12 we scale the original value by a factor of 40 for better visualization purpose) than that of the cohesive surface energy ($2\gamma_s = 10.02$), the ratio between the two is

$$\alpha_{II} = \frac{\gamma_{us}}{2\gamma_s} = 84.16 \gg 1. \tag{18.52}$$

Based on this analysis, one may expect the ductile fracture for Case I, because the unstable stacking fault energy is smaller than the surface energy, the lattice sliding is more susceptible than the lattice cleavage opening along the allowable kinematic failure mode – that is element boundary – whereas in Case II the peak value of unstable stacking energy is about 84 times larger than the surface energy, the lattice is more susceptible to cleavage opening than sliding along the allowable element boundary. Therefore, we would expect brittle fracture for Case II. In the next section, we use both the lattice orientation and mesh fault set-up to simulate crack propagations, compare the results between them, and verify the above analytical predictions.

Above analysis has shown that the element stacking fault energies not only depend on the bulk lattice orientation, but also depend on the microstructure of the interface. This part may be different from the conventional concept of stacking fault energy.

18.3.3 Numerical Examples

Example 18.1 In this example, we apply MCZM method to simulate crack propagations at macroscale. The testing material is a single crystal with hexagonal symmetry. In the simulations, all triangle elements are chosen as equilateral triangles that align their boundary along the directions of the hexagonal lattice in the bulk element. In interphase elements, we choose two sets of hexagonal lattices: (1) the one with the same orientation as in the bulk element and (2) the hexagonal lattice that rotates an angle of $\pi/6$ with respect to the lattice orientation in the bulk element. The two lattice structures are shown in Fig. 18.13(a) and (b). The LJ potential is used

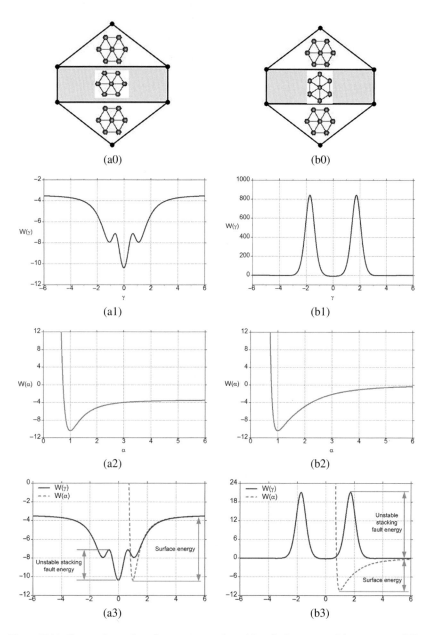

Figure 18.12 Comparison of surface energy and stacking fault energy with respect to different mesh fault orientations: (a1)–(b1) unstable stacking fault energy, (a2)–(b2) cohesive surface energy, and (a3)–(b3) γ_{us} vs. $2\gamma_s$ inside the interphase element: solid line–stacking fault energies, dash line–surface energies.

in the bulk elements, and the depletion potential in Eq. (18.44) is used as the atomistic potential inside the interphase zone. We set $\epsilon = 1$ and $\sigma = 1$ for the bulk and interphase atomistic potentials (see: Eqs. (18.43) and (18.44)).

In the simulation, the test specimen is a 2D plate with dimension (2 mm × 2 mm), which is subjected to unilateral tension in Y-axis, as shown in Fig. 18.13, and there is

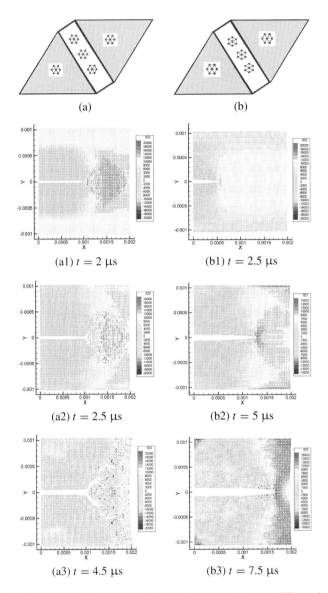

Figure 18.13 Stress distribution for crack propagations vs. different interface lattice orientations: (a) Interface lattice orientation is the same as that in the bulk crystal and the corresponding crack propagation sequence (a1)–(a3), and (b) Interface lattice orientation is different from that in the bulk crystal and the corresponding crack propagation sequence (b1)–(b3).

a pre-crack at the left side of the plate. There are total 9,520 interphase elements and 6,400 triangular bulk elements. The time step is chosen as $\Delta t = 1 \times 10^{-10}$ s. The process of the crack growth is displayed in Fig. 18.13 (a1)–(a3) and Fig. 18.13 (b1)–(b3) for both cases. The mesh stacking fault energies for both cases have been calculated in the previous section. Since in Case I, the unstable element fault energy is smaller than the element surface energy, it exhibits a typical ductile fracture pattern so the crack path is almost along the direction of element boundary where the shear stress is maximum. Whereas in Case II, the mesh stacking fault energy is much larger

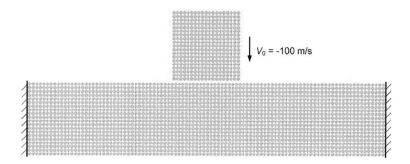

Figure 18.14 The impact problem setting: A rigid square block impact on and penetrate into a plate

than the cohesive surface energy, it exhibits the typical brittle failure pattern with the crack grows along the horizontal line that is aligned with the initial pre-notch direction, or the crack direction is along with the direction of element boundary where the normal stress is maximum. However, void formation at the crack tip can be also observed in simulations with different parameters.

Example 18.2 In this example, numerical simulations have been carried out to simulate high-speed impact-induced spall fractures, which is a very difficult problem that has been elusive to many existing numerical methods (see: Antoun et al. (2003)). The exact problem statement is described in Fig. 18.14. It is a rigid projectile penetrating a deformable plate. The projectile is a (0.38 mm × 0.38 mm) rigid block with impact velocity $v = 100$ m/s, the target is a (2 mm × 0.4 mm) block clamped at the two ends. In this simulation, there are totally 11,880 cohesive elements and 8,000 triangular bulk elements used in the target.

In time integration, the time step is chosen as $\Delta t = 1 \times 10^{-10}$ s. Contact problems are characterized by impenetrability conditions that needs to be enforced during computation. We adopted the exact enforcement of the impenetrability condition in a single time step (see: Hughes et al. (1976)). The simulation results are shown in Fig. 18.15. The wave propagation from the contact point to the opposite boundary has been observed. The phenomena of spall fracture under impacts has been captured (see: Fig. 18.15).

Example 18.3 In this example, we use the multiscale CZ method to simulate the penetration and fragment of a deformable projectile made by aluminum alloy power during high-speed impact and penetration of thin aluminum plate.

We choose the many-body Sutton–Chen (SC) potential (see: Pelaez et al. (2006)) as the bulk material's atomistic potential, which is a special EAM potential designed for aluminum metals. For an chosen atom i in the bulk element,

$$\phi_{bulk}(r) = \epsilon \left\{ \frac{1}{2} \sum_{j \neq i} \left(\frac{a}{r_{ij}} \right)^n - c \sqrt{\rho_i} \right\}, \tag{18.53}$$

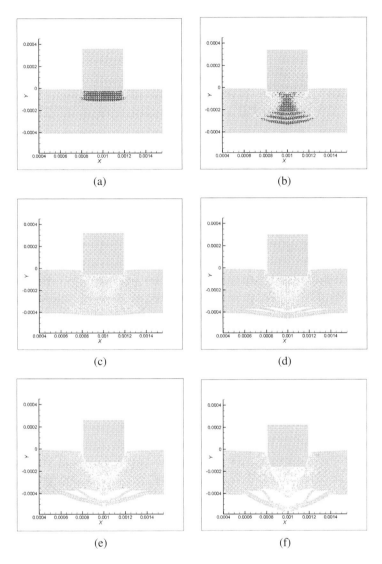

Figure 18.15 The snapshot of stress distribution for the contact impact process: (a) $t = 0.2$ μs; (b) $t = 0.4$ μs; (c) $t = 0.6$ μs; (d) $t = 0.8$ μs; (e) $t = 1.2$ μs; (f) $t = 1.6$ μs (Data source is from Zeng and Li (2010)).

where ρ_i is the electron density at the site of atom i,

$$\rho_i(r) = \sum_{i \neq j} \left(\frac{a}{r_{ij}}\right)^m$$

and ϵ, c, a, m, n are material constants obtained by fitting the atomistic potential (see: Eq. (18.53)) with the results obtained from ab-initio calculation based on a stable bulk configuration. One can see that in EAM potential, there are two distinct part contributions: the pair potential part and embedding electron density part.

Inside interphase elements, we use the general interphase potential expression,

$$\phi_{intp}(r) = \alpha \phi_{bulk}(r) + (1 - \alpha) \phi_{depl}(r). \tag{18.54}$$

However, in our approach, the depletion potential is chosen as the coarse-grain potential of the pair potential part in EAM potential, i.e.,

$$\phi_{depl}(r) = \frac{\epsilon}{2\Omega_0^u} \sum_{j \neq i} \int_{half-space} \left(\frac{a}{\ell}\right)^n dV,$$

with $dV = \pi \ell r^2 dz$, $\ell = \sqrt{r^2 + (z - r_{ij})^2}$

$$= \frac{\epsilon \pi a^3}{\Omega_0^u (n-2)(n-3)} \sum_{j \neq i} \left(\frac{a}{r_{ij}}\right)^{n-3}. \tag{18.55}$$

The parameter for the SC potential for Al are: $\epsilon = 3.31477d - 02 eV$, $c = 16.399$, $a = 4.05$ Å, $m = 6$, and $n = 7$. For the interphase potential we choose the parameter $\alpha = 0.48$ in Eq. (18.54).

The justification of the above approach is based on the argument that significant concentration of vacancy inside a damaged interphase element will greatly affect electron density distribution, and the contribution to material interface strength may be neglected. Therefore, we neglect its contribution in the colloidal crystal approximation. In this example, a dynamic simulation of penetration/fragmentation of a polycrystalline aluminum cylinder through an aluminum plate is carried by using MCZM. The polycrystalline aluminum projectile is 1 mm in diameter and 3 mm in total length, and it impacts a square aluminum plate that has the dimension 5 mm × 5 mm × 0.2 mm (width × length × thickness). The initial velocity of the projectile is: 4000 m/s, shown as Fig. 18.16. The shock wave induced by impact force propagates in both projectile and the target plate, which causes dynamic fracture phenomena in both bodies.

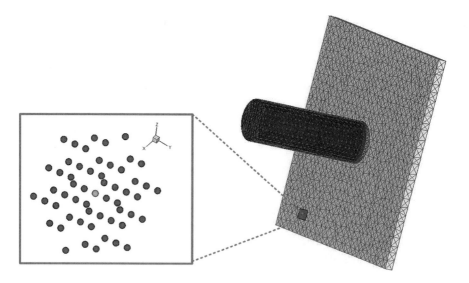

Figure 18.16 MCZM finite element mesh for aluminum projectile/target system with an EAM potential

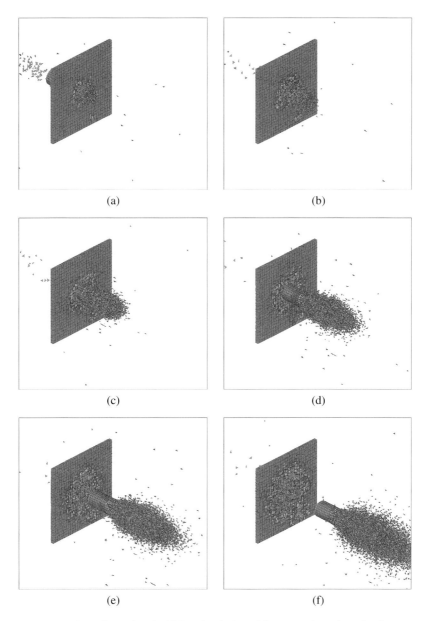

Figure 18.17 Three-dimensional MCZM simulation of fragmentation of an aluminum projectile. (a), (b), (c), (d), (e), and (f) are the tine sequence of the projectile penetration and fragmentation process.(Data source is from Ren and Li (2013)).

In this simulation, the bulk grains are represented by tetrahedron element with various volumes and lattice orientations. Between grains, there are wedge-shaped interphase elements that represent the grain boundary. The computational domain is discrete by 64,608 bulk elements and 124,820 interphase elements. The problem setup and the finite element mesh are shown in Fig. 18.16. The simulation results are shown in Fig. 18.17, which is a time sequence of the deformable project penetration through a thin plate, and then the project starts to disintegrate into many fragments.

The main advantage of MCZM is that the material behaviors inside the CZ or the process zone are determined by the atomistic potential that is related to the atomistic potential of the bulk material. By doing so, the mechanical properties inside the CZ element are consistent with the mechanical properties of the bulk material element, and this is in sharp contrast with conventional CZ method whose empirical cohesive laws have almost no connections to the material properties in the bulk elements (see: Xu and Needleman 1994; Falk et al. 2001).

Here, we only briefly present the results of MCZM simulations on high-speed impact and fragmentation problems. A detailed report on how to use MCZM to simulate 3D spall fracture and fragmentation can be found in Ren and Li (2013).

18.4 Higher-Order MCZM

18.4.1 Hierarchical CB Rule-Based MCZM

In the MCZM, the entire domain of the specimen is discretized by a number of bulk elements as the usual finite element (FEM) discretization, and the cohesion zone is represented by a network of CZ elements among bulk elements, see Fig. 18.18. However, unlike conventional CZ model, the CZ or the process zone in MCZM is represented by an element with extremely thin but finite thickness. In this work, since the tetrahedral element is utilized for the bulk element, the CZ element is represented by a triangular-shaped prism element which shares two triangular facets of the two adjacent tetrahedral (bulk) elements, as shown in Fig. 18.18(b).

Another important characteristic of MCZM is that the strain-stress relation is derived based on the CB rule that is evaluated by deformation of an unit cell, which is embedded in each quadrature point in both bulk and CZ elements.

The key feature of the so-called hierarchical MCZM is that the different order of the CB rule is applied to different order of the CZs. We denote the bulk element as the zeroth order CZ, the interphase element between two bulk elements as the

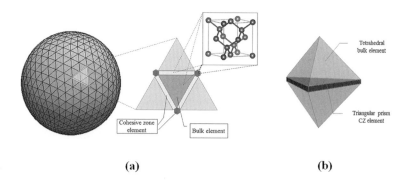

Figure 18.18 Concepts of hierarchical multiscale cohesive zone model: (a) Two-dimensional finite element mesh of multiscale cohesive zone model, and (b) Three-dimensional multiscale cohesive zone finite element setup.

first-order CZ, and void among the multiple bulk elements and interphase elements as the third-order or even fourth-order CZ elements (see: Li et al. (2014) for details).

Consider the finite deformation of a crystalline lattice. We may be able to express any deformed position vector in the current configuration by its undeformed image in the referential configuration through Taylor expansion,

$$\Delta \mathbf{x} = \Delta \frac{\partial \mathbf{x}}{\partial \mathbf{X}}\bigg|_{\mathbf{X}=\mathbf{X}_A} \cdot \Delta \mathbf{X} + \frac{1}{2!}\frac{\partial^2 \mathbf{x}}{\partial \mathbf{X}^2}\bigg|_{\mathbf{X}=\mathbf{X}_A} : \Delta \mathbf{X}_i \otimes \Delta \mathbf{X}$$
$$+ \frac{1}{3!}\frac{\partial^3 \mathbf{x}}{\partial \mathbf{X}^3}\bigg|_{\mathbf{X}=\mathbf{X}_A} \vdots \Delta \mathbf{X} \otimes \Delta \mathbf{X} \otimes \Delta \mathbf{X} + \cdots \quad (18.56)$$

or in a more common notation,

$$\Delta \mathbf{x} = \mathbf{F}_A \cdot \Delta \mathbf{X} + \frac{1}{2!}\mathbf{G}_A : \Delta \mathbf{X} \otimes \Delta \mathbf{X} + \frac{1}{3!}\mathbf{H}_A \vdots \Delta \mathbf{X} \otimes \Delta \mathbf{X} \otimes \Delta \mathbf{X} + \cdots \quad (18.57)$$

where

$$\mathbf{F}_A := \frac{\partial \mathbf{x}}{\partial \mathbf{X}}\bigg|_{\mathbf{X}=\mathbf{X}_A}$$

is the deformation gradient, and

$$\mathbf{G}_A := \frac{\partial^2 \mathbf{x}}{\partial \mathbf{X}^2}\bigg|_{\mathbf{X}=\mathbf{X}_A} \quad \text{and} \quad \mathbf{H}_A := \frac{\partial^3 \mathbf{x}}{\partial \mathbf{X}^3}\bigg|_{\mathbf{X}=\mathbf{X}_A}$$

are the second and the third gradient of $\mathbf{F}$.

The above expression is often referred to as the higher-order CB rule, e.g., Sunyk and Steinmann (2003). The so-called hierarchical multiscale CZ model is that it uses the first-order CB rule,

$$\Delta \mathbf{x} \approx \mathbf{F}_A \cdot \Delta \mathbf{X},$$

to derive the stress–strain relation in the bulk crystal element, i.e., the zeroth order CZ; and it uses the second-order CB rule,

$$\Delta \mathbf{x} \approx \mathbf{F}_A \cdot \Delta \mathbf{X} + \mathbf{G}_A : \Delta \mathbf{X} \otimes \Delta \mathbf{X},$$

to derive the stress–strain relation in the first CZ element, and it uses the third-order CB rule,

$$\Delta \mathbf{x} \approx \mathbf{F}_A \cdot \Delta \mathbf{X} + \mathbf{G}_A : \Delta \mathbf{X} \otimes \Delta \mathbf{X} + \mathbf{H}_A \vdots \Delta \mathbf{X} \otimes \Delta \mathbf{X} \otimes \Delta \mathbf{X}$$

to derive the constitutive relation in the second-order CZ element, and so forth. Here, we restrict our attention only to the bulk element and the interphase element, i.e., the first-order CZ element.

For single silicon, it is not a Bravais lattice, and the unit cell of silicon is composed of a multiple atoms. The silicon lattice structure is a diamond cubic structure. With considering shape change of the unit cell associated with each element deformation, we can evaluate the first Piola–Kirchhoff stress tensor $\mathbf{P}$ and the second-order stress tensor $\mathbf{Q}$ as derivatives of strain energy density W computed with the Tersoff potential as shown previously.

18.4.2 Multiscale Finite Element Formulation for the Second-Order MCZM

To facilitate the discussion, we first briefly outline the Galerkin weak formulation based on virtual work principle for FEM discretization. The detailed procedures can be found in the following references: Zeng and Li (2010); He and Li (2012); Qian and Li (2011); Liu and Li (2012); Fan and Li (2015), and Urata and Li (2017a).

The following presentation is mainly based on Urata and Li (2017a). We first define the total Lagrangian for the continuum medium that is under consideration,

$$\mathcal{L} = \mathcal{K} - \left(\mathcal{W}_{int} + \mathcal{W}_{ext}\right), \tag{18.58}$$

where $\mathcal{W}_{ext}$ is the external potential energy, $\mathcal{W}_{int}$ and $\mathcal{K}$ are the strain energy of the continuum and total kinetic energy, respectively, and defined as,

$$\mathcal{K} = \int_\Omega \frac{1}{2} \rho \dot{\mathbf{u}} \cdot \dot{\mathbf{u}} dV, \tag{18.59}$$

$$\mathcal{W}_{int} = \int_\Omega W(\mathbf{F}, \mathbf{G}) dV, \tag{18.60}$$

where ρ and $\dot{\mathbf{u}}$ are the mass density and velocity field of the continuum, respectively, and W is the strain energy density function as functions of strain and strain gradient.

Then, we employ the Hamiltonian principle to derive the variational weak formulation,

$$\int_{t_0}^{t_1} \left(\delta \mathcal{K} - (\delta \mathcal{W}_{int} + \delta \mathcal{W}_{ext})\right) dt = 0, \tag{18.61}$$

where

$$\int_{t_1}^{t_2} \delta \mathcal{K} dt = \int_{t_1}^{t_2} \int_\Omega \rho \dot{\mathbf{u}} \cdot \delta \dot{\mathbf{u}} dV dt = -\int_{t_1}^{t_2} \int_\Omega \rho \ddot{\mathbf{u}} \cdot \delta \mathbf{u} dV dt, \tag{18.62}$$

$$\delta \mathcal{W}_{int} = \int_\Omega \left[\frac{\partial W}{\partial \mathbf{F}} : \delta \mathbf{F} + \frac{\partial W}{\partial \mathbf{G}} \vdots \delta \mathbf{G}\right] dV = \int_\Omega \left[\mathbf{P} : \delta \mathbf{F} + \mathbf{Q} \vdots \delta \mathbf{G}\right] dV, \tag{18.63}$$

$$\delta \mathcal{W}_{ext} = -\int_\Omega \mathbf{b} \cdot \delta \mathbf{u} dV - \int_{\delta\Omega_t} \bar{\mathbf{T}} \cdot \delta \mathbf{u} dS. \tag{18.64}$$

In Eq. (18.64), $\mathbf{b}$ and $\bar{\mathbf{T}}$ are the body force inside the bulk media and the traction vector on the surface $\delta\Omega_t$, respectively. Consequently, the Galerkin weak formulation can be reformulated in terms of element summation as follows:

$$\mathop{A}_{e=1}^{n_B^e} \left\{\int_{\Omega_B^e} \left(\rho_0 \ddot{\mathbf{u}}^h \cdot \delta \mathbf{u} + \mathbf{P} : \delta \mathbf{F}^h\right) dV\right\} + \mathop{A}_{e=1}^{n_C^e} \left\{\int_{\Omega_C^e} \left(\mathbf{P} : \delta \mathbf{F}^h + \mathbf{Q} \vdots \delta \mathbf{G}^h\right) dV\right\},$$

$$= \mathop{A}_{e=1}^{n_B^e} \left\{\int_{\Omega_B^e} \mathbf{b} \cdot \delta \mathbf{u}^h dV\right\} + \mathop{A}_{e=1}^{n_C^e} \left\{\int_{\Omega_C^e} \mathbf{b} \cdot \delta \mathbf{u}^h dV\right\} + \mathop{A}_{e=1}^{n_B^e} \left\{\int_{\Gamma_t} \bar{\mathbf{T}} \cdot \delta \mathbf{u}^h dS\right\},$$

$$\tag{18.65}$$

where, Ω_B^e and Ω_C^e are the domains of bulk and CZ elements; Γ_t is the traction boundary of the system; n_B^e and n_C^e are number of bulk and CZ elements, respectively;

and superscript h represents kinetic field corresponding to FEM interpolation field. Note that only the bulk element is assumed to have constant deformation gradient, while all cohesive elements are assumed to have up to the second order deformation gradient, i.e., the first gradient of the deformation gradient, which will be discussed in details in later sections.

By considering FEM interpolation approximation, displacement field can be represented by using element shape function matrix $\mathbf{N}$ as follows:

$$\mathbf{u}^h(\mathbf{X}) = \sum_{i=1}^{n_{node}} \mathbf{N}_i(\mathbf{X})\mathbf{d}_i, \quad (18.66)$$

where n_{node} is number of node composing an element and $\mathbf{d}$ is the nodal displacement vector. According to Eqs. (18.65) and (18.66), the discrete equation of motion for FEM procedure is expressed as,

$$[\mathbf{M}][\ddot{\mathbf{d}}] + [\mathbf{f}^{int}(\mathbf{d})] - [\mathbf{f}^{cohe}(\mathbf{d})] = [\mathbf{f}^{ext}], \quad (18.67)$$

where $[\mathbf{M}]$ is the mass matrix. $[\mathbf{f}^{int}]$, $[\mathbf{f}^{cohe}]$, and $[\mathbf{f}^{ext}]$ are force vectors from bulk elements, and CZ elements and external force, respectively. They are defined as follows:

$$[\mathbf{M}] = \underset{e=1}{\overset{n_B^e}{A}} \int_{\Omega_B^e} \rho_0 [\mathbf{N}_e]^T [\mathbf{N}^e] dV, \quad (18.68)$$

$$[\mathbf{f}^{int}] = \underset{e=1}{\overset{n_B^e}{A}} \int_{\Omega_B^e} [\mathbf{B}_e]^T \mathbf{P}(\mathbf{d}) dV, \quad (18.69)$$

$$[\mathbf{f}^{ext}] = \underset{e=1}{\overset{n_B^e}{A}} \left\{ \int_{\Omega_B^e} [\mathbf{N}_e]^T [\mathbf{B}_e] dV + \int_{\partial \Gamma_t} [\mathbf{N}_e]^T \bar{\mathbf{T}} dS \right\}, \quad (18.70)$$

$$\mathbf{f}^{cohe} = \underset{e=1}{\overset{n_{elem}^{CZ}}{A}} \int \left\{ [\mathbf{B}_e]^T \mathbf{P}(\mathbf{d}) + [\mathbf{C}_e]^T \mathbf{Q}(\mathbf{d}) \right\} dV, \quad (18.71)$$

where $[\mathbf{B}_e]$ and $[\mathbf{C}_e]$ are the element strain–displacement matrix and the gradient of the strain–displacement matrix, respectively. The explicit time integration based Newmark-β method with $\beta = 0$ is used in nodal velocity and displacement integrations (see: Belytschko (1983)).

Alternative to Eq. (18.71), the internal cohesive force can be also evaluated by using integration of parts as discussed in Fan and Li (2015),

$$\int_\Omega (\mathbf{P}(\phi) : \delta \mathbf{F} + \mathbf{Q}(\phi) \vdots \delta \mathbf{G}) dV = - \int_\Omega \left(\nabla_X \cdot (\mathbf{P} - \nabla_X \mathbf{Q}) \right) \cdot \delta \mathbf{x} dV,$$

$$+ \int_{\partial \Omega} \left(\mathbf{P} - \nabla_X \cdot \mathbf{Q} \right) : (\mathcal{N} \otimes \delta \mathbf{x}) dS + \int_{\partial \Omega} \mathbf{Q} \vdots (\mathcal{N} \otimes \delta \mathbf{F}) dS, \quad (18.72)$$

where Ω is domain of CZ element, and $\mathcal{N}$ is normal vector of the facet of CZ element. Observing Eq. (18.72), we find that the first term of the right-hand side (RHS) will

Table 18.2 Results of the calculations of fracture stress of the noted specimen (1.0 [μm]) by using the Lennard-Jones potential

	CZ thickness		Yield stress	
	$p_{th}[-]$	Avg. [μm]	σ_y [MPa]	Ratio with Mesh A(3)
Mesh A(1)	0.020	0.095	142	158%
	0.013	0.063	129	144%
	0.007	0.035	134	149%
Mesh A(2)	0.020	0.063	110	122%
	0.011	0.035	100	111%
Mesh A(3)	0.020	0.035	90	–
Mesh B	0.010	0.029	59	66%

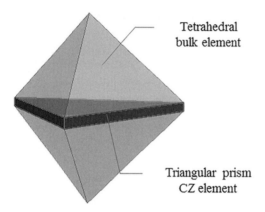

Figure 18.19 Wedge-shaped cohesive zone element with two tetrahedral bulk elements

disappear because it is part of the equation of motion in the CZ. The second and third terms of RHS are boundary conditions, and the third term is the boundary condition for the second-order stress tensor, which we often neglect because of its numerical insignificance. Note that these boundary conditions are in fact the interface boundary conditions between the bulk crystal element and the cohesive zone element (see: Fig. 18.19).

18.4.3 Simulation of Fracture Toughness of Single-Crystal Silicon

In this example, we present the modeling details on simulation of fracture toughness of single-crystal silicon.

For the simulation of a single-crystal silicon specimen, the simulation time interval is set to 1.0×10^{-4}[ns] and uniaxial stretch is at the rate 3×10^5[m/s] in horizontal direction was applied. The crystal orientation in stretched direction is $\langle 110 \rangle$. The MCZM estimated fracture stress σ_y is shown in Table 18.2, and crack patterns are visually illustrated in Figs. 18.20 and 18.21.

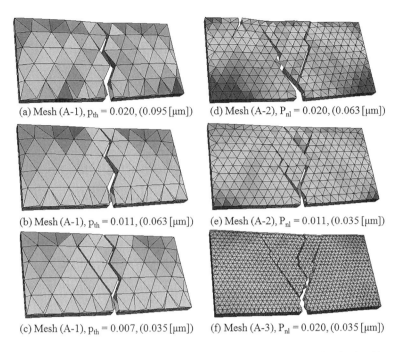

(a) Mesh (A-1), p_{th} = 0.020, (0.095 [μm])
(b) Mesh (A-1), p_{th} = 0.011, (0.063 [μm])
(c) Mesh (A-1), p_{th} = 0.007, (0.035 [μm])
(d) Mesh (A-2), P_{nl} = 0.020, (0.063 [μm])
(e) Mesh (A-2), P_{nl} = 0.011, (0.035 [μm])
(f) Mesh (A-3), P_{nl} = 0.020, (0.035 [μm])

Figure 18.20 Crack paths of the specimen with 1.0[μm] initial notch calculated by using the Lennard-Jones potential to simulate crack growth with different meshes of Mesh type A (see: Table 18.2). Parenthesis in (a), (b), (c), (d), (e) to (f) indicate the average thickness of the cohesive zone element.

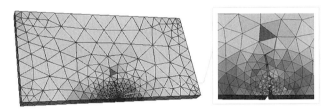

Figure 18.21 Crack initiated in the specimen with 1.0 [μm] initial notch calculated by using the Lennard-Jones potential with finite element Mesh type B (see Table 18.2). (Data source is from Urata and Li (2017a)).

Comparing the critical fracture stress of single-crystal silicon reported in the literature (see: Petersen (1982)) and 6.1 ± 0.8 [GPa] (see: Ericson and Schweitz (1990)). Li and his co-workers (2005) measured fracture stress of the micron size silicon film and it is from 3.0 to 6.4 [GPa]. It is, therefore, reasonable to say that the multiscale CZ method can predict the fracture stress by adjusting the empirical parameter P_{nl} in a reasonable range associated with experimental error for the testing specimen.

Appendix A Crystal Structure

In this Appendix, we briefly introduce some basic concepts and notations of crystal structure.

In general, atoms can be arranged either in a regular periodic array or completely disordered. The former is called crystal structure, which has long range order. The latter is called amorphous structure, which has short range order. The classification of crystal structure is shown in Table A.1.

A.1 Lattice and Basis

An ideal crystal is the periodic arrangement of structural unit, which can be constructed by the infinite repetition of these structural unit in space. The structure of all crystals is described in terms of a lattice with a group of atoms attached to each lattice point, i.e.,

```
Crystal structure = Lattice + Basis
```

In literature, sometimes we also call "*Basis*" as "*Motifs*." A schematic diagram illustrating this definition of crystal structure is shown in Fig. A.1.

As illustrated in Fig. A.1, a lattice is a periodic array of points in space, in which each point has identical surroundings to all others. The lattice basis, or the lattice motifs, is an atom or group of atoms attached to each lattice point in order to generate the crystal structure. Generally speaking, a crystal lattice can be divided into Bravais lattice and non-Bravais lattice, as shown in Fig. A.2. A Bravais lattice is the lattice structure whose basis has only one single atom, whereas for non-Bravais lattices their lattice basis contains several atoms (more than one atom).

Therefore, a Bravais lattice is a lattice array structure with an arrangement and orientation that appears exactly the same, from whichever of the points the array is viewed. That is, a Bravais lattice can always be constructed by the repetition of a fundamental set of translational vectors in real space $\mathbf{a}_i$, i.e., any point in the lattice can be written as,

$$\mathbf{r} = n_1\mathbf{a}_1 + n_2\mathbf{a}_2 + n_3\mathbf{a}_3, \qquad (A.1)$$

where the translational vectors $\mathbf{a}_i$ are the primitive vectors and n_i range through all integral values. Note that the choice for the set $\{\mathbf{a}_i\}$ for any given Bravais lattice is

A Crystal Structure

Table A.1 Classification of crystal structure

Type	Particles	Bonding	Characteristics	Examples
Ionic	Cations and anions	Electrostatic	Strong; hard crystal of high melting point; good thermal and electrical conductors in molten condition	Alkali halides
Molecular	Molecules	Mainly covalent between atoms in molecule; van der Waals or H-bonding between molecules	Soft crystals of low melting point; large coefficient of expansion; insulators	Iodine; ice
Metallic	Metal ions	Metallic	Single crystals are soft; strength depends on structural defects and grain; excellent thermal and electrical conductors	Iron
Covalent	Atoms	Covalent; limited number of electron-pair bonds	Strong; hard crystal of high melting point; insulators	Diamond

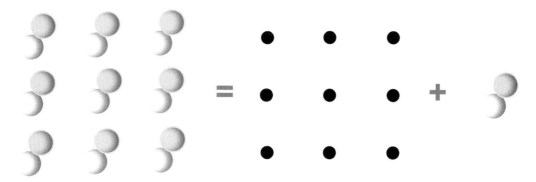

Crystal structure   =   lattice   +   basis

Figure A.1 Schematic of the definition of a crystal structure

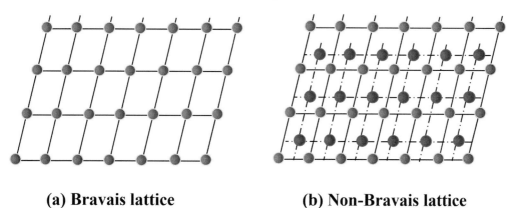

(a) Bravais lattice   **(b) Non-Bravais lattice**

Figure A.2 (a) Bravais lattice and (b) non-Bravais lattice

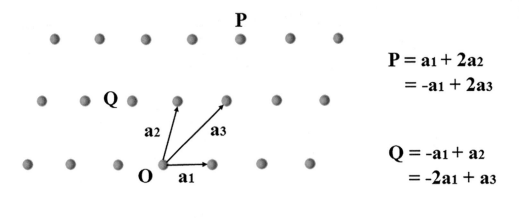

Figure A.3 A general 2D Bravais lattice of no particular symmetry

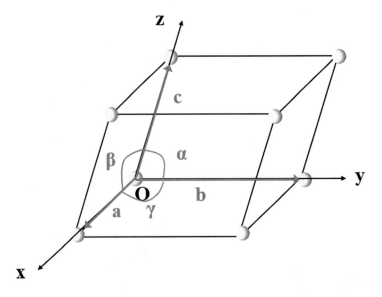

Figure A.4 Schematic of lattice vectors

not unique, as shown in Fig. A.3. In Fig. A.3, the primitive vectors can be chosen as $(\mathbf{a}_1, \mathbf{a}_2)$ or $(\mathbf{a}_1, \mathbf{a}_3)$.

The translational symmetry of a lattice is given by the base vectors or lattice vectors (see: Fig. A.4). Thus, according to the six lattice parameters $a, b, c, \alpha, \beta, \gamma$, there are seven crystal systems, as shown in Table A.2.

According to the definition of Bravais lattice, there are 14 Bravais lattices in 3D space as shown in Table A.3.

To further understand the various types of lattices, it is essential to know elements of group theory. Point group consists of symmetry operations in which at least one

Table A.2 Crystal lattice systems

Crystal system	Lattice parameters	Unit cell	Examples
Cubic	$a = b = c;\ \alpha = \beta = \gamma = 90°$		Fe, Cu, Au
Tetragonal	$a = b \neq c;\ \alpha = \beta = \gamma = 90°$		β–Sn, TiO_2
Orthorhombic	$a \neq b \neq c;\ \alpha = \beta = \gamma = 90°$		Ga, α–Si
Trigonal	$a = b = c;\ \alpha = \beta = \gamma \neq 90°$		Sb, Bi
Hexagonal	$a = b \neq c;\ \alpha = \beta = 90°;\ \gamma = 120°$		Mg, Zn
Monoclinic	$a \neq b \neq c;\ \alpha = \gamma = 90° \neq \beta$		β–S
Triclinic	$a \neq b \neq c;\ \alpha \neq \gamma \neq \beta \neq 90°$		K_2CrO_7

Table A.3 The fourteen Bravais lattices

Bravais lattice	Lattice parameters	Primitive (P)	Body centered (I)	Base centered (C)	Face centered (F)
Cubic	$a = b = c; \alpha = \beta = \gamma = 90°$				
Tetragonal	$a = b \neq c; \alpha = \beta = \gamma = 90°$				
Orthorhombic	$a \neq b \neq c; \alpha = \beta = \gamma = 90°$				

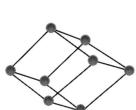

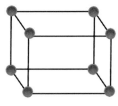

Trigonal $\mathbf{a} = \mathbf{b} = \mathbf{c}; \alpha = \beta = \gamma \neq 90°$

Hexagonal $\mathbf{a} = \mathbf{b} \neq \mathbf{c}; \alpha = \beta = 90°; \gamma = 120°$

Monoclinic $\mathbf{a} \neq \mathbf{b} \neq \mathbf{c}; \alpha = \gamma = 90° \neq \beta$

Triclinic $\mathbf{a} \neq \mathbf{b} \neq \mathbf{c}; \alpha \neq \gamma \neq \beta \neq 90°$

Table A.4 The 32 lattice point groups

Crystal system	Class		Symmetry elements	Crystal system	Class		Symmetry elements
Cubic	23	T	$E\ 4C_3\ 4C_3^2\ 3C_2$	Trigonal	3	C_3	$E\ 2C_3$
	m3	T_h	$E\ 4C_3\ 4C_3^2\ 3C_2\ i\ 8S_6\ 3\sigma_h$		$\bar{3}$	S_6	$E\ 2C_3\ i\ 2S_6$
	$\bar{4}$3m	T_d	$E\ 8C_3\ 3C_2\ 6\sigma_d\ 6S_4$		3m	C_{3v}	$E\ 2C_3\ 3\sigma_v$
	432	O	$E\ 8C_3\ 3C_2\ 6C_2'\ 6C_4$		32	D_3	$E\ 2C_3\ 3C_2$
	m3m	O_h	$E\ 8C_3\ 3C_2\ 6C_2'\ 6C_4\ i\ 8S_6\ 3\sigma_h\ 6\sigma_d\ 6S_4$		$\bar{3}$m	D_{3d}	$E\ 2C_3\ 3C_2\ i\ 2S_6\ 3\sigma_d$
Tetragonal	4	C_4	$E\ 2C_4\ C_2$	Hexagonal	6	C_6	$E\ 2C_6\ 2C_3\ C_2$
	$\bar{4}$	S_4	$E\ 2S_4\ C_2$		$\bar{6}$	C_{3h}	$E\ 2C_3\ \sigma_h\ 2S_3$
	4/m	C_{4h}	$E\ 2C_4\ C_2\ i\ 2S_4\ \sigma_h$		6/m	C_{6h}	$E\ 2C_6\ 2C_3\ C_2\ i\ 2S_3\ 2S_6\ \sigma_h$
	4mm	C_{4v}	$E\ 2C_4\ C_2\ 2\sigma_v'\ 2\sigma_d$		6mm	C_{6v}	$E\ 2C_6\ 2C_3\ C_2\ 3\sigma_v\ 3\sigma_d$
	$\bar{4}$2m	D_{2d}	$E\ C_2\ C_2'\ C_2''\ 2\sigma_d\ 2S_4$		$\bar{6}$m2	D_{3h}	$E\ 2C_3\ 3C_2\ \sigma_h\ 2S_3\ 3\sigma_v$
	422	D_4	$E\ 2C_4\ C_2\ 2C_2'\ 2C_2''$		622	D_6	$E\ 2C_6\ 2C_3\ C_2\ 3C_2'\ 3C_2''$
	4/mmm	D_{4h}	$E\ 2C_4\ C_2\ 2C_2'\ 2C_2''\ i\ 2S_4\ \sigma_h\ 2\sigma_v'\ 2\sigma_h$		6/mmm	D_{6h}	$E\ 2C_6\ 2C_3\ C_2\ 3C_2'\ 3C_2''\ i\ 2S_3\ 2S_6\ \sigma_h\ 3\sigma_d\ 3\sigma_v$
Orthorhombic	2mm	C_{2v}	$E\ C_2\ \sigma_v'\ \sigma_v''$	Monoclinic	m	C_s	$E\ \sigma_h$
	222	D_2	$E\ C_2\ C_2'\ C_2''$		2	C_2	$E\ C_2$
	mmm	D_{2h}	$E\ C_2\ C_2'\ C_2''\ i\ \sigma_h\ \sigma_v'\ \sigma_v''$		2/m	C_{2h}	$E\ C_2\ i\ \sigma_h$
Triclinic	1	C_1	E				
	$\bar{1}$	C_i	$E\ i$				

point remains fixed and unchanged in space. And space group consists of both translational **T** and rotational **R** symmetry operations of a crystal. There are 32 kinds of point group, as shown in Table A.4.

In Table A.4, C_1 only contains the identity E; C_i contains the identity E and a center of inversion i; C_s contains the identity E and a plane of reflection σ; C_n contains the identity E and n-fold axis of rotation; C_{nv} contains the identity E, n-fold axis of rotation, and n vertical mirror planes σ_v; C_{nh} contains the identity E, n-fold axis of rotation, and a horizontal reflection plane σ_h; D_n contains the identity E, n-fold axis of rotation, and n 2-fold rotations about axes perpendicular to the principal axis; D_{nh} contains the same symmetry elements as D_n with the addition of a horizontal mirror plane; D_{nd} contains the same symmetry elements as D_n with the addition of n dihedral mirror planes; S_n contains the identity E and one S_n axis; T_d contains all the symmetry elements of a regular tetrahedron, including the identity E, 4 C_3 axes, 3 C_2 axes, 6 dihedral mirror planes, and 3 S_4 axes; T is same as that of T_d, but do not planes of reflection; T_h is same as that of T, but contains a center of inversion; O_h is the group of the regular octahedron; O is same as that of O_h, but do not contain planes of reflection.

A.2 Unit Cells

A primitive unit cell is a volume of space that, when translated through all the vectors in a Bravais lattice, just fills all the space without either overlapping itself or leaving voids. Like primitive vectors, the choice of primitive unit cell is not unique, as shown in Fig. A.5. Note that each primitive unit cell must contain exactly one lattice point

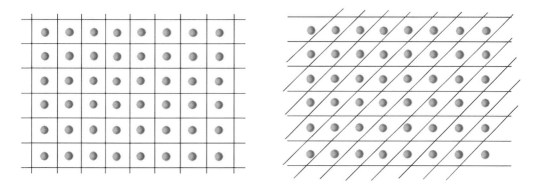

Figure A.5 Two ways of defining primitive cell

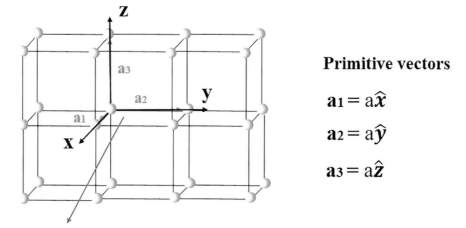

Figure A.6 The primitive cell and conventional unit cell of SC structure

unless it is so chosen that there are lattice points lying on its surface. It then follows that the volume of all primitive unit cells of a given Bravais lattice is the same.

The obvious choice of all points **r** of a primitive cell is,

$$\mathbf{r} = x_1\mathbf{a}_1 + x_2\mathbf{a}_2 + x_3\mathbf{a}_3, \tag{A.2}$$

where x_i are ranging between 0 and 1, and $\mathbf{a}_i$ are primitive vectors.

A conventional unit cell is a region, when translated through some subset of the lattice vectors, can fill up the entire lattice space without voids or overlapping with itself. The conventional unit cell chosen is usually bigger than the primitive cell in favor of preserving the symmetry of the Bravais lattice. Figures A.6–A.9 illustrate the primitive cell and conventional unit cell of simple cubic (SC), body-centered cubic (BCC), face-centered cubic (FCC), and hexagonal close packed cell (HCP), respectively. A primitive unit cell is made of primitive translation vectors $\mathbf{a}_1, \mathbf{a}_2, \mathbf{a}_3$, while the conventional unit cell is determined by six lattice constants $a, b, c, \alpha, \beta, \gamma$.

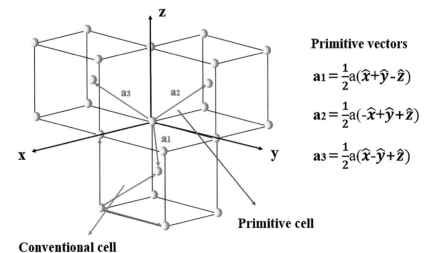

Figure A.7 The primitive cell and conventional unit cell of BCC structure

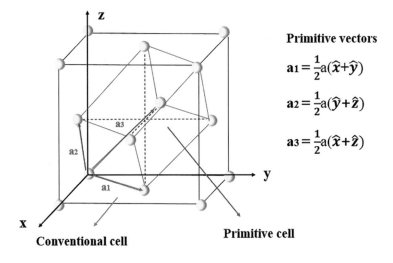

Figure A.8 The primitive cell and conventional unit cell of FCC structure

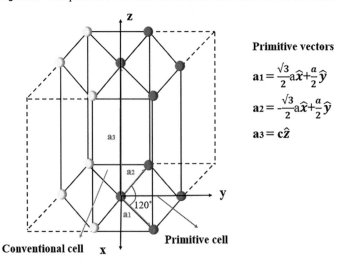

Figure A.9 The primitive cell and conventional unit cell of HCP structure

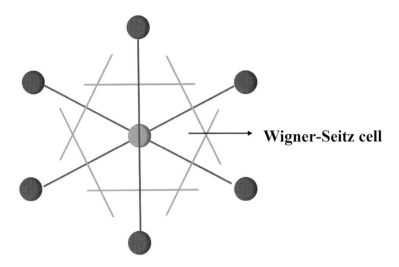

Figure A.10 The Wigner–Seitz cell of a 2D Bravais lattice

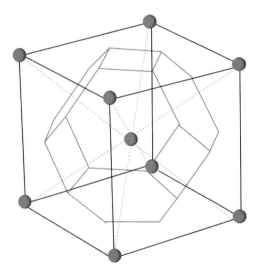

Figure A.11 The Wigner–Seitz cell of a BCC Bravais lattice

The volume of a primitive unit cell can be found by,

$$V = \mathbf{a}_1 \cdot (\mathbf{a}_2 \times \mathbf{a}_3). \tag{A.3}$$

As shown in Fig. A.6, the conventional unit cell of simple cubic structure is the same as that of primitive unit cell. The volume of the conventional unit cell of FCC lattices is four times that of the primitive unit cell, while the conventional unit cell of BCC lattices is two times that of the primitive unit cell. For HCP structure, its conventional unit cell is constructed by two interpenetrating simple hexagonal Bravais lattice.

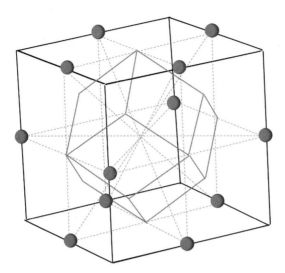

Figure A.12 The Wigner–Seitz cell of a FCC Bravais lattice

There is a simple way to choose a primitive unit cell with the full symmetry of the Bravais lattice, which is using the Wigner–Seitz cell. The Wigner–Seitz cell defines a region of space around a lattice point that is closer to that point than any other lattice points. It is constructed by drawing planes normal to the lines joining nearest lattice points to a particular lattice point, as displayed in Fig. A.10. The smallest polyhedron containing the point bounded by these plane is called the Wigner–Seitz cell.

Figures A.11 and A.12 display the Wigner–Seitz cell for the BCC and FCC lattice, respectively.

A.3 Miller Indices

A.3.1 Miller Indices of Crystal Planes

Since crystals are usually anisotropic, it is useful to regard a crystalline solid as a collection of parallel planes of atoms. Crystallographers and CM physicists use a shorthand notation, i.e., Miller indices to refer to such crystal planes. Miller indices are a symbolic vector representation for the orientation of an atomic plane in a crystal lattice, and they are defined as the reciprocals of the fractional intercepts that a given lattice plane intercepts with crystallographic axes. The following are steps to determine the Miller indices of crystal planes:

- Determine the intercepts l_1, l_2, and l_3 of the plane along each of the three crystallographic directions.
- Take the reciprocals of the intercepts $(\frac{1}{l_1}, \frac{1}{l_2}, \frac{1}{l_3})$.

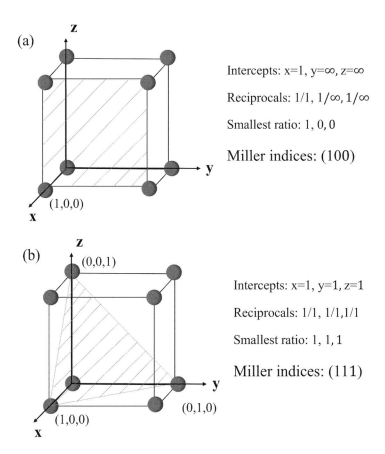

Figure A.13 Miller indices of crystal planes: (a) Lattice plane (100), and (b) lattice plane (111).

- If the results are fractions, multiply each by the smallest constant that make them integers (h, k, l). Note that Miller indices can also be used to label the group of equivalent planes $\{h, k, l\}$.

Figure A.13 shows two examples to find Miller indices of the crystal plane. As shown in Fig. A.13(a), if a plane is parallel to an axis, we say that it cuts at ∞ and $\frac{1}{\infty} = 0$.

While for hexagonal crystals, their Miller indices always be represented as four index notation (h, k, i, l), where $i = (-h + k)$. The use of the four index notation is to bring out the equivalence between crystal equivalent planes and directions. Figure A.14 shows the Miller indices of a crystal plane in hexagonal crystal.

A.3.2 Miller Indices of Crystal Directions

Miller indices are a group of three numbers that describe the orientation of a lattice of atoms or a lattice plane of atoms in a crystal solid. The notation was invented by a British crystallographer, William Hallowes Miller, in nineteen century.

By using Miller indices, crystal directions are specified $[h,k,l]$ as the coordinates of the lattice point closest to the origin along the desired direction. The following are steps to determine the Miller indices of crystal directions:

1. Position the vector so that it passes through the origin (vectors can be translated parallel), or move the origin to the tail of the direction vector.
2. Determine the length of the vector projections on each axis, and then reduce to smallest integer values.
3. Replace negative integers with a bar over the number.
4. Write the indices of crystal directions in $[h,k,l]$. Note that $[h,k,l]$ represent a specific direction, while $<h,k,l>$ denotes a family of symmetry equivalent directions.

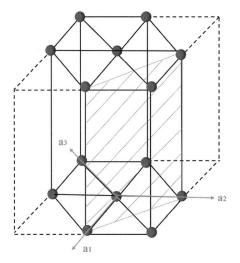

Intercepts: 1, 1, -1/2, ∞

Reciprocals: 1/1, 1/1, 1/(−1/2), 1/∞

Smallest ratio: 1, 1, −2, 0

Miller indices: $(11\bar{2}0)$

Figure A.14 Miller indices of a crystal plane in hexagonal crystal

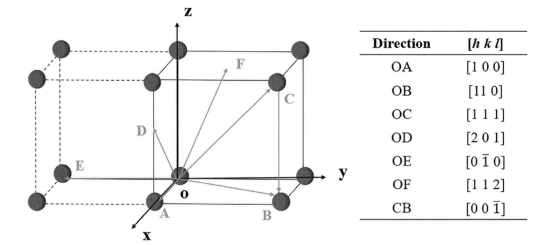

Direction	$[h\ k\ l]$
OA	$[1\ 0\ 0]$
OB	$[1 1\ 0]$
OC	$[1\ 1\ 1]$
OD	$[2\ 0\ 1]$
OE	$[0\ \bar{1}\ 0]$
OF	$[1\ 1\ 2]$
CB	$[0\ 0\ \bar{1}]$

Figure A.15 Schematic illustrations of Miller indices of crystal directions

Here are some examples to find Miller indices of the crystal directions, as shown in Fig. A.15. For cubic lattices, the direction $[h,k,l]$ is perpendicular to the (h,k,l) plane.

A.4 Reciprocal Lattice

The reciprocal lattice is defined as the set of wave vectors $\mathbf{k}$ for which the corresponding plane waves $\Psi_k(\mathbf{r})$ have the periodicity of the Bravais lattice $\mathbf{R}$. The reciprocal lattice vectors are defined as,

$$\left.\begin{aligned} \mathbf{b}_1 &= 2\pi \frac{\mathbf{a}_2 \times \mathbf{a}_3}{\mathbf{a}_1 \cdot (\mathbf{a}_2 \times \mathbf{a}_3)} \\ \mathbf{b}_2 &= 2\pi \frac{\mathbf{a}_3 \times \mathbf{a}_1}{\mathbf{a}_1 \cdot (\mathbf{a}_2 \times \mathbf{a}_3)} \\ \mathbf{b}_3 &= 2\pi \frac{\mathbf{a}_1 \times \mathbf{a}_2}{\mathbf{a}_1 \cdot (\mathbf{a}_2 \times \mathbf{a}_3)} \end{aligned}\right\} \quad (A.4)$$

According to Eq. (A.4), the relation between reciprocal lattice $\mathbf{b}_i$ and real lattice $\mathbf{a}_j$ is,

$$\mathbf{b}_i \cdot \mathbf{a}_j = 2\pi \delta_{ij}, \quad (A.5)$$

where δ_{ij} is the Kronecker delta.

Figure A.16 displays real and reciprocal lattices of a 2D lattice. As shown in Fig. A.16, $\mathbf{b}_1$ is perpendicular to $\mathbf{a}_2$ and $\mathbf{b}_2$ is perpendicular to $\mathbf{a}_1$. It should be noted that the Wigner–Seitz cell of reciprocal lattice is called the first Brillouin zone in solid state physics or just the Brillouin zsone.

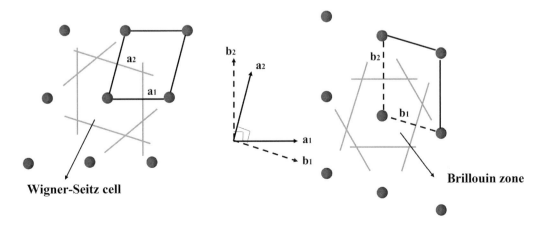

Figure A.16 Illustration of how to construct or identify the Wigner-Seitz cell or the first Brillouin Zone of the reciprocal lattices of 2D lattices.

Appendix A

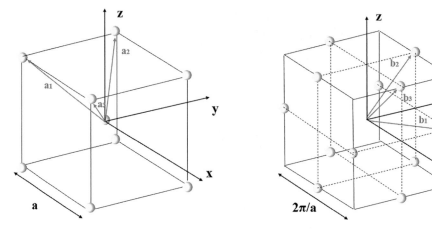

Primitive vectors of a BCC lattice   **Reciprocal lattice of a BCC lattice**

Figure A.17 The real and reciprocal lattices of a BCC lattice

Next, we shall construct the reciprocal lattice of the BCC structure with edge length a, as shown in Fig. A.17. For a BCC lattice, its primitive lattice vectors are (see Fig. A.7) as follows:

$$\left.\begin{array}{l} \mathbf{a}_1 = \tfrac{1}{2}a(\hat{x} + \hat{y} - \hat{z}) \\ \mathbf{a}_2 = \tfrac{1}{2}a(-\hat{x} + \hat{y} + \hat{z}) \\ \mathbf{a}_3 = \tfrac{1}{2}a(\hat{x} - \hat{y} + \hat{z}). \end{array}\right\} \qquad (A.6)$$

Then, according to Eq. (A.4), one can find the reciprocal lattice vectors for BCC structure,

$$\begin{aligned} \mathbf{b}_1 &= \frac{2\pi \mathbf{a}_2 \times \mathbf{a}_3}{\mathbf{a}_1 \cdot (\mathbf{a}_2 \times \mathbf{a}_3)} \\ &= \frac{a^2/4}{a^3/2}(-\hat{x} + \hat{y} + \hat{z}) \times (\hat{x} - \hat{y} - \hat{z}) \cdot 2\pi \\ &= \frac{2\pi}{a}(\hat{x} + \hat{y}) \end{aligned} \qquad (A.7)$$

and

$$\mathbf{b}_2 = \frac{2\pi}{a}(\hat{y} + \hat{z}),$$

$$\mathbf{b}_3 = \frac{2\pi}{a}(\hat{x} + \hat{z}). \qquad (A.8)$$

Thus, the reciprocal lattice of a BCC lattice with edge length a is a FCC lattice with edge length $\frac{2\pi}{a}$. Accordingly, one may find that the reciprocal lattice of a BCC lattice is a FCC lattice.

Figure A.18 The close packed structure of spheres in two-dimensional lattice space

A.5 Some Common Crystal Lattice Structures

A.5.1 Close Packed Structures

The solid structures of crystals can be described by the packing of sphere. In metallic crystals, atoms are assumed to be spherical to explain the bonding and the crystal structures. These spherical particles can be packed into different arrangements. In close packed structures, the arrangement of the spheres are densely packed in order to take up the greatest amount of space possible. For 2D structure, one can easily see that the closest packing of spheres is realized by a hexagonal structure, as shown in Fig. A.18.

In 3D, when the first layer of spheres is laid down, a second layer may be placed to cover the trigonal holes from the first layer. Then, the third layer can be either exactly above the first one or shifted with respect to both the first and the second one. So, there are three relative positions of these layers possible (denoted by A, B, and C). These could, in principle, be combined arbitrarily but in nature the sequences A-B-A-B-A-B and A-B-C-A-B-C are most common. Figure A.19 illustrates the close packed structure of spheres in 3D. In a first layer the spheres are arranged in a hexagonal pattern, where each sphere is surrounded by six others (A). A second layer with the same structure is added. But this layer is slightly shifted and hence just filling the gaps of the first layer (B). Then, another equivalent layer (a third layer) is added filling the gaps just as before. But there are two options: either this layer lies exactly above the first one (A) or it is shifted with respect to both A and B and thus has its own position C. In the following, we will introduce two close packed lattice structures: FCC, which is the cubic close packed lattice structure, and hexagonal close packed (HCP) lattice structure.

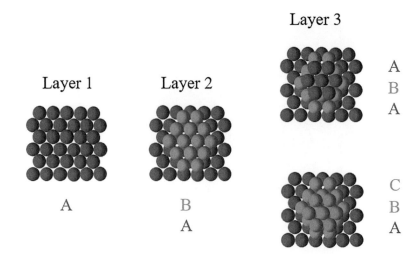

Figure A.19 The close packed structure of spheres in 3D.

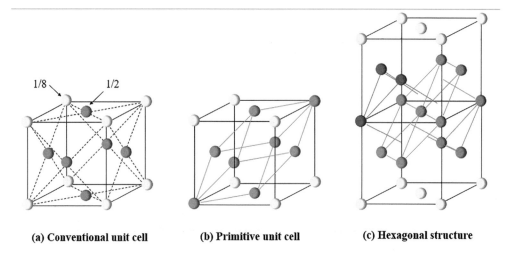

(a) Conventional unit cell (b) Primitive unit cell (c) Hexagonal structure

Figure A.20 The unit cell of the FCC lattice

A.5.2 FCC Crystal Structure

As displayed in Fig. A.20(a), the conventional unit cell of FCC lattice is a cube with edge length a, where eight lattice sites at the corners and six additional ones at the faces of the cube are present. Each corner atom is shared by eight neighboring unit cells and contribute, therefore, only with $\frac{1}{8}$ each. Face-centered atoms only belong to two unit cells and therefore contribute with $\frac{1}{2}$, respectively. Thus, the total number of atoms per FCC unit cell is,

$$n = 8 \cdot \frac{1}{8} + 6 \cdot \frac{1}{2} = 4. \tag{A.9}$$

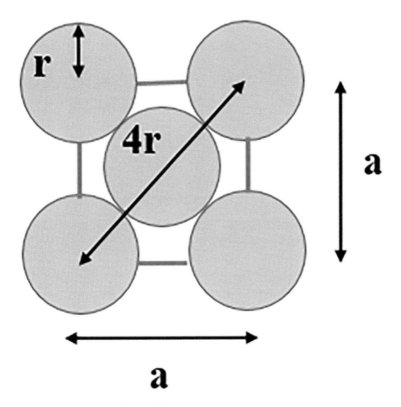

Figure A.21 The APF of the FCC crystal structure

Figure A.20(c) illustrates the hexagonal structure of FCC lattices. It shows that three hexagonal layers are stacked while being shifted against each other, i.e., the type of layer is A-B-C-A-B-C.

Next, we will introduce a basic concept: atomic packing factor (APF), which is defined as the volume of atoms within the unit cell divided by the volume of the unit cell. In FCC lattices, as displayed in Fig. A.21, there are $n = 4$ atoms per unit cell with a volume of $V_{sphere} = \frac{4}{3}\pi r^3$. Thus, the atomic packing factor for FCC lattice is,

$$APF = \frac{n \cdot V_{sphere}}{V_{cell}}$$

$$= \frac{4 \cdot \frac{4}{3}\pi \cdot (\frac{\sqrt{2}}{4})^3 a^3}{a^3} = \frac{\sqrt{2}\pi}{6} \approx 0.74. \qquad (A.10)$$

The APF of FCC structure is the highest possible for spherical objects.

The number of points in a Bravais lattice that are closest to a given point (nearest neighbors) is called the coordination number. As shown in Fig. A.22, each atom in FCC crystal structure has 12 nearest neighbors that are neighboring face atoms, and 6 next-nearest neighbors that are located along the vertices of the lattice. There are

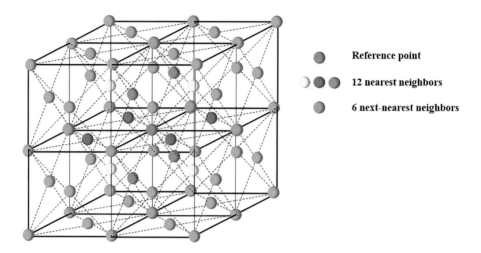

Figure A.22 The nearest neighbors and next-nearest neighbors for FCC crystal structure

$c_1 = 12$ nearest neighbors for FCC structure, so that the coordination number of FCC structure is 12. Its distance is

$$d_{c1} = \frac{a}{\sqrt{2}}. \tag{A.11}$$

For FCC lattice structures, the next-nearest neighbors $c_2 = 6$, and it has the distance of

$$d_{c2} = a. \tag{A.12}$$

A.5.3 HCP Crystal Structure

As shown in Fig. A.23, the HCP structure is constructed by two nested hexagonal lattice that are shifted by the vector $(\frac{2}{3}, \frac{1}{3}, \frac{1}{2})$ against each other (see Fig. A.23(a)). Clearly, the HCP structure is not a Bravais lattice, but it can be regarded as a hexagonal Bravais lattice with a two-atomic basis with the atoms sitting at the positions $(0, 0, 0)$ and $(\frac{2}{3}, \frac{1}{3}, \frac{1}{2})$. And the coordination number of HCP structure is 12. Note that even though the term HCP is used for all those lattices, a close-packing of equal atoms is only obtained for a certain ratio $\frac{c}{a} = \sqrt{\frac{8}{3}}$ of the lattice constant.

A.5.4 BCC Crystal Structure

The body-center cubic (BCC) crystal structure is a non-close packed structure. As shown in Fig. A.7, the spheres are located in each corner of a cube and one sphere in the middle of the cube for the conventional unit cell of BCC lattice with edge length a. Different from the FCC lattice, there is a lattice point that is completely inside the conventional unit cell. So the total number of atoms per BCC unit cell is,

$$n = 8 \cdot \frac{1}{8} + 1 = 2. \tag{A.13}$$

A Crystal Structure   555

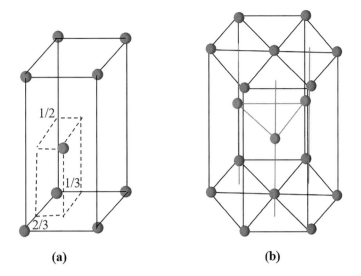

Figure A.23 HCP crystal structure

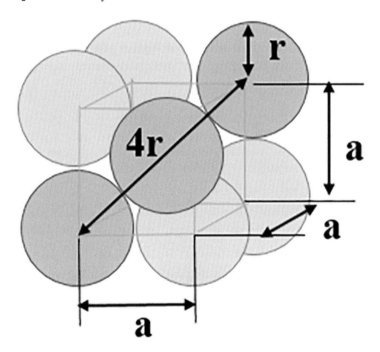

Figure A.24 The APF of the BCC structure

The atomic packing factor for BCC lattice is given as follows (see Fig. A.24):

$$APF = \frac{n \cdot V_{sphere}}{V_{cell}}$$

$$= \frac{2 \cdot \frac{4}{3}\pi \cdot \left(\frac{\sqrt{3}}{4}\right)^3 a^3}{a^3} = \frac{\sqrt{3}\pi}{8} \approx 0.68. \qquad (A.14)$$

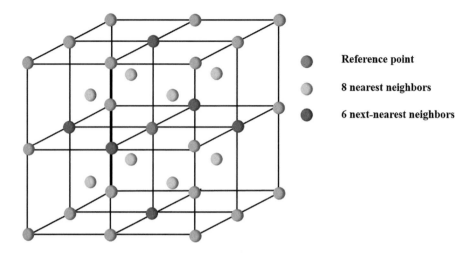

Figure A.25 The nearest neighbors and next-nearest neighbors for the BCC structure

In the BCC structure, as displayed in Fig. A.25, each lattice point has 8 nearest neighbors, which are in the centers of the neighboring cubes, and 6 next-nearest neighbors, which are located at the neighboring vertices of the lattice. The coordination number of BCC lattice $c_1 = 8$, i.e., each atom has 8 nearest neighboring atoms, and the lattice distance is,

$$d_{c1} = \frac{\sqrt{3}}{2}a. \tag{A.15}$$

For BCC lattices, the coordination number $c_2 = 6$, i.e., each atom in BCC lattices has 6 second nearest neighbors, and they are at a distance,

$$d_{c2} = a. \tag{A.16}$$

A.5.5 Diamond Structure

In this section, we shall introduce a classical lattice: diamond structure, such as carbon, germanium, and silicon, which is formed by many elements of the fourth main group of the periodic table.

Figure A.26 shows the conventional unit cell of diamond structure. It shows that the diamond structure is not a Bravais lattice by itself because there are two types of lattice points with different environments. But the underlying structure is actually two interpenetrating FCC lattice structures. Thus, there are two atoms attached to each FCC lattice point: one located just at the position of the lattice point $(0,0,0)$ and one being shifted by the vector $(\frac{1}{4}, \frac{1}{4}, \frac{1}{4})$.

As discussed previously, the APF is defined as the ratio of the volume filled by the spheres to the total volume. In diamond structure, there are $n = 4$ lattice points per unit cell with $N = 2$ atoms sitting on each such lattice point. Neighbored atoms are

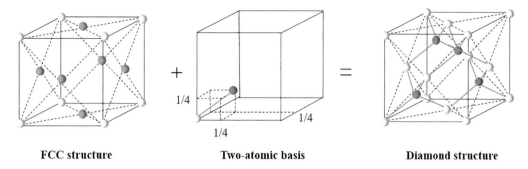

Figure A.26 The conventional unit cell of diamond structure.

shifted by a vector of length $d = \frac{\sqrt{3}a}{4}$. Thus, the atoms are assigned a radius of $r = \frac{d}{2}$. Then the APF of diamond structure can be written as,

$$APF = \frac{n \cdot N \cdot V_{sphere}}{V_{cell}}$$

$$= \frac{2 \cdot 4 \cdot \frac{4}{3}\pi \cdot \left(\frac{\sqrt{3}}{8}\right)^3 a^3}{a^3} = \frac{\sqrt{3}\pi}{16} \approx 0.34. \tag{A.17}$$

The atoms in the diamond structure have $c_1 = 4$ nearest neighbors at a distance of

$$d_{c1} = \frac{\sqrt{3}}{4}a, \tag{A.18}$$

where a is the lattice constant.

The next-nearest neighbors $c_2 = 12$ are located at the neighbored faces of the cube with a distance of

$$d_{c2} = \frac{1}{\sqrt{2}}a. \tag{A.19}$$

A.6 Bloch Theorem

In quantum mechanics, particles in a lattice have a periodic potential function. This is known as the Bloch theorem. In this section, we shall state and prove the Bloch theorem, which is attributed to a Swiss-American physicist Feliz Bloch (1905–1983).

For a given lattice structure, the underlying translational periodicity of the lattice can be defined by the primitive lattice

$$\mathbf{T} = \ell_1 \mathbf{a}_1 + \ell_2 \mathbf{a}_2 + \ell_3 \mathbf{a}_3, \tag{A.20}$$

where ℓ_1, ℓ_2, and ℓ_3 are integers and $\mathbf{a}_1, \mathbf{a}_2$, and $\mathbf{a}_3$ are three lattice vectors. The potential function $V(\mathbf{r})$ inside the lattice has to be periodic, i.e.,

$$V(\mathbf{r} + \mathbf{T}) = V(\mathbf{r}). \tag{A.21}$$

Appendix A

The periodic nature of $V(\mathbf{r})$ ensures that $V(\mathbf{r})$ can be expressed as a Fourier series, which may be written as

$$V(\mathbf{r}) = \sum_{\mathbf{G}} V_{\mathbf{G}} \exp(i\mathbf{G} \cdot \mathbf{r}), \qquad (A.22)$$

where $\mathbf{G}$ are a set of wave number vectors and $V_{\mathbf{G}}$ are the corresponding Fourier coefficients.

The periodic condition implies that

$$\exp(i\mathbf{G} \cdot \mathbf{T}) = 1 \quad \rightarrow \quad \mathbf{G} \cdot \mathbf{T} = 2p\pi,$$

where p is an integer.

Bibliography

Abraham, F. F., Broughton, J. Q., Bernstein, N., and Kaxiras, E. 1998. Spanning the length scales in dynamic simulation. *Computers in Physics*, **12(6)**, 538–546.

Andersen, H. C. 1980. Molecular dynamics simulations at constant pressure and or temperature. *Journal of Chemical Physics*, **72**, 2384–2393.

Antoun, T., Curran, D. R., Seaman, L., Kanel, G. I., Razorenov, S. V., and Utkin, A. V. 2003. *Spall fracture*. Springer Science & Business Media.

Arndt, M., Nairz, O., Vos-Andreae, J., Keller, C., Van der Zouw, G., and Zeilinger, A. 1999. Wave–particle duality of C 60 molecules. *Nature*, **401(6754)**, 680–682.

Baskes, M. I. 1999. Many-body effects in fcc metals: A Lennard-Jones embedded-atoms potential. *Physical Review Letters*, **83(13)**, 2592.

Belytschko, T. 1983. An overview of semidiscretization and time integration procedures. In *Computational methods for transient analysis (A 84-29160 12-64)*, 1–65. North-Holland.

Born, M. and Huang, K. 1954. *Dynamical theory of crystal lattice*. Oxford University Press.

Car, R. and Parrinello, M. 1985. Unified approach for molecular dynamics and density-functional theory. *Physical Review Letters*, **55(22)**, 2471.

Chandler, D. 1987. *Introduction to modern statistical mechanics*. Oxford University Press.

Chialvo, A. A. and Debenedetti, P. G. 1990. On the use of the Verlet neighbor list in molecular dynamics. *Computer Physics Communications*, **60(2)**, 215–224.

Clatterbuck, D. M., Chrzan, D. C., and Morris, J. W., Jr. 2003. The ideal strength of iron in tension and shear. *Acta Materials*, **51**, 2271–2283.

Clausius, R. J. E. 1870. On a mechanical theorem applicable to heat. *Philosophical Magazine Series 4*, 122–127.

Dodson, B. W. 1987. Development of a many-body Tersoff-type potential for silicon. *Physical Review B*, **35(6)**, 2795.

Einstein, A. 1905. On the motion of small particles suspended in liquids at rest required by the molecular-kinetic theory of heat. *Annalen der Physik*, **17**, 208, 549–560.

Elder, R. M., Mattson, W. D., and Sirk, T. W. 2019. Origins of error in the localized virial stress. *Chemical Physics Letters*, **731**, 136580.

Elices, M. G. G. V., Guinea, G. V., Gomez, J., and Planas, J. 2002. The cohesive zone model: Advantages, limitations and challenges. *Engineering Fracture Mechanics*, **69(2)**, 137–163.

Ericson, F. and Schweitz, J. A. 1990. Micromechanical fracture strength of silicon. *Journal of Applied Physics*, **68**, 5840.

Eringen, A. C. and Maugin, G. A. 2012. *Electrodynamics of continua I: Foundations and solid media*. Springer Science & Business Media.

Español, P. and Warren, P. 1995. Statistical mechanics of dissipative particle dynamics. *Europhysics Letters*, **30(4)**, 191–196.

Evans, D. J. and Morriss, G. 2008. *Statistical mechanics of nonequilibrium liquids*. Cambridge University Press.

Falk, M. L., Needleman, A., and Rice, J. R. 2001. A critical evaluation of cohesive zone models of dynamic fracture. *Journal de Physique IV, France*, **11**, 43–50.

Fan, H. and Li, S. 2015. Multiscale cohesive zone modeling of crack propagations in polycrystalline solids. *GAMM-Mitteilungen*, **38**, 268–284.

Finnis, M. W. and Sinclair, J. E. 1984. A simple empirical N-body potential for transition metals. *Philosophical Magazine, A*, **50(1)**, 45–55.

Finnis, M. W. and Sinclair, J. E. 1986. Erratum: A simple empirical N-body potential for transition metals. *Philosophical Magazine, A*, **53(1)**, 161.

Fleck, N. A. and Hutchinson, J. W. 1997. Strain gradient plasticity. *Advances in Applied Mechanics*, **33**, 296–361.

Fock, V. 1930. Näherungsmethode zur Lösung des quantenmechanischen Mehrkörperproblems. *Zeitschrift für Physik*, **61(1–2)**, 126–148.

Fokker, A. D. 1914. Die mittlere Energie rotierender elektrischer Dipole im Strahlungsfeld. *Annalen der Physik*, **348(5)**, 810–820.

Frenkel, D. and Smit, B. 1996. *Understanding molecular simulation from algorithm to applications*. Academic Press.

Gear, C. W. 1966. *The numerical integration of ordinary differential equations of various orders*. Technical Report 7126. Argonne National Laboratory.

Gear, C. W. 1971. *Numerical initial value problems in ordinary differential equations*. Prentice-Hall.

Groot, R. D. and Warren, P. B. 1997. Dissipative particle dynamics: Bridging the gap between atomistic and mesoscopic simulation. *The Journal of Chemical Physics*, **107(11)**, 4423–4435.

Hardy, R. J. 1982. Formulas for determining local properties in molecular-dynamics simulations: Shock waves. *The Journal of Chemical Physics*, **76(1)**, 622–628.

Hartree, D. R. 1957. *The calculation of atomic structures*. Wiley & Sons.

Hartree, D. R. 1928. The wave mechanics of an atom with a non-coulomb central field. Part II. Theory and method. *Mathematical Proceedings of the Cambridge Philosophical Society*, **24(1)**, 89–110.

He, M. and Li, S. 2012. An embedded atom hyperelastic constitutive model and multiscale. *Computational Mechanics*, **49**, 337–355.

Heisenberg, W. 1985. *Über den anschaulichen Inhalt der quantentheoretischen Kinematik und Mechanik*. Springer. In Original Scientific Papers Wissenschaftliche Originalarbeiten (pp. 478–504).

Hogenberg, P. and Kohn, W. 1964. Inhomogeneous electron gas. *Physical Review*, **136(3B)**, B864.

Hoogerbrugge, P. J. and Koelman, J. M. V. A. 1992. Simulating microscopic hydrodynamic phenomena with dissipative particle dynamics. *Europhysics Letters*, **19(3)**, 155.

Hoover, W. G. 1985. Canonical dynamics: Equilibrium phase-space distributions. *Physical Review A*, **31**, 1695.

Hu, H., Liu, M., Wang, Z. F., Zhu, J., Wu, D., Ding, H., Liu, Z., and Liu, F. 2012. Quantum electronic stress: Density-functional-theory formulation and physical manifestation. *Physical Review Letters*, **109(5)**, 055501.

Hughes, T. J. 2012. *The finite element method: Linear static and dynamic finite element analysis*. Courier Corporation.

Hughes, T. J., Taylor, R. L., Sackman, J. L., Curnier, A., and Kanoknukulchai, W. 1976. A finite element method for a class of contact-impact problems. *Computer Methods in Applied Mechanics and Engineering*, **8(3)**, 249–276.

Irving, J. H. and Kirkwood, J. G. 1950. The statistical mechanical theory of transport processes. IV. The equations of hydrodynamics. *The Journal of Chemical Physics*, **18(6)**, 817–829.

Israelachvili, J. N. 2011. *Intermolecular and surface forces*. Academic Press.

Izumi, S. and Sakai, S. 2004. Internal displacement and elastic properties of the silicon Tersoff model. *JSME International Journal Series A Solid Mechanics and Material Engineering*, **47**, 54–61.

Kadanoff, L. P. 2000. *Statistical physics: Statics, dynamics and renormalization*. World Scientific Publishing Company.

Kelchner, C. L., Plimpton, S. J., and Hamilton, J. C. 1998. Dislocation nucleation and defect structure during surface indentation. *Physical Review B*, **58(17)**, 11085.

Khoei, A. R. and DorMohammadi, H. 2012. Validity and size-dependency of Cauchy-Born hypothesis with Tersoff potential in silicon nano-structures. *Computational Materials Science*, **63**, 168–177.

Khoei, A. R., DorMohammadi, H., and Aramoon, A. 2014. A temperature-related boundary Cauchy-Born method for multi-scale modeling of silicon nano-structures. *Physics Letter A*, **378**, 551–560.

Kinjo, T. and Hyodo, S. A. 2007. Equation of motion for coarse-grained simulation based on microscopic description. *Physical Review E*, **75(5)**, 051109.

Koelman, J. M. V. A. and Hoogerbrugge, P. J. 1993. Dynamic simulations of hard-sphere suspensions under steady shear. *Europhysics Letters*, **21(3)**, 363.

Kohn, W. and Sham, J. L. 1965. Self-consistent equations including exchange and correlation effects. *Physical Review*, **140 (4A)**, A1133–1138.

Kubo, R. 1966. The fluctuation-dissipation theorem. *Reports on Progress in Physics*, **29(1)**, 255.

Lehoucq, R. B. and Sears, M. P. 2011. Statistical mechanical foundation of the peridynamic nonlocal continuum theory: Energy and momentum conservation laws. *Physical Review E*, **84(3)**, 031112.

Li, S. and Liu, W. K. 1999. Reproducing kernel hierarchical partition of unity. Part I: Formulation and theory. *International Journal of Numerical Methods for Engineering*, **45**, 251–288.

Li, S., Ren, B., and H. Minaki, H. 2014. Multiscale crystal defect dynamics: A dual-lattice process zone model. *Philosophical Magazine*, **94**, 1414–1450.

Li, S. and Sheng, N. 2010. On multiscale non-equilibrium molecular dynamics simulations. *International Journal for Numerical Methods in Engineering*, **83**, 998–1038.

Li, S. and Tong, Q. 2015. A concurrent multiscale micromorphic molecular dynamics. *Journal of Applied Physics*, **117**, 154303.

Li, S. and Urata, S. 2016. An atomistic-to-continuum molecular dynamics: Theory, algorithm, and applications. *Computer Methods in Applied Mechanics and Engineering*, **306**, 452–478.

Li, S., Zeng, X., Ren, B., Qian, J., Zhang, J., and Jha, A. J. 2012. An atomistic-based interphase zone model for crystalline solids. *Computer Methods in Applied Mechanics and Engineering*, **229–232**, 87–109.

Li, X., Kasai, T., Nakao, S., Tanaka, H., Ando, T., Shikida, M., and Sato, K. 2005. Measurement for fracture toughness of single crystal silicon film with tensile test. *Sensors and Actuators A: Physical*, **119**, 229–235.

Liu, L. and Li, S. 2012. A finite temperature multiscale interphase zone model and simulations of fracture. *Journal of Engineering Materials and Technology*, **134**, 31014.

Lyu, D. and Li, S. 2017. Multiscale crystal defect dynamics: A coarse-grained lattice defect model based on crystal microstructure. *Journal of Mechanics and Physics of Solids*, **107**, 379–410.

Marsden, J. and Hughes, T. 1983. *Mathematical foundations of elasticity*. Prentice-Hall.

Martínez, L., Andrade, R., Birgin, E. G., and Martínez, J. M. 2009. Packmol: A package for building initial configurations for molecular dynamics simulations. *Journal of Computational Chemistry*, **30(13)**, 2157–2164.

Martyna, G. J., Tuckerman, M. E., Tobias, D. J., and Klein, M. L. 1996. Explicit reversible integrators for extended systems dynamics. *Molecular Physics*, **87(5)**, 1117.

Milstein, F. and Farber, B. 1980. Theoretical fcc-bcc transition under [100] tensile loading. *Physical Review Letters*, **44**, 277–280.

Mishin, Y., Mehl, M. J., Papaconstantopoulos, D. A., Voter, A. F., and Kress, J. D. 2001. Structural stability and lattice defects in copper: Ab initio, tight-binding, and embedded-atom calculations. *Physical Review B*, **63(22)**, 224106.

Murashima, T., Urata, S., and Li, S. 2019. Coupling finite element method with Large Scale Atomic/Molecular Massively Parallel Simulator (LAMMPS) for hierarchical multiscale simulations. *The European Physical Journal B*, **9**, 211–215.

Murdoch, A. I. 1983. The motivation of continuum concepts and relations from discrete considerations. *The Quarterly Journal of Mechanics and Applied Mathematics*, **36(2)**, 163–187.

Murdoch, A. I. 2007. A critique of atomistic definition of the stress tensor. *Journal of Elasticity*, **88**, 113–140.

Nielsen, F. 2010 (September). *Legendre transformation and information geometry*. Technical Report CIG-MEMO2. www.informationgeometry.org.

Nielsen, O. H. and Martin, R. M. 1983. First-principles calculation of stress. *Physical Review Letters*, **50(9)**, 697.

Nielsen, O. H. and Martin, R. M. 1985. Quantum-mechanical theory of stress and force. *Physical Review B*, **32(6)**, 3780.

Nosé, S. 1984. A unified formulation of the constant temperature molecular dynamics methods. *Journal of Chemical Physics*, **81**, 511.

Onsager, L. 1944. Crystal statistics. I. A two-dimensional model with an order-disorder transition. *Physical Review*, **65(3–4)**, 117.

Park, H. S. and Klein, P. A. 2008. A surface Cauchy-Born model for silicon nanostructures. *Computer Methods in Applied Mechanics and Engineering*, **197**, 3249–3260.

Parrinello, M. and Rahman, A. 1980. Crystal structure and pair potentials: A molecular dynamics study. *Physical Review Letters*, **14**, 1196–1199.

Parrinello, M. and Rahman, A. 1981. Polymorphic transitions in single crystals: A new molecular dynamics method. *Journal of Applied Physics*, **12**, 7182–7190.

Pelaez, S., Garcia-Mochales, P., and Serena, P. A. 2006. A comparison between EAM interatomic potentials for Al and Ni: From bulk systems to nanowires. *Physica Status Solidi (A)*, **203(6)**, 1248–1253.

Petersen, K. E. 1982. Silicon as a mechanical material. *Proceedings of the IEEE*, **70**, 420–457.

Pilar, F. L. 1990. *Elementary quantum chemistry*. Dover Publication.

Planck, M. 1917. Über einen Satz der statistischen Dynamik und seine Erweiterung in der Quantentheorie. *Sitzungsberichte der Preussischen Akademie der Wissenschaften zu Berlin*, **24**, 324–341.

Podio-Guidugli, P., 2010. On (Andersen–) Parrinello–Rahman molecular dynamics, the related metadynamics, and the use of the Cauchy–Born rule. *Journal of Elasticity*, **100(1)**, 145–153.

Qian, J. and Li, S. 2011. Application of multiscale cohesive zone model to simulate fracture in polycrystalline solids. *ASME Journal of Engineering Materials and Technology*, **133**, 011010.

Ren, B. and Li, S. 2013. A three-dimensional atomistic-based process zone finite element simulation of fragmentation in polycrystalline solids. *International Journal for Numerical Methods in Engineering*, **93**, 989–1014.

Roussas, G. G. 2003. *An introduction to probability and statistical inference*. Elsevier.

Rudd, R. E. and Broughton, J. Q. 1998. Coarse-grained molecular dynamics and the atomic limit of finite elements. *Physical Review B*, **58(10)**, R5893–R5896.

Rudd, R. E. and Broughton, J. Q. 2005. Coarse-grained molecular dynamics: Nonlinear finite elements and finite temperature. *Physical Review B*, **72(14)**, 144104.

Rudzinski, J. F. 2019. Recent progress towards chemically-specific coarse-grained simulation models with consistent dynamical properties. *Computation*, **7(3)**, 42.

The LAMMPS Developers. 2022. LAMMPS Documentation, Sandia National Laboratories (SNL), Albuquerque, NM, https://docs.lammps.org/Manual.pdf

Sauer, R. and Li, S. 2007. A contact mechanics model for quasi-continua. *International Journal for Numerical Methods in Engineering*, **71**, 931–962.

Silling, S. A. 2000. Reformulation of elasticity theory for discontinuities and long-range forces. *Journal of the Mechanics and Physics of Solids*, **48(1)**, 175–209.

Silling, S. A. and Askari, E. 2005. A meshfree method based on the peridynamic model of solid mechanics. *Computers & Structures*, **83(17–18)**, 1526–1535.

Silling, S. A., Epton, M., Weckner, O., Xu, J., and Askari, E. 2007. Peridynamic states and constitutive modeling. *Journal of Elasticity*, **88(2)**, 151–184.

Silling, S. A. and Lehoucq, R. B. 2010. Peridynamic theory of solid mechanics. *Advances in Applied Mechanics*, **44**, 73–168.

Slater, J. C. 1928. The self consistent field and the structure of atoms. *Physical Review*, **32(3)**, 339–348.

Subramaniyan, A. K. and Sun, C. T. 2008. Continuum interpretation of virial stress in molecular simulations. *International Journal of Solids and Structures*, **45**, 4340–4346.

Sunyk, R. and Steinmann, P. 2003. On higher gradients in continuum-atomistic modelling. *International Journal of Solids and Structures*, **40(24)**, 6877–6896.

Tadmor, E. B., Ortiz, M., and Phillips, R. 1996. Quasicontinuum analysis of defects in solids. *Philosophical Magazine A*, **73**, 1529–1563.

Tang, S., Hou, T. Y., and Liu, W. K. 2006. A pseudo-spectral multiscale method: Interfacial conditions and coarse grid equations. *Journal of Computational Physics*, **213(1)**, 57–85.

Taylor, G. I. 1938. Plastic strain in metals. *Journal of the Institute of Metals*, **62**, 307–324.

Tersoff, J. 1988. Empirical interatomic potential for silicon with improved elastic properties. *Physical Review B*, **38**, 9902.

Todd, B. D. and Daivis, P. J. 2017. *Nonequilibrium molecular dynamics: Theory, algorithms and applications*. Cambridge University Press.

Tong, Q. and Li, S. 2015. From molecular systems to continuum solids: A multiscale structure and dynamics. *Journal of Chemical Physics*, **143**, 064101.

Tong, Q. and Li, S. 2016. Multiscale coupling of molecular dynamics and peridynamics. *Journal of Mechanics and Physics of Solids*, **95**, 169–187.

Tong, Q. and Li, S. 2020. A concurrent multiscale study of dynamic fracture. *Computer Methods in Applied Mechanics and Engineering*, **366**, 113075.

Tretiakov, K. V. and Scandolo, S. 2004. Thermal conductivity of solid argon from molecular dynamics simulations. *The Journal of Chemical Physics*, **120(8)**, 3765–3769.

Tsai, D. H. 1979. The virial theorem and stress calculation in molecular dynamics. *The Journal of Chemical Physics*, **70**, 1375.

Urata, S. and Li, S. 2017a. Higher order Cauchy-Born rule based multiscale cohesive zone model and prediction of fracture toughness of Silicon thin films. *International Journal of Fracture*, **203(1)**, 159–181.

Urata, S. and Li, S. 2017b. A multiscale model for amorphous materials. *Computational Materials Science*, **135**, 64–77.

Urata, S. and Li, S. 2018. A multiscale shear-transformation-zone (STZ) model and simulation of plasticity in amorphous solids. *Acta Materialia*, **155**, 153–165.

Wagner, G. J. and Liu, W. K. 2003. Coupling of atomistic and continuum simulations using a bridging scale decomposition. *Journal of Computational Physics*, **190**, 249–274.

Wang, S., Li, Z., and Pan, W. 2019. Implicit-solvent coarse-grained modeling for polymer solutions via Mori-Zwanzig formalism. *Soft Matter*, **15(38)**, 7567–7582.

Wang, S., Ma, Z., and Pan, W. 2020. Data-driven coarse-grained modeling of polymers in solution with structural and dynamic properties conserved. *Soft Matter*, **16(36)**, 8330–8344.

Warren, T. L., Silling, S. A., Askari, A., Weckner, O., Epton, M. A., and Xu, J. 2009. A non-ordinary state-based peridynamic method to model solid material deformation and fracture. *International Journal of Solids and Structures*, **46(5)**, 1186–1195.

Wiener, N. 1964. *Time series*. MIT Press.

Xu, X. P. and Needleman, A. 1994. Numerical simulations of fast crack growth in brittle solids. *Journal of the Mechanics and Physics of Solids*, **42(9)**, 1397–1434.

Zeng, X. 2011. Application of an atomistic field theory to Nano/Micro materials modeling and simulation. *Computer Modeling in Engineering and Sciences*, **74**, 183–201.

Zeng, X. and Li, S. 2010. A multiscale cohesive zone model and simulations of fracture. *Computer Methods in Applied Mechanics and Engineering*, **199**, 547–556.

Zhou, M. 2003. A new look at the atomic level virial stress: On continuum-molecular system equivalence. *Proceedings of the Royal Society of London. A: Mathematical, Physical and Engineering Sciences*, **459(2037)**, 2347–2392.

Author Index

Andersen, Hans C., 206, 225, 487

Bloch, Feliz, 557
Bohr, Niels, 28
Boltzmann, Ludwig E., 124, 125
Born, Max, 10, 166, 434
Brenner, Michael P., 258
Broughton, Jeremy Q., 462

Callaway, David J. E., 237
Car, Roberto, 165
Cauchy, Augustin-Louis, 409, 434
Chandler, David, 378, 395
Chialvo, Ariel A., 187
Clausius, Rudolf J. E., 413
Compton, Arthur H., 6

Davis, Peter J., 373
de Broglie, Louis, 3
Debenedetti, Pablo G., 187
Dirac, Paul A. M., 19, 27

Ehrenfest, Paul, 162
Einstein, Albert, 5, 381
Español, P., 359
Evans, Denis J., 221, 373

Feynman, Richard, 164
Fleck, Norman A., 73
Fock, Vladimir A., 38
Fokker, Adriaan, 352

Güttinger, Paul, 164
Gear, C. William, 200
Gibbs, Josiah W., 128
Green, Melville S., 350, 380

Hardy, Robert J., 427
Hartree, Douglas R., 38
Heisenberg, Werner, 28
Hellmann, Hans, 164
Hertz, Heinrich R., 4

Hohenberg, Pierre, 40
Hoogerbrugge, P. J., 359
Hoover, William G., 208, 214, 223
Hu, Hao, 65
Hutchinson, John W., 73
Hyodo, Shi-aki, 359

Irving, John H., 419, 426
Izumi, Saroshi, 511

Kadanoff, Leo P., 353
Kelchner, Cynthia L., 284
Khinchin, Aleksandr, 393
Kinjo, Tomoyuki, 359
Kirkwood, John G., 419, 426
Koelman, J. M. V. A., 359
Kohn, Walter, 40, 46
Kubo, Ryogo, 350, 380

Lagrange, Joseph-Louis, 105
Langevin, Paul, 350
Lennard-Jones, John E., 171
Li, Shaofan, 453, 472, 495
Liouville, Joseph, 426
London, Fritz, 170

Markov, Andrey, 325
Martin, Richard M., 54
Martinez, L., 366
Martyna, Glenn J., 218
Miller, William H., 547
Millikan, Robert A., 5
Mishin, Y., 445
Mori, Hajime, 401
Morriss, Gary P., 373
Morse, Philip M., 175
Murdoch, A. Ian, 427

Neumann, John von, 315
Nielsen, Ole H., 54
Nosé, Shuichi, 208, 214

Onsager, Lars, 337, 399
Oppenheimer, J. Robert, 166

Parrinello, Michele, 165, 225, 487
Pauli, Wolfgang, 164
Pearson, Karl, 375
Planck, Max, 124, 352
Podio-Guidugli, Paolo, 246

Rahman, Aneesur, 225, 237, 487
Rudd, Robert E., 462

Saint-Venant, Jean Claude, 411
Sakai, Shinsuke, 511
Scandolo, Sandro, 389
Schödinger, Erwin, 10, 12, 31
Sham, Lu Jeu, 40
Silling, Stewart, 499

Slater, John C., 38

Tersoff, Jerry, 258, 510
Thomson, Joseph J., 4
Todd, Billy D., 373
Tong, Qi, 472, 495
Tretiakov, Konstantin V., 389
Trotter, Hale F., 218

Urata, Shingo, 453, 472

Verlet, Loup, 191

Warren, P., 359
Warren, Thomas L., 501
Wiener, Norbert, 393

Young, Thomas, 6

Zhou, Min, 485
Zwanzig, Robert, 401

Subject Index

Ab-initio molecular dynamics, 162
Acoustic wave speed, 468
Adiabatic condition, 116
Amorphous solid, 453
Andersen Lagrangian, 241
Andersen thermostat, 204, 206
Andersen–Parrinello–Rahman molecular dynamics, 225, 487
Angle interaction, 281
Angular momentum operator, 19
Aperiodicity, 328
Atomistic potential, 169
Autocorrelation function, 374, 480

Backward Euler method, 190
Balance of linear momentum, 423, 483
Basis atom, 261
Berendsen thermostat, 204, 206
Biased potential, 323
Binning algorithm, 274
Biological membrane, 366
Bivariate Gaussian distribution, 341
Bloch solution, 466
Bloch's theorem, 47, 55, 466, 557
Block diagonal matrix, 465
Body-centered cubic lattice, 260
Boltzmann constant, 170
Boltzmann distribution, 140
Boltzmann entropy, 128
Bond interaction, 281
Bond order potential, 170
Bonded interaction, 259
Born–Oppenheimer approximation, 35, 162
Born–Oppenheimer molecular dynamics, 162
Bosons, 30, 469
Bra–ket notation, 19
Bravais lattice, 262, 435
Brenner potential, 170
Brownian motion, 353
Brownian particle, 350
Buffer zone, 186

Calculus, 190
Canonical ensemble, 138, 210, 314, 345, 462
Canonical transformation, 114
Car–Parrinello molecular dynamics, 165
Cauchy relation, 442
Cauchy stress, 58, 409, 436
Cauchy–Born rule, 52, 60, 433, 463
Cauchy–Schwartz inequality, 23
Cell list, 186
Cell shape tensor, 454
Center of mass (COM), 472
Central limit theorem, 143, 150, 314
Centrosymmetruc parameter (CSP), 284
Charge equilibrium condition, 63
CHARMM (Chemistry at Harvard Macromolecular Mechanics), 177
CHARMM force field, 292
Chemical equilibrium, 117
Clausius virial theorem, 412
Coarse-grained model, 345, 358
Coarse-grained molecular dynamics (CGMD), 462
Coarse-scale deformation gradient, 473
Coarse-scale dynamics, 236
Commutator, 22
COMPASS (class 2) bond, 259
Compton effect, 5
Concurrent multiscale coupling approach, 495
Conditional probability, 325
Conjugate gradient algorithm, 85
Conservation of energy, 115
Conservation of linear momentum, 419
Conservation of mass, 112
Conservation of probability, 112
Conservative force, 167, 359
Convergence criteria, 281
Copenhagen interpretation, 31
Correlation function, 374
Corresponding principle, 29
Cotangent space, 242
Coulombic interaction, 281
Couple stress, 72

Subject Index

Covalent bond, 169
Crack branching, 503
Critical temperature, 337
Cumulative distribution function, 315
Curse of dimensionality, 319
Cutoff energy, 77
Cutoff method, 172

Damping coefficient, 206
DCD trajectory file, 293
De Broglie relations, 3
Debye force, 170
Deformation gradient, 52, 69, 434
Density functional theory (DFT), 35, 60
Detailed balance, 329
Deterministic random bit generator (DRBG), 315
Diamond cubic lattice, 260
Dielectric constant, 170
Diffusion process, 381
Dihedral bond, 177
Dihedral interaction, 281
Dipole polarizability, 170
Dipole–dipole interaction, 170
Dipole–induced dipole interaction, 170
Dirac comb, 423
Dirac delta function, 210, 316
Dirac equation, 27
Dirac sampling, 422
Discrete element method (DEM), 174
Discrete gradient operator, 484
Dislocation, 503
Dispersion relation, 466
Dissipative force, 345
Dissipative particle dynamics (DPD), 258, 345, 358
Divergence theorem, 111, 221
Domain decomposition, 472
Double-slit experiment, 6
Drag force, 345, 359
Dynamic loading condition, 275

Effective mass, 208
Ehrenfest molecular dynamics, 162
Eigenvalue problem, 461
Einstein relation, 382
Elastic energy density, 435
Elastic tensor, 436, 444
Electric current, 373
Electric potential, 373
Electrical conductivity, 373
Electromagnetic force, 169
Electron density, 36, 443
Electron localization function (ELF), 102
Electronic band structure, 46, 93
Embedded atom method, 170, 258, 443
Embedding function, 443
Empirical distribution, 316

Energy level, 469
Energy minimization, 281
Enthalpy, 123
Entropy, 116, 470
Equal partition theorem, 145
Equilibrium state, 281, 397
Ergodic hypothesis, 134
Ergodic Markov chain, 327
Euler–Lagrange equation, 106
Evans–Hoover thermostat, 220
Exchange-correction quantum stress, 63
Exchange-correlation energy functional, 45
Exchange-correlation higher-order quantum stress, 72
Exchange-correlation quantum couple stress, 75
Excitation, 462
Excited state, 462

Face-centered cubic lattice, 260
FCC lattice, 445
FENE (finite-extensible nonlinear elastic) bond, 259
Fermions, 30
Ferromagnetism, 337
FFT mesh, 77
Fick's law, 381
Fine-scale velocity, 236
First Brillouin zone (BZ), 48, 92
First law of thermodynamics, 115
First Piola–Kirchhoff stress (PK-I), 58, 435, 470
First-principles Lagrangian, 239
Fixed boundary condition, 257
Fluctuation force, 345
Fluctuation–dissipation theorem, 349, 374, 392
Fokker–Planck equation, 350
Force field, 167
Force state, 499
Forward Euler method, 190
Fourier law, 373
Fourier transform, 55
Fugacity, 154

Gaussian distribution, 143, 330, 350
Gaussian function, 475
Gaussian isokinetic thermostat, 221
Gear's predictor–corrector algorithm, 199
Generalized gradient approximation, 96
Gibbs entropy, 128
Gibbs free energy, 123
Global truncation error, 190
Grand canonical ensemble, 152
Gravitational force, 169
Green–Kubo formula, 378
Green–Kubo relation, 350, 374
Green–Lagrangian strain tensor, 242
Griffith critical fracture stress, 503
Ground state, 17

Hamilton principle, 166
Hamilton's principle, 105
Hamilton–Jacobi equation, 161
Hamiltonian (Hybrid) Monte Carlo method, 340
Hamiltonian mechanics, 106
Hamiltonian operator, 10
Hamiltonian system, 121
Hand-shaking, 495
Hard sphere model, 174
Hardy stress, 427
Harmonic approximation, 460
Harmonic bond, 259
Harmonic oscillator, 17, 462
Harmonic potential, 324
Hartree higher-order quantum stress, 72
Hartree potential, 39
Hartree quantum couple stress, 75
Hartree quantum stress, 62
Hartree–Fock approximation, 38, 43
Heat capacity, 337
Heat conductivity coefficient, 373
Heat flux, 373
Heaviside function, 394
Hellmann–Feynman theorem, 163
Helmholtz free energy, 140, 324, 463
Hermitian operator, 23
Hexagonal close packed lattice, 260
Hexagonal lattice, 260
Higher-order Cauchy–Born rule, 446
Higher-order multiscale cohesive zone model, 530
Higher-order quantum stress, 69
Hit-and-miss algorithm, 319
Hohenberg and Kohn theorems, 39
Homogenization model, 358
Horizon, 499
Hydrogen bond, 169

Image atoms, 181
Importance sampling, 323
Improper dihedral bond, 177
Improper interaction, 281
Induced dipole–induced dipole interaction, 170
Infinitesimal invariant, 202
Initialization, 254
Integration algorithm, 189
Interference of matter waves, 6
Intermediate equilibrium configuration, 476
Intermolecular force, 345
Internal degrees of freedom, 496
Invariant distribution, 326
Inverse transform technique, 315
Ion–electron higher-order quantum stress, 71
Ion–electron quantum couple stress, 74
Ion–electron quantum stress, 61
Ion–ion quantum couple stress, 74
Ion–ion quantum stress, 62

Ion-ion higher-order quantum stress, 71
Ionic bond, 169
Irving–Kirkwood formalism, 419
Isentropic process, 116
Isochoric deformation transformation, 225
Isokinetic thermostat, 204
Isolated system, 116
Iteration method, 281
Ito lemma, 361
Ito process, 361

Jacobian matrix, 201
John von Neumann's middle square method, 315

K-points mesh, 81
Keesom force, 170
Kinematic condition, 455
Kinetic energy operator, 19
Kinetic higher-order quantum stress, 71
Kinetic quantum couple stress, 74
Kinetic quantum stress, 63
Kohn–Sham equations, 43
Kronecker delta, 166
Kubo expression, 400

Lagrangian mechanics, 105
Lagrangian multiplier, 166
Lagrangian system, 121
Lagrangian volumetric strain, 242
LAMMPS, 249
Langevin dynamics, 348
Langevin equation, 345
Langevin thermostat, 204, 207
Lattice constants, 260
Lattice curvature, 73
Lattice gauge theory, 237
Lattice motif, 262
Lattice pattern, 260
Lattice space, 264
Lattice vector, 80
Lattice vibrations, 460
Leapfrog algorithm, 197, 341
Least action (Hamilton's) principle, 105
Least squares principle, 220
Left Cauchy–Green tensor, 436
Legendre transformation, 107, 120
Lennard-Jones embedded atom potential, 444
Lennard-Jones potential, 170
Lie–Trotter product formula, 218
Liouville theorem, 111, 112, 221, 350, 420
Lipschitz condition, 191
Local defect, 284
Local density approximation, 61
Local truncation error, 190
Logarithmic scale, 277
London force, 170
Long-range order, 477

Macroscale deformation gradient, 477
Macroscale traction boundary condition, 482
Macroscopic relaxation, 397
Magnetic susceptibility, 337
Major symmetry, 443
Many-body Schrödinger equation, 35
Markov chain, 325
Markov Chain Monte Carlo (MCMC) method, 325
Markov process, 346
Mass density distribution, 110
Mass-reduced units, 460
Maxwell velocity distribution, 353
Maxwell–Boltzmann distribution, 147, 206
Mean value theorem, 190, 417
Mechanical equilibrium, 117
Memorylessness, 325
Message Passing Interface (MPI) parallel execution, 253
Metallic bond, 169
Metropolis–Hastings method, 329
Micro deformation tensor, 473
Microcanonical ensemble, 131, 210
Micromorphic continuum, 488
Micromorphic deformation gradient, 477
Micromorphic multiplicative decomposition, 473
Microscale right Cauchy–Green tensor, 481
Microstate, 324
Middle point rule, 199
Miller indices, 547
Minimum image criterion, 182
Minor symmetry, 443
Mishin embedded atom potential, 445
Mixed higher-order elastic tensor, 449
Molecular dynamics, 158
Moment of inertia tensor, 479
Monkhorst–Pack's technique, 82
Monte Carlo method, 313
Mori-Zwanzig formalism, 359, 401
Morse potential, 175, 258
Multi-thread parallel execution, 252
Multiplicative multiscale decomposition, 477
Multiscale kinematic decomposition, 497
Multiscale method, 433
multiscale micromorphic molecular dynamics (MMMD), 472, 481, 486
Multiscale multiplicative decomposition, 229
Multivariate normal distribution, 329

Neighbor list, 185, 274
Neighbor shell, 445
Newton second law, 162
Newton–Raphson method, 451
Newtonian fluid, 374
Newtonian mechanics, 158
Nośe Hamiltonian, 208
Non-Bravais lattice, 262, 450

Nonequilibrium molecular dynamics, 373
Nonequilibrium physical process, 373
Nonlocal balance equation, 500
Nonlocal deformation gradient, 499
Normal distribution, 318, 350
Normal mode, 463
Normal mode coordinate, 461
Normalization factor, 467
Nosé–Hoover chain, 213
Nosé–Hoover molecular dynamics, 216
Nosé–Hoover thermostat, 204, 208

Observable, 21
Occupation number, 124, 469
Onsager's regression hypothesis, 399
Open Visualization Tool (OVITO), 305
Operator, 22
Orientation vector, 261

PACKMOL, 366
Pair potential, 167, 258
Parallel universe, 31
Paramagnetism, 337
Parrinello–Rahman closure condition, 239
Parrinello–Rahman extended Lagrangian, 244
Parrinello–Rahman Lagrangian, 241
Parrinello–Rahman molecular dynamics (PR MD), 453
Partition function, 140, 462
Pauli's exclusion principle, 30
PDB topology file, 293
Pearson product moment correlation coefficient, 375
Peculiar velocity, 236
Perdew–Burke–Ernzerhof (PBE) functional, 96
Peridynamics (PD), 495, 499
Periodic boundary condition, 180, 257
Permittivity, 170
Phase separation, 337, 362
Phase space compression factor, 221
Phonon, 462
Photoelectric Effect, 4
Planck constant, 469
Poisson distribution, 207
Poisson's bracket, 110, 219
Posterior distribution, 325
Power spectrum, 392
Predictor–corrector method, 199
Primitive vector, 261
Principal coordinate, 461
Principle of corresponding state, 179
Probability, 110
Probability distribution, 143
Probability distribution function, 315
Probability distribution function (PDF), 317
Probability mass function, 316

Subject Index

Projector-augmented wave (PAW) method, 76
Proposal distribution, 329
Pseudopotential, 50, 82
Pseudorandom number generator (PRNG), 315
PSF topology file, 292
Pull back, 243
Push forward, 242

Quadratic form, 241
Quadratic polynomial, 324
Quadrature integration, 314
Quantum couple stress, 72
Quantum effect, 469
Quantum electron stress, 66
Quantum fluid dynamics, 160
Quantum mechanics, 163
Quantum state, 124
Quantum stress, 52, 60
Quantum virial theorem, 53
Quasi-harmonic approximation, 459

Random force, 345, 359
Random walk proposal distribution, 332
Rayleigh wave speed, 503
Real atoms, 181
Real space lattice vector, 55
Real variable, 209
ReaxFF potential, 258
REBO potential, 258
Reciprocal lattice vector, 47, 55
Reciprocal space, 55
Recurrence, 328
Reduced unit, 178
Reduced-order model, 358
Reducibility, 327
Referential equilibrium configuration, 476
Reproducing kernel particle method, 473
Response function, 392
Reversible Markov chain, 329
Reversible process, 116
Reynolds transport theorem, 53
Right Cauchy–Green tensor, 246, 436
Runge–Kutta method, 199

Sample mean method, 318
Sampling, 316
Scatterplot, 374
Schödinger cat, 31
Schrödinger equation, 8, 162
Schrödinger equation, 469
Second law of thermodynamics, 117
Second Piola–Kirchhoff stress (PK-II), 61, 436
Self-consistent charge density, 83
Self-consistent field (SCF) method, 47
Separation of variable, 159
Serial execution, 252

Shape tensor, 473
Shear viscosity, 374
Short-range order, 477
Shrink-wrapped boundary condition, 257
Simple cubic lattice, 260
Simple function, 324
Size effect, 180, 453
Slater determinant, 39
Smearing method, 79
Soft sphere model, 174
Solvation method, 207
Spatial configuration, 476
Spatial distribution, 473
Spatial location, 324
Spectral decomposition, 479
Spherical Bessel function, 59
Spin, 26
Spin angular momentum, 26
Spin tensor, 247
Spinodal decomposition, 362
Spontaneous (irreversible) process, 116
Square lattice, 260
Stömer–Verlet formula, 192
State function, 115
State variable, 115
State-based peridynamics, 473, 499
Static correlation, 377
Static standard deviation, 352
Stationary condition, 160
Stationary distribution, 326
Stationary wave, 468
Statistical conditions, 478
Statistical configuration, 476
Statistical ensemble, 124
Statistical mechanics, 105
Statistical parametric configuration, 473
Stillinger–Weber potential, 170, 176, 258
Stochastic collision frequency, 207
Stochastic differential equation, 361
Stochastic Langevin dynamics, 207
Stochastic thermostat, 206
Stopping tolerance, 281
Strain energy, 121
Strain gradient, 69
Stress–strain relation, 52, 102
Stretch gradient, 73
Strong nuclear force, 169
Structural optimization, 84
Susceptibility, 399
Symplectic algorithm, 201
Symplectic condition, 114
Symplectic integrator, 343

Subject Index

Tangent space, 243
Taylor expansion, 192, 446
Tensorial correlation function, 241
Tensorial kinetic energy, 414
Tensorial virial theorem, 414
Tersoff potential, 170, 258, 510
Thermal coefficient, 464
Thermal conductivity, 207, 385
Thermal diffusivity, 386
Thermal equilibrium, 117
Thermal expansion constant, 464
Thermal fluctuations, 214, 462, 464
Thermal reservoir, 346
Thermodynamic phase transition, 362
Thermodynamic states, 117
Thermodynamical flux, 374
Thermostat, 204
Three-body atomistic potential, 170
Time reversible, 214
Time-dependent Schrödinger equation, 10
Time-independent Schrödinger equation, 12
Transformation matrix, 461
Transition probability, 325
Transition zone, 495
Transport coefficient, 373
Trotter algorithm, 218
Truncation error, 192
Two-dimensional (2D) lattice, 263

Ultrasoft Vanderbilt pseudopotentials (US PP), 76
Umbrella potential, 322
Umbrella sampling, 322
Unbiased estimator, 322
Unbiased variance, 314
Uncertainty principle, 8, 21, 28, 135
Urey–Bradley bond, 177

Van der Waals force, 169
Van der Waals interaction, 170, 281
Velocity Verlet algorithm, 195, 217
Verlet algorithms, 191
Verlet list, 186
Vienna Ab-inito Simulation Package (VASP), 76
Virial pressure, 235
Virial stress, 232, 411, 433, 485
Virial theorem, 412
Virtual variable, 209
Viscosity coefficient, 383
Viscous force, 347
Visual molecular dynamics (VMD), 291

Wave particle duality, 3
Weak nuclear force, 169
Weeks–Chandler–Andersen theory, 227
White noise, 346
Wiener–Khinchin theorem, 393
Wigner–Seitz cell, 48, 262, 435, 450
Window function, 474